T0319285

Genetic and Genomic Resources for Grain Cereals Improvement

Genetic and Genomic Resources for Grain Cereals Improvement

Edited by

Mohar Singh

Hari D. Upadhyaya

AMSTERDAM • BOSTON • HEIDELBERG • LONDON
NEW YORK • OXFORD • PARIS • SAN DIEGO
SAN FRANCISCO • SINGAPORE • SYDNEY • TOKYO
Academic Press is an Imprint of Elsevier

Academic Press is an imprint of Elsevier
125, London Wall, EC2Y 5AS, UK
525 B Street, Suite 1800, San Diego, CA 92101-4495, USA
225 Wyman Street, Waltham, MA 02451, USA
The Boulevard, Langford Lane, Kidlington, Oxford OX5 1GB, UK

British Library Cataloguing-in-Publication Data
A catalogue record for this book is available from the British Library

Library of Congress Cataloging-in-Publication Data
A catalog record for this book is available from the Library of Congress

ISBN: 978-0-12-802000-5

For information on all Academic Press publications
visit our website at http://store.elsevier.com/

Publisher: Nikki Levy
Acquisition Editor: Nancy Maragioglio
Editorial Project Manager: Billie Jean Fernandez
Production Project Manager: Julie-Ann Stansfield
Designer: Mark Rogers

Typeset by Thomson Digital

Printed and bound in the United States of America

Contents

List of contributors

Shephalika Amrapali Indian Council of Agricultural Research, Directorate of Floriculture Research, College of Agriculture Campus, Shivaji Nagar, Pune, India

Ahmad Amri Department of Wheat Breeding/Genetics, International Center for Agricultural Research in the Dry Areas (ICARDA), Rabat, Morocco

Banisetti Kalyana Babu Indian Council of Agricultural Research, Indian Institute of Oil Palm Research (IIOPR), Pedavegi, Andhra Pradesh, India

Michael Baum Department of Wheat Breeding/Genetics, International Center for Agricultural Research in the Dry Areas (ICARDA), Rabat, Morocco

Maja Boczkowska Department of Functional Genomics, Plant Breeding and Acclimatization Institute (IHAR), National Research Institute, Radzików, Poland

Ismail Dweikat Department of Horticulture, University of Nebraska, Lincoln, Nebraska, USA

Sangam Lal Dwivedi International Crops Research Institute for the Semi-Arid Tropics (ICRISAT), Genebank, Patancheru, Telangana, India

Lakshmi Kant Crop Improvement Division, Indian Council of Agricultural Research, Vivekananda Pravatiya Krishi Anusandhan Sansthan (Vivekanada Institute for Hill Agriculture), Almora, Uttarakhand, India

Bogusław Łapiński National Centre for Plant Genetic Resources, Plant Breeding and Acclimatization Institute (IHAR), National Research Institute, Radzików, Poland

Trilochan Mohapatra Indian Council of Agricultural Research, Indian Agricultural Research Institute, New Delhi, India

Umakanta Ngangkham Crop Improvement Division, Indian Council of Agricultural Research, National Rice Research Institute, Cuttack, Odisha, India

Francis C. Ogbonnaya Grains Research and Development Corporation (GRDC), Australia

Bhaskar C. Patra Crop Improvement Division, Indian Council of Agricultural Research, National Rice Research Institute, Cuttack, Odisha, India

Santosh K. Pattanashetti International Crops Research Institute for the Semi-Arid Tropics (ICRISAT), Genebank, Patancheru, Telangana, India

Wiesław Podyma Organic Farming Section, Plant Breeding and Acclimatization Institute (IHAR), National Research Institute, Radzików, Poland; Laboratory of Gene Bank, Polish Academy of Sciences Botanical Garden, Center for Biological Diversity Conservation in Powsin, Warsaw, Poland

Soham Ray Crop Improvement Division, Indian Council of Agricultural Research, National Rice Research Institute, Cuttack, Odisha, India

Kothapally Narsimha Reddy International Crops Research Institute for the Semi-Arid Tropics (ICRISAT), Genebank, Patancheru, Telangana, India

Miguel Sanchez-Garcia Department of Wheat Breeding/Genetics, International Center for Agricultural Research in the Dry Areas (ICARDA), Rabat, Morocco

Shailesh Kumar Singh International Crops Research Institute for the Semi-Arid Tropics (ICRISAT), Genebank, Patancheru, Telangana, India

Mohar Singh Indian Council of Agricultural Research, National Bureau of Plant Genetic Resources Regional Station, Shimla, Himachal Pradesh, India

Quahir Sohail Department of Wheat Breeding/Genetics, International Center for Agricultural Research in the Dry Areas (ICARDA), Rabat, Morocco

Wuletaw Tadesse Department of Wheat Breeding/Genetics, International Center for Agricultural Research in the Dry Areas (ICARDA), Rabat, Morocco

Hari D. Upadhyaya International Crops Research Institute for the Semi-Arid Tropics (ICRISAT), Genebank, Patancheru, Telangana, India

Mani Vetriventhan International Crops Research Institute for the Semi-Arid Tropics (ICRISAT), Genebank, Patancheru, Telangana, India

Yi-Hong Wang Department of Biology, University of Louisiana at Lafayette, Lafayette, Louisiana, USA

Preface

Grain cereals mainly comprised of rice, wheat, maize, barley, oat, sorghum, and millets (pearl, finger, foxtail, proso, barnyard, little, and kodo) are the members of grass family and are very important to human diet because of their role as staple food crops in many parts of the world. These cereals are also used to produce animal feed, oils, starch, flour, sugar, and processed foods including malts and alcoholic beverages. Further, about 50% of the world's calories are being provided by wheat and maize and in several parts of Africa and Asia, people rely on grains such as sorghum and millets. The increasing human population and enhanced standard of living are placing greater demands on food-related requirements in terms of quality, quantity as well as diversity. As a basic raw material for future crop breeding, genetic resources are the key to future food security. An excellent performance has been achieved by applying contemporary approaches for germplasm characterization and evaluation to manage the crop genetic resources effectively. In parallel, use of genomic resources and specialized germplasm sets such as minicore collections and reference sets will facilitate identification of trait-specific germplasm, trait mapping, and allele mining for resistance to various biotic and abiotic stresses and also for useful agromorphologic traits.

The book entitled "Genetic and Genomic Resources for Grain Cereals Improvement" comprises a total of eight chapters contributed by eminent researchers around the world. The first introductory chapter highlights the landmark research on genetic and genomic resources of grain cereals improvement. Subsequently, Chapters 1–8 deal with aspects related to genetic and genomic resources of grain cereals improvement. Each chapter provides a comprehensive account of information on the origin, distribution, diversity and taxonomy, erosion of genetic diversity from the traditional areas, status of germplasm resource conservation, germplasm characterization and evaluation, use of germplasm in crop improvement, and integration of genetic and genomic resources in crop improvement. The editors are grateful to all chapter contributors for their outstanding efforts in the preparation of this book and we had very cordial relations during the entire process of development of this manuscript. The editors are also thankful to the Academic Press staff for shepherding the book through the editorial process with a complete academic approach. The edited multiauthored book describing the problem of genetic and genomic resources of grain cereals improvement will facilitate students, faculty, researchers, and policy makers, effectively managing and utilizing the genetic resources for the benefit of humankind.

Editors

Introduction

*Mohar Singh**, *Hari D. Upadhyaya*[†]
*Indian Council of Agricultural Research, National Bureau of Plant Genetic Resources Regional Station, Shimla, Himachal Pradesh, India; [†]International Crops Research Institute for the Semi-Arid Tropics (ICRISAT), Genebank, Patancheru, Telangana, India

Grain cereals, including rice, wheat, maize, barley, oat, sorghum, and millets (pearl millet, finger millet, foxtail millet, proso millet, barnyard millet, little millet, and kodo millet), are members of the grass family and occupy a considerable area under commercial cultivation worldwide. These cereals are also used to produce animal feed, oils, starch, flour, sugar, and processed foods including malts and alcoholic beverages. The increasing human population and enhanced standard of living are placing greater demands on food-related requirements in terms of quality, quantity, as well as genetic variability. As a base material for future crop improvement, genetic resources are the key to future food and nutritional security. An excellent performance has been obtained by applying contemporary approaches for germplasm characterization and evaluation to manage the crop genetic resources effectively and efficiently. Use of genomic resources and specialized germplasm sets, such as minicore collection and reference sets, will facilitate identification of trait-specific germplasm, trait mapping, and allele mining for resistance to major prevailing biotic and abiotic stresses and also for useful agronomic traits of interest. Here we conclude brief details on the genetic and genomic resources research on important grain cereals.

1 Rice

Rice is the staple food crop of more than half the world's population. Asia accounts for more than 90% of the world's total rice production and the balance is divided almost equally between Africa and Latin America, where the demand for rice is increasing. Rice has been cultivated in Asia since ancient times, and for generations farmers have maintained thousands of different local landraces for their subsistence agriculture (Jackson, 1995). Most countries in Asia maintain rice germplasm collections, and the largest are in China, India, Thailand, and Japan (FAO, 2013). The International Rice Research Institute (IRRI) holds the largest collection and is also the most genetically diverse and complete world rice collection. Africa contains a diversity of both cultivated and wild/weedy rice species. The region has 8 species representing 6 of the 10 known genome types. Genetic resources of these species are conserved in various global germplasm repositories, but they remain under collected and hence underrepresented in germplasm collections. The lack of *in situ* germplasm conservation programs further exposes them to possible genetic erosion or extinction. In order to obtain maximum benefits from these resources, it is imperative that they are collected, efficiently preserved, and optimally utilized. High-throughput molecular approaches, such as genome sequencing, could be employed to study their genetic diversity and precise value and thereby enhance their use in rice genetic improvement.

2 Wheat

Wheat is the most important grain cereal for ensuring food security worldwide. Total demand for wheat has been growing with the increasing human population pressure globally. The production of wheat has increased substantially from 218.5 million tons in 1961 to 732 million tons in 2013 (www.fao.org) primarily due to the adoption of semidwarf high-yielding and input-responsive cultivars. Likewise, wheat genetic resources have played a pivotal role in genetic improvement by contributing potential gene sources for yield, wider adaptation, short stature plant height, improved grain quality, and resistance/tolerance to major prevailing biotic and abiotic stresses. In view of climate change and genetic erosion associated with many natural and anthropogenic factors as well as rapid expansion and domination of mega wheat cultivars across the major wheat agroecologies, efforts have been made to collect and preserve wheat genetic resources in *ex situ* collection. The center of genetic diversity for wild wheat relatives includes Egypt, Israel, Jordan, Lebanon, Syria, Turkey, Armenia, Azerbaijan, Iraq, Iran, Afghanistan, and the Turkic Republics of Central Asia. The range of distribution of wheat relatives occurs from the Canary Islands to western China and from southern Russia to northern Pakistan and India. To-date more than 900,000 wheat accessions (wild/weedy relatives, landraces, synthetic wheats, advance breeding lines, genetic stocks) are conserved in different gene banks worldwide. The wheat genetic resource center (WGRC) maintains 2500 wheat accessions including cytogenetic stocks, developed by wheat researchers across the globe. Genes for host-plant resistance to viral, bacterial, fungal, and insect pests and major abiotic stresses have been identified and introgressed into agronomically elite genetic backgrounds. Effective utilization of a large number of genetic resources, however, is a big challenge. Application of modern tools and techniques, such as focused identification of germplasm strategy (FIGS), effective gene introgression methods, and genomics, are essential in improving genetic resource utilization and improving breeding efficiency.

3 Barley

Barley belongs to the genus *Hordeum*, and all species have the basic chromosome number of $n = x = 7$. Furthermore, cultivated barley, *Hordeum vulgare* ssp. *vulgare*, and its immediate wild progenitor *H. vulgare* ssp. *spontaneous* (K. Koch.) Asch. & Graebn. are true diploid species with $2n = 2x = 14$ chromosome numbers. Likewise, other *Hordeum* species are diploid, tetraploid ($2n = 4x = 28$), or hexaploid ($2n = 6x = 42$). According to Harlan's gene pool concept, all barley species have been classified into three different gene pools. The primary gene pool includes elite breeding materials, commercial cultivars, landraces, and the wild ancestor of cultivated barley. The secondary gene pool includes only one species, *Hordeum bulbosum* L., which shares the basic *Hordeum* genome. The tertiary gene pool of barley is very large and comprises all other remaining wild species (Bothmer et al., 1991). Genetic diversity of any crop species is defined as genetic variation within and between populations, landraces, and cultivars, arising due to recombination, mutations, and introgression. The use of highly diverse germplasm increases the chances for success in developing wider populations through introgression. Globally, more than 400,000 barley accessions are available for research and breeding purposes at different gene banks. Total gene bank collections represent landraces (44%), breeding lines (17%), crop wild relatives (CWR) (15%), commercial cultivars (15%), and other genetic stocks (9%).

4 Oat

Oat is one of the minor cereals used as feed, food, and industrial feedstock purposes. Common oat (*Avena sativa*) is the cultivated species grown under diverse agroecologic conditions. Globally, germplasm collection of *Avena* species consist of approximately 131,000 accessions preserved by more than 63 countries. Further distribution of total germplasm holding revealed that only 14 countries held more than 80% genetic resources. The largest collections are held in Canada (~40,000), the United States (~22,000), and Russia (~12,000). In Canada and Russia, cultivated species and several wild species are preserved, while in the United States emphasis was placed on *A. sativa* and its wild relatives from the primary gene pool. The genetic reserve conservation is defined as management and monitoring of genetic diversity of natural populations of CWR in specific areas for the long-term preservation. *On-farm* conservation is focused on cultivated species and in particular on landraces and traditional cultivars, and consists of agrobiodiversity preservation in a dynamic agroecosystem that is self-supporting and favoring evolutionary processes. The issue of *in situ* conservation of genetic resources in the genus *Avena* has been specifically targeted at the framework of the European project (Frese et al., 2013). It is aimed at the creation of conservation strategies for CWR and landraces and to transfer them into elite backgrounds.

5 Sorghum

Worldwide, a quarter million sorghum accessions have been collected and maintained by several national and international gene banks and the biggest sorghum germplasm holders are the US Department of Agriculture (USDA) and the International Crops Research Institute for the Semi-Arid Tropics (ICRISAT), India. The majority of collections in the United States gene bank are from Ethiopia, Sudan, Yemen, Mali, India, and the United States (http://www.ars-grin.gov/cgi-bin/npgs/html/tax_stat.pl). About 16% of the world collection of sorghum (235,711 accessions) is conserved in ICRISAT's gene bank in India (FAO, 2009). This collection of 37,949 accessions from 92 countries comprises 32,578 landraces, 4,814 advanced breeding lines, 99 cultivars, and 458 wild and weedy relatives (Upadhyaya et al., 2014). Most of the accessions were characterized and evaluated for several traits of interest including trait-specific germplasm. Sorghum researchers can access these useful germplasm accessions to meet their research needs. More importantly, core and minicore collections or genotype-based reference sets, representing diversity available in the whole germplasm have been formed and using these subsets new sources of variations have been identified for use in sorghum genetic improvement. Furthermore, the ICRISAT collection is divided into active and base collections (Upadhyaya et al., 2014). More than 30,000 sorghum accessions have also been conserved in the Svalbard Global Seed Vault, Norway (Upadhyaya et al., 2014). Furthermore, molecular-marker development, genome mapping, and tagging of agronomically important traits have been taken well into consideration. A large number of single-nucleotide polymorphisms were identified through whole genome resequencing (Morris et al., 2013; Mace et al., 2013).

6 Pearl millet

Pearl millet is an important staple crop in the semi-arid tracts of Asia and Africa. Globally, 66,682 accessions of pearl millet are conserved in 97 gene banks, in which, ICRISAT has the largest collection. Tremendous genetic diversity has been observed in the cultivated gene pool for morphoagronomic traits and resistance to abiotic and biotic stresses, including nutritional traits. Core and minicore collections developed at ICRISAT would facilitate extensive evaluation and identification of trait-specific diverse germplasm accessions. Interspecific crosses were also developed within the primary gene pool for widening the genetic base of elite genetic background. A large number of germplasm accessions have been characterized at ICRISAT for several morphoagronomic traits using pearl millet descriptor states (IBPGR and ICRISAT, 1993). These accessions showed large phenotypic diversity for almost all qualitative and quantitative traits. Substantial variation was also reported for morphologic traits among landraces and wild relatives from India, west and central Africa, Cameroon, Yemen, and Ghana (Dwivedi et al., 2012). Among abiotic stresses, high-temperature stress at seedling and reproductive stages has an impact on crop establishment and yield of pearl millet. Genetic variation has been observed for heat tolerance at seedling and reproductive stage among germplasm. A recent finding for reproductive stage heat tolerance over 3–4 years could identify tolerant breeding and germplasm lines (Gupta et al., 2015). Low-temperature stress at vegetative stage causes increased basal tillering and grain yield; at elongation stage, it leads to reduced spikelet fertility, inflorescence length, and decreased grain yield; at grain development stage, it leads to increase in grain yield (Fussell et al., 1980). Pearl millet germplasm tolerant to salinity have also been reported. At ICRISAT, characterization and evaluation of a large number of germplasm accessions has led to the identification of resistant/tolerant gene sources for downy mildew, smut, ergot, and rust (Upadhyaya et al., 2007). Several germplasms with multiple disease resistance to major prevailing diseases have also been identified (Dwivedi et al., 2012). Enormous variability has been reported in pearl millet germplasm collection for protein (up to 24.3%) among 260 accessions and micronutrient concentrations among 191 accessions (Rai et al., 2015). Genomic resources are expected to increase with pearl millet genome sequence due for release and faster developments in next-generation sequencing technologies, which would enhance germplasm management and crop improvement.

7 Finger and foxtail millets

Finger and foxtail millets are important ancient crops of dry-land agriculture and the climate-resilient crops for food and nutritional security. Assessing genetic variability of germplasm collections, development and use of genetic and genomic resources for breeding high-yielding cultivars, developing crop production and processing technologies, value addition for improving consumption, public–private partnerships, and policy recommendations are needed to upscale these crops to make them more remunerative to the farming community.

These crops are highly nutritious with diverse usage, well adapted to marginal lands, and mostly grown by resource-poor farmers. Worldwide more than 46,000 foxtail millet and about 37,000 finger millet germplasm accessions have been preserved and the largest collections of finger and foxtail millets are in India and China, respectively. Considerable variation exists for various biotic and abiotic stresses, and for quality including important agronomic traits. Entire genetic diversity of these crops has been captured in the form of core and minicore collections and is being used in genetic and genomic studies for identification of new sources of variation. Genomic resources are available in foxtail millet, while in finger millet these resources are being developed. Furthermore, use of genetic and genomic resources need to be accelerated to assist in developing improved cultivars of these crops.

8 Proso, barnyard, little, and kodo millets

Proso, barnyard, little, and kodo millets are highly nutritious crops and have climate-resilient traits. Globally, about 50,000 germplasm accessions of these crops have been conserved, and the largest collections of proso millet are in the Russian Federation and China, barnyard millet in Japan, and kodo millet and little millet in India. These crops have larger variation for yield and its component traits including stress tolerance related characters. Core collections representing diversity of entire collections of these crops have been developed for identification of new sources of variation for major prevailing biotic and abiotic stresses, and for quality as well as important agronomic traits. Globally, more than 29,000 accessions of proso millet, 8,000 accessions each of barnyard and kodo millet, and more than 3,000 accessions of little millet have been conserved. The ICRISAT gene bank in India conserves 849 accessions of proso millet,

749 accessions of barnyard millet, 665 accessions of kodo millet, and 473 accessions of little millet under medium- and long-term storage. Limited research works have been done on germplasm characterization and evaluation of various agronomic traits, nutritional traits, and biotic and abiotic stresses. A few studies on germplasm characterization and evaluation were conducted by Upadhyaya et al. (2011) and identified important gene sources including trait-specific germplasm in these crops. Genomic resources are limited and efforts to develop such resources through high-throughput genotyping are in progress.

References

Dwivedi, S., Upadhyaya, H., Senthilvel, S., Hash, C., Fukunaga, K., Diao, X., et al., 2012. Millets: genetic and genomic resources. In: Janick, J. (Ed.), Plant Breeding Reviews, vol. 35, AVI Publishing Company, Inc., Germany, pp. 247–375.

FAO, 2009. Commission on Genetic Resources for Food and Agriculture. Draft Second Report on the World's Plant Genetic Resources for Food and Agriculture (Final Version). Rome, 330 p. Available from: http://www.fao.org/3/a-k6276e.pdf (accessed September 2014.).

FAO, 2013. Commission on Genetic Resources for Food and Agriculture. Draft Second Report on the World's Plant Genetic Resources for Food and Agriculture. Available from: http://www.fao.org/3/a-k6276e.pdf

Frese, L., Henning, A., Neumann, B., Unger, S., 2013. CWR *In Situ* Strategy Helpdesk created and managed by S. Kell (University of Birmingham). Available from: http://www.agrobiodiversidad.org/aegro (accessed 5.10.2013.).

Fussell, L.K., Pearson, C.J., Norman, M.J.T., 1980. Effect of temperature during various growth stages on grain development and yield of *Pennisetum americanum*. J. Exp. Bot. 31 (121), 621–633.

Gupta, S.K., Rai, K.N., Singh, P., Ameta, V.L., Gupta, S.K., Jayalekha, A.K., et al., 2015. Seed set variability under high temperatures during flowering period in pearl millet (*Pennisetum glaucum* (L.) R. Br.). Field Crops Res. 171, 41–53.

IBPGR, ICRISAT, 1993. Descriptors for Pearl Millet [*Pennisetum glaucum* (L.) R. Br.]. IBPGR/ICRISAT, Rome.

Jackson, M.T., 1995. Protecting the heritage of rice biodiversity. Geo J. 35, 267–274.

Mace, E.S., Tai, S., Gilding, E.K., Li, Y., Prentis, P.J., Bian, L., et al., 2013. Whole-genome sequencing reveals untapped genetic potential in Africa's indigenous cereal crop sorghum. Nat. Commun. 4, 2320.

Morris, G.P., Ramu, P., Deshpande, S.P., Hash, C.T., Shah, T., Upadhyaya, H.D., et al., 2013. Population genomic and genome-wide association studies of agroclimatic traits in sorghum. Proc. Natl. Acad. Sci. USA 110, 453–458.

Rai, K.N., Velu, G., Govindaraj, M., Upadhyaya, H.D., Rao, A.S., Shivade, H., et al., 2015. Iniadi pearl millet germplasm as a valuable genetic resource for high grain iron and zinc densities. Plant Genet. Resour. 13 (1), 75–82.

Upadhyaya, H.D., Reddy, K.N., Gowda, C.L.L., 2007. Pearl millet germplasm at ICRISAT genebank – status and impact. J. SAT Agric. Res. 3 (1), 1–5.

Upadhyaya, H.D., Sharma, S., Dwivedi, S.L., Singh, S.K., 2014. Sorghum genetic resources: conservation and diversity assessment for enhanced utilization in sorghum improvement. In: Wang, Y.H., Upadhyaya, H.D., Kole, C. (Eds.), Genetics, Genomics and Breeding of Sorghum. CRC Press, New York, pp. 28–55.

Upadhyaya, H.D., Sharma, S., Gowda, C.L.L., Reddy, V.G., Singh, S., 2011. Developing proso millet (*Panicum miliaceum* L.) core collection using geographic and morpho-agronomic data. Crop Pasture Sci. 62, 383–389.

von Bothmer, R., Jacobsen, N., Jørgensen, R.B., Linde-Laursen, I., 1991. An ecogeographical study of the genus *Hordeum*. Systematic and Ecogeographic Studies on Crop Genepools 7 International Board for Crop Genetic Resources, Rome.

Rice

1

Bhaskar C. Patra, Soham Ray*, Umakanta Ngangkham*,*
Trilochan Mohapatra[†]
*Crop Improvement Division, Indian Council of Agricultural Research, National Rice Research Institute, Cuttack, Odisha, India; [†]Indian Council of Agricultural Research, Indian Agricultural Research Institute, New Delhi, India

1.1 Introduction

Plant genetic resources (PGR) constitute the basic raw material for any crop improvement program. It may consist of seed or vegetative propagules (tuber, sucker, rhizome, cutting, seedling, etc.) of plants and also include pollen, cell, DNA, or any other component, which contains the functional units of heredity. They are generally referred to as germplasm or genetic resource material. Sir Otto Frankel coined the word "Genetic Resources."

The green revolution in India was fueled by increasing productivity in staple food production. The rapid yield growth in the 1970s and 1980s was built on a solid foundation of systematic development of genetic resources. By 2030, the production of rice must increase by at least 25% in order to keep up with population growth and demand in the country. Accelerated genetic gains in rice improvement are needed to mitigate the effects of climate change and loss of arable land, as well as to ensure a stable global food supply. The enormous rice genetic diversity available in the gene banks will be the foundation of the genetic improvement of the crop through unraveling the new genes and traits that will help rice-producing farmers who are facing the challenges brought about by climate change, pests and diseases, and other unfavorable conditions. It is well known that the traditional rice varieties and their wild relatives constitute an invaluable gene pool in terms of resistance/tolerance to biotic and abiotic stresses, which can be exploited for developing modern varieties having enough resilience to sustain adverse climatic changes.

1.2 Origin, distribution, and diversity

Carl von Linneaus described the genus *Oryza* with a single species *Oryza sativa* in his *Species Plantarum* in 1753. Steudel (1855) and Bentham and Hooker (1861–1883) attempted the enumeration of the species of *Oryza* and Baillon (1894) provided the first classification of its species. Prodohel (1922) was the first to describe different species of genus *Oryza*, though her effort was imperfect. Later, Roschevicz (1931), a Russian scientist, wrote the first comprehensive and reliable monograph on the genus *Oryza*. Most major graminaceous crops are closely related to another domesticated grain crop; for example, wheat and barley are both in Triticeae, while *Setaria* and *Panicum* millets are in Paniceae, sorghum and maize both in Andropgoneae. However, rice is the only major crop in the tribe Oryzeae (Vaughan, 1994).

Genetic and Genomic Resources for Grain Cereals Improvement. http://dx.doi.org/10.1016/B978-0-12-802000-5.00001-0

Rice is cultivated as far north as the banks of the Amur River (53°N) on the border between Russia and China, and as far south as central Argentina (40°S). It is grown in cool climates in the mountains of Nepal and India, and under irrigation in the hot deserts of Pakistan, Iran, and Egypt. It is an upland crop in parts of Asia, Africa, and Latin America. At the other environmental extreme are floating rice, which thrive in seasonally deeply flooded areas such as river deltas – the Mekong in Vietnam, the Chao Phraya in Thailand, the Irrawady in Myanmar, and the Ganges–Brahmaputra in Bangladesh and eastern India. Rice can also be grown in areas with saline, alkali, or acid sulfate soils. Clearly, it is well adapted to diverse growing conditions.

The center of origin of any crop plant is decided on the basis of the following.

1. Distribution of the progenitor species;
2. Genetic diversity of the crop;
3. Antiquity of its cultivation in a region;
4. Diverse usages of its products and by-products;
5. Presence of a large number of words for its various products in the language of a region;
6. Use in traditions, customs, and folklores of the people;
7. On the basis of archeologic findings.

Based on these criteria, Asian rice could have originated anywhere in the region from China to India.

The origin of cultivated rice has been debated and discussed for quite a long time. The plant is of such antiquity that precise time and place of its first development will perhaps never be known (Huke and Huke, 1990). A clear understanding of the origin of cultivated rice very much depends upon a good knowledge of the taxonomy of the genus *Oryza*, the phylogenetic relationships among its species, and the genetic variability within the two cultivated species *O. sativa* of Asia and *Oyrza glaberrima* of Africa. It was assumed that *Oryza perennis* subsp. *balunga* (*Oryza rufipogon*) has given rise to *O. sativa* in Asia and *O. perennis* subsp. *barthii (Oryza longistaminata*) has given rise to *O. glaberrima* in Africa thereby proposing a monophyletic hypothesis about the origin of both cultivated species. It was also assumed that the progenitor species hybridize in nature with the cultivated species and give rise to hybrid swarms. The recent view (since 1965) about the origin of cultivated rice considers that the Asian annual wild species *Oryza nivara* has given rise to the Asian cultivated species *O. sativa* and the African annual wild species *Oryza barthii* (=*O. breviligulata*) to the African cultivated species *O. glaberrima*. These progenitor species hybridize in nature with the cultivated species and give rise to various intergrades through introgressive hybridization.

The people responsible for the origin of cultivated rice, as proposed by Prof. H. Hamada are the proto-Australoids who were widespread in South and Southeast Asia up to the second millennium BC. The migration of Aryans into the Indian subcontinent and of the Tibeto-Burmans into Southeast Asia later very much altered the ethnic picture of this region but not the spectrum of the basic ecotypic diversity of the region.

According to the available archeologic evidences, China has the oldest rice remains and richest rice culture. The rice could have originated at many sites in South Asia, Southeast Asia, and South and Southwest China. Archeologists have also found evidence that rice was an important food in Lothal and Rangpur in Gujarat during the Harappan civilization as early as 2500 BC and in the Yangtze Basin 8000 years BP during the late Neolithic period (Chang, 1976). This would support the hypothesis of "diffused origin" proposed by Harlan (1975). With the development of puddling and transplanting, rice became truly domesticated. In China, the history of rice in river valleys and low-lying areas is longer than its history as a dry land crop. In Southeast Asia, by contrast, rice was originally grown in uplands using slash and burn (shifting cultivation) practice. Migrants from south China or perhaps northern Vietnam carried the traditions of wetland rice cultivation to the Philippines during 2000 BC, and Deutero-Malayas carried the practice to Indonesia around 1500 BC. From China or Korea, the crop was introduced to Japan no later than 100 BC.

Movement to Sri Lanka was also accomplished as early as 1000 BC. The crop may well have been introduced to Greece and the Middle East by Alexander the Great's expedition to India ca. 344–324 BC. From Sicily Island, rice spread throughout the southern portion of Europe and to a few locations in North Africa. Rice cultivation was introduced to the New World by early European settlers. The Portuguese carried it to Brazil and the Spanish introduced its cultivation to several locations in Central and South America. The first record from North America dates to 1685, when the crop was produced on the coastal lowlands and islands of what is now South Carolina. The crop may well have been carried to that area by slaves brought from Madagascar. In the eighteenth century, it spread to Louisiana and by twentieth century it spread to California (Fig. 1.1).

Figure 1.1 Origin and spread of cultivated rice.

The nonshattering types were already evolved by 5000 BC at the Hemudu site in the Taifu area of eastern China. The earliest and most convincing evidence has come from ^{14}C and thermoluminescence test of the pottery shreds bearing imprints of grains and husks of *O. sativa* discovered at Non Nok Tha in the Korat area of Thailand (Higham and Kijngam, 1984) dating back to 4000 BC. This evidence not only pushed back the documented origin of cultivated rice but when viewed in conjunction with plant remains of 10,000 BC discovered at Spirit Cave on the Thailand–Myanmar border suggests that agriculture itself is much older than what it was thought of earlier.

Linguistic evidence also points to the early origin of cultivated rice in parts of Southeast Asia, which is considered as the heartland of rice cultivation. In several regional languages, the general terms for rice and food, or for rice and agriculture, are synonymous. Such is not the case in any other part of the world. Hindu and Buddhist scriptures make frequent reference to rice as staple food, as well as the grain being used as a major offering to the gods and goddesses. In contrast, there is no reference of rice in Jewish scripture of the Old Testament and Egyptian records as well.

The center of origin and centers of diversity of two cultivated species *O. sativa* and *O. glaberrima* have been identified using genetic diversity, historical and archeologic evidences, and geographical distribution. It is generally agreed that river valleys of the Yangtze and Mekong rivers could be the primary center of origin of *O. sativa*, while the Delta of Niger River in Africa is the primary center of origin of *O. glaberrima* (Porteres, 1956). The foothills of the Himalayas, Chhattisgarh, Jeypore tract of Odisha, northeastern India, northern parts of Myanmar and Thailand, Yunnan Province of China, and so on, are some of the centers of diversity for Asian cultigens. The inner delta of Niger River and some areas around the Guinean coast of Africa are considered to be the center of diversity of the African species of *O. glaberrima* (Chang, 1976; Oka, 1988).

Varieties of the same group when grown in different seasons and different cultural managements are named as different ecotypes (crop growing time):

1. Boro: Nov–Dec/April–May: In water-stagnated areas or with irrigation; cold tolerant at seedling stage (spring or summer rice).
2. Aus: April–Aug: Autumn rice, broadcast (*aus*) or transplanted (*ahu*).
3. Broadcast Aman: April–Dec: Broadcast, deep-water rice (also called *bao*), shallow-water rice (also called *asra*).
4. Transplanted Aman: July–Dec: Winter rice, transplanted, photoperiod sensitive (*sali, kharif*).

The natural hybridization between *aus* and *O. rufipogon* gives rise to *aman* in eastern India; whereas *japonica* and *O. rufipogon* give rise to *sali* in Brahmaputra valley, *boro* is the intermediate between *aus* and *aman*. The *aman* ecotype migrated to Southeast Asia and spread there very fast. It gave rise to the *tjereh* or *bulu* ecotype in Indonesia. The order of ecotypes with respect to its mean sterility value are *aus*, *aman*, *boro*, *tjereh*, *sali*, and *japonica*.

The *bulu* types of Indonesia could have been the progenitor of *javanica* rice. The closer relationships between *japonica* and *javanica* ecotypes could be attributed to the possible closer genetic relationship between the populations of *O. nivara* of South China and Southeast Asia (Glaszmann, 1986; Chang, 1985). The *aman* ecotype was evolved from the *aus* ecotype as a result of introgression of *O. rufipogon* genes into the *aus* ecotype in the lower Gangetic valley. According to Ramiah and Ghose (1951) and Chang (1976), the deepwater rice cultivars are the product of introgression of *O. rufipogon* characters into *O. sativa*. Phylogenetic analyses based on single-nucleotide polymorphism (SNP) data confirmed differentiation of the *O. sativa* gene pool into five varietal groups – *indica*, *aus/boro*, basmati/sadri, tropical *japonica*, and temperate *japonica*.

Vaughan (1989) reported a new species of *Oryza* (*Oryza rhizomatis*) from Sri Lanka. It is a diploid species closely related to *Oryza eichingeri* of Sri Lanka and *Oryza officinalis* of South and Southeast Asia. The genome of *O. rhizomatis* was determined by crossing with other known genomes of *Oryza* species and studying the pairing of chromosomes at the meiotic stage in the F_1 hybrids. Dhua (1994) worked on the genome analysis of *O. rhizomatis* and found that it has DD genome. The spikelets of *O. rhizomatis* are characterized by a wash of purple pigmentation. The largest size of spikelets is seen in *Oryza australiensis* followed by that of *Oryza punctata*, *Oryza grandiglumis*, and *Oryza alta*. The smallest size of spikelets is seen in *O. minuta* and *O. eichingeri*. The longest awns are reported in *O. brachyantha*, an annual African species.

Unfortunately, there is no unanimity among the rice researchers regarding delimitation of species in the Asian complex of *O. sativa*. Whereas some recognize three species, namely, *O. rufipogon*, *O. nivara*, *O. sativa* in Asia in this complex (Sharma and Shastry, 1965; Chang, 1976, 1985; Vaughan, 1994), others recognize only two species (*O. rufipogon* and *O. sativa*) in this complex and treat the annual species as variations within *O. rufipogon* (Morishima, 1984). Besides, the nomenclature of these elements has also been changing. The perennial wild species was earlier referred to as *O. perennis* by all rice workers until Bor (1960) identified it as *O. rufipogon*. The annual wild species was earlier known as *Oryza fatua* Koenig (a *nomen nudum*). Sharma and Shastry (1965) assigned it a new name (*O. nivara*) as *O. fatua* was not a validly published name for the annual wild species (Fig. 1.2).

Watt (1891), Roschevicz (1931), Chatterjee (1951), Oka (1964, 1988), Morishima (1984), and Sharma (2003) have discussed the probable progenitor species from which *O. sativa* could have originated. According to them, more than one wild species have played a role in the origin of Asian cultivated rice. According to Roschevicz (1931), *O. sativa* originated from *O. sativa* f. *spontanea*. According to him, *O. sativa* f. *spontanea* "indubitably represents a complex of several species" and is "indisputably the ancestor of the majority of varieties of cultivated rice." He believed that *O. officinalis* and *O. minuta* have also played a role in the origin of small-grained rice varieties and *Oryza coarctata* in the origin of saline-resistant rice varieties. The taxonomic delimitation of the various elements of *O. sativa* complex was not very clear at that time. Besides, the chromosome numbers or genomic constitutions of these species were also not known at that time. Therefore, the role of *O. minuta* and *O. coarctata* has been ruled out as these are tetraploid ($2n = 48$) species.

According to Chatterjee (1951), the annual wild species (*O. nivara*) has played the major role in the origin of cultivated rice though *O. officinalis* has also played a role especially in the origin of small-grained rice varieties. In other words, Roschevicz (1931) and Chatterjee (1951) proposed a polyphyletic origin of *O. sativa*. *O. officinalis* is a diploid species ($2n = 24$) with genomic constitution CC. The F_1 hybrid between

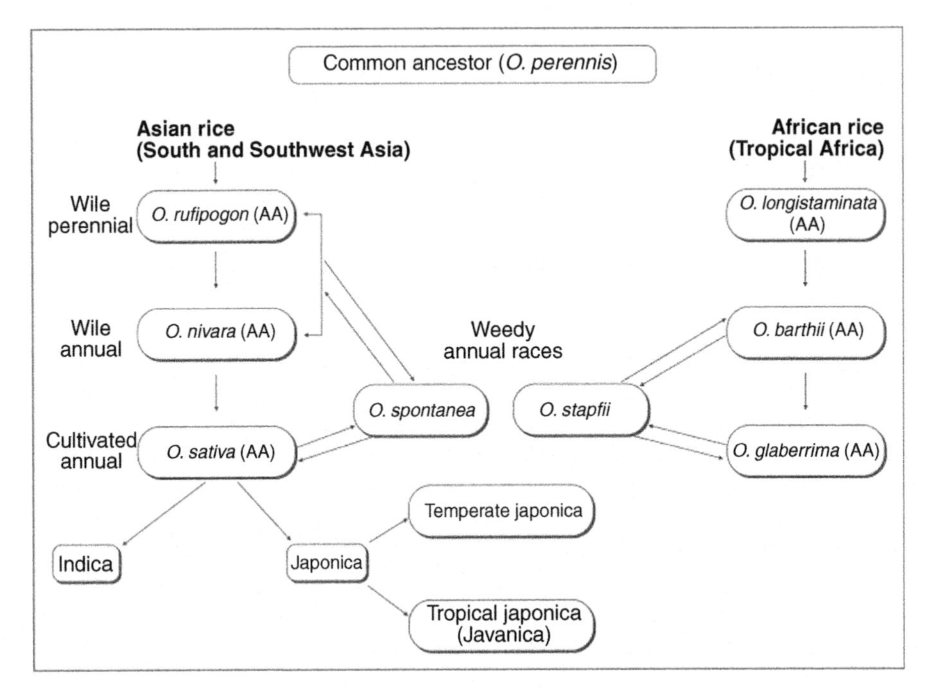

Figure 1.2 Schematic representation of the evolutionary pathways of Asian and African cultivated rice.

O. sativa and *O. officinalis* is completely sterile and the chromosomes of the two species do not pair at meiotic metaphase I. Besides, the two species grow in distinctly different ecologies and hence there is hardly any chance of their hybridization in nature. There is, therefore, hardly any probability that *O. officinalis* could have played a role in the origin of Asian cultivated rice.

Sampath and Rao (1951) treated the perennial wild species of Asia (*O. rufipogon),* Africa (*O. longistaminata*), and America (*Oryza glumaepatula*) as a single species *O. perennis* Moench., as suggested by Chatterjee (1948). They proposed that the perennial form of *O. perennis* in Africa (*O. longistaminata*) has given rise to *O. glaberrima* in tropical West Africa and the perennial form of *O. perennis* in Asia (*O. rufipogon*) has given rise to *O. sativa* in South and Southeast Asia. According to them, the two perennial species hybridize in nature with their cultivated counterparts to form natural hybrids. They considered that the annual wild species (*O. nivara* of Asia and *O. barthii* of Africa) occurring in the two continents are fixed forms of such hybridization. In other words, Sampath and Rao (1951) proposed a monophyletic origin for cultivated rice of Africa as well as Asia. Their view was further elaborated by Richharia (1960). Later, Sampath (1962) modified his hypothesis and recognized the Asian *O. rufipogon* and African *O. longistaminata*, the perennial rice, as two distinct and different species and, in this sense, demolished his own hypothesis of monophyletic origin of cultivated rice. Besides, he recognized the existence of an annual wild species in South and Southeast Asia but defined it as the fixed form of natural hybrids between the perennial (*O. perennis*) and the cultivated species *O. sativa* (Sampath, 1964).

Ramiah and Ghose (1951) recognized three species in the *O. sativa* complex of Asia: (1) a perennial wild species (*O. perennis* Moench., *O. rufipogon*), (2) an annual wild species (*O. fatua* Koenig, *O. nivara*), and (3) the cultivated rice (*O. sativa*). According to them, the annual wild species (their *O. fatua*, our *O. nivara*) is the progenitor of the Asian cultivated rice. According to Chang (1976, 1985), the broad belt from the foothills of the Himalayas and the Gangetic belt up to the southern China and northern Vietnam could be the "homeland" of cultivated rice.

The archeologic evidence suggests that agriculture originated in 10,000 BC on the Thailand–Myanmar border. Rice originated around 5000 BC at the Hemudu site in Taifu area of eastern China. The evidences from pottery shreds suggest that rice originated around 4000 BC at Non Nok Tha in Korat in northern Thailand (Higham and Kijngam, 1984) and was confirmed by ^{14}C test. According to Morishima (1984), "all biologic evidence indicates that the homeland of *O. sativa* must be the large area between eastern India and southern China."

The diversity within the Asian cultivated rice (*O. sativa*) is enormous. Some controversy exists over when and where rice was domesticated (Sweeney and Mc Couch, 2007; Huang et al., 2012). It is fairly safe to say that rice was being cultivated at least 10,000 years ago and that it was domesticated from its wild ancestor *O. rufipogon* (Khush, 1997). Two major subgroups of rice, *indica* and *japonica*, led rice genetic resource specialists to conclude that there were two centers of origin. One was thought to be in the tropical regions of South Asia where *indica* rice varieties dominated and the other near Central China where *japonica* rice dominated (Londo et al., 2006; Vaughan et al., 2008). It has generally been recognized that genetically the *japonica* (*sensu stricto*) is a fairly homogeneous group whereas the *indica* is a highly heterogeneous group (Jennings, 1966). With the discovery that there are tropical *japonica* traditional varieties and another major group in South Asia called *aus* rice, things have become less clear. A recent study based on extensive DNA sequence analysis of *indica* and *japonica* varieties has concluded that there may have been only one center of domestication. Because rice was domesticated from *O. rufipogon*, repeated crosses occurred between increasingly domesticated rice and its wild ancestor, resulting, along with natural mutations, in increasing genetic diversity until modern times. This out-crossing continues even today, with weedy rice being an irritant to most farmers at one time or other. This probably explains the extraordinary diversity in rice germplasm. Weedy rice belongs to the same genus and species as cultivated rice but with different forma. It appears as hybrid swarms due to introgression of genes between wild and cultivated species in nature. In Asian rice, it is known as *Oryza spontanea* whereas in the African context it is known as *Oryza stapfii*. It grows faster; produces more tillers, panicles, and biomass; makes better use of available N; shatters earlier; has better resistance to adverse conditions; and possesses longer dormancy in soil. Because of its high competitive ability, it becomes a serious threat to rice growers worldwide. Great morphologic variability, similar growth behavior, and high biologic affinity with cultivated varieties make its control difficult.

The cultivated rice (*O. sativa* L.) is said to have originated in the southeast part of Asia. India forms a major part of this region and is traditionally rich in the genetic diversity including the wild progenitors. Rice research started in India a little more than a hundred years ago with the establishment of the Imperial Agricultural Research Institute in Pusa, Bihar in 1905. Real rice breeding work started at Paddy Breeding Station, Coimbatore in 1912 mainly through pure line selection of landraces. This was further geared up when the country was facing widespread drought and famine. Several million people suffered a setback due to the Bengal famine in 1943, which was caused by the brown spot disease of rice in an epidemic form. Since then, systematic research in developing biotic (disease/insect) and abiotic stress (drought/submergence/salinity) resistant/tolerant varieties were undertaken through pure line selection from a vast range of landraces available in the country. Thus, keeping a comprehensive account of rice germplasm collection through exploration and acquisition, its conservation, characterization, evaluation, documentation, and utilization have always remained priority for any rice research.

In India, during the early part of the twentieth century and more specifically in the years between 1910 and 1920, Mr F.R. Parnell, Mr G.P. Hector, and Mr Graham initiated the collection of rice germplasm from the Madras Presidency, Bengal, and the Central provinces, respectively (Table 1.1). At the same time, in 1911 the first Agricultural Research Station devoted to rice research was established in Dacca (now Dhaka in Bangladesh) and the paddy breeding station was established at Coimbatore in 1912. Subsequently, Karimganj (1913) and Titabar (1923) in Assam; Pattambi in Kerala and Nagina in Uttar Pradesh in 1927; Chinsurah, West Bengal in 1932; Habiganj (now in Bangladesh) in 1934; Sabour, Bihar in 1936; and many more research stations were established in different agroclimatic zones of the country (Table 1.2). Before the formation of Orissa province in 1936, two rice research stations were established for southern Odisha at Berhampur in 1932 under the paddy specialist of Madras Presidency at Coimbatore and the other at Cuttack for developing rice varieties in Mahanadi delta area under the control of the paddy specialist of Bihar and Orissa province at Sabour with financial support from the Indian Council of Agricultural Research (ICAR). When Orissa became a separate province in 1936, the Cuttack centre became the main rice research station of the state and Berhampur became its substation. Later, the Cuttack center was shifted to Bhubaneswar. However, the Cuttack farm was transferred to the Government of India to establish a central institute for rice research. Dr K. Ramiah was invited as Special Officer to select a suitable location for establishing a research center on rice and he chose this farm. Accordingly, the Govt. of India, with the help of the then premier of Orissa, the Maharaja of Paralakhemundi and Dr P.K. Parija, the then Director of Agriculture, Government of Orissa, established the Central Rice Research Institute (CRRI) at Cuttack in April 1946 with Dr Ramiah as its founding director. All these rice research stations collected traditional rice varieties from their respective regions/localities, practiced pure line selection to identify higher-yield potential lines, and recommended them for general cultivation by the farmers.

Table 1.1 **Germplasm acquisitions at CRRI**

Year	Source		CRRI accessions	Remarks
1964–1983	D.P. Bhattacharya	Plant Physiology Division, CRRI, Cuttack	AC 8601-8672	Drought-tolerant lines
1969–1983	S. Govindaswami and A.K. Ghosh	CRRI, Cuttack	AC 8901-8932	Scented types
1972–1983	R.K. Bhattacharya	Central Soil Salinity Research Institute, Substation Canning, 24 Parganas, WB	AC 8001-8573	Salt-tolerant lines
1973	S.D. Sharma	IARI Regional Station, Hyderabad	AC 16785-16849	
1974	R.K. Arora	Plant Introduction Division, IARI, New Delhi	AC 16850-16881	Meghalaya
1974	T.A. Thomas	Plant Introduction Division, IARI, New Delhi	AC 16882	Madhya Pradesh
1974	T.A. Thomas	Plant Introduction Division, IARI, New Delhi	AC16883-16887	Uttar Pradesh
1983	A. Panda	CRRI, Cuttack	AC 16888-16960	Western Odisha
1976	D.P. Srivastava	CRRI, Cuttack	NCS 1-152 (AC 17001-17152)	Ranchi and Chainbasa districts of Bihar
1976	D. Choudhury	CRRI, Cuttack	NCS 153-212 (AC17153-17212)	Gonda, Bahraich, Azamgarh, Mirzapur, Gorakhpur, and Faizabad districts of UP, India
1976	R.N. De and S. Biswal	CRRI, Cuttack Rice specialist, Chinsurah	NCS 213-335 (AC 17213-17335)	Bankura, Purulia, and New Jalpaiguri districts of WB
1976	J.K. Roy	CRRI, Cuttack	NCS 336-401 (AC 17336-17401)	Sambalpur, Dhenkanal, Sundargarh and Bolangir districts of Odisha
1976	M. Nagaraju	CRRI, Cuttack	NCS 402–413 (AC 17402-17413)	Andhra Pradesh

(Continued)

Table 1.1 Germplasm acquisitions at CRRI *(cont.)*

Year	Source		CRRI accessions	Remarks
1976	D.P. Ghorai	CRRI, Cuttack	NCS 414-620 (AC 17414-17620)	Rewa and Raipur districts of MP
1976	S.N. Ratho	CRRI, Cuttack	NCS 621-784 AC 17621-17784)	Karnataka
1976	P.J. Jachuck	CRRI, Cuttack	NCS 785-1038 (AC 17785-17999	Pune, Satara, Gangabarda, Ajra, Ratnagiri, Nipanni, Patan, Koyna valley of Maharashtra
1976	S.D. Sharma	IARI Regional Station, Hyderabad	ARC 18001-18623 (AC 18001-18623)	Derivatives of ARC types
1977	K.D. Sharma	HPAU, Palampur	NCS 1127-1453 (AC 19127-19453)	Himachal Pradesh
1976–1978	N.D. Desai	GAU, Nawagam Gujarat	NCS 1454-1594 (AC 19454-19594)	Gujarat
1977	J.K. Roy	NDUAT, Faizabad	NCS 1595-1610 (AC 19595-19610)	Deep-water rice from Ghaghara ghat, UP
1978	N.P.S. Varde and R.C. Mandal	ICAR Research Complex, Goa and CPCRI, MARGAO	NCS 1611-1646 (AC19611-19646)	Goa
1978–1979	S.M. Ahmed	Agril. Res. Station, Banswara, Rajasthan	NCS 1751-1954 (AC 19751-19954)	Four districts of Rajasthan
1979	Ashutosh Roy	Rice specialist, Chiplima, Sambalpur	NCS 1996-2185 (AC 19996-19999 and AC 22000-22185)	Odisha
1979	Sh. Mansoor Kazim	NBPGR, New Delhi	NCS 2186-2340 (AC 22186-22340)	Agril. Research Station, Vadgaon (Maval), Dist. Pune, Maharashtra
1978–1979	D.M. Maurya	NDUAT, Faizabad	NCS 2341-2563 (AC 22341-22563)	Uttar Pradesh

1978–1979	A.T. Roy	Rice Specialist, Chiplima, Sambalpur	NCS 2564-2670 (AC 22564-22670)	Odisha
1978–1979	J.L. Dwivedi	Asstt. Rice Breeder, Rice Research Station, Titabar, Assam	NCS 2671-2689 (AC 22671-22689)	Assam
1979–1980	M.S. Patel	Main Rice Research Station, Gujarat Agril. University, Nawagam	NCS 2814-2873 (AC 22814-22873)	Kheda district, Gujarat
1979	IRRI	Manilla, Philippines	AC 16016-16025	Exotic varieties
1980	K. Maruyama	National Hokiriku Agri. Expt. Sta.1-2-1, Indane, JOETSU-SHI, NIIGATA-KEN, Japan	AC 16026-16073	Leading varieties of Japan
1980	M.N. Koppar	NBPGR, New Delhi	AC-16077-16190	Landraces from Himachal Pradesh
1980	T.A. Thomas	NBPGR, New Delhi	AC 16091-16196	Sikkim
1980	T.A. Thomas	NBPGR, New Delhi	AC 16197-16779	Exotic collections from Nigeria, West Africa
1980	T.A. Thomas	NBPGR, New Delhi	AC16780-16784	–
1978–1980	R.M. Singh	BHU, Varanasi	NCS 2889-2960 (AC22889-22960)	Uttar Pradesh
1980	S.A. Dadlani	IARI, New Delhi	AC 22961-23043	Collections maintained at Rice Research Station, Kapurthala, Punjab
1980	A.N. Asthana	ICAR Research complex for NEH Region, Shillong	AC 23113-23414	NE Hill region, India
1980	S.R. Das	College of Agriculture, OUAT, Bhubaneswar	AC 23415-23916	Odisha
1980	R.C. Chaudhary	Agril. Res. Institute (Rajendra Agril. University), Mithapur, Patna	AC 23917-24006)	Bihar
1980	A.T. Roy	Chiplima, Sambalpur	AC 24007-24066	Odisha
1980	J.K. Roy	CRRI, Cuttack	AC 24067-24152	Coastal areas of Odisha
1980	G.J.N. Rao	CRRI, Cuttack	AC 24153-24212	Choudwar area of Cuttack, Odisha
1980	S.C. Prasad	Agril. College, Kanke, Ranchi	AC 24213-24513	Chotanagpur, Bihar

(Continued)

Table 1.1 Germplasm acquisitions at CRRI *(cont.)*

Year	Source		CRRI accessions	Remarks
1980–1981	H.K. Nagar	NDUAT, Faizabad	AC 24514-24554)	Collections from Uttar Pradesh
1980–1981	J.K. Roy	CRRI, Cuttack	AC 24555-24678	Chilka and Basudevpur areas (coastal saline) of Odisha
1980–1981	J.K. Roy	CRRI, Cuttack	AC 24679-24685	Alkaline resistant materials from Uttar Pradesh
1980–1981	J.K. Roy	CRRI, Cuttack	AC 24686-24697	Saline tolerant materials from Kerala
1980–1981	J.K. Roy through AICRIP	CRRI, Cuttack	AC 24698-24700	Saline tolerant materials from Odisha
1980–1981	J.K. Roy through AICRIP	CRRI, Cuttack	AC 24701-24707	Tamil Nadu
1980–1981	J.K. Roy through AICRIP	NBPGR	AC 24708-24720	Monkompu, Kerala
1981	R.K. Arora	NBPGR, New Delhi	AC 24721-24784	Sikkim
			AC 24785-24900	Trichur Dist, Kerala
1981	J.K. Roy	CRRI, Cuttack	AC 24901-24960	Nayagarh, Daspalla, Narsinghpur areas of Puri district, Odisha
1981	N.K.C. Patnaik	CRRI, Cuttack	AC 24961-24964	24 Parganas district, WB
1981	A. Panda	CRRI, Cuttack	AC 24965-24972	Balasore district, Odisha
1981	J.K. Roy	CRRI, Cuttack	AC 24973-24985	Nirgundi of Cuttack district, Odisha
1981	R.K. Arora	NBPGR, New Delhi	AC 24986-25019	Ratnagiri and Pune districts of Maharashtra
1981	M.N. Koppar	NBPGR, New Delhi	AC 25020-25037	Champaran district of Bihar
1981	M.N. Koppar	NBPGR, New Delhi	AC 25499-25506	UP, HP, Punjab, and Haryana
1981	K.P.S. Chandel	NBPGR, New Delhi	AC 25507-25510	J&K
1982	R.K. Arora	NBPGR, New Delhi	AC 25038-25055	Darjeeling district of WB
1982	J.K. Roy	CRRI, Cuttack	AC 25056-25102	Puri district of Odisha

1982	R.H. Richharia and M.N. Srivastava	Zonal Agril. Res. Station, Raipur, MP	AC 25103-25264	Madhya Pradesh
1982	J.K. Roy	CRRI, Cuttack	AC 25487-25497	Paradip port of Cuttack, Odisha
1984	Siva Subramaniyan	Tamil Nadu Agril. University, Coimbatore	AC 26500-26626	Tamil Nadu
1984	R.C. Chaudhury	Agril. Res. Instt. Mithapur, Patna (Rajendra Agril. University, Bihar)	AC 26662-26694	Bihar
1984	M. Gopal Krishnan	Regional Res. Station, VC Farm, Mandya	AC 26996-26720	Karnataka
1984	H.K. Gardoo	Rice Research Station, Khudwani, Anantnag	AC 26733-26740	J&K
1984	S. Sardan	ICAR Res. Complex for NEH Region, Tripura Centre, Lembiechera	AC 26766-26769	Tripura
1984	S.C. Pradhan	Technical Asst. CRRI, Cuttack	AC 26770-26777	Amiyajhar, Cuttack, Odisha
1984	P.N. Sreedharan	CRRI, Cuttack	AC 26778-26782	Paradip port area, Cuttack, Odisha
1984	P.N. Sreedharan	CRRI, Cuttack	AC 26783-26786	Rajnagar, Coastal Saline areas, Odisha
1984	Sr. Rice Breeder	Birsa Agril. University, Kanke, Ranchi	AC 26787-26821	Bihar
1984	P.P. Khanna	NBPGR, New Delhi	AC 26884-26897	Lalitpur, Jhansi, Banda districts of UP
1984	S. Devadata	CRRI, Cuttack	AC 26898-26908	Odisha
1984	R. Sreedhar	CRRI, Cuttack	AC 26909-26920	Odisha
1984	J.S. Nanda	G.B. Pant University of Agric. and Tech., Pantnagar, UP	AC 26921-26931	Uttar Pradesh
1984	D.M. Maurya	NDUAT, Faizabad, UP	AC 26932-26945	Uttar Pradesh
1984	S.D. Sharma	CRRI, Cuttack	AC 26946-26950)	Sundargarh district of Odisha
2000	B.N. Singh	West African Rice Development Board (WARDA), CGIAR	96 accessions	Exotic collections

(Continued)

Table 1.1 Germplasm acquisitions at CRRI *(cont.)*

Year	Source		CRRI accessions	Remarks
2000	B.C. Patra	CRRI, Cuttack	92 accessions	Assam (12 districts namely Kamrup, Marigaon, Nagoan, Golaghat, Jorhat, Sonitpur, Darrang, Nalbari, Barpeta, Karimganj, Hailakandi, and Cachar
2001	B.C. Patra	CRRI, Cuttack	40 accessions	North Bihar (14 districts namely Katihar, Purnea, Saharsa, Supaul, Madhubani, Darbhanga, Samastipur, Vaishali, Muzaffarpur, Begusarai, Patna, Khagaria, Rohtas, and Bhojpur)
2001	B.C. Patra	CRRI, Cuttack	25 accessions	North Bengal (seven districts, namely, Murshidabad, Malda, Dakshin Dinajpur, Uttar Dinajpur, Darjeeling, Jalpaiguri, and Cooch Bihar
2001	OIC	RRURRS, Hazaribag	742 accessions	Upland rice germplasm
2002	B.C. Patra	CRRI, Cuttack	24 accessions	Eastern UP (12 districts, namely, Balia, Ghazipur, Sant Rabi Das Nagar, Varanasi, Basti, Sant Kabir Nagar, Siddharth Nagar, Gorakhpur, Maharajganj, Deoria, Mau, and Kushi Nagar)
2005	Directorate of Rice Research	Hyderabad	1224 accessions	Indian germplasm
2005	NBPGR	New Delhi	2262 accessions	Indian germplasm
2006	Directorate of Rice Research	Hyderabad	1224 accessions	Indian germplasm
2006	NBPGR	New Delhi	2005 accessions	Indian germplasm
2007	NBPGR	New Delhi	2556 accessions	Indian germplasm

2008	NBPGR	New Delhi	1911 accessions	Indian germplasm
2009	B.C. Patra	CRRI	127	Breakfast rice from Jajpur, Cuttack, Balasore, Mayurbhanj, Keonjhar, Dhenkanal, Angul, Jagatsinghpur, Puri, and Nayagarh of Odisha
2010	B.C. Patra	CRRI, Cuttack	56 accessions	Cold tolerant rice from the districts of East Siang, Pasighat, Itanagar, Papum Pare, and West Kamang Arunachal Pradesh
2011	D. Swain	CRRI, Cuttack	74 accessions	Collections from Tripura
2012	B.C. Patra and B.C. Marndi	CRRI, Cuttack	409 accessions	Collections from Assam, Arunachal Pradesh, Odisha
2012	B.C. Patra and D. Pani	CRRI, Cuttack NBPGR, CRRI campus, Cuttack	125 accessions	Coastal Odisha (submergence tolerant landraces)
2014	B.C. Patra and B.C. Marndi	CRRI, Cuttack	87 accessions	Wild/Weedy rice germplasm from Jharkhand and Chhattisgarh states

Table 1.2 **Establishment of rice research stations in India**

1911	Dacca (Dhaka, Bangladesh)
1912	Coimbatore (Paddy Breeding Station)
1913	Karimganj, Assam and Anakapalli, AP
1914	Nagercoil, TN
1917	Karjat, Maharashtra
1921	Aduthurai, Thanjavur, TN and Mugad, Karnataka
1923	Titabar, Assam
1925	Maruteru, AP
1927	Nagina, UP
1927	Pattambi, Kerala
1932	Chinsurah, WB; Nagina (Bijnor), UP; Cuttack, Odisha (later shifted to Bhubaneswar); and Berhampur, Odisha
1934	Habiganj (now in Bangladesh)
1936	Sabour, Bihar
1937	Ambasamudram, Tirunelveli, TN
1946	Central Rice Research Institute, Cuttack
1960–onward	Agricultural Universities in different states (GBPUAT-1961, OUAT, and PAU-1962) and other 82 SAUs

1.3 Germplasm exploration and collection

Nikolai Ivanovich Vavilov (1887–1943), a Russian scientist was the foremost plant geographer of the world. He had organized over 100 collecting missions. His major foreign expeditions included those to Iran (1916), the United States, Central and South America (1921, 1930, and 1932), the Mediterranean, and Ethiopia (1926–1927). In India, Sardar Harbhajan Singh was one of the most distinguished plant explorers of India and was referred by many as the Indian Vavilov. In 1971, he was conferred with the Padma Shri for his outstanding contribution in making promising germplasm introductions from all over the world. Later, R.H. Richharia, the former Director of CRRI, had introduced 67 varieties from Taiwan and tested them at CRRI. Two or three cultures were dwarf types and one of them was identified as Taichung Native 1 (TN 1), which laid the foundation for the Green Revolution in the country.

In the past, the scientists involved with the crop improvement programs at different research stations undertook the evaluation of germplasm and were choosing the varieties through pure line selection. Several donors were identified and utilized for the crop improvement programs. This led to the recommendation of 394 varieties for general cultivation, as pure line selections, from the collected germplasm. In 1955, when Dr N. Parthasarathy was Director, the CRRI undertook its first planned exploration and collection mission of rice germplasm in the Jeypore tract (now Koraput district of Odisha). The collection program continued for 5 years (1955–1959) by a team of scientists led by Dr S. Govindaswami and supported by a scheme sanctioned by the ICAR. This mission was popularly known as the Jeypore Botanical Survey (JBS) and was the first of its kind, ever organized in the world to collect rice germplasm

(Chang, 1989). The team explored about 27,000 km^2 and collected a total of 1,745 cultivated rice and 150 wild rice accessions (Govindaswami and Krishnamurty, 1959). Later when Dr R.H. Richharia became Director of CRRI, he initiated the exploration and collection of rice germplasm from Manipur and Nagaland during 1965–1969. This mission was spread over a period of 5 years and a total of 874 accessions were collected. Simultaneously, a PL-480 project on collection of rice germplasm was operative during 1967–1972 with Dr M.S. Swaminathan and Dr S.V.S. Shastry at the Indian Agricultural Research Institute (IARI), New Delhi. In this program, during a period of 5 years, Dr S.D. Sharma and his associates collected a total of 6630 accessions from all the districts of Arunachal Pradesh, Nagaland, Manipur, Tripura, Meghalaya, and North Lakhimpur, Guwahati, and Goalpara districts of Assam. This mission was popularly known as the Assam Rice Collection (ARC). During 1970–1979, a special program was undertaken to collect rice germplasm from all the rice-growing districts of Madhya Pradesh by Dr R.H. Richharia after he left CRRI in 1969. He explored 42 districts and collected a total of 19,226 accessions, which formed the Raipur Collection. A special drive for upland paddy varieties under cultivation in Andhra Pradesh, Karnataka, Maharashtra, Madhya Pradesh, Uttar Pradesh, Odisha, and West Bengal resulted in the collection of 1938 cultivars. In 1975, under the leadership of Dr J.K. Roy, a comprehensive exploration and collection program was drawn for the whole country especially for the traditional rice-growing areas of Karnataka, Maharashtra, Madhya Pradesh, Uttar Pradesh, Bihar, West Bengal, and Odisha covering 30 districts of 7 states. This program is popularly known as the National Collection from States (NCS) and resulted in the collection of 1038 accessions. Similarly, another collaborative program was in operation at CRRI with agricultural universities, state departments of agriculture, and ICAR centers for collection of rice germplasm all over the country during 1978–1980. The materials were collected from over 100 districts of 14 states and a total of 6349 accessions were collected. The National Bureau of Plant Genetic Resources (NBPGR) augmented the collections during 1983–1989 by 4862 accessions. In addition, joint explorations by NBPGR in collaboration with state agricultural universities (SAUs) during 1978–1980 and CRRI during 1985 resulted in the collection of about 7000 and 447 accessions (from Sikkim, South Bihar, and parts of Odisha), respectively. The ICAR-VPKAS, Almora explored the hilly regions of Uttar Pradesh and 1247 accessions of primitive cultivars were collected. Increased interest in herbal medicines during the last decade has necessitated the collection of rice germplasm with special emphasis on their medicinal value from Bastar region of Chhattisgarh and the Palakkad area, the northern part of Kerala for the world famous "njavara" rice. Recently, landraces/farmers' varieties from Assam have been found to have high levels of protein (14–15%). Traditional landraces, like Bindli, are now reported to have high Zn (>50 ppm) in brown rice. A detailed characterization of landraces available in the gene bank therefore bears a paramount importance. This will enable adequately addressing the demand for specialty rice like medicinal rice, high protein rice, and high micronutrient rice.

In the new millennium, trait-specific collection programs were undertaken. To mention a few are the boro rice germplasm from Assam, north Bihar, north Bengal, and eastern Uttar Pradesh (UP); saline-tolerant rice from Pokkali region of

Kerala; bao rice from deep-water areas of Assam and Meghalaya; aman rice from West Bengal; Kalanamak (scented) rice from eastern UP, medicinal rice from Kerala and Chhattisgarh; high-altitude cold-tolerant germplasm from Sikkim and Tripura, and so on. Attempts are being made to collect specialty rices such as aromatic rice, soft rice, wine rice (country liquor known as *Handia* in eastern India), glutinous or waxy rice, colored rice (brown, black, red), beaten rice, pop rice, organic rice, and nutritional rice. During 2011–2012, the Department of Agriculture, Government of Odisha in collaboration with CRRI collected about 851 farmers' varieties for the purpose of conserving and providing support to farmers' rights under the PPV&FR Act, 2001.

The diversity in nondomesticated rice is also not less in our country. *O. rufipogon* is a perennial wild rice that grows abundantly in the roadside ditches and margins of ponds. Similarly, *O. nivara*, an annual wild rice, grows quite frequently in the seasonal pools. *O. sativa* f. *spontanea* often infests the cultivated fields and its eradication poses a serious problem for the farmers. *O. officinalis* and *Oryza granulata* have been reported from the forest shades. *O. coarctata* (*Porteresia*) grows in abundance in the tidal creeks in the seacoast especially in the Bhitarkanika mangrove areas in the mouth of Brahmani River and in the Sunderbans. *O. rufipogon*, *O. nivara*, and *O. officinalis* have already been utilized in the genetic improvement of cultivated rice.

During 1984–1989, a total of 204 accessions of wild species, namely, *O. nivara*, *O. rufipogon*, and *O. officinalis*, and 3697 accessions of cultivated rice were collected from north Odisha, south Bihar, West Bengal, Sikkim, Mizoram, Maharashtra, Karnataka, Goa, and Gujarat in collaboration with NBPGR. The importance of wild rice was felt when resistance to the grassy stunt virus biotype-1 (Khush and Ling, 1974) was identified in one of the accessions of *O. nivara* (IRGC Acc. 101508) collected from Basti district in Uttar Pradesh. Since then efforts have been ongoing to collect wild and weedy rice from different regions of the country by the CRRI (Table 1.3). In 1984, the Division of Genetic Resources was created in CRRI to look after all the

Table 1.3 Collection of wild/weedy species of rice by CRRI

Year	Species	No. of collections	Area/region
1984–1989	*O. nivara* *O. rufipogon* *O. officinalis*	204	Odisha, West Bengal, Maharashtra, Karnataka, Goa, Gujarat
1994–1995	*P. coarctata*	9	Bhitarkanika (Odisha)
1999–2003 (under NATP)	*O. nivara* *O. rufipogon* *O. sativa* f. *spontanea*	161 151 6	Odisha
1999–2003	*O. nivara* *O. rufipogon* *O. sativa* f. *spontanea*	78 84 1	West Bengal
2012–2014	*O. nivara* *O. rufipogon* *O. sativa* f. *spontanea*	223	Odisha, Assam, Jharkhand, and Chhattisgarh

aspects of rice germplasm but unfortunately again it was merged with the creation of a new Division of Crop Improvement in 2001. During October 1999, the Odisha coast faced a devastating super cyclone through the Bay of Bengal causing huge loss of life, property, and natural resources. In order to salvage the genetic resources that regenerated/survived after the natural calamity, a special rescue mission was organized. The ICAR-NRRI was recognized as the National Active Collection Centre (now National Active Germplasm Site) by the ICAR-NBPGR. During 1990–2000, the scientists of CRRI collected about 2400 accessions of cultivated and wild rice germplasm in collaboration with NBPGR. With the introduction of a mission mode NATP project on plant biodiversity (by the NBPGR as its lead center), a new thrust was made on the exploration and collection of rice germplasm with special reference to wild rice.

1.3.1 Germplasm introduction/acquisition

The CRRI was established at Cuttack in the 1946 with Dr K. Ramiah as its first Director. He brought with him a nucleus set of about 2400 accessions of rice germplasm from Coimbatore, which was being maintained at the Paddy Breeding station. This became the starting point of the building up of the National Germplasm Collection for rice at CRRI Cuttack, India. Subsequently, many exploration and collection programs, introduction and acquisition through exchange activities have helped to enrich the gene pool (Table 1.1).

1.3.2 Collection of wild relatives

The collections of wild rice germplasm were initiated at CRRI during 1948–1955. Subsequently, collections were made from western, northern, central, and eastern India. Later, variability in *O. coarctata (Porteresia)* was collected from the coastal regions. Many joint explorations with other countries have also been undertaken. The first such trip was mad by Dr H.I. Oka and his team from the National Institute of Genetics, Mishima, Japan in 1957. They explored the eastern and southern parts of India. In 1959, Dr H. Kihara and Dr S. Nakao of the same institute visited India and collected wild rice from Assam. In 1971, Prof Watabe of Tottori University of Japan visited India with his team and collected *O. nivara* and *O. rufipogon* from almost all the rice-growing regions of India. They also had collected samples of rice embedded in the bricks of ancient buildings to check if there has been any shift in the type of rice grown in the country during the last several centuries. In 1984, the Institut de Recherche Agronomiques Tropicales (IRAT) and Office de Recherche Scientifique et Technique en Outre Mer (ORSTOM) of France collected 79 accessions of *O. nivara*, *O. rufipogon*, and *O. officinalis* from Goa, Karnataka, Maharashtra, and Gujarat. Additional 75 samples comprising *O. granulata*, *O. rufipogon*, *O. nivara*, *O. officinalis*, and *O. spontanea* were collected from coastal, midland, and mountainous areas of Kerala in 1987. From 1987–1989, ICAR and International Rice Research Institute (IRRI) scientists undertook intensive collection for wild rice in South India and West Bengal (Table 1.4).

Table 1.4 **Cultivated rice and its wild and related species as represented in Central National Herbarium in India**

S. No.	Species	Name of collector (s)	Collector's no.	Period of collection	Locality	Herbarium ACC no.	Remarks
1	*Hygroryza aristata*	Mokim	10	1894	Gaya		Verified by N.L. Bor at RBG, Kew
2	*Leersia hexandra*	J.D. Naskar	1234	2.10.1894	Howrah		
3	*Leersia japonica*	Makino		23.10.1911	Matasudo-cho, Higashikatsugh-ikagum, Japan		
4	*Leersia lenticularis*	Charles C. Deam		21.9.1918	Indiana, Northwest of Mt. Vernon, Posey country		
5	*Leersia orisoides*	N. Vy-hodewski		22.9.1954	Bulgaria	569844	
6	*Leersia sayanuka*	M. Tagawa		18.10.1950	Japan		
7	*O. coarctata*	Robert Ellis		20.8.1891	Sundarbans, WB		
8	*O. coarctata*	Dr. Watt.		23.7.1896	Karachi, Pakistan	525952	Commonly called as Nanoi grass
9	*O. grandiglumis*	N.P. Singh and U.R. Deshpande		25.9.1970	Satrem, Goa	124555 BSC, WC	
10	*O. granulata*	J.D. Hooker			Rangial Valley, Sikkim	525936	
11	*O. granulata*	C.C. Calder and M.S. Ramas-wami	1607	4.5.1935	Travancore, Madras		
12	*O. granulata*	M.S. Ramas-wami		8.8.1914	Rawpa hill, Godavari, AP	525953	Corrected as *O. meyeriana* by O. Fischer, 22.3.1932

13	*O. granulata*	A.N. Henry and M. S. Swaminathan		4.9.1976	Veerapulli R.F, river side Kanyakumari, TN	48235 BSI, SC	
14	*O. granulata*	K. Vivekanathan		18.11.1975	Vannathiparai, Madurai, TN	46721 BSI, SC Coimbatore	Herb to 60 cm; Glumes white, grains green, in the undergrowth of forest, not common
15	*O. granulata*	G.V. Subba Rao and G.R. Kumari		21.9.1980	Maredumilli to Kakur, East Godavari, AP	67590 BSI, Sc	
16	*O. granulata*	R.T. Balakrishnan		31.5.1984	Tholpetty, Wynaad dist. Kerala – Karnataka border	40249 Cali. Uni.	
17	*O. granulata*	C.B. Clarke		11.10.1883	Parasnath, Hazaribag,Bihar	33832	
18	*O. granulata*	A.C. Chatterjee	564	July 1902	Hathogaon jungle, Assam		
19	*O. granulata*	N.C. Nair	50965	10.9.1977	Paravallypatha, Punalur, Quilon dt, Kerala		Glumes greenish, not common
20	*O. granulata*	N.C. Nair and P. Bhargavan		13.3.1984	Nedumgayam, Malappuram, Kerala		
21	*O. granulata*	M. Mohanan	65186	22.11.1979	Kollar, Trivandrum, Kerala	MH 117844	
22	*O. granulata*	S. Kurz		21.11.1890	Rajmahal hill, Bihar	525934	
23	*O. granulata*	C.B. Clarke		18.11.1894	Parasnath, Hazaribag, Bihar	525935	
24	*O. granulata*	S. Kurz	10	1868	Sikkim Himalayas, Little Rangiel valley	525937	

(Continued)

Table 1.4 Cultivated rice and its wild and related species as represented in Central National Herbarium in India *(cont.)*

S. No.	Species	Name of collector (s)	Collector's no.	Period of collection	Locality	Herbarium ACC no.	Remarks
25	*O. granulata*	A.S. Rao	38783	13.6.1964	Garbhanga forest, South Kamrup, Assam		
26	*O. granulata*	G.V. Subbarao	28107	25.8.1966	Madgole to Paderu, Visakhapatnam, AP	53818	
27	*O. granulata*	G.V. Subbarao	30087	7.6.1968	Chintapalli–Nararipatnam, Visakhapatnam, AP	57884	
28	*O. granulata*	M.S. Ramaswami	1524	8.8.1914	Rumpa hill, Godavari, AP		
29	*O. granulata*	A.C. Chatterjee		May, 1902	Takeshari hills, Assam		
30	*O. granulata*	A.C. Chatterjee		1902	Hathegaon, Assam		
31	*O. granulata*	A.S.Rao	38783	13.6.1964	Garbhanga forest, South Kamrup, Assam		
32	*O. granulata*	J.S. Gamble	883	11.10.1883	Parasnath, Hazaribag, Bihar		
33	*O. granulata*	M.S. Calder	294	September, 1913	Travancore, Kerala		
34	*O. granulata*	J.L. Ellis	20411	15.8.1964	Pavagada, Calicut, Kerala	32510	
35	*O. granulata*	K. Vivekanathan	50552	27.8.1977	Idukki to Mangala temple, Kerala	MH 103703	
36	*O. granulata*	R.R Beddome		1866	Malabar, Kerala		
37	*O. granulata*	Wight		August, 1835	Courtallum, TN		
38	*O. granulata*	S. Malhi		1904	HBC, WB		

39	*O. latifolia*	D.B. Deb	2552	31.8.1954	Lowsipak, Assam	559831	
40	*O. latifolia*	J.D. Hooker	1841		Jenkins, Upper Assam	325921	
41	*O. latifolia*	G. Schaller	205		Kanha National Park, Mandla, MP	525922	
42	*Oryza meyeriana*	J.L. Ellis	20411	15.8.1964	Calicut, Kerala		900 m, common
43	*O. meyeriana*	G.V.S. Rao	23188	6.1.1961	Birbhum, Bihar		
44	*O. meyeriana*	Rolla Sesha-giri Rao	8876	27.8.1957	Tripura		Fairly abundant, herb on moist soil along rice field, on way to Cherulan village
45	*O. meyeriana*	V. Narayan-swamy	434	20.2.1947	Dharakonda, Vizag, AP (Fl. of Madras)		
46	*O. minuta*	S.R. Rolla	32794	8.1.1962	Experimental Garden, CRRI, Cuttack, Odisha		
47	*O. minuta*	D.B. Deb	2549	31.8.1954	Lousipath, Agartala, Tripura		Probably not *Oryza* by SVS Shastri 29.8.1963
48	*O. officinalis*	J.D. Hooker			Sikkim	525925	
49	*O. officinalis*	D.C.S Raju	377	22.9.1962	Blavaram Agency, Devara goudi, AP		
50	*O. officinalis*	K.M. Sebas-tine	6672	23.9.1958	Narsapur, Medak dt, AP		
51	*O. officinalis*	R.S. Rao	17530	28.9.1959	Subansin frontier, Khartan, Arunachal Pradesh		

(Continued)

Table 1.4 Cultivated rice and its wild and related species as represented in Central National Herbarium in India *(cont.)*

S. No.	Species	Name of collector (s)	Collector's no.	Period of collection	Locality	Herbarium ACC no.	Remarks
52	*O. officinalis*	R.S. Rao	10600		Shoehang to Paya, Lahit frontier, Arunachal Pradesh		
53	*O. officinalis*	R.S. Rao	15067	2.9.1958	Niusa to Wanu, Arunachal Pradesh		
54	*O. officinalis*	G. Panigrahi	11548	23.11.1957	Kokri reserve forest, Assam		
55	*O. officinalis*	I.H. Burkill			Rotong, Assam		
56	*O. officinalis*	B.B Pramanik	707	20.9.1965	Champaran dt, Bihar		
57	*O. officinalis*	G.V. Subbarao	23188	6.1.1956	Birbhum, Bihar		
58	*O. officinalis*	J.H. Lace	1112	9.10. 1881	Simla, HP		
59	*O. officinalis*	N.C. Nair	642556	4.10.1979	Palghat, Kerala		
60	*O. officinalis*	G. Panigrahi	2192	18.1.1964	Sindri, MP		
61	*O. officinalis*	S.K. Jain	5102	24.9.1962	Raipur, MP		
62	*O. officinalis*	S.D. Mahajan	6688	22.8.1956	Khandesh, Maharastra		
63	*O. officinalis*	D.B. Deb	29252	6.9.1962	Garo hills, Meghalaya		
64	*O. officinalis*	G. Panigrahi	45171	29.10.1956	Road. to Jerin, Meghalaya		
65	*O. officinalis*	G. Panigrahi	430	30.10.1959	Jowai, Garampani, Meghalaya		
66	*O. officinalis*	H. Deka	19230	8.11.1959	Hongpoh, Meghalaya		
67	*O. officinalis*	S. Nakao and H. Kihara		November, 1959	Around Shillong, Meghalaya		
68	*O. officinalis*	G. Panigrahi	20510	29.10.1959	Sambalpur, Odisha		
69	*O. officinalis*	P.M. Debbarman	972	25.10.1975	Agartala, Tripura		

70	*O. officinalis*	G. Panigrahi	6576	22.11.1964	Gorakhpur, UP		
71	*O. officinalis*	K.K. Biswas			Hooghly, West Bengal		
72	*O. officinalis*	M.K. Ghosh	2499	10.11.1964	Near Hadyangram college, WB		
73	*O. officinalis*	P.K. Hazra	112	22.111961	Arandi, Hooghly, WB		
74	*O. officinalis*	R.M. Dutta	385	21.8.1966	Manik Chowkghat, WB		
75	*O. officinalis*	S. Sen		8.12.1914	Chinsurah, WB		
76	*O. officinalis*	J.A. Lorzing	1690	10.6.1914	Pemarrang, Karang Baling, Pembliangan, Indonesia		
77	*O. perennis*	D.B. Deb	1944	29.11.1959	Jirania, Agartala, Tripura		Corrected by S.V.S. Shastri as *O. rufipogon*
78	*O. rufipogon*	R. K. Basak	1261	18.12.1969	Birbhum, WB		
79	*O. rufipogon*	G. Panigrahi	12664	15.3.1970	Mirzapur, UP		
80	*O. rufipogon*	N.C. Nair	64590	13.10.1979	Palghat, Kerala		
81	*P. coarctata*	A.K. Mukherjee	8702	7.10.1971	Kakdathip point bar, 24 Parganas,WB	10078	
82	*P. coarctata*	D. Prain		6.8.1902	Athara banki, Fl. of Sunderbans, WB	525950	
83	*Zizania aquatica*	H.H. Babcock		8.8.1874	Chicago, USA	525728	
84	*Zizania latifolia*	Abdul Kalil	1896		Upper Burma	525718	
85	*Z. latifolia*	Tomitaro Makino	8040	1919	Koiwa, Tokyo		
86	*Zizania texana*	Neil Hotchkiss		14.1.1936	Texas, San marcos, in San marcos river, USA	565727	American grass from the US National Herbarium, Smithsonian Institute

1.3.3 Wild species and its importance

Wild species are an excellent reservoir of variability for several traits including resistance to biotic and abiotic stresses and quality and productivity traits. The first use of a wild species for the genetic improvement of cultivated rice was made at the Paddy Breeding Station, Coimbatore. Here, *O. sativa* was crossed with *O. rufipogon* (then identified as *O. longistaminata*) and the drought-resistant varieties CO 15 and CO 16 were developed (Ramiah, 1953). It is suspected that this wild species was *O. rufipogon* only but misidentified as *O. longistaminata*, which is an African rhizomatous perennial species. Both these wild species are, however, diploid ($2n = 24$) and have the same genome (AA) as that of *O. sativa*.

Realizing the importance of wild and weedy relatives of the crop plants, several accessions of wild rice species, namely, *O. nivara*, *O. rufipogon*, and *O. coarctata* (*Porteresia*) were collected from Odisha and West Bengal, which were hitherto untapped and remained as the largest repository of rice genetic resources in the country. *O. nivara* and *O. rufipogon* have the same chromosome number and genomic constitution with that of *O. sativa* ($2n = 24$ and AA). *O. nivara* is an annual species with gregarious habit, synchronous flowering of tillers, and bold seeds. It is photoperiod insensitive and occurs frequently and abundantly in seasonal pools in the eastern part of the Deccan plateau and in the central Gangetic plain of India. It also occurs in the western and southern part of the Deccan plateau though less frequently. It further grows in the plateau regions of Myanmar, Indochina, and southern China. *O. nivara* is conspicuously missing in the lower Gangetic valley and in northeastern India. *O. rufipogon* is a photoperiod-sensitive perennial species with nonsynchronous flowering of tillers and, compared with *O. nivara* has slender grains and low seed productivity. While screening about 400 accessions of both the wild species, which showed high variability, 24 lines were found resistant to bacterial blight and 5 others were found resistant to brown planthopper (BPH). Two accessions of *O. nivara* (AC-100374 and AC-100476) collected from West Bengal were found to have high degree of tolerance to drought with an SES score of 0 and 1, respectively.

The genus *Oryza* consists of about 20 wild (Sampath, 1961; Tateoka, 1964a; Vaughan, 1989) and 2 cultivated species. The common cultivated rice *O. sativa* is distributed worldwide while the African cultivated rice *O. glaberrima* is confined to West Africa only. Most of the wild species have similar genomes (AA) as that of *O. sativa* and *O. glaberrima*; other distantly related species have BB, CC, BBCC, CCDD, EE, FF,GG, HHJJ, HHKK, and KKLL genomes (Khush, 2004). A comprehensive list of potential wild species available in the genus *Oryza*, their genome, distribution, along with their reaction to biotic and abiotic stresses are presented to reinstate the genetic diversity of the species into the rice-breeding program (Table 1.3). There are four distinct nomenclatural valid wild species of rice in India, which are *O. nivara*, *O. rufipogon*, *O. granulate*, and *O. officinalis*. Three other species (*Oryza indandamanica*, *Oryza malampuzhaensis*, and *Oryza jeyporensis*) described earlier from the Indian peninsula were later merged with *O. granulata*, *O. officinalis*, and *O. sativa*, respectively. Apart from the species of genus *Oryza*, several allied weedy species of other genera, like *Porteresia coarctata*, *Leersia hexandra*, and *Hygroyza aristata*, have also shown importance in having different abiotic stress-tolerant genes and forms the tertiary gene pool. Exploration and collection of *P. coarctata* to one of the biodiversity-rich mangrove forests of Bhitarkanika in three different seasons have shown tremendous variability (Patra and Dhua, 1996). *O. coarctata* is a tetraploid species ($2n = 48$) and shares none of its genomes with *O. sativa*. Natural hybridization between *O. sativa* and *O. coarctata* has never been reported and success in artificial hybridization has been rare (Jena, 1994). *Oryza schlechteri*, first found by Richard Schlechter in 1907 from Papua New Guinea and described in 1910, is one of the rarest species in the collection. It was recollected as living species from its type locality in 1991 (Vaughan and Sitch, 1991). It is a stoloniferous species of unstable stony soil, such as river banks, and grows in full or semishade and does not flower; attempts to induce flowering also have failed to date.

Hooker (1897), in his Flora of British India, has mentioned that the genus *Oryza* had only five species (*O. sativa*, *Oryza latifolia*, *O. granulata*, *Oryza ridleyi*, and *O. coarctata*) whereas Haines (1921–1925) in the book "The Botany of Bihar and Orissa," has mentioned only two species (*O. sativa* and *O. granulata*) available in the region.

A herbarium plays an important role not only in preserving the specimens but also in providing a lot of passport information with respect to locality, habit, habitat, flowering, and fruiting time and many more, which help researchers in confirming and determining the identity of the taxon. It also serves as a reference tool for botanists and conservationists interested in the ecology and diversity study. The genetic diversity and distribution pattern of different species in a particular genus is also ascertained by studying the herbarium sheets. The rarity and existence of *O. schlechteri* can be cited as an example. This is an endemic and very rare wild rice species described by Schlechter in 1910 from a few sites of Papua New Guinea and was not available for study either in living form or in a herbarium thereafter. However, after visiting major world herbaria and studying the passport information, only one or two plants could be collected from the same sites in Madang and Asway after more than eight decades (Vaughan, 1994). This gives the idea of the importance of passport information in a herbarium. A detailed account of the herbarium sheets (exsiccates) of cultivated/wild/weedy rice species studied (Patra, 2008) in the Central National Herbarium (CNH, CAL), Kolkata is presented in Tables 1.3 and 1.4.

1.3.4 Collection of trait-specific germplasm

1. Medicinal rice: Seventy-two accessions of rice germplasm were collected from the Bastar region of Chhattisgarh in which some medicinal rices, namely, Gudmatia, Bhejari, Danwar, Baisur, and Gathuwan, were also reported. Also, the world famous medicinal "njavara" rice was collected from the northern part of Kerala.
2. Saline-tolerant rice: Fifty-one accessions of saline-tolerant rice mostly from Pokkali region of Kerala (one of the potential regions for geographical indication (GI)) were collected.
3. Basmati rice: Eighty-eight accessions of long slender basmati rice germplasm were collected from eight districts of western UP and six districts of Haryana state.
4. Aromatic short grained rice: Sixty-seven accessions of short-grained scented rice "Kalanamak" germplasm were collected from eastern UP.

5. Boro rice: A total of 208 accessions of Boro rice germplasm were collected from Assam, north Bihar, north Bengal, and eastern UP.
6. Bao rice: About 126 accessions of Bao rice germplasm were collected from deep-water areas of Assam and Meghalaya.
7. Aman rice: A set of 69 accessions of Aman rice germplasm were collected from West Bengal.
8. Cold-tolerant rice: A set of 116 accessions of cold-tolerant rice germplasm were collected from hilly regions of Arunachal Pradesh.
9. Wild and weedy rice: About 495 accessions of wild rice germplasm (*O. nivara*, *O. rufipogon*, and *O. coarctata (P. coarctata)* were collected in 12 exploration trips from Odisha and West Bengal under the National Agricultural Technology Project. Apart from this about 223 accessions of weedy rice (*O. sativa* f. *spontanea*) have also been added to the gene pool.
10. Specialty rices: Attempts are ongoing to collect aromatic rice, soft rice, wine rice, glutinous or waxy rice, colored rice (brown, black, red), beaten rice, pop rice, organic rice, nutritional rice, and so on.

1.3.4.1 Nerice rice

West Africans domesticated *O. glaberrima about* 3500 years ago. The Asian species *O. sativa* reached Africa about 450–600 years ago and slowly displaced the native rice because of low harvest. By the 1990s, native African rice was reduced to a few pockets on scattered farms. Then Sierra Leonean plant breeder Monty Jones and his colleagues found a way to create a fertile hybrid between African and Asian rice, called "*Nerica*" (New Rice for Africa), it could yield a bumper harvest like its Asian parent, but it was as tough as its African side, resistant to drought, pests, and diseases. Scientists have bred many varieties of Nerica and farmers have started growing them. This new rice, descended from an endangered species, is helping Africa to feed itself, yet this opportunity would have been lost if *O. glaberrima* had gone extinct.

1.3.4.2 Medicinal (njavara) rice

The documentation of indigenous traditional knowledge on the medicinal and nutritional significance of red rice is another aspect that is gaining momentum due to recognition of njavara rice of Kerala as one of the regions of GI of Goods Act, 1999 under the Intellectual Property Right. Two varieties of *njavara* – black glumed and golden yellow glumed – cultivated in parts of northern Kerala confined to Palakkad district have secured registration. This was the first time that a rice variety of Kerala received a Geographical Indication Registry in 2008. It is estimated that about 300 tons of this medicinal rice is used annually for Ayurveda treatment in the state. Studies found that *njavara* has increased levels of protein and amino acid in the organically grown seeds; thus, it should be developed as baby food and a health product to save this wonder rice from extinction. This unique medicinal rice is now cultivated only on about 30 acres of land in the state because farmers are not taking up its cultivation in spite of high demand owing to poor marketing network and the absence of a remunerative price. Therefore, this region and Pokkali region should be worked out as prospective sites of *in situ* conservation.

1.4 Germplasm introduction

When the International Rice Commission (IRC) recognized CRRI as a center for the maintenance of world genetic stocks of rice varieties, especially of the *indica* types, many rice varieties of South and Southeast Asian countries were introduced to the country for their maintenance at CRRI. This provided an opportunity for Indian rice scientists to test and recommend a few of them for general cultivation. Again, when CRRI was recognized as the main center for the inter-racial hybridization program between *japonicas* and *indicas* during 1950–1964, many exotic *japonica* rice germplasm were introduced to India. The participants of the southeast and south Asian countries came with their own rice varieties for hybridization. This further provided opportunity to Indian rice scientists to grow and study the rice varieties of other countries. Some of the *japonica*s when tried in temperate hilly regions were found suitable for direct introduction. During the operation of this interracial hybridization project, many *japonica* varieties (Aikoku, Asahi, Fukoku, Gimbozu, Norin 1, Norin 6, Norin 8, Norin 17, Norin 18, Norin 20, Rikuu 132, Taichu 65) were crossed with the popular varieties of Odisha (T 90, T 812, T 1145, BAM 9) and the progenies were grown at three rice research stations (Bhubaneswar, Berhampur, and Jeypore) for further selections. Very limited success was achieved in this project. In all, 192 improved local varieties were selected and a total of 710 different *indica* × *japonica* crosses were made. The F_1 seeds were distributed to many countries for further crop improvement programs. Only four varieties were released: Malinja and Mahsuri released in Malaysia, ADT 27 in Tamil Nadu state of India, and Circna in Australia.

Among the *japonica–indica* hybrids, Mahsuri, which was bred in Malaysia from a cross between Myang Ebos 80 × Taichu 65, was introduced to India and became very popular and was spread in many parts of the country. Later, this project involving two subspecies was abruptly ended in 1966 with the advent of semidwarf rice varieties. However, in later years, a few popular high-yielding varieties, like CR 1014, CR 2002, Annada, and Utkal Prabha, were released using the introduced *japonica* and *javanica* parents.

The *indica* cultivars were tall and nonresponsive to nitrogenous chemical fertilizer. Hence, efforts were made to introduce some exotic germplasms from China, Japan, Taiwan, and Russia, which are semitall and have nitrogen-responsive character. Some cultivars from Chinese origin, like CH 4, CH 45, CH 55, CH 62, and CH 63, proved to be very good donors for better yield and early maturity duration. Even before independence, varieties like CH 1039, CH 27, CH 47, CH 962, CH 971, and CH 972 were grown in Kashmir valley. Contrary to the success of the Chinese varieties, the Japanese and Russian germplasms were found unsuitable under Indian conditions.

The introduction of Taichung (Native) 1 from the semidwarf mutant Dee-geo-woo-gen ("*Dee-geo*" means dwarf and "*woo-gen*" means brown apiculus) of Taiwan in 1964 was the most significant landmark in the history of rice breeding in India. This was the single variety adaptable to the whole of India starting from Kashmir to Kanyakumari. Due to its superior response to high levels of nitrogen fertilizer, this variety out-yielded significantly in *kharif* as well as in *rabi* seasons. Ratna was the first miracle rice variety of CRRI developed from the cross TKM 6 × IR 8 and released by Central Variety Release Committee in 1972. It had a wider coverage throughout the country.

More than 500 varieties including landraces and exotic collections were evaluated for selecting a suitable donor for breeding work in the high-altitude areas of Kashmir and other Himalayan temperate regions. Four Chinese varieties, including CH 1039, and three Russian varieties, including R 3073, were selected for the valley and hilly regions. CH 1039 was found to be a high yielder until 1954 when the *indica* × *japonica* hybridization program was started and one of the few successful crosses, namely, Rikku 137, out-yielded CH 1039 in many locations. In another attempt, after a thorough multilocation screening and evaluation of a large number of rice germplasm, only two Chinese varieties CH 13 and CH 45 were found resistant to brown spot disease. This disease was the root cause of the Bengal Famine in 1943. Two Chinese varieties, namely, CH 47 and CH51, were found resistant to yellow stem borer. Earlier, exotic *Oryza* spp., like *O. longistaminata*, *O. barthii*, *Oryza schweinfurthiana*, *Oryza minuta*, and *O. latifolia*, were introduced to India during the 1930s especially to study their relationships with the cultivated rice. The African cultivated species *O. glaberrima* was also introduced (Ramiah, 1953).

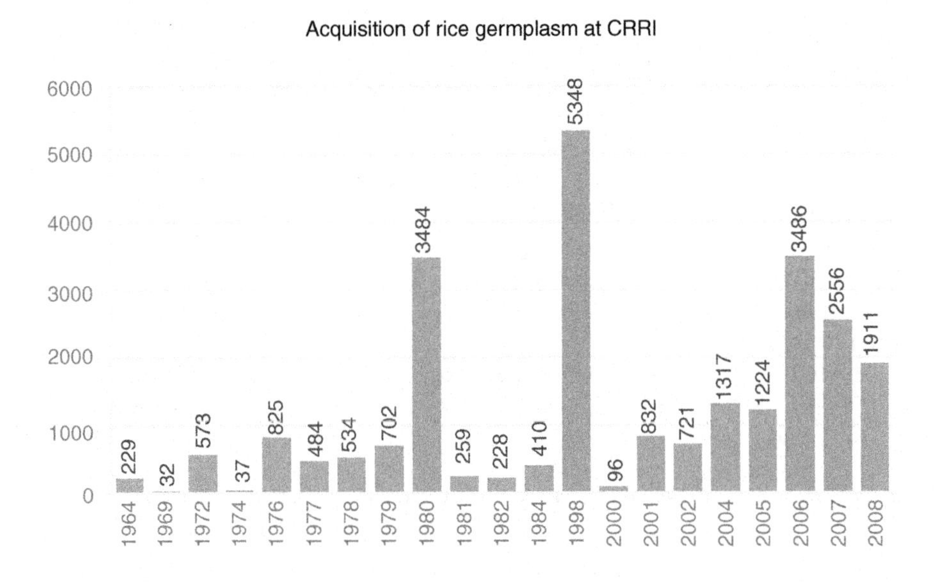

Rice is a much easier crop to work with genetically. A diploid, it has by far the smallest genome among all the cereals in contrast to bread wheat, a hexaploid, which has a genome 40 times larger than rice. Significant achievements were made to create viable progeny from crosses between species comprising the 10 genomes of genus *Oryza*, with the exception of *O. schlechteri*, which never flowers. The gene pool of submergence-tolerant rice cultivars is scanty. Some submergence-tolerant rice cultivars, like FR 13A and FR 43B, were identified awhile back and they were similar in the sense that all the cultivars did possess the same submergence-tolerant gene (*Sub 1*) in chromosome number 9. Moreover, the existing tolerant cultivars had very poor combining ability and agronomic characters. Hence, development of high-yielding submergence-tolerant cultivars using these cultures as one of the parents was not successful. Identification of new donors (gene resources) thus became imperative to achieve success. Some landraces were identified as tolerant to complete submergence. They were Khoda, Khadara, Kusuma, Gangasiuli, Atiranga, Ande Karma, Nahng tip, and Kalaputia. In some areas crop suffers from floods when it is submerged under water for up to 10 days. Rice cultivars cannot survive such prolonged submergence. A few rice cultivars have been identified, which survive submergence up to 80 cm water depth, for 10–12 days at the early vegetative stage of the crop. Genetic analysis of one such cultivar, FR 13A, revealed that tolerance to submergence is controlled by one major gene. Using FR 13A as a donor, an improved rice cultivar, Swarna *Sub1*, has been developed and released in India and is gaining popularity among the farmers. Odisha is the first state in the country to release the Swarna *Sub1* variety, which was developed by the IRRI and the University of California-Davis by deriving the flood-tolerant gene *Sub1* from traditional rice FR 13A, grown in the Odisha State.

Similarly, drought is a major abiotic stress that adversely affects the crop leading to low productivity. The need for other stress-tolerant varieties is now felt as the paddy cultivation in the region is largely affected by extreme natural calamities like drought, cyclones, and most recently, after rapid climate change, through a fickle monsoon. After repeated screening some landraces were identified as tolerant to vegetative stress drought; these are Mahulata, Sunamani, Naliakhura, Ranganatha Bao, Bhuta, Bibhisal, Brahman nakhi, Salkain, Gauranga, Karinagin, and Kiaketi. At Rice Research Station, Nagina, Uttar Pradesh, N 22 was selected through pure line selection from a landrace "Rajbhog," which is internationally known as a very good donor for high-temperature tolerance, drought tolerance, and grain dormancy.

Direct seeding is common in rainfed lowlands. In eastern India, sometimes, early rain causes water stagnation in the field just after sowing, which results in poor crop establishment. Two cultivars, namely, Panikekoa and T 1471, were identified as anaerobic seeding tolerant rice germplasms.

A number of resistant germplasms were short-listed for major pests like BPH, such as Salkathi, Panidubi, Dhoiya Bankoi, Banspati, Dhobanumberi, and Jalakanthi. Similarly, a set was screened against major disease "blast" under simulated epiphytotic condition in a uniform blast nursery as well as in a net house under controlled conditions by artificial inoculation. Two landraces, namely, Ambika 2 and Khalvi, were found to be highly resistant to leaf blast with the score of 1 in the SES scale.

1.5 Germplasm conservation

There are about 124,000 (*O. sativa*), 1,651 (*O. glaberrima*), and 4,508 (wild relatives) accessions of rice germplasm conserved at the IRRI in the Philippines. China has several rice germplasm collections but the collection at Beijing has over 50,000 accessions. The Japanese national rice collection contains about 26,000 accessions. The USDA-ARS rice collection has 17,279 accessions from 110 countries. The IRRI collection is probably the most genetically diverse rice collection in the world because the acquisition and field collection efforts were implemented in the appropriate places and at opportune time before advanced genetic erosion occurred (Chang, 1989). However, it has been estimated that more than 100,000 rice cultivars exist in Asia (Chang, 1985). In India alone, there are 104,427 accessions of rice germplasm conserved at −18°C and with 3–4% relative humidity (RH) in the national gene bank of NBPGR, New Delhi as of April 2015. Therefore, the collection and acquisition of rice germplasm must be a continuous effort. Moreover, the important point is that only a small proportion of the total genetic diversity of rice has been utilized (Chang, 1989).

Several categories of germplasms are conserved for different purposes:

1. Working collection: A collection of germplasms maintained and used by a breeder or other scientist for their own breeding or research, without taking any specific measures to conserve. The collection may have a short life span and the composition of the collection may vary greatly during its lifetime.
2. Active collection: A collection maintained by a gene bank and used as the source of seeds for active use, including distribution, characterization, and regeneration. It is usually conserved under short- or medium-term storage conditions.
3. Base collection: A collection of seeds ideally prepared and held in ideal conditions for long-term conservation. The seeds should be conserved and never used except for
    a. Periodic germination tests,
    b. Regeneration of samples conserved in long-term storage when their viability decreases below threshold,
    c. Regeneration to replace stocks in an active collection after accumulating three successive generations of regeneration from an active collection, and
    d. As the primary point of rescue when the accession is accidentally lost from all active collections.
4. Seed file: A small sample of original seeds, set aside when a seed sample first arrives at the gene bank, to serve as the definitive reference sample. The seed file should be maintained dry under conditions preventing disease or pest damage, although not necessarily alive. Other seed samples of the same accession, for example, for every new harvest, should be visually cross-checked with the seed file.
5. Safety back-up: Duplicate samples of the base collection, stored in a different gene bank, preferably in a different continent. The storage conditions in the safety back-up should be at least as good as those in the corresponding long-term collection. The holder of the safety back-up has no rights to use or distribute the seed in any way or to monitor seed health or viability. Additional duplication of the base collection to the Svalbard Global Seed Vault (SGSV) provides definitive safety back-up in case of large-scale loss of crop diversity. The SGSV commissioned on the Arctic island of Svalbard on the North Pole in 2008 conserves about 0.8 million germplasms. It is managed by Norway's Department of Agriculture and the Global Crop Diversity Trust (GCDT) under the International Treaty on Plant Genetic Resources for Food and Agriculture (ITPGRFA) and supported by the Bill & Melinda Gates Foundation. India has deposited 25 accessions of pigeon pea in April 2014 as the fifty-ninth nation.

The importance of genetic resources is widely recognized. Activities related to genetic resources, such as germplasm introduction, exchange, collection, characterization, evaluation, documentation, and conservation, are characterized by high cost and long-term return. Until the recent past, conservation of rice germplasm was synonymous with repeated rejuvenation in the field. This process of maintenance subjected the germplasm to the threat of losing their identity because of random and nonrandom processes due to sampling. Also, loss due to unforeseen natural calamities, such as the super-cyclone and flood that affected Odisha, devastating the native germplasm cannot be ruled out as far as on-farm *ex situ* conservation is concerned. Therefore, realizing the importance of genetic diversity, Jeypore tract of Odisha, the Palakkad area of Kerala, and Apatani valley of Arunachal Pradesh deserve to be protected as on-farm *in situ* conservation sites. In 2010, the FAO recognized Koraput as a geographically important agriculture heritage site.

Due to the danger of such unforeseen genetic erosion, the effort of developing a cold storage system for rice germplasm was initiated at CRRI in 1984. Meanwhile, during 1986, it was decided to conserve all the germplasms of CRRI at the National Gene Bank. Since then, more than 30,000 rice germplasm accessions have been deposited in the long-term storage of NBPGR. Under the aegis of the Indo-USAID collaborative project, a cold module was gifted to CRRI. The facility became operative in 1998 with a controlled temperature of 4 ± 2°C and 33 ± 5% RH and found to be rather dependable. The gene bank facility thus created is meant for medium-term storage (MTS) and the seeds are kept viable for 6–8 years. When accessions in the MTS working collection drop below 50 g or if seed viability falls below 85%, then the accession is increased (rejuvenated). The *japonica* varieties are monitored more frequently than *indica* rice as they have an inherently shorter storage life than *indica* varieties.

The seeds of each of the accessions were dried to reduce the moisture content up to 10–12% and kept in three-layered aluminum foil pouches for medium-term storage. The outer layer of the pouch is polyester of 12 μm; intermediate layer is aluminum of 12 μm, and the innermost layer is polythene of 250 gauge. These aluminum foil pouches are stored in a cold module at a regulated temperature of 4°C and 33% RH.

At CRRI about 12,000 accessions of rice germplasm are conserved as an active collection and about 30,000 accessions of rice germplasm are deposited in the National Gene Bank as base collection (Fig. 1.3). Now the stage is set for streamlining and systematizing the conservation of rice germplasm in the country. In this endeavor, the first step is to allot a national accession number (IC No.) to all these collections, which is being allotted by the NBPGR, New Delhi. This has to be followed by seed increase in sufficient quantity at the National Active Germplasm Site (NAGS) for further deposit in the long-term storage and also in the medium-term storage of the active collection in a phase-wise manner. At the same time they need to be characterized and documented. Further evaluation and utilization has to be followed up for the purpose of crop improvement. However, the possible area/source of resistance and tolerance for some important biotic and abiotic traits has been identified over the years.

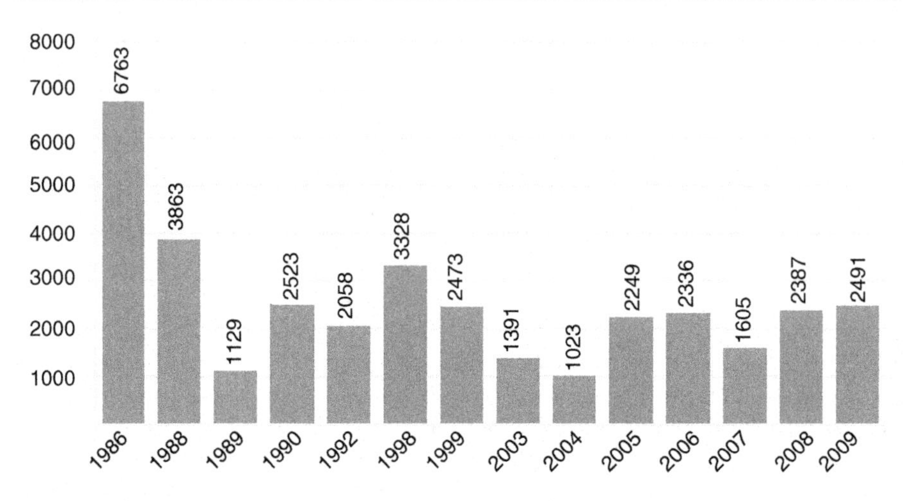

Figure 1.3 Rice germplasm deposited by CRRI for long-term storage at NBPGR, New Delhi.

1.6 Germplasm evaluation and utilization

Genetic erosion has been very fast in recent years due to the rapid modernization of society, and genetic diversity has been replaced by the introduction of a few high-yielding varieties. Farmers are leaving their own traditional varieties and growing the improved cultures thereby many of the landraces have become extinct. The need for *in situ* and *ex situ* conservation is now felt as the paddy cultivation in the country is largely affected by extreme natural calamities after rapid climate change, through an erratic monsoon. Earlier, the biggest challenge was flooding, but subsequently other factors, such as salinity after frequent cyclones and sea water surge, temperature rise, and drought-like situations in many parts of the country, have put the challenge before rice researchers to incorporate these genetic factors in the plant.

Activities related to genetic resources are characterized by high cost and long-term return. Introduction and germplasm exchange, collection, characterization, evaluation, documentation, and conservation are essential steps that cannot be overemphasized. An appropriate synchronism among these activities is required for the bank to be effective in maintaining genetic variability and to assure germplasm utilization.

The importance of genetic resources is widely recognized. Activities in germplasm banks demand qualified researchers in several areas of knowledge. Besides the conservation of genetic variability for the future, the actual utilization of available accessions is another important goal. The main factors responsible for the low utilization of PGR are the lack of documentation and adequate description of collections, lack of the desired information by breeders, accessions with restricted adaptability, insufficient plant breeders particularly in developing countries, and lack of systematic evaluations of the collections. Low seed availability due to inadequate seed regeneration programs is another barrier to their use (Dowswell et al., 1996).

Furthermore, breeder-to-breeder exchange materials are very common and constitute a reasonable alternative to extend genetic variability in breeding programs. In general, it seems that breeders are satisfied with the available genetic variability among agronomically advanced materials (Duvick, 1984; Paterniani, 1987; Peeters and Galwey, 1988).

In the past, the scientists involved in crop improvement programs at different research stations undertook the evaluation of germplasm and were choosing the varieties through pure line selection. The rice research in the country started between 1911 and 1929 when 21 rice research stations were established in different agroclimatic zones and as many as 37 rice varieties were bred. During this period, the economic botanists (later designated as paddy specialists) of these research stations were collecting the local varieties from their respective regions to study their morphology, floral biology, physiology, traditional indigenous agronomic practices, response to organic manuring (since there was no chemical fertilizer in those days), yield potential, and other traits. During the collection of landraces, which used to be the admixtures of varieties, the rice workers were purifying and maintaining them for hybridization purposes. Government Economic Botanist 24 (GEB 24) was the first rice variety recommended by the Paddy Breeding Station, Coimbatore for cultivation in the whole of Madras Presidency as well as in central provinces. It was a short-duration variety and was a selection from Konamani, a landrace. During the early 1920s, some improved strains of rice, namely, Benibhog (autumn), Cuttack 1 (early winter), Cuttack 2 (midwinter), Cuttack 3, Cuttack 5 (late winter), Kujang 1, and Kujang 2 (for the saline tract) were developed for cultivation in the erstwhile province of Bihar (including Jharkhand) and Odisha (Ramiah, 1953). In Odisha, T 141 was selected from Saruchinamali of Cuttack district; it had fine translucent grains and good cooking qualities, 150 days duration, and had wide adaptability. Some other very popular pure line selections were T 90 (a selection from Machhakanta of Balasore district), T 442 (a selection from Kalakartika of Sambalpur district), T 812 (a selection from Rangalata), T 1145 (a selection from Ussa of Puri district), T 1242 (a selection from Magura), SR 26B (a selection from Kalambanka, a salinity tolerant variety), and FR 13A (a selection from Dhalaputia, a flood tolerant variety). Similarly, Latisail was released in West Bengal in 1948 through selection from a bulk collection of Bangladesh. FR 13A released in West Bengal in 1953 from Orissa collections; FR 43 B and SR 26 B released in 1953 in Orissa from local germplasm; NC 1281 released in West Bengal in 1960 through selection from materials collected from low-lying areas of 24 Parganas district of West Bengal; CB1 released in 1952 from boro germplasm collected in Hooghly district of West Bengal; CB 2 released in 1956 from a collection of Midnapore district of West Bengal; Adday released in Sikkim in 1955 through selection from a local germplasm of Darjeeling, and so on. Some of the varieties remained quite popular for a long time in Odisha. Of these, T 90, T 141, T 1242, BAM 6, FR 13A, FR 43B, and SR 26B continued to be popular among the farmers and hence the Department of Agriculture used to produce and supply pure seeds of these varieties. As a result, these varieties had to be officially notified as per the Seed Act.

However, in the absence of a national numbering system and proper conservation of such evaluated materials, the information generated often suffered from deficient identification and thus utilization of such information remained inadequate.

1.6.1 Disease resistance

Research for identifying the source of resistance from local landraces is a continuous process and donors for several important diseases have been screened at different research stations. Germplasm identified as possible source of major diseases, like blast, brown spot, bacterial blight, and RTV, are depicted in Tables 1.5 and 1.6.

1.6.2 Insect resistance

Before 1960 the efforts on the identification of resistance source were confined mainly to field tolerance. Such efforts helped in the identification of gall midge resistant donors such as HR 13, HR 14, and Ratnachudi. Systematic screening of 3600 accessions at CRRI in the early 1950s resulted in the identification of about 246 resistant varieties against gall midge. Also, efforts in the endemic area of Warangal, Andhra Pradesh during 1954–1964 resulted in the identification of Eswarakora, HR 12, HR 42, HR 63, and so on.

Insects, like gall midge, stem borer, brown planthopper, white-backed planthopper, leaf folder, and green leaf hopper, are of major concern causing considerable yield loss. Therefore, the identification of a resistance source has been one of the major objectives and this has resulted in the identification of several donors (Table 1.7)

Table 1.5 Sources of resistance/tolerance for some important traits

Traits	Possible sources
Resistance against different pests and diseases	Landraces of insect or disease-specific endemic areas
Drought tolerance and recovery at vegetative stage	Dry seeded upland germplasm and landraces grown in the hilly region under shifting (Jhum) cultivation
Cold tolerance at vegetative stage	Landraces of Boro cultivation
Low light intensity	Landraces of traditional Sali, Asra, and Bao cultivation areas of Assam and traditional Aman varieties of Bengal and Odisha
Flash flood	Coastal areas of Odisha and West Bengal. Also Sali and Asra germplasm of the Brahmaputra valley of Assam
Deep-water situation	Bao germplasm of Assam and landraces from deep-water areas of Eastern UP, Bihar, and West Bengal
Earliness	Ahu germplasm of Assam and Aus group of materials from West Bengal
Aromatic rice	Joha varieties of Assam, Kalanamak group of varieties of eastern UP, Small grain scented varieties of Sikkim, West Bengal, and Odisha
Nonaromatic fine grain	Lahi group of materials from Assam, Balami grain type of Odisha and short slender grain types of West Bengal

Table 1.6 Germplasm identified as possible source of resistance against major diseases

Diseases	Possible donor for resistance	
	Landraces	**Improved varieties**
Blast (*Pyricularia oryzae*)	Tetep, Tadukan, Carreon, Gampai, Dawn, Peta, Sigadis, Saleem, Yerravadlu, Bailvadlu, Dular, Kalamdani, Tarabali, Mugisali, Madrisali, Soll-pona, Rongaahu, Manoharsali, Andrewsali, Gajepsali, Sail Sali, Beganbisi 2, Rongagutia, Kolimekuri, Rikhojoi 2, ARC 7098	Zenith, Raminad Str.3, ADT 29, ADT 25, CO 43, MTU 3626, MTU 6203, MTU 7014, MTU 9992, WGL 26889, WGL 47969, WGL 47970, MTU 993, BJ-1, NLR 145, PTB 10, Ceysron, Milyang 83, NLR 36, Hi-malaya 2, TK 1, CO 4, CO 25, CO 29, CO-30, MTU 9993, Saleem, Thikkana, Kotha Molagoli, Kulu-72, Pinakini, Swarna muli
Bacterial leaf blight (*Xan-thomonas oryzae*)	Dholamula, Moinasail, Kartik kolma, Japorisali, Gajepsali, Jatiosali, Kola ahu, Ahusuri, ARC 5827	Dular, Vajram, Prahlad, Lacrose-Zenith-Nira, BJ1, Karjat, Pallavi, Zenith M. Sung Song, Syntha, TKM 6, Sigadis, DZZ 192, DV 85, Pelita 1, MTU 15, MTU 16, ASD 5, T 1069, ARC 18562, MTU 4870, MTU 2400, MTU 3626, MTU 6203, MTU 7014, MTU 9992, Swarna, PLA 9180, BPT 3291, Yerra vadlu, Mahsuri, RNR 10786, PNR 2736, RNR 4970, RNR 10208, AS 330
Rice Tungro Virus (*Nephotettix malayanus*)	Kataribhog, Latisail, Sigadis, Ambemohar, Habiganj, ARC 14766	BJ 1, PTB 18, W 1263, Gampai 15, Pankhari 203
Helminthosporium oryzae	Bhatta Dhan	Ch 13, Ch 45, BAM 10, AC 2550, ADT 29, CO 29
False smut (*Ustilaginoidea virens*)	Sugandha, Sabari, Karna, Deepa, Sona, ARC 5378	Udaya, CO 9, IR 62, MNP 85, BR 16, IR 24, IR 29
Stem rot (*Hel-minthosporium sigmoideum*)	–	Basmati 370, Bara 62
Ragged stunt virus	–	PTB 21, PTB 33
Grassy stunt virus		*O. nivara*
Sheath rot (*Saro-cladium oryzae*)	Bhatta Dhan	–

Table 1.7 Important donors identified against major insect pests

Insect	Donors	
	Landraces	Improved varieties
Brown planthopper (BPH) (*Nilaparvata lugens*)	ARC 6650, ARC 5984, Rathuhenati, Suali, ARC 7080, ARC 14766, NCS 91, NCS 131, NCS 707, Milyang-55, PTB 43, PTB 21, Tribeni, Sinna sivappu, Payur-3, Pasimidikuppam, Suil-phul, Palghar-1, Velluthacheera, W-1263, HKR 50, ARC 6248, ARC 6605, ARC 6619, ARC 5757, ARC 6158, ARC 6102	Ptb 33, Leb Mue Nahng, Udaya, CR 1009, Ptb 18, Ptb 21, Vajram, Pratibha, Nandi, Chait-anya, Krishna veni
Gall Midge (*Orseolia oryzae*)	ARC 5984, ARC 10660, ARC 6605, Leuang 152, ARC 5959, ARC 13516, ARC 14787	Eswarakora, Siam 29, OB 677, CR 94-1512-6, Shakti, Ptb 18, Ptb 21,W 1263, AC 5, AC 6, AC 7, Oru-mundakam, Surekha, Velluthicheera, W 1263, WGL 20471, WGL 47970, Pothana, Phalguna, RP 140, ORS 677, CR 157 -212, CR 157- 303, Kakatiya, Di-vya, Erramalelu, Kavya, Orgallu, Dhauya laxmi
White Backed Plant Hopper (*Sogatella furcifera*)	ARC 5803, ARC 6064, ARC 7138, ARC 7318, ARC 10340	IET 6288
Stem Borer	(*Chilo suppressalis*)	ARC 6158, ARC 10386, ARC 10443, NCS 266, NCS 336, NCS 464, ARC 5500, W 1263
Green Leaf Hopper (GLH) (*Nephotettix* spp.)	ARC 6606	Ptb 2, Ptb 21, ADT 14, Vijaya, ADR 52
Leaf Folder (*Cnaphalo-crocis medinalis*)	ARC 1129, Gorsa, Darukasali	Ptb 12

1.6.3 Abiotic stress tolerance

Besides biotic stresses, the rice crop frequently faces problems of drought, low temperature, submergence, waterlogging, salinity/alkalinity, and so on. These abiotic stress situations cause drastic reduction in yield and thus varieties with in-built resistance to such stresses are desirable. The germplasms having resistance to such stress situations have been identified (Tables 1.5 and 1.8).

Table 1.8 **Important donors identified against abiotic stress**

Stress	Donors	
	Landraces	**Improved varieties**
Drought tolerance	Mahulata, Brahman nakhi, JBS 508, Saria, Sathchali, Nepalikalam, Kodibudama, NC 487, Dagaranga, Mettamolagolukulu, NC 488, ARC 10372, NC 492, AS 180, Hasakumra, Maibi, Koijapori, Bairing, Ahu joha, ARC 10372, Noga ahu, Soraituni, Lakhi, Pera vanga, Bodat Mayang, Prabhabati,	N 22, MTU 17, Lalnakanda 41, Kalakeri, CR 143-2-2, Dular, Janaki, AS 313/11, AS 47, Aditya, Tulsi
Cold tolerance	AC 540, Siga, Rajai, CB 1, Dholiboro, Dunghansali, Raja Sanula	Boro 33, IRGC 100081, 10114, 10028, Barkat, Kalinga 2, Tella Hamsa, Satya, Gavinda
Submergence tolerance	S 24, S 25, S 28, Solpona, Sail badal, Dhola badal, Kolasali, Boga bordhan, Rongasali, Khajara, Dhusara, Nali Baunsagaja, FR 13A, FR 43B, Chakia 59, CN 540, S 22, Madhukar, Khoda, Khadara	–
Deep water	Nagari bao, Kekoa bao, HBJ 1, Jalamagna, Jaladhi 1, Jaladhi 2	–
Coastal saline/alkaline	Pokkali, Kala farm, Rahas Panjar, Nona Bokra, Katla, Bhura ratha	SR 26B, Getu, Dasal, Patnai 23, Pokkali, Hamilton, CSR 10, CSR 13, CSR 18, Vikas, Co 43
Waterlogging	Tilakkachari, NC 496, Kalakhersail	Jhingasail, Patnai 23

The All India Coordinated Rice Improvement Project was launched in 1965 and thereafter, more systematic evaluation against major biotic stress situations was undertaken with multilocation field screening followed by greenhouse evaluation.

1.6.4 Donors identified for different traits

1. Bacterial blight: AC-33523 (Tarical), AC-33557 (Dulla karma), AC-33562 (Kangpui), AC-3094 (TKM-6), AC-8368 (BJ)-1, AC-26903 (DV-85), Chinsurah boro-II, Somera mangga, Wase Aikoku 3, Malagkit, Sung son.
2. Blast: AC-55 (CH-55) AC-8368 (BJ-1), AC-8369 (S-67), SM-6, SM-8, SM-9, CP-6, AC-293 (AKP-8), AC-294 (AKP-9), AC-360 (PTB-10), AC-26904 (Tetep), Fukunishike CO-4, CO-29 Tadukan, Zenith, Carreon.
3. Tungro virus: ARC-7125, ARC-7149, DW-8, AC-368 (PTB-18), AC-5079 (Kataribhog), Bhagirathi, Boitalpakhia, AC-34558 (Nalini) AC-17933 (Kamod-153), AC-34650 (Usha), AC-273 (ADT-20), AC- 297 (ASD-1), AC-304 (CO-1), AC-315 (CO-12), AC-351 (PTB-1), AC-360 (PTB-10), AC-3094 (TKM-6), AC-8396 (CB-1), and so on.

4. False smut: AC-26570 (ADT-33), AC-40119 (PTB-23), AC-40124 (PTB-26), AC-3070 (PTB-32), and so on.
5. Sheath rot: AC- 26904 (Tetep), Ram tulasi.
6. Brown spot: Katak tara, Bhut muri, BAM 10, SR-26B, CH-45, CO-20.
7. Stem borer: AC-3094 (TKM-6), AC-392 (Slo-12), AC-267 (ADT-14), AC-8396 (CB-1), AC-344 (MTU-15), AC-20006 (JBS-1638), Tepa-1, and so on.
8. Gall midge: AC-35 (Ningar small), AC-39 (CNAB white rice), AC-210 (Bhadas-79), AC-391 (Bikiri sannam), ARC-5984 (Suto syamara), ARC-10660, ARC-12508 (Khauji), ARC-12586 (Vale matse), ARC-12588 (Amamma matse), ARC-12670 (Nien sah), ARC-13166 (Jaksa), ARC-13210 (Yangbelok), ARC-14915 (Maich dol), ARC-14967 (Galong), AC- 352 (PTB-2), AC-362 (PTB-12), AC- (PTB-18), AC-371 (PTB-21), PTB-24, AC-26704 (Phalguna), and Leaung-152.
9. Brown planthopper: ARC-6650 (Gomiri bora), AC-34969 (Baidya raj), AC-34993 (Ghusuri), AC-34997 (Jhupjhupa), AC-35014 (Nal dhan), AC-371 (PTB-21), AC-40634 (PTB-33), AC-30300 (MR-1523), AC-35181 (Salkathi), AC-35184 (Dhoba numberi), AC-35228 (Jalakanthi), AC-35066 (Banspati), AC-35070 (Panidubi), AC-35108 (China bali), AC-17912 (Ganga sagar), AC-20363 (Kalachudi), Tarapith, Haldi ganthi, and so on.
10. Leaf folder: AC-33849 (Bundei), AC-35034 (Hari sankar), AC-33831 (Sunakathi), AC-33832 (Surjana), Juli, AC-35338 (Saru chinamali), and so on.
11. Yellow stem borer: AC-33515, AC-33526, AC-33538, AC-33563, ARC-5984, AC-30300 (MR-1523), and AC-30349 (Aganni).
12. Nematode: AC-26594 (TKM-6), AC-40083 (MTU-17), AC-467 (Lalnakanda-41), Hasma, Bahagia, AC-40509 (Manoharsali), Amla, AC-17134 (Sathia), AC-22899 (Anang), AC-23652 (Kalakeri), Kanyakaprashant, and so on.
13. Drought: AC-254, AC-263, AC-304, AC-511, AC-2298, AC-3035, AC-3111, AC-3577, AC-9066, AC-9387, ARC-7063, AC-45 (CH-45), AC-40083 (MTU-17), W-691, AC-467 (Lalnakanda-41), AC-35207 (Dular), AC-37077 (Dhan gora), AC-37127 (Black gora), AC-37291 (Kalakeri), AC-8205 (Surjamukhi), AC-34440 (Salumpikit), AC- 34256 (Kabiraj Sal), AC-34296 (Bombay murgi), AC-34992 (Salkiann), AC-35021 (Kalon dani), AC-35038 (Godhi akhi), AC-35046 (Nadi tikar), AC-35059, (Phutki bari), AC-35060 (Bhuska), AC-35143 (Baihunda), AC-35452 (Karama) AC- 100374 (*O. nivara*), AC-100476 (*O. nivara*), and so on.
14. Submergence: AC-24682 (FR-13A), AC-35741(Telgri), AC-35323 (Chaula pakhia), AC-35675 (Biesik), AC-36107 (SL276), AC-36470 (Khoda), Khadara, Kalaputia, AC-26670 (Janki), AC-40844 (Manasarovar), Sarumuli, AC-40916 (Jalamagna), AC-40604 (Jaladhi-1), Kanawar, and so on.
15. Salinity: AC-2405 (SR-26B), AC-8532 (Pokkali), Pateni-2, AC-41360 (Nonabokra), AC-35255 (Rahaspanjar), Canning-7, Ravana, and so on.

1.7 Limitations in germplasm use

The low utilization of PGR conserved in the National Gene Bank of India (about 102,000) is due to lack of documentation and adequate description of collections, accessions with restricted adaptability, and insufficient rice breeders in the country. There is also a gap between available genetic resources and breeding program activities. While germplasm banks try to conserve as much as possible the genetic variability to be used by breeders, breeding programs do not explore efficiently the available diversity, relying almost exclusively on their working collection.

The search for superior genotypes regarding yielding ability, disease and pest resistance, stress tolerance, or better nutritional quality is very hard, competitive, and expensive. This is why breeders tend to concentrate on adapted and improved materials, avoiding wild parents, landraces, and exotics that are available in germplasm banks, which would require a long time and high financial support besides the difficulty to identify potentially useful genes. Marshall (1989) emphasized that the difficulty to identify useful genes is the main factor responsible for the low utilization of these accessions. Evidently, there is a gap between available genetic resources and breeding program activities.

1.8 Germplasm enhancement through wide crosses

Generally, a breeder avoids making a wide cross for transferring any desirable character from a wild species to the cultivated one if the same desirable character is available within the cultivated species because, in the process, many undesirable characters of wild rice also get transferred and then the breeder has to make efforts to eliminate these undesirable traits. It is generally the resistance to insect pests and diseases or tolerance to some rare physiologic traits that the wild species are screened for and transferred to the cultivated species.

With regard to utilization of wild species for the genetic improvement of a crop plant, Harlan and DeWet (1971) classified the species into three classes: (1) primary gene pool, (2) secondary gene pool, and (3) tertiary gene pool based on their isolation barriers. In the case of *O. sativa*, it is easier to transfer a trait from a wild species that is diploid and has the same genome (AA) as that of *O. sativa.* Transferring a trait from a diploid species having a different genome or from a tetraploid species is comparatively difficult. Again the BB, CC, and DD genomes are considered homoeologous as they have some affinity with AA genome species. It is therefore easier to transfer a trait from, say, *O. officinalis* (CC) to *O. sativa* (AA) than transferring a trait from, say, *O. granulata* (GG) or *Oryza brachyantha* (FF). Transferring the characters from tetraploid ($2n = 48$) wild species to diploid ($2n = 24$, *O. sativa*) involves the additional problem of bringing the hybrid to a diploid level retaining mostly the *sativa* chromosomes with a few introgressed characters of the tetraploid wild species.

The plant breeders have utilized the variability in landraces for the selection and improvement of crops. The variability and germplasm resources available for many cultivated varieties are becoming extremely limited (Harlan, 1976). As additional genetic resources are required to enrich the germplasm, unique and imaginative procedures are required to exploit fully the potential of crop plants. Utilization of wild species (Sen et al., 2005; Nayak et al., 1996; Bose, 2005; Bose et al., 1990), therefore, is one method designed to introduce additional germplasm into cultivated varieties (Stalker, 1980).

The wild species of rice are an important reservoir of useful genes for resistance to major diseases and pests, and tolerance to abiotic stresses. Wide hybridization in *Oryza* is normally difficult to achieve because many are difficult to cross with cultivated rice because of the difference in chromosome number or genetic constitution.

Fertilization may occur, but the embryo is aborted. Embryo rescue technique is used to maintain the hybrid embryos and F_1's through several cycles of back-crosses until fertility is restored. Interspecific hybridization has been attempted to study the phylogeny and taxonomy but with limited success in transferring desirable traits from wild to cultivated species. However, the transfer of grassy stunt virus resistance from one accession of *O. nivara* was achieved successfully. The resistance from this *O. nivara* was transferred to the cultivated rice and varieties like IR28, IR29, IR30, IR32, IR34, and IR36 were developed and released for cultivation in countries affected with this disease (Jena and Khush, 1990).

Another species belonging to the AA genome, *O. longistaminata* has been exploited for transfer of bacterial blight resistance to cultivated rice. The resistance of BLB in one of the accessions of *O. longistaminata* (earlier designated as *O. barthii*) was first identified at CRRI (Devadath, 1983). After knowing about the resistance of the plant from A.R. Panda, Dr R.C. Choudhury took it to the Rice Research Station, Patna during 1980–1981. He planted a few stubbles in the backyard of his house, which proliferated from the rhizomes. Later in 1982, an international symposium on rainfed rice was organized at Patna by R.C. Choudhury where Dr Khush from IRRI was attending and came to know about the plant. Then Khush took it to IRRI, from where it was taken to the University of California-Davis where the *Xa21* gene was identified and cloned for developing BLB-resistant rice varieties. Wild species with genomes nonhomologous to the AA genome of *O. sativa*, such as *O. officinalis* (CC), *O. australiensis* (EE), and *O. minuta* (BBCC) possessing resistance to brown planthopper, white-backed planthopper, bacterial blight, and blast, have been used to transfer these desirable alien traits to cultivated rice. Therefore, wide hybridization is one of the key components in the program aiming at transferring alien genes from diverse sources surmounting sexual barriers. Advances in embryo rescue, anther culture, chromosome (genetic) engineering, and overall genetics have facilitated in the transfer of genes and in precise monitoring and characterization of alien introgression from different genomes of *Oryza* into cultivated rice. By integrating conventional breeding with advanced methods of alien introgression, disease and insect pest resistant varieties can be developed.

Advances made in biotechnology have largely broken down barriers to the transfer of genes among organisms. Thus, a broad-based knowledge of the relationships among organisms provides a framework for understanding diversity, evolution, and appropriate approaches to gene transfer among organisms. Recently, molecular markers, such as restriction fragment length polymorphism (RFLP) and genome-specific DNA probes, have become available, which offer the potential to facilitate monitoring of alien gene introgression. The choice of a particular method like alien addition lines, anther culture, somaclonal variation, and recombinant DNA technology for alien gene transfer through wide hybridization depends upon genomic relationship, extent of chromosome pairing, and recombination between the genomes of the alien wild and cultivated species. If there is no restriction on chromosome pairing and recombination in wide hybrids, direct hybridization and back-crossing to the recurrent cultivated parent are followed. When the chromosomes of wide cross-hybrids do not pair certain chromosome manipulation techniques such as alien addition lines, alien substitution lines, induced homologous pairing, and amphiploids are utilized.

More than 15,000 accessions of rice germplasm have so far been evaluated systematically for biotic and abiotic stress situations at CRRI during the last decade. Several donors have been identified and they are being utilized for crop improvement programs.

Rice plants, many a times, are exposed to stress-prone environments that adversely affect the crop leading to low productivity. Among the more often experienced stresses, drought brings maximum damage to the rice physiology and causes setbacks to its production potentiality. Various USDA statistical analyses identified drought as the single most yield-reducing factor. Drought at vegetative stage causes irreparable loss of canopy, whereas at flowering stage, water deficit hampers anthesis and seed setting leading to higher spikelet sterility and lower yields. Keeping this in view, about 12,000 rice germplasms collected from uplands, medium lands, lowlands, and 97 accessions of wild rice (*O. nivara*) were screened against vegetative-stage drought under field condition. Dry seeding was done in the first week of February and the crop was grown with adequate soil moisture for 30 days with a sprinkler irrigation system. Thirty-one days after sowing, the irrigation was suspended until the susceptible check showed permanent wilting and then reirrigated for plant recovery. The soil moisture content during the drought period decreased from 34% to 3% of the soil profile of 45 cm. Data on leaf rolling and recovery were recorded using SES of IRRI. The lines with early leaf rolling after suspension of sprinkler irrigation showed higher score for drought tolerance (7–9). Few germplasm lines with delayed leaf rolling recovered faster after reirrigation. Two accessions of *O. nivara* collected from West Bengal having the identity numbers (SRD01-17, AC-100374, IC-330470) and (BCPW-30, AC-100476, IC-330611) were found to be tolerant to drought. They are being processed for registration at NBPGR. While characterizing these accessions for their agromorphologic traits as per IRRI-IPGRI descriptor, it was found that both are morphologically the same except in a few characters such as ligule, collar, internode and auricle color, culm and flag leaf angle, culm strength, and panicle exsertion. Days to 50% flowering also differ from each other being 107 days for AC-100374 and 118 days for AC-100476. It bears significance because the habitat of this species is usually swampy and marshy places along the roadside seasonal ditches.

The genetic pool of submergence-tolerant rice cultivars is very small. For screening against submergence tolerance, the seeds were direct seeded in specially designed field cemented tanks (40 × 8 m) with known submergence-tolerant (e.g., FR 13A) and susceptible (e.g., IR 42) cultivars as checks. Twenty-one-day-old seedlings were submerged for 12 days under 80 cm of water. Survival count was taken visually on the tenth day of desubmergence. The characteristics of the floodwater in terms of light transmission (%) were measured at 1200 h and water temperature and oxygen concentration were determined at 0600 and 1630 h. Light intensity at 60 cm water depth varied between 37.6% and 41.2% of the incident irradiance above the floodwater. The oxygen concentration at the same water depth was 4.4–7.4 mg/L at 0600 h and 6.2–11.6 mg/L at 1630 h. The temperature did not vary greatly, being 29.1–33.3°C throughout the period of experiment. After thorough screening of 11,000 germplasm lines, a few landraces have been identified as tolerant to complete submergence. They are Khoda, Khadara, Kusuma, Matiaburush, Gangasiuli, Atirang, and Kalaputia.

Direct seeding is common in rainfed lowlands. In eastern India, sometimes, early rain causes water stagnation in the field just after sowing, which results in poor crop establishment. Two cultivars, namely, Panikekoa and T 1471, have been identified from the huge genetic stock available in the gene bank. They establish very well under water in anaerobic condition. This work has immense value for irrigated conditions also in reducing the cost of cultivation.

Agromorphologic characterization of rice germplasm is another important aspect for any crop improvement project. The germplasm accessions of cultivated rice are characterized with the help of descriptors developed by IRRI-IPGRI. Every year the characterized data of 45 morphoagronomic characters are documented and published as a catalog. The wild species of cultivated rice represent an untapped reservoir of genetic variability for agronomic traits. They harbor significantly higher genetic diversity than does the cultivated species. Therefore, an attempt was made under NATP to collect, evaluate, and conserve the wild rice germplasm from eastern India where it is found abundantly. About 483 accessions of wild rice species, namely, *O. nivara* and *O. rufipogon*, were collected and multiplied for seed increase, then characterized, evaluated, and conserved. The wild rices show a strong dormancy, so to facilitate quick germination and raise seedlings the seeds were dehusked and put on germination paper in petri dishes. Once the seeds sprouted, these were planted in small earthen pots containing sterilized soil.

1.9 Rejuvenation of cultivated germplasm

Samples sent to the rice gene bank for long-term conservation must be multiplied to produce sufficient seeds for the active and base collections. Accessions are also regenerated if seed viability falls below 85%, or seed stocks in the active collection are 50 g or less. Every effort is made to produce seeds of the highest viability with the minimum number of multiplication or regeneration cycles. Accessions are harvested individually at 21–28 days after anthesis. Panicles are threshed in the field, and within an hour or so, seeds can be placed in the drying room. The changes to field operations, combined with the controlled drying regime ensure the production of high-quality seeds with high initial viability. This even extends to the *japonica* rice that are notoriously difficult to grow successfully under tropical conditions and that are less well adapted to the environment. They have an inherently shorter storage life than *indica* varieties.

When a variety is collected either through exploration or through correspondence/exchange, it is given an accession number in the CRRI germplasm register, which runs serially from AC-1 onward. After recording the passport characters such as place of collection, farmer's name, and other preliminary plant characteristics, the seeds are kept in the cold storage module until the cropping season starts in the month of June. At the onset of monsoon, the kharif season starts and newly acquired seeds are taken out along with other rice germplasm, which are to be regenerated. To avoid weed and other agronomic management problems, usually seedbed sowing followed by transplanting is done. The seedbeds are of 1 m width and 20–25 m long; in each bed 120–130 lines are drawn and in each line one accession of germplasm is sown in dry condition. When the seeds are properly conditioned in the soil with temperature and if there is no rain immediately, then one irrigation is given in the channels between two beds and the seeds start sprouting in 3–4 days time. To minimize mistakes, human errors, or to know if any line has not germinated, one purple rice variety is sown as a marker in every 20-line interval; also, at the edge of each line a pinch of the same purple marker is sown to keep the line distinct from another when the plants grow up in competition of the weeds.

When the seedlings become 3–4 weeks old, they become ready for transplanting. The 40-cent (40 m × 40 m) plots are prepared with thorough ploughing, followed by irrigation, then leaving it for 6–7 days so that whatever rice or other weed seeds that have fallen earlier from the previous season have germinated. Then the plot is filled with water up to 30–40 cm. Next day with a tractor-drawn rotavator, the experimental field is puddled and all the undesired sprouted seedlings perish. A basal dose of fertilizer is applied before leveling so that the chemical fertilizer is uniformly mingled in the soil. Then through a leveler, drawn by the bullocks, the plot is leveled manually.

Meanwhile in the nursery seedbeds, for every variety a bamboo peg of about a meter length is put with an aluminum label or waxed label so that it becomes easy to transport and arrange in the puddled field. Furthermore, care is taken to see that the label remains intact until harvesting stage. Then sufficient water is given in the channels between the seedbeds so that it becomes easy to uproot the seedlings and wash the roots making them clean from soil. Every variety is uprooted and tied with the bamboo pegs with label and kept in a tray. Then they are arranged in the puddled field serially for line transplanting. In each plot 10 ranks are made giving about a 1 m gap between the ranks so that it becomes easy to monitor the plants as well as for wind flow to the culm base of the varieties. In each rank, 50–60 varieties are planted. Each variety is grown in three rows of 25–30 plants in each row, with a spacing of 20 cm between plants and 30 cm between lines. A blank line or gap between varieties is left as a measure to check cross-pollination and leaving a distinct mark between two accessions. After every 20 varieties one line of purple variety, namely, Crossa, is planted as a number counting mark. Soon after panicle initiation stage, rogueing is done to discard the off types. The agromorphologic observations are recorded at vegetative and reproductive stages of the plant growth. The postharvest grain characteristics are taken in the laboratory. A single plant from the middle row is selected and the panicles are harvested in a paper packet for next season's sowing, to ensure genetic purity. Depending upon the maturity duration the varieties are harvested when the seeds become physiologically matured.

1.10 Sharing of germplasm

Seed supply is another important aspect for any crop improvement program. It is provided to different researchers of the country as and when they need it. In the past CRRI was supplying the germplasm even to foreign agencies. But in the context of the present IPR regime, the sharing of germplasm has been restricted to researchers within the country only. Furthermore, a Standard Material Transfer Agreement (SMTA) has been introduced before supplying any germplasm to any researcher. The major achievement in seed supply was made when the RRS, Chinsurah lost their entire genetic stock due to heavy flooding in 2000. CRRI supplied about 815 germplasm lines and replenished the genetic stock to start their research program afresh. Likewise, HPAU, Palampur also lost their germplasm over the years; fortunately, they had deposited a set at CRRI, which later was restored to them. Apart from this, every year at least hundreds of germplasms are being supplied to different researchers in the country and thousands are screened within the institute for identifying donors for different biotic and abiotic stress, and nutritional aspects of the germplasm.

1.11 Registration of germplasm (Table 1.9)

Table 1.9 Unique rice germplasm of CRRI registered by Plant Germplasm Registration Committee of ICAR in India

S. No.	National identity	Trait of interest
1	INGR No. 04001, Khoda (PD 27)-2004	Tolerance to complete submergence
2	INGR No. 05001, T-1471 (Kodiyan)-2005	Tolerance to anaerobic seeding
3	INGR No. 08108, Khadara (PD33)-2008	Tolerance to complete submergence
4	INGR No. 08109, Atiranga (RM5/232)-2008	Tolerance to complete submergence
5	INGR No.08110, Kalaputia (PCP-01)-2008	Tolerance to complete submergence
6	INGR No. 08111, Gangasiuli (PB-265)-2008	Tolerance to complete submergence
7	INGR No.08113, Kusuma (PD75)-2008	Tolerance to complete submergence
8	INGR No.08112, Mahulata (PB-294)-2008	Tolerance to vegetative-stage drought stress
9	INGR No. 10147, Medinapore (RM5/AK-225;IC-0258990) -2010	Tolerance to complete submergence
10	INGR No.10148, Andekarma (JBS-420;IC-0256801)-2010	Tolerance to complete submergence
11	INGR No.10149, Champakali (IC-0258830)-2010	Tolerance to complete submergence
12	INGR No.10150, Brahman Nakhi (DPS-3)-2010	Tolerance to vegetative-stage drought stress
13	INGR No.08112, Sal kaiin (PB-78;IC-0256590)-2010	Tolerance to vegetative-stage drought stress
14	INGR 14025, Bhundi (JRS-9;IC0575277;AC42091)-2014	Tolerance to complete submergence and having shoot elongation ability
15	INGR 14026,Kalaketki (JRS-4;IC0575273;AC42087)-2014	Tolerance to 20 days complete submergence

1.12 Integration of genomic and genetic resources in crop improvement

1.12.1 Structural and functional genomic resources

Rice (*O. sativa* L.) is one of the most important cereal crops providing more than 50% of the world's staple food with 20% of the world's dietary energy source (Schatz et al., 2014). Considering the future food demand of the world population, which has been predicted to increase by 25% or more (Seck et al., 2012) by 2030, and the current rate of population growth with the rapid decline of arable land area worldwide and unpredictable environmental factors, such as drought, flood, and ill-soils, increasing the yield/productivity of rice by adopting innovative technology to develop new high-yielding cultivars with resistances to multiple stresses is the utmost vital means for the future world population to survive. However, this aim requires more complete knowledge of the genetic diversity of rice and systematic exploitation of this rich genetic diversity by associating with the traits of interests through innovative breeding strategies. Rice has been well known to be the most diverse genus among food crops (Calpe, 2003) by a large collection of 120,000 rice accessions (Anacleto et al., 2015). This rich genetic diversity has been the driving force for rice improvement in the past, and will be so for the future. Despite such within-species genetic diversity and varietal group differentiation being collected and maintained, less than 5% of these accessions have been used in breeding programs. To understand the complete genetic information and knowledge of rice traits, rice genomic research is necessary. This addresses the present yield bottlenecks by designing efficient ways to tap into the wealth of rice genome sequence information through genes or multilocus combinations reaping the maximum yield potential and desirable quality.

The whole genome of rice was sequenced for the first time almost a decade ago. In fact, it is the first crop genome to be sequenced completely. Here we briefly review the global effort on the whole-genome sequencing of rice and associated structural genomic studies.

1.12.2 Rice genome

The rice genome is a haploid set of 12 chromosomes and its mitochondrial and plastid genomes. Genomics is the study of genomes and its annotation, which includes identifying the genes and location along the chromosomes, gene function, and regulation. The genus *Oryza* is small and comprises two cultivated and 22 wild species accounting for about 24 species that categorize them into 10 genome types (Zhang et al., 2014) represented by six diploid (AA, BB, CC, EE, FF, GG genomes) and four tetraploid (BBCC, CCDD, HHJJ, HHKK genomes) groups (Ge et al., 1999) that belongs to the family Poaceae and tribe Oryzeae. Both the cultivated rices carry the same genome type (AA), which are classified as Asian rice *O. sativa* L. and African rice *O. glaberrima* Steud. This diverse species of rice is a huge rich source of genetic resources providing an enormous gene pool for the genetic improvement of rice cultivars and for studying many biologic processes involving comparative and functional genomics, polyploid evolution, speciation and biogeography, as well as ecologic adaptation and domestication (Ge et al., 1999; Wing et al., 2005; Sang and Ge, 2007; Ammiraju et al., 2010; Zang et al., 2011). Unlike cultivated rice, wild rice consists of 22 species typically displaying long awns and severe shattering for seed dispersal. All members of the *Oryza* genus have $n = 12$ chromosomes and while interspecific crossing is possible within each complex, it is difficult to recover fertile offspring from crosses across complexes (Vaughan et al., 2003). A considerable number of useful genetic resources are preserved in the wild species and can be utilized to improve cultivated rice.

Asian cultivated rice was reported to be domesticated from its wild progenitor *O. rufipogon* and/or *O. nivara* around 8000–9000 years ago (Ouyang and Zhang, 2013). Based on hybridization studies and genetic differentiation, Asian cultivated rice is further divided into two subspecies: *indica* and *japonica.* Based on many morphologic and physiologic traits, Japonica is further subdivided into temperate and tropical *japonica* ecotype (Kato et al., 1928; Oka, 1953, 1958; Glaszmann, 1987). These two subspecies show a wide range of morphologic and physiologic distinctiveness in characteristics such as seed size, abiotic and biotic stress responses with distinct patterns of adaptation to environmental conditions (Oka, 1988), and they are considered as independently domesticated from wild ancestral species (Yamanaka et al., 2003; Londo et al., 2006; Kwon et al., 2006; Kawasaki et al., 2007). However, Huang et al. (2012) suggested another model in which *japonica* rice was first domesticated and then *indica* rice was generated from the cross between *japonica* and wild rice. Molecular markers as well as whole-genome comparisons between the reference genome Nipponbare (*japonica*) and 93-11 (*indica*) analyses revealed that *indica* and *japonica* have profound genetic differentiation and divergence happened (Ma and Bennetzen, 2004) around 0.44 million years ago, which clearly confirmed the distinction between them at the whole-genome level too. Rice is considered to have been cultivated since more than 7000 years ago (Agrama et al., 2007) and these two groups have gone through many human artificial selections throughout the domestication process since its divergence. The progenitor *O. rufipogon* is divided into two ecotypes: *O. nivara* (annual) and *O. rufipogon* (perennial), which are morphologically, physiologically, and ecologically distinct. The perennial *O. rufipogon* is adapted to the stable, deep-water habitat, predominantly out-crossing, and photoperiod sensitive, whereas *O. nivara* is considered to be derived/descended from perennial *O. rufipogon* adapted to the seasonally dry habitat, annual, self-fertilized, and photoperiod-insensitive species and is rather similar to the cultivated rice as compared to perennial *O. rufipogon.*

The African cultivated rice, *O. glaberrima* Steud., is endemic to African countries and was found to be independently domesticated from the wild progenitor *O. barthii* ~3000 years ago, which is around 6000–7000 years after the domestication of Asian rice (*O. sativa*) (Vaughan et al., 2008). It is well adapted for cultivation in Africa and possesses traits for increased tolerance to biotic and abiotic stresses, including drought, soil acidity, iron and aluminum toxicity, as well as weed competitiveness.

1.12.3 Rice whole-genome sequencing

Among cereals, rice has the smallest genome size at an estimated 400–430 Mb as compared to the significantly large cereal genome sizes of sorghum, maize, barley, and wheat. Unlike other cereal crops, rice shows remarkable diverse ecologic adaptations due to huge genetic resources and shares a large degree of synteny among the grass genomes, which makes the best option for studying comparative genomics and evolutionary biology. The 2C (twice the gametic) value for *O. sativa* ssp. *japonica* ranges from 0.86 pg to 0.91 pg (Arumuganathan and Earle, 1991). It is also well mapped and characterized with the easiest genome to transform genetically thus providing powerful tools such as reverse genetics and insertion mutagenesis for functional genomics. Extensive linkage mapping has been already reported in rice with several available molecular markers such as RFLP, amplified fragment length polymorphism (AFLP), random amplified polymorphic DNA (RAPD), microsatellite or simple sequence length polymorphism (SSLP), and cleaved amplified polymorphic sequence (CAPS) markers (Mohan et al., 1997; McCouch et al., 1988; Nagamura et al., 1997), which can be used as foundation stones for reference genome sequencing through clone-by-clone approach. The availability of reference genome sequences are the basis of crop genetic studies, which is considered as the backbone for functional and comparative genomics studies. A high-quality reference sequence is essential to fully understand the biology of the organism, which greatly affects subsequent research depending on the sequencing strategy and the genome of the crop to be sequenced (Pan et al., 2014). This can be achieved through clone-by-clone sequencing approach, which provides a way to achieve high-quality sequence assemblies for genomes as compared to *de novo* assembly from whole-genome shotgun (WGS) sequencing, which yields a large number of sequence gaps. It has been used to guide the assembly of newly sequenced genomes with lower coverage and short sequence reads (Schneeberger et al., 2010), resequencing of diverse varieties through next-generation sequencing (NGS) technologies, or to identify missing sequences and insertions in individual genomes (Kidd et al., 2010). Some of the landmark rice genome sequencing that can be used as reference rice genomes are summarized as follows.

1.12.3.1 Cultivated rice

O. sativa (ecotype japonica): In 1998, the International Rice Genome Sequencing Project (IRGSP), a publicly funded consortium from 11 countries, initiated the sequencing of rice, and Nipponbare, a temperate *japonica* variety, was selected as a representative of the whole rice plants for sequencing. The main reasons for selecting Nipponbare was due to its availability of high-density and robust molecular genetic map (Harushima et al., 1998), ESTs and PCR-based markers, large-sized rice genomic DNA fragments ligated in yeast artificial chromosome (YAC), bacterial artificial chromosome (BAC), and P1-derived artificial chromosome (PAC) vectors, with easy regeneration from callus for genetic transformation. The consortium adopted a map-based, clone-by-clone approach using BAC clones similar to the human genome sequencing to yield the best quality reference genome sequence. In this approach, a robust BAC physical map was first constructed and anchored to a dense genetic map to provide high accuracy for location of the BAC contigs (Harushima et al., 1998; Chen et al., 2002). Fingerprinting and physical mapping were used to make minimal tiling paths, which was followed by end sequencing, connecting, and extending of contigs to obtain the genome sequence (Eckardt, 2000; Sasaki and Burr, 2000; Chen et al., 2002). Then, a minimal tile of BAC clones was individually sequenced, finished, and assembled and the whole-genome sequence was assembled from the completed individual BAC sequences. In 2004, the IRGSP succeeded in completely decoding the rice genome sequence with 99.99% accuracy. The resulting Nipponbare reference sequence is of high quality and is considered a "gold standard." The Nipponbare genome assembly was updated by revising and validating the minimal tiling path of clones with the optical map for rice (Kawahara, 2013). In the latest update of IRGSP1.0, the sequence was improved by a revision that used optical map data and by whole-genome resequencing using next-generation sequencing (Kawahara et al., 2013). Subsequently, two genome assemblies were independently produced: one by the Rice Genome Annotation Project initially located at The Institute for Genomic Research and now at Michigan State University (MSU) and another by the Rice Annotation Project (RAP) (Ouyang et al., 2007; Tanaka et al., 2008). The annotation of the Rice Annotation Project Database (RAP-DB) was updated for Os-Nipponbare-Reference- IRGSP-1.0. As a result, 37,872 loci, of which 35,681 have protein-coding potential, were determined. The current release of the MSU Rice Genome Annotation Project (Release 7) based on the Os-Nipponbare-Reference-IRGSP-1.0 pseudomolecules contains 56,081 loci encoding 66,433 gene models. Of these, 39,102 loci (49,110 gene models) are nontransposable element related. Comparison of the RAP and MSU annotation datasets show a high degree of concordance between the two annotations. These genome sequences provide a solid foundation for integrating biologic information, including genetics, gene expression, development, and physiology.

O. sativa (*ecotype indica*): *Indica* rice varieties account for about 90% of total rice production worldwide. Therefore, due to divergence from the japonica genome, cracking the genome sequence information of an *indica* rice variety is very important for rice genetic improvement. The 93-11 cultivar, an elite representative of *indica* varieties, was selected due to its involvement as a parental line of the popular super-hybrid rice, LYP9, which accounts for more than 70% of the world's rice production. Obtaining a high-quality 93-11 genome sequence as a reliable *indica* reference sequence is imperative for world food security (Pan et al., 2014). In contrast to the Nipponbare genome sequencing, sequencing of the 93-11 genome was conducted by BGI and adopted the WGS approach in which no BAC library and physical map were used. In this approach, the genome is randomly broken down into smaller pieces, which are then sequenced and subsequently assembled thereby significantly reducing the preparation time and cost. However, the short sequence data are more difficult to assemble resulting in relatively lower quality genome sequence as compared to the Nipponbare genome sequence.

Oryza glaberrima: The African cultivated rice genome is especially important due to many traits that make African rice resistant to environmental stress, such as long periods of drought, high salinity in the soils, and flooding, unlike the Asian cultivated rice. For this, the CG14 accession, which is one of the parental lines used in the generation of the "New Rice for Africa" (NERICA) cultivars that revolutionized rice cultivation in west Africa by combining the high-yielding traits of Asian rice with the adaptive traits of west African rice, was selected for whole-genome sequencing adopting BAC clone and WGS combined approach and assembled around 316 Mb genome sequences using the high-quality reference genome Nipponbare (389 Mb). The genome annotation yielded around 33,164 gene models as compared to that of the gold standard, *O. sativa* Nipponbare (41,620 genes). Transposable elements constitute about 104 Mb as compared to 156 Mb of the *O. sativa* genome. The estimated divergence time between *O. sativa* and *O. glaberrima* lineages, that is, ~600,000 years, provides a unique opportunity to study the dynamics of TE-driven structural changes through a genome-wide comparative approach. Genome-wide introgression analysis of *O. glaberrima* and *O. sativa* suggested a strong evidence for domestication independently from *O. barthii* (Wang et al., 2014).

1.12.3.2 Wild rice

Oryza brachyantha: This wild rice is considered the most diverse genome from the others and is placed on the basal lineage in *Oryza* genus with the FF genome type. Because of this behavior, the genome is likely to be more static compared with other *Oryza* genomes. This provides an opportunity to explore the signatures of gene and genome evolution of *Oryza* by comparing with the rice genome. The whole genome was sequenced through WGS approach combined with the BAC-based physical map to assemble 261 Mb genome size, annotated into 32,038 genes, which is much lower than in cultivated rice, implying a massive amplification of gene families in the domesticated rice genome. It was found to be a compact genome composed of less than 30% of repeat elements. The genomic comparison revealed that only 35% of the *O. brachyantha* genome was conserved with the rice genome due to differences in the lineage-specific evolution of intergenic sequences, of which LTR retrotransposons alone contributed to 50% of the size difference.

1.12.4 The Oryza map alignment project

To unlock the genetic potential of wild rice, the International Oryza Map Alignment Project (OMAP) was initiated with the goal of completing the sequencing and assembly of all 23 species in the *Oryza* genus (Jacquemin et al., 2013). It has already completed the construction of BAC libraries and BAC-based physical maps for 17 of 23 *Oryza* species representing all the 10 genome types (Ammiraju et al., 2006, 2010; Kim et al., 2008; Jacquemin et al., 2013). Landraces and wild species of rice (genus *Oryza*) possess an underused source of novel alleles that have great potential for crop improvement of cultivated rice species (*O. sativa* and *O. glaberrima*), since they possess new genes that could be exploited for yield increases and for developing resistance to biotic stresses and tolerance of abiotic stresses. Apart from providing insights into the evolution of the *Oryza* genus, other expected outcomes are the identification of new genes and quantitative trait loci (QTL) that could be subsequently incorporated into adapted rice varieties. So far, the *de novo* assembly genome sequence of

Table 1.10 Description of rice genome databases

Databases	URL links	Applications
RiceVarMap	http://ricevarmap.ncpgr.cn/	Rice variation map, SNPs and InDels
SNP-Seek	http://oryzasnp.org/iric-portal/	Rice SNPs database from 3000 genomes
RAP-DB	http://rapdb.dna.affrc.go.jp/gb1/index.html	Rice genome and annotation: *Japonica*
MSU-RGAP	http://rice.plantbiology.msu.edu/	Rice genome and annotation: *Japonica*
Gramene	http://www.gramene.org/	Plant genome database and comparative genomics
Phytozome	http://phytozome.jgi.doe.gov/pz/portal.html	Plant genome including rice
OryzaSNP	http://oryzasnp.plantbiology.msu.edu/	SNP database from 20 rice genomes
OMAP	http://www.omap.org/nsf.html	Comparative genomics of rice
3K RGP	http://dx.doi.org/10.5524/200001	3000 Rice genome sequences
RIS-BGI	http://rice.genomics.org.cn/rice/index2.jsp	Rice genome and annotation: *Indica*
PlantGDB	http://www.plantgdb.org/	Plant genome sequence including rice
RGKbase	http://rgkbase.big.ac.cn/RGKbase/	Rice comparative genomics and evolutionary
Rice-Map	http://www.ricemap.org/	Rice genome browser
Grassius	http://grassius.org/links.html	Comparative regulatory genomics of grasses
PlantGM	http://www.niab.go.kr/nabic/PlantGM	Genetic markers for rice and Chinese cabbage
RIMD	http://202.120.45.71/	Rice InDels markers

O. brachyantha (genome FF) and *O. rufipogon* (genome AA) have been published, and the sequence assembly and genome annotation of other wild rice species are under way. Gramene, a rice genus-level resource, includes 13 of the estimated 24 species within the *Oryza* genus, complete reference assemblies for *O. glaberrima*, its wild progenitor *O. barthii*, and the distantly related wild species *O. punctata* and *O. brachyantha*. An additional eight *Oryza* species, including one polyploid, out group species *Leersia perrieri*, are available as chromosome short-arm assemblies. Gramene currently incorporates an SNP database also and/or SV datasets for *japonica*, *indica* rice, and African rice. Some of the other important genome databases are listed in Table 1.10.

1.12.5 Rice genome resequencing project (3000 accs.)

Studies of genome variation in rice may reveal the origin of cultivated rice and provide clues about the domestication processes in rice that can be achieved through the resequencing of thousands of cultivated and wild rice genomes as high-quality reference is available in rice. Therefore, an international effort has been made to understand the total genetic diversity within *O. sativa* and to create a public rice database containing genetic and genomic information suitable for advancing rice breeding technology gene by resequencing a total of 3000 germplasm accessions from 89 different countries/regions (3K RGP 2014). The project had generated an average depth of ~14×, average genome coverage of 94.0%, and yielded a combined total of approximately 17 TB of high-quality sequence data, which are now deposited in the GigaScience database (GigaDB). These data will be helpful in exploring the within-species diversity and genome-level population structure of *O. sativa* in great detail thereby leading to a more thorough understanding of the molecular, cellular, and physiologic mechanism responsible for the growth and development of rice plants and their responses to various abiotic and biotic stresses. The phylogenetic analyses of 3000 accessions revealed clear differentiation into two major groups – *indica* and *japonica*, indicating the genome distinctness between this subspecies.

1.12.6 Plastome genome: plastid and mitochondrial genomes

Chloroplast and mitochondrial genomes are an integral part of the rice genome. The chloroplast is the center for photosynthesis whereas the mitochondrion is the center for oxidative respiration; both supply energy within the cells and are regarded as the powerhouse of the cells thereby playing an important role in plant growth and development processes. Mitochondria and chloroplasts of rice are considered as originating from alpha-proteobacteria and cyanobacteria endosymbionts, respectively (Tian et al., 2006). Both the organelles have their own genomes similar to prokaryotes, which are distinct from the nuclear genome.

For the first time among the monocot plants, the complete mitochondrial genome sequence of Nipponbare, a *japonica* rice, was reported by Notsu et al. (2002). The mitochondrial genome was comprised of 490,520 bp with the average 43.8 GC% with a total of 35 genes for known proteins, three ribosomal RNAs (rrn5, rrn18, rrn26), two pseudoribosomal protein genes, 17 kinds of tRNAs, and five pseudo tRNAs. Another group, Tian et al. (2006), reported the complete mitochondrial genome sequence of two cultivars, 93-11 (*indica* variety) and *PA64S* (*indica*-like variety with maternal origin of *japonica*) with 491,515 bp and 490,673 bp mitochondrial genomes assembled, respectively. On comparison of available *indica* and *japonica* mitochondrial genomes, they identified 96 SNPs, 25 indels, and 3 segmental sequence variations, which led to the detection of divergence of *indica* and *japonica* mitochondrial genomes, which could have occurred approximately 45,000–250,000 years ago.

The first chloroplast genome sequence of rice was published by Hiratsuka et al. (1989) from Nipponbare cultivar with an assembly of 134,525 bp genome size. Another group also sequenced the plastid genome of *indica* (93-11) and *PA64S* and compared the interspecies variation. They reported that the intersubspecific variations of 93-11 (*indica*) and *PA64S* (*japonica*) chloroplast genomes consisted of 72 SNPs and 27 insertions or deletions, which help to estimate the divergence of *indica* and *japonica* chloroplast genomes as occurring approximately 86,000–200,000 years ago.

The comparison of the whole chloroplast genome sequence of *O. rufipogon* from Asia and Australia and *Oryza meridionalis* and *O. australiensis* emphasized the genetic distinctness of the Australian populations and revealed their potential as a source of novel rice germplasm (Waters et al., 2012).

1.12.7 Resequencing of rice genome

As the high-quality reference genome sequence for both the cultivated species of rice is available in the public domain, resequencing of new rice genotype has been feasible and affordable to many rice researchers due to the low cost of genome sequencing and next-generation sequencing technology. Several groups have adopted this resequencing technology to identify genetic variations within the species, evolution process, discovery of important agronomic QTLs/genes using biparental mapping or GWAS, domestication process, and so on.

Genetic diversity analysis of 950 cultivated rice collected worldwide through resequencing leads to the classification of cultivated rice into five divergent groups: *indica, aus, temperate japonica, tropical japonica*, and intermediate (Huang et al., 2012). In another study, resequencing of 517 *indica* subspecies led to the identification of 3,625,200 SNPs, which were used to construct a high-density SNP map called HapMap (Huang et al., 2010). In order to tap the novel alleles present in the wild rice, several groups are also resequencing the wild rice to identify genetic variation and selection sweep (Xu et al., 2012; Huang et al., 2012). Phylogenetic analysis using genome variation from 1529 rice genome resequencing data revealed the progenitor of cultivated rice *O. sativa* and subspecies as *O. rufipogon – I* and *O. rufipogon – III* of the three types of *rufipogon.* They also suggested that cultivated rice originated from southern China and *japonica* was first domesticated followed by generation of *indica* by crossing it with wild rice. Xu et al. (2012) also reported a similar hypothesis that *japonica* was domesticated directly from *rufipogon* and that *indica* and *japonica* were domesticated independently by resequencing of 50 accessions. Resequencing is also adopted for the identification of QTLs/genes for agronomically important rice. Gao et al. (2013) identified 43 yield-associated QTL while resequencing 132 core RILs derived from two *indica* rice. While resequencing another RILs population, 49 QTLs for 14 agronomic traits were identified by Wang et al. (2011).

Weedy rice (*O. sativa* f. *spontanea* Rosh.) or red rice is a member of the *Oryza* genus with weedy characters mainly due to its mixture of domesticated and undomesticated traits and widely grown in rice-planting areas all over the world (Ferrero et al., 1999; Noldin and Cobucci, 1999; Mortimer et al., 2000; Sun et al., 2013). It harbors characteristics of undomesticated *Oryza* species, including seed dispersal/shattering mechanisms and seed dormancy and domesticated traits, such as rapid growth resembling domesticated rice during the seedling stage, which promotes its invasiveness in the agroecosystem (Sun et al., 2013). Several hypotheses have been established about its evolution such as due to ongoing selection and adaptation of wild rice (Harlan, 1992), dedomestication or reversion of crop species to a wild or feral form from abandoned domesticated rice (Bres-Patry et al., 2001), or hybridization between cultivated rice and its progenitor type (Tang and Morishima, 1996). Therefore, genome sequencing of weedy rice will elucidate the origin and evolution of problematic weedy species through dissection of the underlying genomic information. Genome resequencing of three weedy rice accessions from the lower Yangtze region in Jiangsu Province were conducted and the results suggest that this weedy rice originated from hybridization between *indica* and temperate *japonica* rice (Qiu et al., 2014). Ishikawa et al. (2005) also performed isozyme and morphophysiologic analyses, and proposed that weedy rice originated from *indica–japonica* hybridization. Xiong et al. (2012) also demonstrated that some percentage of weedy rice-like plants could have been segregated from the offspring of intersubspecies and intervarietal hybridization based on seven *indica/japonica* and four *indica/indica* combinations. Several rice-breeding programs adopted the intersubspecies hybridization between *indica* and *japonica* to improve rice cultivars and the introduction of germplasm of *indica* subspecies into temperate *japonica* has been reported to be largely responsible for the increased rice yield in northern China (Chen et al., 2007; Sun et al., 2012), which might be the source for the evolution of weedy rice. Therefore, there should be a stringent breeding process that involves interspecies hybridization to avoid such evolution (Chen et al., 2007; Sun et al., 2012).

1.12.8 Structural genomics

With the completion of sequencing, tremendous advances have been made in the rice genetic and genomic research. Characterization and sequencing of the genome facilitated the identification of genes corresponding to phenotypes and genome variation, which differentiates one species from another within the same genus or other closely related genera. Since the whole-genome sequencing of rice has predicted/annotated about 37,000 genes, the next challenge is to determine the function and to utilize agronomically important genes for genetic improvement of cultivated rice. These include genes related to yield potential, biotic and abiotic stress tolerance, heading date, and other agronomically important traits, which are either orthologous among cereal crops or unique to each crop. Genome-wide variations are considered to drive important phenotypic variation within the crop plant species. The variation within and between species is most commonly quantified by molecular markers such as RAPD, RFLP, AFLP, microsatellite repeats, and SNPs. However, there has been a recent introduction of genome variation in terms of structural variation (SV) including CNV caused by insertion, deletion, or rearrangement like inversions and translocations,

1.12.8.1 Molecular marker

One immediate output from genome sequencing is the development of DNA markers. DNA markers and genetic linkage maps are prerequisite tools for performing molecular genetic studies of plants. Different types of DNA markers have been developed and used for constructing linkage maps, which are used for genetic mapping of the loci onto chromosomes and identification of gene(s) of interests for agronomic importance traits. Later, marker-assisted selection (MAS) in which DNA markers are used to infer trait of interests/phenotypes, which can be easily transferred to the desired cultivars thereby accelerating the development of new improved crop varieties. Sequence-tagged DNA markers provide opportunities to perform intra- or interspecies comparisons of genome structure. During the pregenome sequence of rice, RFLP, AFLP, and RAPD markers were commonly used for rice breeding research. With the advent of polymerase chain reaction (PCR) technology and the availability of the genome sequence of rice in the public domain, PCR-based markers called sequence-tagged site (STS) markers, a technically simple and reliabile technology, have revolutionized the rice genetic studies. Simple sequence repeats (SSR or "microsatellites") are the short consecutive repetitive DNA sequence motifs with 1–6 bp in length. It became the most widely used marker in rice due to its abundance, high reliability, codominant in inheritance, highly polymorphic, and transferable between mapping populations. McCouch et al. (2002) identified around 2200 validated SSRs using publicly available BAC and PAC clones of rice and followed by 18,828 SSRs that were released after the completion of the Nipponbare genome sequence in 2005 (International Rice Genome Sequencing Project, 2005). Presently, around 19,480 SSR markers covering the entire genome are available at the Gramene website (http://archive.gramene.org/markers/microsat/). Zhang et al. (2007) have also developed a new database that contains 52,845 polymorphic SSRs between *indica* and *japonica*. Parida et al. (2009) also identified 19,555 SSR markers from the genic region of rice and named as genic but noncoding microsatellites. Therefore, due to the availability of high-density SSR markers in the rice genome, it is the most commonly used PCR-based markers in the rice genetic studies and MAS. Another postgenome sequencing era marker, SNPs became the most abundant forms of genetic variation compared to the other markers among individuals of the same species, which are ubiquitous and amenable to high- and ultra-high-throughput automation. In rice, SNPs can be readily identified by direct comparisons of genomes with one or both reference sequences. Recently, it has been widely used for QTLs mapping, GWAS analysis, diversity, and evolutionary studies in rice. Next-generation sequencing technology also accelerated the identification of more SNPs by resequencing the diverse genotypes of rice and discovering genome-wide SNPs. Whole-genome sequence comparison of the reference high-quality complete sequence of *indica* and *japonica* leads to the generation of 5.41 million SNP loci polymorphisms (Feltus et al., 2004; Shen et al., 2004). Another set of 158,000 high-quality SNPs was identified from 20 accessions by sequencing of 100 Mb gene-rich genomic sequences, which are available for public use at the OryzaSNP database (McNally et al., 2009). Huang et al. (2012) also sequenced 950 rice accessions and identified 4,109,366 SNPs. The Alexandrov et al. (2014) group also identified about 20 million rice SNPs by aligning short reads from the 3000 rice genomes project with the Nipponbare genome reference, which is now available at SNP-Seek database. Zhao et al. (2014) also identified 6,551,358 SNPs from 1479 rice accessions, and constructed a comprehensive database of rice genomic variations, which is available at the RiceVarMap database. Such data would be highly useful for exploring genetic variations and evolution studies of rice, GWAS, and genetic mapping of important agronomic traits in rice. Another robust PCR-based codominant marker is the insertion–deletion length polymorphisms (InDels), which is abundant, simple, and cost effective. It can also be easily identified by genome sequence comparison of desired genotypes. Such variation is usually observed more in introns compared to exons due to the high tolerability of nucleotide insertion or deletions or low selective pressure in the noncoding regions of the genome, and such polymorphisms have been exploited by the development of a new class of intron length polymorphic (ILP) markers. Yamaki et al. (2013) has developed several InDels markers that can differentiate all the rice species. Zhao et al. (2014) also identified 1,214,627 InDels from 1479 rice accessions, and constructed a comprehensive database of rice genomic variations, which is available at RiceVarMap database. Lu et al. (2015) developed about 205,659 InDels from *japonica* and *indica* cultivars of which 2681 InDels showed subgroup-specific pattern and they are available at the Rice InDel Marker Database.

1.12.8.2 Gene identification and characterization

After completion of genome sequencing and availability to the public domain for both the subspecies of rice, tremendous advances have been made in rice genetic and genomic research. These include genetic mapping of large numbers of genes/QTLs that control many important traits and its cloning, characterization, and introgression in the cultivars through MAS. Identification of genes/QTLs that control desirable traits such as yield, quality, and biotic and abiotic stresses is the next step to be carried out after completing the rice genome sequencing. Since the whole-genome annotation predicted about 44,755 gene loci, excluding transposable elements (TEs) and ribosomal protein or tRNA loci, in RAP (Rice Annotation Project 2008; http://rapdb.dna), 702 were functionally characterized accounting for only 1.6% of the total loci annotated till 2012 (Yamamoto et al., 2012). Therefore, the next challenge is to determine the function and characterization of the remaining >98% annotated genes to utilize immediately for genetic improvement of cultivated rice. To map and characterize a gene locus for a particular trait of interest, the methodology that is commonly adopted in rice can be explained/summarized in two broad approaches.

1.12.8.2.1 Conventional approaches: map-based cloning

The most common approach for identifying and characterizing a gene controlling a particular trait of interest is map-based cloning. In this approach, two individual lines carrying contrasting traits of interest are identified and crossed to develop a mapping population, such as F_2, RILs, doubled haploids, and NILs, and depending on the time and cost factors, an appropriate mapping population is selected. Using molecular markers, the polymorphic DNA markers between the parental lines are conducted and through which a genetic map is constructed to identify the genomic region by searching for linkages to markers to locate the candidate genes that are located near closely linked flanking markers. Then, fine genetic mapping to narrow down the candidate genomic region is followed by identifying the linked flanking markers using more polymorphic molecular markers from the same genome region in the large number of mapping population. Finally, the small and narrowed-down genomic region is scanned through chromosome walking or targeted sequencing in order to detect the DNA variation in the genic regions between the parental line, which might show the candidate gene. Some examples identified through this approach are semidwarf genes in rice, *Gn1a* and *DENSE AND ERECT PANICLE1 (DEP1), DEP1 GS3, GW2, qSW5/GW5,* and *GIF1* for grain size, *MOC1, LAZY1, IPA1,* and *Prog1* for plant architecture, and so on (from a review of Ikeda et al., 2013).

1.12.8.2.2 Next-generational sequencing technology approach

The conventional methods for functional characterization of rice genes related to important agronomic traits are labor intensive and time-consuming. The advent of advanced DNA sequencing technology has brought down the cost of DNA sequencing and made the sequencing genomes feasible and cost effective, which has greatly accelerated and facilitated genotyping procedures (Dhanapal, 2015). With the availability of high-quality reference genome sequences and the rapid development of genome sequencing technologies, an alternative strategy for identifying a gene has developed by increasing the sequencing throughput, thereby allowing the simultaneous sequencing of numerous samples using a multiplexed sequencing strategy (Craig et al., 2008; Cronn et al., 2008). This method identifies genes by combining whole-genome resequencing with high-throughput genotyping, which generates a huge amount of short read in a high-throughput manner at relatively low cost that is assembled into whole-genome sequences by aligning these high-quality reference genome sequences. Unlike the conventional approach, bulk segregants with extreme phenotype coupled with pooling resequencing can be used for rapid gene isolation/identification for simple qualitative/quantitative traits or mutant mapping through next-generation sequencing technologies such as MutMap (Abe et al., 2012), MutMap-Gap (Takagi et al., 2013), and QTL-seq (Takagi et al., 2013). MutMap is used for the identification of mutant genes created through induced mutagenesis. In this, the mutant is allowed to cross with the wild parental line to generate F_2 mapping population followed by whole-genome resequencing of pooled DNA from bulk F_2 segregants with extreme phenotype. In the case of QTL-seq, the parental lines are allowed to cross to generate segregating mapping population followed by bulking segregants with extreme phenotypes exhibiting the highest and lowest values and resequencing (Takagi et al., 2013). An advantage of this method is that any segregating population can be used. This next-generation DNA sequencing technology also provides the opportunity to speed up genotyping for effective genetic mapping and genome analysis. The first use of high-throughput genotyping technology also known as microarray-based genotyping facilitates direct scanning of allelic variation across the genome with relatively easy data analysis, covering hundreds to thousands of SNPs by hybridizing DNA to oligonucleotides spotted on the chips. High-resolution genotyping technologies, such as the 44 K Affymetrix array and 50 K Infinium array (RiceSNP50), have been developed for rice SNP genotyping (Chen et al., 2013). Another 50 K SNP chip has been designed by incorporating 50,051 SNPs from 18,980 different genes spanning 12 rice chromosomes including 3710 single-copy (SC) genes conserved between wheat and rice, 14,959 SC genes unique to rice, 194 agronomically important cloned rice genes, and 117 multicopy rice genes (Singh et al., 2015). Using this microarray-based genotyping approach, several groups have conducted GWAS in rice to study the genome-wide genetic diversity analyses and identification of agronomically important genomic loci. The low-coverage whole-genome resequencing approach was also used in the genotyping of natural populations to study the genetic diversity and characterization of important traits. In rice, 1083 cultivated *O. sativa* ssp. *indica* and *O. sativa* ssp. *japonica* varieties and 446 wild rice accessions (*O. rufipogon*) were collected and sequenced with low genome coverage. Another genotyping through NGS is genotyping by sequencing (GBS), which is efficient for large-scale, low-cost genotyping. Huang et al. (2009) conducted genotyping by resequencing 150 rice RILs, derived from a cross between the *indica*, 93-11 and the *japonica*, Nipponbare to construct SNP-based ultra-high-density linkage maps (Huang et al., 2010). In another study, a sequencing-based GWAS was developed based on the direct resequencing of 517 diverse rice landraces capturing more of the common sequence variation in cultivated rice than has any other dataset to date (Huang et al., 2010). Using such SNP-based ultra-high-density linkage maps, 80 robust association signals were identified for 14 agronomically relevant traits. This study provides results that can be immediately followed up to gain new biologic insights in rice and demonstrates the utility of this type of approach. Therefore, this technology provides a promising tool for fine-mapping of the major QTLs underlying complex traits from untapped germplasm (Bandillo et al., 2013).

1.12.9 Functional genomics

The availability of the whole-genome sequence in the public domain coupled with the amenability to genetic manipulation has opened up a vast field for functional genomic research in rice. During the last decade a strong foundation for rice functional genomics has been established as a result of a global effort that offers various resources, tools, and techniques for the identification of genes/alleles and pathways in a high-throughput manner. In fact, rice has come up as the model for monocot functional genomic research.

Rice functional genomics caught pace with the completion of the IRGSP on the International Rice Genome Sequencing Project (2005), which offers high-quality genome sequence data. High-quality sequence data are also available in the forms of BAC libraries, expressed sequence tags (EST), cDNA libraries, and unigenes in the public domain, for example, at the National Centre for Biotechnology Information (NCBI) database. The Rice Genome Annotation Project database (maintained by MSU) offers annotation of rice genes. Several web-based programs have integrated these available genomic resources as their database to perform data analysis.

Analysis of mutants, traditionally, has been the basis for studying gene function. Rice has a rich gene pool distributed over more than 230,000 germplasm accessions worldwide (Li et al., 2014) containing enormous diversity. This wide range of natural variation can be efficiently exploited in functional genomics studies. In the absence of natural variation, an efficient way for gene function analysis is to study the artificially induced mutants. Several mutant resources developed via physical, chemical, or biological means are available for rice (Hirochika et al., 2004).

With the advent of high-throughput sequencing technologies the cost and time required for sequencing has reduced by many folds. Rice, being a crop of immense global interest but having a relatively small genome (~400 Mb), has been resequenced widely (Huang et al., 2010, 2012; Xu et al., 2011). Recently, 3000 rice accessions have been sequenced in a collaborative effort of IRRI, the Chinese Academy of Agricultural Science (CAAS), and Beijing Genomics Institute (BGI) (The 3000 Rice Genomes Project, 2014). These resequencing studies have also generated very useful resources to decipher the structure as well as function of the rice genome.

Here in this chapter we will briefly discuss the different resources used for the functional analysis of the rice genome. Germplasms are a very important resource for conducting research on functional genomics as these are the repository of natural mutation accumulated over ages. Since germplasms have been dealt separately earlier in this chapter, here we will avoid repetition of the same and initiate our discussion with mutant resources of rice, which are developed through artificial mutation. We will also discuss several sequence databases available for rice functional genomics and finally conclude by discussing in brief the rice system biology, which integrates all available resources to investigate genome function.

1.12.10 Mutant resources of rice

Mutation results in changes in the phenotype and thus provides a direct causal relationship between the structure and biologic function of a gene. Induction of artificial mutation and their analysis is a vividly used technique to study gene function since Morgan and his group demonstrated the power of this technique way back in 1910 using *Drosophila* as a model system. Since rice has come up as a model for basic and applied research in cereal biology, several mutant resources have also been developed to decipher the function of its genes. Several mutagens, such as physical (γ-ray, fast nutron, etc.), chemical (ethyl methane sulfonate, *N*-methyl-*N*-nitrosourea, etc.), and biologic (transposons, retrotransposons, T-DNA, etc.), have been used to create these mutant resources. RNA interference (RNAi) has also been used for creating functional knockouts. Table 1.11 provides the web-link of some important rice mutant databases, which are shared extensively between global rice researchers.

Table 1.11 Web information of some important mutant databases

Database	Websites
POSTECH RISD	http://www.postech.ac.kr/life/pfg/risd/
RMD	http://rmd.ncpgr.cn/
Zhejiang University	http://www.genomics.zju.edu.cn/ricetdna.html
TRIM	http://trim.sinica.edu.tw
Oryza Tag Line (OTL) Genoplante	http://oryzatagline.cirad.fr/
SHIP	http://ship.plantsignal.cn/index.do
NIAS (RTIM)	http://tos.nias.affrc.go.jp
RDA-Genebank	http://genebank.rda.go.kr/dstag
SIRO	http://www.csiro.au/science/Rice-Functional-Genomics-Project
EU-OSTID	http://orygenesdb.cirad.fr/
Sundaresan Lab	http://www-plb.ucdavis.edu/Labs/sundar/
NRIMD	http://www.niab.go.kr/RDS/
IR64 deletion mutant population	http://www.iris.irri.org/cgibin/MutantHome.pl
UC Davis TILLING population	http://tilling.ucdavis.edu/
RiceFOX	http://ricefox.psc.riken.jp/
CSIRO	http://www.csiro.au/pi
NIAS	https://tos.nias.affrc.go.jp/

1.12.10.1 Mutants developed through physical and chemical mutagenesis

γ-Ray and fast neutron are the most commonly used physical mutagens in rice, which are also known as clastogen as they induce double-strand DNA breakage. Hence, use of these mutagenic agents introduces deletions in the genome of 50–700 kb size range. The resultant mutants are particularly helpful for characterizing genes present as tandemly repeated gene families. Of the rice genes, 22% are present as tandem repeats (Droc et al., 2013). Physical mutagen-induced deletion mutants are also important from another aspect. They create knockouts in accession where insertional mutagenesis is not so easy due to the basic difficulty of nonresponsiveness of the accessions to tissue culture, for example, many of the *indica* varieties. These deletion mutants can be characterized by hybridizing fluorescently labeled mutant DNA to the Affymetrix Rice GeneChip®, which can rapidly identify the deleted region (Bruce et al., 2009). Deleat-a-gene® is another technique for characterizing fast neutron-based deletion mutants, which preferentially amplifies and subsequently enriches deletion-induced smaller fragments from the PCR products amplified from DNA samples of mutagenized populations pooled together (Li et al., 2001, 2002).

Apart from physical mutagen, chemical mutagens are also frequently used for producing mutants in rice. Ethyl methanesulfonate (EMS), methyl nitrosourea (MNU), diepoxybutane (DEB), and sodium azide (SA) are some of the frequently used chemical mutagens. Unlike physical mutagens discussed earlier, these chemical mutagens mostly induce point mutation through base transition or transversion. Chemical mutagens create a high density of point mutations, which are distributed randomly throughout the genome (Sega, 1984; Vogel and Natarajan, 1995). Targeting-induced local lesions in genome (TILLING) is a reverse genetic approach that can be employed to find mutations in the target genes. The University of California-Davis has applied this approach to screen 10 target genes and identified 27 and 30 nucleotide changes in the EMS and SA-MNU-derived mutant population of Nipponbare, respectively. The throughput of TILLING can further be increased by coupling NGS along with it – a procedure termed as "TILLING by sequencing" (Tsai et al., 2011), which allows screening of large mutagenized populations within a short time. The power of forward genetics-based mutant analysis for gene discovery has also been greatly enhanced by coupling NGS with balk segregant analysis (BSA). The technique termed as MutMap helps in quick identification of causal SNP related to gene function from the mutant population (Abe et al., 2012).

Physical and chemical mutagen-induced mutant libraries have been widely used for gene discovery in rice. For example, a large mutant collection (>60,000) of *indica* rice IR64 has been developed by inducing physical (fast neutron, and γ-ray) and chemical (DEB and EMS) mutations (Wu et al., 2005). Phenotyping of these mutants for altered morphology and resistance/tolerance to several biotic and abiotic stresses are underway. Twenty-one lesion mimic mutants were recently isolated from these IR64-derived mutant populations, which have been comprehensively studied in order to elucidate disease resistance pathways (Wu et al., 2008a). With the availability of high-throughput screening technologies, the analysis of mutants is gradually becoming less cumbersome. This in turn holds great promise for taking rice functional genomics to the next level.

1.12.10.2 Mutants developed through insertional mutagenesis

Exogenous DNA elements, like T-DNAs, transposons, and retrotransposons, have the ability to cause mutation by getting inserted into different sites of the genome. The resultant, hence, mutation is termed as insertional mutagenesis. Insertional mutagenesis has been extensively utilized to generate rice mutant populations mainly because of two reasons. First, characterization of the mutants is easy as the sites of insertion in the mutants can be identified easily by using simple PCR-based techniques such as inverse PCR or TAIL PCR. Second, there are situations where knockout mutants (developed through physical and chemical mutagenesis) cannot be used and insertional mutagenesis is the only available option to overcome the problem. For example, knockout mutants cannot be used for identifying vital genes or the genes having redundancy in the genome. In such cases alternative approaches like activation tagging and gene trap can be used, which are only possible through insertional mutagenesis. Nevertheless, knockout mutants can also be developed through insertional mutagenesis. The development of highly efficient transformation and regeneration protocols in rice has enabled researchers to generate large numbers of mutants using insertional mutagenesis (Hiei et al., 1994; Lin and Zhang, 2005; Toki et al., 2006). Next, we will discuss briefly the mutant resources developed by using three of the most commonly used insertional mutagenesis techniques.

1.12.10.3 T-DNA insertion mutagenesis

T-DNA is a tumor inducing (Ti) plasmid-born DNA fragment of *Agrobacterium tumifaciens* and *Agrobacterium rhizogenes* that has the natural ability to get transferred and integrated in the genome of the host plant. Integration of T-DNA in the host genome results in random insertion mutagenesis with low copy numbers but having stable inheritance (Wu et al., 2003). Since the transformation and regeneration protocol of *japonica* rice lines are more standardized, *japonica*-derived T-DNA mutant lines are mostly available worldwide. For example, about 129,000 independent rice enhancer trap mutant lines are available at the RMD database (http://rmd.ncpgr.cn/) in the background of Zhonghua 11. More than 45,000 flanking sequence tags have been developed so far using these lines (http://signal.salk.edu/RiceGE/RiceGE_Data_Source.html) (Zhang et al., 2006). These mutants are very much helpful to characterize gene function. Many important rice genes, like *RID1*, which is a master switch for flowering induction (Wu et al., 2008b); *JMJ706*, which is involved in flower development (Sun and Zhou, 2008), and *ILA1*, another key gene involved in mechanical tissue formation at leaf lamina joint, were identified through T-DNA mutant screening.

1.12.10.4 Transposon mutagenesis

Transposons are the mobile genetic elements that have the ability to shift position in the genome by employing cycles of transient breakage and subsequent fusion of double-stranded DNA. Though T-DNA mutagenesis is most frequently used because of the ease of transformation, there are situations where transposon mutagenesis provides greater advantage over the former. For example, a large number of transposon-tagged population can be produced from a smaller number of parental lines. And unlike T-DNA insertion mutants where complementation to produce the wild type is difficult, wild revertants in transposition-derived mutants can be produced with relative ease by simply eliminating the transposon tags through transposon mobilization. Enhancer/suppressor-mutator (En/Spm-dSpm) (Greco et al., 2004; Kumar et al., 2005) and activator/dissociation (Ac/Ds) are two-component transposon systems from maize (*Zea mays*), which have been used to generate large-scale insertion mutants in rice. Phenotype screening of Ds insertional lines (Jiang et al., 2005) helped in identifying mutants like *oscyp96b4*, a novel rice semidwarf mutant (Ramamoorthy et al., 2011), and *Osnop*, a mutant producing pollen-less flowers (Jiang et al., 2005). Apart from maize transposons, which have been traditionally used in genetic screening, there are rice-based transposons that can be used for this purpose. A miniature inverted-repeat transposable element (MITE) named miniature Ping (mPing) (Jiang et al., 2003; Kikuchi et al., 2003b; Nakazaki et al., 2003) and a two-component system named nonautonomous/autonomous DNA-based active rice transposon (nDart/aDart) (Nishimura et al., 2008), are the two rice-derived transposons that can be potentially used for transposon mutagenesis.

1.12.10.5 Retrotransposon mutagenesis

Retrotransposons, on the other hand, are the specialized transposons that transposes through producing RNA copy involving a cycle of transcription followed by a reverse transcription. An endogenous retrotransposon of rice named Tos17 is reported to get activated only when tissue culture is performed (Hirochika et al., 1996). Hence, it can produce permanent insertion in tissue culture regenerated plants. Since the induction of Tos17 mutation needs only tissue culture and does not involve recombinant DNA technology, field testing of the mutant lines become easy as these do not come under the GM regulation. This gives Tos17-derived mutants an edge over T-DNA- or transposon-derived mutants. Tos17 has been extensively used for creating insertion mutants in rice. More than 47,000 Tos17-induced insertion mutants have been developed by Miyao et al. (2003) in the Nipponbare background.

Besides these genomic knockout mutants, functional knockouts can be produced by utilizing RNAi. In rice RNAi has been found to be efficient too. But there are not many reports available where a genome-wide RNAi-based mutant library has been established for rice. Wang et al. (2013) have established a hairpin RNA (hpRNA) library for the identification of rice gene function in genome-wide scale using the rolling circle amplification-mediated hpRNA (RMHR) method.

1.12.11 Transcriptomic resources

The term "transcriptome" refers to the collection of RNA molecules produced by a particular cell or tissue at a specific developmental stage or in response to specific environmental conditions. Hence, unlike the genome, which remains unaltered for a species, transcriptome composition varies spatially as well as temporally based on the set of genes expressed at the particular site at a particular point of time. Regulation of gene expression is a fascinating topic of research in biology. There are genes that are expressed only at specific tissue and/or growth stage in response to specific

Table 1.12 Web addresses of some important transcriptome databases

Database	Websites
ROAD	http://www.ricearray.org/index.shtml
Genevestigator	https://www.genevestigator.com/gv/plant.jsp
PlexDB	http://www.plexdb.org/
RiceXPro V3	http://ricexpro.dna.affrc.go.jp/
RiceGE	http://signal.salk.edu/RiceGE/RiceGE_Data_Source.html

endogenous or exogenous stimuli. These genes are termed as differentially expressed genes and can be identified through transcriptome profiling. There are two aspects of studying the transcriptome – one is its composition and the other is its dynamics. Table 1.12 provides web-links of some important transcriptome databases that are commonly used by the global rice research community.

1.12.11.1 Study of transcriptome composition

Transcriptome composition analysis refers to the identification of the set of genes expressed to constitute the transcriptome. The commonly used methods for this purpose are described next.

1.12.11.1.1 Sequencing of ESTs and FL-cDNA

Transcriptome studies in many organisms initiated with the collection and sequencing of a large number of ESTs. ESTs are useful for gene discovery and genome analysis. About 1.2 million rice ESTs have already been registered in the NCBI dbEST database (http://www.ncbi.nlm.nih.gov/dbEST/). However, due to their fragmented nature, often ESTs do not contain the whole open reading frame (ORF) of an expressed gene, thus posing a difficulty in the precise identification of a gene structure. The evolution of standard methods for the preparation of full-length cDNA (FL-cDNA) libraries helped to overcome the demerits of ESTs and facilitated the identification of complete gene structures including ORFs as well as 5′ and 3′ untranslated regions. FL-cDNA sequences are useful not only for transcriptome composition analysis (Kikuchi et al., 2003a; Yamada et al., 2003) but also for knowledge-based identification of conserved protein motifs or domains accurate annotation of sequenced genomes (Kikuchi et al., 2003a; Yu et al., 2005).

In 2000, the Rice Full-Length cDNA Consortium was launched in Japan under which EST and FL-cDNA libraries were constructed from 50 different tissue types by inducing different stresses to the rice (*japonica*) plants along with proper controls (Kikuchi et al., 2003a). Random sequencing from 5′ and 3′ ends of about 380,000 clones from the libraries could characterize and annotate 2800 FL-cDNA clones. This collection of FL-cDNA clones from the *japonica* rice has eventually expanded to 578,000 FL-cDNA clones and FL-ESTs, among which 35,000 cDNA clones are completely sequenced and annotated (Satoh et al., 2007; http://cdna01.dna.affrc.go.jp/cDNA/). In another effort, over 20,000 FL-cDNAs were isolated from two *indica* rice varieties – Guangluai 4 and

Minghui 63 (Lui et al., 2007; Lu et al., 2008; http://www.ncgr.ac.cn/ricd) – as part of the National Rice Functional Genomics Project of China. Sequencing of 10,096 Guangluai 4 FL-cDNAs helped in identifying 1,200 new transcription units (Lui et al., 2007). The utility of FL-cDNAs in genome annotation has been demonstrated by Satoh et al. (2007) where they used FL-cDNA sequences to validate the expression of *in silico* predicted genes of rice genome. In their effort, they were able to validate expression of 28,500 predicted genes in TIGR rice genome assembly as these were supported by FL-cDNA sequences (Ouyang et al., 2007).

1.12.11.1.2 Genomic tiling microarray analysis

Genomic tiling microarrays were used to systematically identify the transcriptome components. The first genomic tiling microarray for chromosome 4 of rice was made from PCR-generated overlapping probes that were on an average 3 kb long. The investigators could identify 80% of the annotated genes along with 1643 novel transcribed loci expressed at different stages of plant development (Jiao et al., 2005). Subsequently, *indica* (Li et al., 2006) and *japonica* genomes (Li et al., 2007) were also analyzed using oligonucleotide tiling microarrays developed on the NimbleGen platform. This led to the identification of 25,352 transcriptionally active regions (TARs) in *japonica* and 27,744 TARs in *indica* rice encoded by novel exons (Li et al., 2007). A portion of these TARs were found to be derived from duplicated gene fragments and the novel TARs compositionally resembled the exonic regions.

1.12.11.2 Study of transcriptome dynamics

Transcriptome dynamics refers to the comparative analysis of the number of each transcripts present in the transcriptome. This involves mapping as well as quantification of transcriptome either by cDNA hybridization using microarray or by RNA sequencing using RNA-seq technique.

1.12.11.2.1 Microarray analysis

Microarray is a probe-based technique used to study expression dynamics of annotated genes in a high-throughput manner. The first expression array of rice was developed by the Rice Microarray Project of Japan using 1265-element PCR amplicon array of EST clones (Yazaki et al., 2000). Subsequently, several developmental stage-, trait-, and treatment-specific arrays were developed for the identification and quantification of differentially expressed genes and to analyze the level of their expression (Kawasaki et al., 2001; Wang et al., 2005; Lian et al., 2006; Huang et al., 2006; Yazaki et al., 2004; Furutani et al., 2006). The early microarray-based transcriptome-profiling efforts provided an excellent tool for the functional analysis of predicted genes and for the verification of annotated gene models. Whole-genome microarrays are useful for the expression study of mutants, understanding specific biologic responses, and for identification of underlying regulatory networks. Whole-genome oligonucleotide arrays are available for *japonica* and *indica* rice varieties. A near whole genome oligonucleotide array consisting of 43,312 oligonucleotide probes (designed on the basis of annotated gene models) corresponding to 44,974 *japonica* rice transcripts was developed (Jung et al., 2008) and used to compare expression profiles of light- and dark-grown rice leaf tissues to understand the biologic functions of genes and pathways. In case of *indica* rice, a 70-mer microarray covering 41,754 annotated *indica* genes was developed to analyze the transcriptome in six representatives (Ma et al., 2005). The OryzaExpress database (Hamada et al., 2011), Rice Oligonucleotide Array Database, and so on, provide an excellent interface for performing genome annotations, microarray-based transcriptome data analysis, and metabolic pathways studies in rice. The Affymetrix GeneChip platform has developed rice microarrays that have been used for the transcriptome profiling (Zhu et al., 2003; Hazen et al., 2005), expression pattern analysis (Bethke et al., 2006), characterization of transcriptional programs (Xue et al., 2012), identification of regulatory genes along with their signaling pathways (Xue et al., 2012), and even for the study of expression level in case of quantitative traits (Wang et al., 2010) for elucidating the regulatory networks. However, the spatial resolution of the previous microarray studies was limited by the extraction of RNA from biologic samples consisting of a mixture of cell types. The advent of two major techniques – fluorescence activated cell sorting (Birnbaum et al., 2005) and laser capture microdissection (Asano et al., 2002) – have facilitated the isolation of specific cell types prior to RNA isolation and improved the precision of the expression studies. Jiao et al. (2009) produced a cell-specific transcriptome atlas in rice. This atlas included 40 distinct cell types from shoot, root, and germinating seed at several developmental stages.

1.12.11.2.2 RNA-Seq analysis

Microarray is a technique purely for studying transcriptome dynamics. It can only measure the expression levels of the previously annotated genes for which probe sets are present in the microarray chip but it is unable to identify the presence of any novel transcripts. Powered by next-generation sequencing technologies, it is now possible to sequence the transcriptome using the RNA sequencing (RNA-Seq) approach. This has provided a new avenue for assaying the presence and prevalence of transcripts simultaneously. So, over microarray or other presently existing transcriptome profiling techniques, RNA-Seq has a clear advantage – it can study transcriptome composition and transcriptome dynamics, together at the same time. Coupled with deep sequencing, RNA-Seq is able to produce high resolution for every sequenced nucleotide. This is crucial to precisely reveal the location of important genomic features like transcription start sites, intron–exon boundaries, polyadenylation sites, and events of alternative splicing and their sites. In a pioneering study Lu et al. used RNA-Seq to analyze rice transcriptome where 83% of the rice predicted gene models were validated, and it was also demonstrated that about 6228 gene models could be extended at least by 50 base pairs at the ends of the transcripts. Besides, they also found that nearly half of rice genes are capable of alternative splicing. As mentioned earlier, RNA-Seq renders the critical advantage of detecting novel transcripts. Global sampling of the transcripts using RNA-Seq approach in the *indica* and *japonica* subspecies by Lu et al. (2010) could confirm the expression of more than 60% of novel transcripts, which were identified by tilling array analysis (Li et al., 2007). Mizuno et al. (2010) were also able to identify 2795 shoot-derived and 3082 root-derived novel transcripts and also demonstrated differential expression of a few of these transcripts in response to salinity. There are databases that acquire and provide RNA-Seq data. RNA-Seq data for *indica* and *japonica* are available at the Rice Functional Genomic Express (RiceGE) Database (http://signal.salk.edu/RiceGE/RiceGE_Data_Source.html).

Comparative transcriptomics of related species is another area where RNA-Seq can really be valuable. Comparative transcriptomics can help in studying the expression of ortologous genes of the domesticated species in their wild species counterpart. Vis-a-vis it has the potential to throw light on the evolution of different gene regulatory networks in the related species. Microarray can be used for this purpose as was done by Peng et al. (2009) while analyzing six lineages of cultivated and wild rice. They capitalized on the close relatedness among the rice lineages used in their study, but similar success could not have been achieved if distant lineages were included in the study due to the basic difficulties associated with probe designing. RNA-Seq can be a boon in such situations as it measures the transcriptomes of different species independently based on their sequences. For example, Davidson et al. (2012) used RNA-Seq to compare reproductive tissues derived transcriptomes of *Brachypodium*, sorghum, and rice and showed that only a fraction of orthologous genes exhibit conserved expression patterns.

1.12.12 Proteomic resources of rice

The proteome is the collection of all the proteins present in a specific cell or cell-type at a particular time point in a given set of conditions. Like the transcriptome it is also spatial and temporal in nature. Proteomics is the study of the proteome, which is comparatively more tedious than genomics and transcriptomics owing to the fact that proteins, unlike nucleic acids, cannot be amplified *in vitro*. The techniques involved in studying rice proteomics can broadly be classified in three groups, namely, gel-based (1-DE, 2-DE, and 2-DIGE) and gel-free (LC-MS/MS, MudPIT, iTRAQ); besides there are a few that combine both (Agrawal et al., 2010, 2013). In recent times, gel-free quantitative approaches have been found promising for large-scale and high-throughput protein identification in rice. In fact, the advent of matrix-assisted laser desorption/ionization coupled with mass spectroscopy has revolutionized the field of proteomics by taking it to the high-throughput era. However, two-dimensional gel electrophoresis (2-DE) still continues to be the method of choice worldwide in the case of rice proteomics. The Rice Proteome Database (http://gene64.dna.affrc.go.jp/RPD/) has cataloged a set of rice proteins extracted from various tissues and organelles using 2-DE among which a few are functional characterizations. Reference maps were constructed comprising 13,129 rice proteins among which the amino acid sequences of 5092 proteins are present in the database. Significant progress has been made in the field of rice proteomics during the last decade. Proteomic resources have been generated for several tissues and organs by growing rice plants under ambient growth condition as well as under the challenge of various abiotic and biotic stresses (Komatsu and Tanaka, 2004; Agrawal and Rakwal, 2011; Kim et al., 2014). The Rice Proteome Database also has cataloged the major proteins involved in growth or stress responses, which have been identified by a proteomics approach. There are also a few other databases, namely Plant Protein Phosphorylation Database (http://p3db.org/), ARAMEMNON (http://aramemnon.botanik.uni-koeln.de/), Plant Phosphoproteome Database (http://metadb.riken.jp/), and OryzaPG-DB (http://www.iab.keio.ac.jp/en/content/view/406/133/), which provide useful information regarding rice proteomics.

1.12.13 Rice systems biology

Heaps of "omics" data generated globally as a result of intensive research efforts carried out to understand rice biology during the last 25 years have flooded the public domain. In order to fetch meaningful information out of these enormous data there is an urgent need to integrate this multifaceted information into biologic networks and models. This has given birth to the field of systems biology, which studies the interactions among biologic components by using models and/or networks. It includes and integrates genome data (developed through whole-genome sequencing and resequencing), transcriptome and proteome data, interactomes and reactomes, gene network, and gene-indexed mutant populations. The first challenge in this regard is to construct a unified platform that can facilitate easy retrieval of primary and secondary data. Several genome browsers, like the Rice Annotation Project Database (http://rapdb.dna.affrc.go.jp/), Rice Genome Annotation Project (http://rice.plantbiology.msu.edu/), Gramene (http://www.gramene.org/), and Rice Functional Genomic Express Database (http://signal.salk.edu/cgi-bin/RiceGE), have been developed for this purpose, which allows data retrieval. There are browsers as well for cross-referencing the retrieved data with other species – a study technically known as orthology analysis. GreenPhylDB (http://www.greenphyl.org/v2/cgi-bin/index.cgi), Phytozome (http://www.phytozome.net/), Plaza (http://bioinformatics.psb.ugent.be/plaza/), InParanoid (http://inparanoid.sbc.su.se/cgi-bin/index.cgi), and so on, are the few well-known browsers for this purpose. Spatial and temporal expression of the genes is governed by the promoters of the genes. Promoters are the *cis*-regulatory elements generally located upstream of the structural gene and can be identified by DNA sequence analysis. There are programs, like PLACE (http://www.dna.affrc.go.jp/PLACE/), Grassius (http://grassius.org/grasspromdb.html) PlantCARE (http://bioinformatics.psb.ugent.be/webtools/plantcare/html/), Osiris (http://www.bioinformatics2.wsu.edu/cgi-bin/Osiris/cgi/home.pl), PPDB (http://ppdb.agr.gifu-u.ac.jp/ppdb/cgi-bin/index.cgi), PlantPAN (http://plantpan.mbc.nctu.edu.tw/), etc. that aid in the identification of plant promoter elements. Genome is expressed through the transcriptome and proteome. Databases that allow analysis of transcriptome and proteome have already been discussed earlier. Response of an organism to different endogenous and exogenous stimuli is governed by the intricate network of interactions between coexpressed proteins taking place at the cellular level. There are also several softwares and browsers that facilitate analysis of the coexpression network (Table 1.13) and protein–protein interaction analysis (Table 1.14), among which some are specific for rice. Finally, there are different databases and softwares that allow pathway analysis (Table 1.15) in rice and thus help in interpreting the available data in the context of biologic processes, pathways, and networks.

Table 1.13 Web addresses of some important coexpression databases

Database	Websites
Oryzaexpress	http://bioinf.mind.meiji.ac.jp/Rice_network_public/script/
Ricefrend	http://ricefrend.dna.affrc.go.jp/
PLANEX	http://planex.plantbioinformatics.org/
STARNET	http://vanburenlab.medicine.tamhsc.edu/starnet2.html
ATTED-II	http://atted.jp/
Plant ArrayNet	http://arraynet.mju.ac.kr/arraynet/
PlaNet	http://aranet.mpimp-golm.mpg.de/
CoP	http://webs2.kazusa.or.jp/kagiana/cop0911/

Table 1.14 Web addresses of some important protein–protein interaction databases

Database	Websites
STRING	http://string-db.org/
DIPOS	http://csb.shu.edu.cn/dipos/?id=5
PRIN	http://bis.zju.edu.cn/prin/
Rice Interaction Viewer	http://bar.utoronto.ca/interactions/cgi-bin/rice_interactions_viewer.cgi

Table 1.15 Web addresses of some important protein–protein interaction databases

Database	Websites
Mapman	http://mapman.gabipd.org/web/guest
KEGG	http://www.genome.jp/kegg/
PANTHER	http://www.pantherdb.org/pathway/
Reactome	http://www.reactome.org/
Ricecyc	http://pathway.gramene.org/gramene/ricecyc.shtml

1.13 Conclusions

The donors identified over the years at CRRI and other rice research stations of the country have been utilized for rice improvement programs and more than 1000 high-yielding varieties have been released so far. The genetic erosion has been very fast in recent years due to rapid modernization of society and genetic diversity has been replaced by the introduction of high-yielding varieties. Farmers are leaving their own traditional varieties and growing the improved cultures; therefore, many of the landraces have become extinct. The loss of genetic resources has resulted in major concerns about future food and nutrition security. The need for *in situ* (on-farm) and *ex situ* (gene bank) conservation is now felt as the paddy cultivation in the country is largely affected by extreme natural calamities after rapid climate change, through an erratic monsoon. Earlier, the biggest challenge was flooding, but subsequently other factors, like salinity after frequent cyclones and sea-water surge, temperature rise, and drought-like situations in many parts of the country, have put the challenge before rice researchers. Various programs have to be addressed for the conservation and sustainable use of agrobiodiversity. The role of farmers and communities toward on-farm *in situ* conservation can only be realized through the empowerment of farming communities by making them learn the sustainable use of genetic diversity to adapt to such adverse changes.

Ever since rice was established as a model crop for studying cereal biology, the combined efforts of rice researchers' globally have accelerated the functional annotation of the rice genome. The large collection of germplasms and mutant resources enables gene identification by employing forward and reverse genetic approaches. Powered by the NGS-driven "omics" technologies, the process of gene identification has been precise and less time-consuming. Rather than focusing on a single gene as is done through the conventional approach, new generation techniques have enabled simultaneous analyses of many genes at the system level, which can provide new insights about the genome function as a whole. More than 600 rice genes have been cloned by the end of 2010 among which 36% are related to yield trait and 31% related to resistance/tolerance to biotic and abiotic stresses (Jiang et al., 2012). But still there is room for further improvement. A centralized resource containing high-quality data that are required will combine bioinformatics predictions with experimental validations. With this objective, the International Rice Functional Genomics steering committee had proposed the "RICE2020" project in 2008 with the goal of assigning biologic function to every rice gene by the year 2020. If this is achieved rice could become the first crop where design-directed targeted breeding can be practiced with maximum precision and virtually no futile effort.

References

Abe, A., Kosugi, S., Yoshida, K., 2012. Genome sequencing reveals agronomically-important loci in rice from mutant populations. Nat. Biotechnol. 30, 174–178.

Agrama, H.A., Eizenga, G.C., Yan, W., 2007. Association mapping of yield and its components in rice cultivars. Mol. Breed. 19, 341–356.

Agrawal, G.K., Rakwal, R., 2011. Rice proteomics: a move toward expanded proteome coverage to comparative and functional proteomics uncovers the mysteries of rice and plant biology. Proteomics 11, 1630–1649.

Agrawal, G.K., Jwa, N.S., Lebrun, M.H., Job, D., Rakwal, R., 2010. Plant secretome: unlocking secrets of the secreted proteins. Proteomics 10, 799–827.

Agrawal, G.K., Sarkar, A., Righetti, P.G., Pedreschi, R., Carpentier, S., 2013. A decade of plant proteomics and mass spectrometry: translation of technical advancements to food security and safety issues. Mass Spectrom. Rev. 32, 335–365.

Alexandrov, N., Tai, S., Wang, W., Mansueto, L., Palis, K., Fuentes, R.R., et al., 2014. SNP-Seek database of SNPs derived from 3000 rice genomes. Nucleic Acids Res. 49, D1023–D1027.

Ammiraju, J.S., Luo, M., Goicoechea, J.L., Wang, W., Kudrna, D., et al., 2006. The *Oryza* bacterial artificial chromosome library resource: construction and analysis of 12 deep-coverage large-insert BAC libraries that represent the 10 genome types of the genus *Oryza*. Genome Res. 16, 140–147.

Ammiraju, J.S.S., Song, X., Luo, M., Sisneros, N., Angelova, A., Kudrna, D., et al., 2010. The *Oryza* BAC resource: a genus-wide and genome scale tool for exploring rice genome evolution and leveraging useful genetic diversity from wild relatives. Breed. Sci. 60, 536–543.

Anacleto, R., Cuevas, R.P., Jimenez, R., Llorente, C., Nissila, Eero., Henry, R., et al., 2015. Prospects of breeding high-quality rice using post-genomic tools. Theor. Appl. Genet. 8, 1449–1466.

Arumuganathan, K., Earle, E.D., 1991. Nuclear DNA content of some important plant species. Plant Mol. Biol. Rep. 3, 208–218.

Asano, T., Masumura, T., Kusano, H., 2002. Construction of a specialized cDNA library from plant cells isolated by laser capture microdissection: toward comprehensive analysis of the genes expressed in the rice phloem. Plant J. 32, 401–408.

Baillon, H., 1894. Histoire des Plantes. vol. XII. Paris.

Bandillo, N., Raghavan, C., Muyco, P.A., Sevila, M.A.L., Lobina, I.T., Dilla-Ermita, C.J., 2013. Multi-parent advanced generation inter-cross (MAGIC) populations in rice: progress and potential for genetics research and breeding. Rice 6, 11.

Bentham, G., Hooker, J.D., 1862–1883. Genera Plantarum. London.

Bethke, P.C., Hwang, Y.S., Zhu, T., Jones, R.L., 2006. Global patterns of gene expression in the aleurone of wild-type and dwarf1 mutant rice. Plant Physiol. 140, 484–498.

Birnbaum, K., Jung, J.W., Wang, J.Y., 2005. Cell type-specific expression profiling in plants via cell sorting of protoplasts from fluorescent reporter lines. Nat. Methods 2, 615–619.

Bor, N.L., 1960. The Grasses of Burma, Ceylon, India and Pakistan (Excluding Bambuseae). Pergamon Press, Oxford.

Bose, L.K., 2005. Broadening gene pool of rice for resistance to biotic stresses through wide hybridization. Iran. J. Biotechnol. 3 (3), 140–143.

Bose, L.K., Panigrahi, A., Prusty, S., Nayak, P., Misra, R.N., 1990. Wide hybridization to transfer alien gene in rice. In: Proceedings of International Symposium on Rice Research: New Frontiers. November 15–16, 1990. Hyderabad, India: Directorate of Rice Research, pp. 125–126.

Bres-Patry, C., Bangratz, M., Ghesquiere, A., 2001. Genetic diversity and population dynamics of weedy rice in Camargue area. Genet. Sel. Evol. 33, S425–S440.

Bruce, M., Hess, A., Bai, J., Mauleon, R., Diaz, M.G., Sugiyama, N., et al., 2009. Detection of genomic deletions in rice using oligonucleotide microarrays. BMC Genomics 10, 129.

Calpe, C., 2003. Status of the world rice market in 2002. In: Proceedings of the FAO Twentieth Session of the International Rice Commission, Bangkok, Thailand, July 23–26, 2002. Food and Agriculture Organization of the United Nations, Rome.

Chang, T.T., 1976. The origin, evolution, cultivation dissemination and diversification of Asian and African rices. Euphytica 25, 435–441.

Chang, T.T., 1985. Crop history and genetic conservation: rice – a case study. Iowa State J. Res. 59 (4), 425–456.

Chang, T.T., 1989. Domestication and spread of the cultivated rices. In: Harris, D.R., Hillman, G.C. (Eds.), Foraging and Farming – The Evaluation of Plant Exploration. Unwin Hyman, London.

Chatterjee, D., 1948. A modified key and enumeration of the species of *Oryza* Linn. Indian J. Agric. Sci. 18, 185–192.

Chatterjee, D., 1951. Note on the origin and distribution of wild and cultivated rices. Indian J. Genet. 11, 18–22.

Chen, M., Presting, G., Barbazuk, W.B., Goicoechea, J.L., Blackmon, B., Fang, G., et al., 2002. An integrated physical and genetic map of the rice genome. Plant Cell 14, 537–545.

Chen, W., Xu, Z., Zhang, L., Zhang, W., Ma, D., 2007. Theories and practices of breeding japonica rice for super high yield. Sci. Agric. Sin. 40, 869–874, (in Chinese).

Chen, H., He, H., Zhou, F., Yu, H., Deng, X.W., 2013. Development of genomics-based genotyping platforms and their applications in rice breeding. Curr. Opin. Plant Biol. 16, 247–254.
Craig, D.W., Pearson, J.V., Szelinger, S., Sekar, A., Redman, M., Corneveaux, J.J., et al., 2008. Identification of genetic variants using bar-coded multiplexed sequencing. Nat. Methods 5, 887–893.
Cronn, R., Liston, A., Parks, M., Gernandt, D.S., Shen, R., Mockler, T., 2008. Multiplex sequencing of plant chloroplast genomes using Solexa sequencing-by-synthesis technology. Nucleic Acids Res. 36, e122.
Davidson, R.M., Gowda, M., Moghe, G., 2012. Comparative transcriptomics of three Poaceae species reveals patterns of gene expression evolution. Plant J. 71 (3), 492–502.
Devadath, S., 1983. A strain of *Oryza barthii*, an African wild rice immune to bacterial blight of rice. Curr. Sci. 52 (1), 27–28.
Dhanapal, A.P., Govindaraj, M., 2015. Unlimited thirst for genome sequencing, data interpretation, and database usage in genomic era: the road towards fast-track crop plant improvement. Genet. Res. Int. 2015, 684321.
Dhua, S.R., 1994. Genome analysis of *Oryza rhizomatis* Vaughan. PhD Thesis, Visva Bharati, Sriniketan, India.
Dowswell, C.R., Paliwal, R.L., Cantrell, R.P., 1996. Maize in the Third World. Westview Press, Boulder, USA.
Droc, G., An, G., Wu, C.Y., Hsing, Y.I., Hirochika, H., Pereira, A., Sundaresan, V., Han, C.D., Upadhyaya, N., Ramachandran, S., Comai, L., Leung, H., Guiderdoni, E., 2013. Mutant resources for functional analysis of the rice genome. In: Zhang, Q., Wing, R. (Eds.), Genetics and Genomics of Rice: Plant Genetics and Genomics: Crops and Models, vol. 5. Springer, New York, pp. 81–115.
Duvick, D.N., 1984. Genetic diversity in major farm crops on the farm and in reserve. Econ. Bot. 38, 161–178.
Eckardt, N.A., 2000. Sequencing the rice genome. Plant Cell 12, 2011–2017.
Feltus, F.A., Wan, J., Schulze, S.R., Estill, J.C., Jiang, N., Paterson, A.H., 2004. An SNP resource for rice genetics and breeding based on subspecies *indica* and *japonica* genome alignments. Genome Res. 14, 1812–1819.
Ferrero, A., Vidotto, F., Balsari, P., Airoldi, G., 1999. Mechanical and chemical control of red rice (*Oryza sativa* L. var. sylvatica) in rice (*Oryza sativa*) pre-planting. Crop Prot. 18, 245–251.
Furutani, I., Sukegawa, S., Kyozuka, J., 2006. Genomewide analysis of spatial and temporal gene expression in rice panicle development. Plant J. 46, 503–511.
Gao, Z.Y., Zhao, S.C., He, W.M., Guo, L.B., Peng, Y.L., Wang, J.J., et al., 2013. Dissecting yield-associated loci in super hybrid rice by resequencing recombinant inbred lines and improving parental genome sequences. Proc. Natl. Acad. Sci. USA 110, 14492–14497.
Ge, S., Sang, T., Lu, B.R., Hong, D.Y., 1999. Phylogeny of rice genomes with emphasis on origins of allotetraploid species. Proc. Natl. Acad. Sci. USA 96, 14400–14405.
Glaszmann, J.C., 1986. A varietal classification of Asian cultivated rice (*O. sativa* L.) based on isozyme polymorphism. In: Rice Genetics. IRRI, Philippines, pp. 83–90.
Glaszmann, J.C., 1987. Isozymes and classification of Asian rice varieties. Theor. Appl. Genet. 74, 21–30.
Govindaswami, S., Krishnamurty, A., 1959. Genetic variability among cultivated rices of Jeypore tract and its utility in rice breeding. Rice News Teller 7, 12–15.
Greco, R., Ouwerkerk, P.B., Taal, A.J., Sallaud, C., Guiderdoni, E., Meijer, A.H., et al., 2004. Transcription and somatic transposition of the maize En/Spm transposon system in rice. Mol. Genet. Genomics 270, 514–523.
Haines, H.H., 1921–1925.The Botany of Bihar and Orissa. London (Reprint, 1961, Calcutta).

Hamada, K., Hongo, K., Suwabe, K., 2011. *Oryza* express: an integrated database of gene expression networks and omics annotations in rice. Plant Cell Physiol. 52, 220–229.

Harlan, J.R., 1975. Crops and Man. American Society of Agronomy, Madison, WI.

Harlan, J.R., 1976. Genetic resources in wild relatives of crops. Crop Sci. 16, 329–333.

Harlan, J., 1992. Crops and Man. American Society of Agronomy – Crop Science Society of America, Madison, WI, pp. 117–130.

Harlan, J.R., DeWet, J.M.J., 1971. Toward a rational classification of cultivated plants. Taxonomy 20 (4), 509–517.

Harushima, Y., Yano, M., Shomura, A., Sato, M., Shimano, T., Kuboki, Y., et al., 1998. A high-density rice genetic linkage map with 2275 markers using a single F_2 population. Genetics 148, 479–494.

Hazen, S.P., Pathan, M.S., Sanchez, A., 2005. Expression profiling of rice segregating for drought tolerance QTLs using a rice genome array. Funct. Integr. Genomics 5, 104–116.

Hiei, Y., Ohta, S., Komari, T., Kumashiro, T., 1994. Efficient transformation of rice (*Oryza sativa* L.) mediated by *Agrobacterium* and sequence analysis of the boundaries of the T-DNA. Plant J. 6, 271–282.

Higham, C.F.W., Kijngam, A., 1984. Prehistoric Excavations in Northeast Thailand: Excavations at Ban Na Di, Ban Chiang Hian, Ban Muang Phruk, Ban Sangui, Non Noi and Ban Kho Noi, British Archaeological Reports, International Series 231 (i–iii), Oxford.

Hiratsuka, J., Shimada, H., Whittier, R., Ishibashi, T., Sakamoto, M., Mori, M., et al., 1989. The complete sequence of the rice (*Oryza sativa*) chloroplast genome: intermolecular recombination between distinct tRNA genes accounts for a major plastid DNA inversion during the evolution of the cereals. Mol. Gen. Genet. 217, 185–194.

Hirochika, H., Sugimoto, K., Otsuki, Y., Tsugawa, H., Kanda, M., 1996. Retrotransposons of rice involved in mutations induced by tissue culture. Proc. Natl. Acad. Sci. USA 93, 7783–7788.

Hirochika, H., Guiderdoni, E., An, G., Hsing, Y.I., Eun, M.Y., Han, C.D., et al., 2004. Rice mutant resources for gene discovery. Plant Mol. Biol. 54, 325–334.

Hooker, J.D., 1897. Flora of British IndiaVIIReeve & Co.Ltd, London.

Huang, Y., Zhang, L., Zhang, J., 2006. Heterosis and polymorphisms of gene expression in an elite rice hybrid as revealed by a microarray analysis of 9198 unique ESTs. Plant Mol. Biol. 62, 579–591.

Huang, X., Feng, Q., Qian, Q., Zhao, Q., Wang, L., Wang, A., et al., 2009. High-throughput genotyping by whole-genome resequencing. Genome Res. 19, 1068–1076.

Huang, X., Wei, X., Sang, T., Zhao, Q., Feng, Q., Zhao, Y., et al., 2010. Genome-wide association studies of 14 agronomic traits in rice landraces. Nat. Genet. 42, 961–967.

Huang, X., Kurata, N., Wei, X., Wang, Z., Wang, A., Zhao, Q., et al., 2012. A map of rice genome variation reveals the origin of cultivated rice. Nature 490, 497–501.

Huke, R.E., Huke, E.H., 1990. Rice: Then and Now. International Rice Research Institute, Los Baños, Philippines.

Ikeda, M., Miura, K., Aya, K., Kitano, H., Matsuoka, M., 2013. Genes offering the potential for designing yield-related traits in rice. Curr Opin. Plant Biol. 16 (2), 213–220.

International Rice Genome Sequencing Project, 2005. The map-based sequence of the rice genome. Nature 436, 793–800.

Ishikawa, R., Toki, N., Imai, K., Sato, Y.I., Yamagishi, H., Shimamoto, Y., et al., 2005. Origin of weedy rice grown in Bhutan and the force of genetic diversity. Genet. Resour. Crop Evol. 52, 395–403.

Jacquemin, J., Bhatia, D., Singh, K., Wing, R.A., 2013. The International Oryza Map Alignment Project: development of a genus-wide comparative genomics platform to help solve the 9 billion-people question. Curr. Opin. Plant Biol. 16, 147–156.

Jena, K.K., 1994. Production of intergeneric hybrid between *Oryza sativa* L. and *Porteresia coarctata* (Roxb.) Tateoka. Curr. Sci. 67, 744–746.

Jena, K.K., Khush, G.S., 1990. Introgression of genes from *O. officinalis* Wall. *ex* Watt to cultivated rice, *O. sativa* L. Theor. Appl. Genet. 80, 737–745.

Jennings, P.R., 1966. Evaluation of Partial Sterility in *indica* × *japonica* Rice Hybrids. IRRI, Philippines, The Bulletin No. 5.

Jiang, N., Bao, Z., Zhang, X., Hirochika, H., Eddy, S.R., McCouch, S.R., et al., 2003. An active DNA transposon family in rice. Nature 421, 163–167.

Jiang, S.Y., Cai, M., Ramachandran, S., 2005. The *Oryza sativa* no pollen (Osnop) gene plays a role in male gametophyte development and most likely encodes a C2-GRAM domain-containing protein. Plant Mol. Biol. 57, 835–853.

Jiang, Y., Cai, Z., Xie, W., Long, T., Yu, H., Zhang, Q., 2012. Rice functional genomics research: progress and implications for crop genetic improvement. Biotechnol. Adv. 30, 1059–1070.

Jiao, Y., Jia, P., Wang, X., 2005. A tiling microarray expression analysis of rice chromosome 4 suggests a chromosome-level regulation of transcription. Plant Cell 17, 1641–1657.

Jiao, Y., Tausta, S.L., Gandotra, N., 2009. A transcriptome atlas of rice cell types uncovers cellular, functional and developmental hierarchies. Nat. Genet. 41, 258–263.

Jung, K.H., Dardick, C., Bartley, L.E., 2008. Refinement of light-responsive transcript lists using rice oligonucleotide arrays: evaluation of gene redundancy. PLoS ONE 3, e3337.

Kato, S., Kosaka, H., Hara S., 1928. On the Affinity of Rice Varieties as Shown by the Fertility of Rice Plants, vol. 2. Central Agricultural Institute of Kyushu Imperial University, pp. 241–276.

Kawahara, Y., de la Bastide, M., Hamilton, J.P., Kanamori, H., McCombie, W.R., Ouyang, S., et al., 2013. Improvement of the *Oryza sativa* Nipponbare reference genome using next tide generation sequence and optical map data. Rice 6, 1–10.

Kawasaki, S., Borchert, C., Deyholos, M., 2001. Gene expression profiles during the initial phase of salt stress in rice. Plant Cell 13, 889–905.

Kawasaki, S.I., Ebana, K., Nishikawa, T., Sato, Y.I., Vaughan, D.A., Kadowaki, K.I., 2007. Genetic variation in the chloroplast genome suggests multiple domestication of cultivated Asian rice (*Oryza sativa* L.). Genome 50, 180–187.

Khush, G.S., 1997. Origin, dispersal, cultivation and variation of rice. Plant Mol. Biol. 35 (1–2), 25–34.

Khush, G.S., 2004. Harnessing science and technology for sustainable rice based production system. Paper presented at the Conference on Rice in Global Markets and Sustainable Production Systems, February 12–13, 2004. Food and Agriculture Organization of the United Nations (FAO), Rome, Italy.

Khush, G.S., Ling, K.C., 1974. Inheritance of resistance to grassy stunt virus and its vector in rice. J. Hered. 65, 134–136.

Kidd, J.M., Sampas, N., Antonacci, F., Graves, T., Fulton, R., Hayden, H.S., et al., 2010. Characterization of missing human genome sequences and copy-number polymorphic insertions. Nat. Methods 7, 365–371.

Kikuchi, S., Satoh, K., Nagata, T., 2003a. Collection, mapping, and annotation of over 28,000 cDNA clones from *japonica* rice. Science 300, 1566–1569.

Kikuchi, K., Terauchi, K., Wada, M., Hirano, H.Y., 2003b. The plant MITE mPing is mobilized in anther culture. Nature 421, 167–170.

Kim, H., Hurwitz, B., Yu, Y., Collura, K., Gill, N., SanMiguel, P., et al., 2008. Construction, alignment and analysis of twelve framework physical maps that represent the ten genome types of the genus *Oryza*. Genome Biol. 9, R45.

Kim, S.T., Kim, S.G., Agrawal, G.K., Kikuchi, S., Rakwal, R., 2014. Rice proteomics: a model system for crop improvement and food security. Proteomics 14, 593–610.

Komatsu, S., Tanaka, N., 2004. Rice proteome analysis: a step toward functional analysis of the rice genome. Proteomics 4, 938–949.

Kumar, C.S., Wing, R.A., Sundaresan, V., 2005. Efficient insertional mutagenesis in rice using the maize En/Spm elements. Plant J. 44, 879–892.

Kwon, S.J., Lee, J.K., Hong, S.W., Park, Y.J., McNally, K.L., Kim, N.S., 2006. Genetic diversity and phylogenetic relationship in AA *Oryza* species as revealed by Rim2/Hipa CACTA transposon display. Gen. Genet. Syst 81, 93–101.

Li, X., Song, Y., Century, K., Straight, S., Ronald, P., Dong, X., et al., 2001. A fast neutron deletion mutagenesis-based reverse genetics system for plants. Plant J. 27, 235–242.

Li, X., Lassner, M., Zhang, Y., 2002. Deleteagene: a fast neutron deletion mutagenesis-based gene knockout system for plants. Comp. Funct. Genomics 3 (2), 158–160.

Li, L., Wang, X., Stolc, V., 2006. Genome-wide transcription analyses in rice using tiling microarrays. Nat. Genet. 38, 124–129.

Li, L., Wang, X., Sasidharan, R., 2007. Global identify cation and characterization of transcriptionally active regions in the rice genome. PLoS ONE 2, e294.

Li, J.Y., Wang, J., Zeigler, R.S., 2014. The 3000 Rice Genome Project: opportunities and challenges for future rice research. GigaScience 3, 8.

Lian, X., Wang, S., Zhang, J., 2006. Expression profiles of 10,422 genes at early stage of low nitrogen stress in rice assayed using a cDNA microarray. Plant Mol. Biol. 60, 617–631.

Lin, Y.J., Zhang, Q., 2005. Optimising the tissue culture conditions for high efficiency transformation of indica rice. Plant Cell Rep. 23, 540–547.

Londo, J., Chiang, Y., Hung, K., Chiang, Y., Schall, B., 2006. Phytogeography of Asian wild rice, *Oryza rufipogon*, reveals multiple independent domestications of cultivated rice, *Oryza sativa*. Proc. Natl. Acad. Sci. USA 103, 9578–9583.

Lu, T., Huang, X., Zhu, C., 2008. RICD: a rice indica cDNA database resource for rice functional genomics. BMC Plant Biol. 8, 118.

Lu, T., Lu, G., Fan, D., 2010. Function annotation of the rice transcriptome at single-nucleotide resolution by RNA-seq. Genome Res. 20, 1238–1249.

Lü, Y., Cui, X., Li, R., Huang, P., Zong, J., Yao, D., Li, G., Zhang, D., Yuan, Z., 2015. Development of genome-wide insertion/deletion markers in rice based on graphic pipeline platform. J. Integr. Plant Biol. doi: 10.1111/jipb.12354.

Lui, X., Lu, T., Yu, S., 2007. A collection of 10,096 indica rice full-length cDNAs reveals highly expressed sequence divergence between *Oryza sativa* indica and japonica subspecies. Plant Mol. Biol. 65, 403–415.

Ma, J., Bennetzen, J.L., 2004. Rapid recent growth and divergence of rice nuclear genomes. Proc. Natl. Acad. Sci. USA 101, 12404–12410.

Ma, L., Chen, C., Liu, X., 2005. A microarray analysis of the rice transcriptome and its comparison to *Arabidopsis*. Genome Res. 15, 1274–1283.

Marshall, D.R., 1989. Limitations to the use of germplasm collections. In: Brown, A.D.H., Frankel, O.H., Marshall, D.R., Williams, J.T. (Eds.), The Use of Plant Genetic Resources. Cambridge University Press, Cambridge, UK, pp. 105–120.

McCouch, S.R., Kochert, G., Yu, Z.H., Wang, Z.Y., Khush, G.S., Coffman, W.R., et al., 1988. Molecular mapping of rice chromosomes. Theor. Appl. Genet. 76, 815–829.

McCouch, S.R., Teytelman, L., Xu, Y., Lobos, K.B., Clare, K., Walton, M., et al., 2002. Development and mapping of 2240 new SSR markers for rice (*Oryza sativa* L.). DNA Res. 9, 199–207.

McNally, K.L., Childs, K.L., Bohnert, R., Davidson, R.M., Zhao, K., Ulat, V.J., et al., 2009. Genomewide SNP variation reveals relationships among landraces and modern varieties of rice. Proc. Natl. Acad. Sci. USA 106, 12273–12278.

Miyao, A., Tanaka, K., Murata, K., Sawaki, H., Takeda, S., Abe, K., et al., 2003. Target site specificity of the *Tos17* retrotransposon shows a preference for insertion within genes and against insertion in retrotransposon-rich regions of the genome. Plant Cell 15 (8), 1771–1780.

Mizuno, H., Kawahara, Y., Sakai, H., 2010. Massive parallel sequencing of mRNA in identification of unannotated salinity stress-inducible transcripts in rice (*Oryza sativa* L.). BMC Genomics 11, 683.

Mohan, M., Nair, S., Bhagwat, A., Krishna, T.G., Yano, M., Bhatia, C.R., et al., 1997. Genome mapping, molecular markers and marker-assisted selection in crop plants. Mol. Breed. 3, 87–103.

Morishima, H., 1984. Species relationships and the search for ancestors. In: Tsunoda, S., Takahashi, N. (Eds.), Biology of Rice. Elsevier, Science Society Press, Amsterdam, Tokyo, Japan, pp. 3–30.

Mortimer, M., Pandey, S., Piggin, C., 2000. Weedy rice: approaches to ecological appraisal and implications for research priorities. In: Baki, B.B., Chin, D.V., Mortimer, M. (Eds.), Proceedings of Wild and Weedy Rice in Rice Ecosystems in Asia – A Review (pp. 97–105). Limited proceedings no. 2, Los Baños, Philippines, International Rice Research Institute.

Nagamura, Y., Antonio, B.A., Sasaki, T., 1997. Rice molecular genetic map using RFLPs and its applications. Plant Mol. Biol. 35, 79–87.

Nakazaki, T., Okumoto, Y., Horibata, A., Yamahira, S., Teraishi, M., Nishida, H., et al., 2003. Mobilization of a transposon in the rice genome. Nature 421, 170–172.

Nayak, D., Bose, L.K., Dikshit, N., Ranga Reddy, P., Misra, R.N., 1996. Alternate source of resistance for bacterial leaf blight disease of rice. Indian J. Plant Genet. Resour. 9, 143–146.

Nishimura, H., Ahmed, N., Tsugane, K., Iida, S., Maekawa, M., 2008. Distribution and mapping of an active autonomous aDart element responsible for mobilizing nonautonomous nDart1 transposons in cultivated rice varieties. Theor. Appl. Genet. 116, 395–405.

Noldin, J.A., Cobucci, T., 1999. Red rice infestation and management in Brasil. Report of the global workshop on red rice control, August 30–September 3, Varadero, Cuba. pp. 9–13.

Notsu, Y., Masood, S., Nishikawa, T., Kubo, N., Akiduki, G., Nakazono, M., et al., 2002. The complete sequence of the rice (*Oryza sativa* L.) mitochondrial genome: frequent DNA sequence acquisition and loss during the evolution of flowering plants. Mol. Genet. Genomics 268, 434–445.

Oka, H.I., 1953. Phylogenetic differentiation of the cultivated rice plant. VI. The mechanism of sterility in the intervarietal hybrid of rice. Jpn. J. Breed. 2, 217–224.

Oka, H.I., 1958. Intervarietal variation and classification of cultivated rice. Indian J. Genet. 18, 79–89.

Oka, H.I., 1964. Pattern of interspecific relationships and evolutionary dynamics in *Oryza*. In Rice Genetics and Cytogenetics. Elsevier Publishing Company, Amsterdam, pp. 71–90.

Oka, H.I., 1988. Origin of Cultivated Rice. Elsevier Publishing Company, Amsterdam.

Ouyang, Y., Zhang, Q., 2013. Understanding reproductive isolation based on the rice model. Annu. Rev. Plant Biol. 64, 111–113.

Ouyang, S., Zhu, W., Hamilton, J., 2007. The TIGR Rice Genome Annotation Resource: improvements and new features. Nucleic Acids Res. 35, D883–D887.

Pan, Y., Deng, Y., Lin, H., Kudrna, D.A., Wing, R.A., Li, L., et al., 2014. Comparative BAC-based physical mapping of *Oryza sativa* ssp. indica var. 93-11 and evaluation of the two rice reference sequence assemblies. Plant J. 77, 795–805.

Parida, S.K., Kalia, S.K., Kaul, S., Dalal, V., Hemaprabha, G., Selvi, A., et al., 2009. Informative genomic microsatellite markers for efficient genotyping applications in sugarcane. Theor. Appl. Genet. 118, 327–338.

Paterniani, E., 1987. An evaluation of the genetic diversity in the varieties currently utilized. In: Plant Breeding Research Forum; Report 1985. Caracas, 1987. pp. 45–58.

Patra, B.C., 2008. Diversity study of rice *(Oryza sativa* Linn.) in herbarium. Environ. Ecol. 26 (1), 250–255.

Patra, B.C., Dhua, S.R., 1996. Genetic diversity in *Porteresia coarctata* collected from Bhitarkanika mangrove forest, Orissa. J. Econ. Taxon. Bot. 20 (1), 241–244.

Peeters, J.P., Galwey, N.W., 1988. Germplasm collections and breeding needs in Europe. Econ. Bot. 42 (4), 503–521.

Peng, Z.Y., Zhang, H., Liu, T., 2009. Characterization of the genome expression trends in the heading-stage panicle of six rice lineages. Genomics 93, 169–178.

Porteres, R., 1956. Taxonomic agrobotanique des ris cultives *O. sativa* Linn. *et O. glaberrima* Steud. Journal d'Agriculture Tropicale et de Botanique Appliquée, Paris 3, pp. 343–384, 541–580, 627–700, 822–856.

Prodohel, A., 1922. Oryzeae monographice describintur. Bot. Arch. 1, 221–224, 231–255.

Qiu, J., Zhu, J., Fu, F., Ye, C.Y., Wang, W., Mao, L., 2014. Genome re-sequencing suggested a weedy rice origin from domesticated *indica–japonica* hybridization: a case study from southern China. Planta 240, 1353–1363.

Ramamoorthy, R., Jiang, S.Y., Ramachandran, S., 2011. *Oryza sativa* cytochrome P450 family member OsCYP96B4 reduces plant height in a transcript dosage dependent manner. PLoS ONE 6, e28069.

Ramiah, K., 1953. Rice Breeding and Genetics. ICAR, New Delhi.

Ramiah, K., Ghose, R.L.M., 1951. Origin and distribution of cultivated plants of South Asia-Rice. Indian J. Genet. 11, 7–13.

Richharia, R.H., 1960. Origin of cultivated rices. Indian J. Genet. 20, 1–14.

Roschevicz, R.J., 1931. Critical botanical review of species of rice. Bull. Appl. Bot. Genet. Plant Breed. 27 (4), 3–133, (in Russian).

Sahu, S.C., Bose, L.K., Pani, J., Misra, R.N., Misra, C.D., 1994. Genetic evaluation of wild *Oryza* species for resistance against the rice root knot nematode, *Meloidogyne graminicola*, Golden and Birhfield, 1968. Ann. Plant Prot. Sci. 2, 90–91.

Sampath, S., 1961. Notes on taxonomy of genus *Oryza*. Shokubutsugaku Zasshi 74, 269–270.

Sampath, S., 1962. The genus *Oryza*: its taxonomy and species relationships. Oryza 1, 1–29.

Sampath, S., 1964. The species ancestral to cultivated rice. Curr. Sci. 33 (7), 205–207.

Sampath, S., Rao, M.B.V.N., 1951. Inter-relationship between species in the genus *Oryza*. Indian J. Genet. 11, 14–17.

Sang, T., Ge, S., 2007. Genetics and phylogenetics of rice domestication. Curr. Opin. Genet. Dev. 17, 533.

Sasaki, T., Burr, B., 2000. International rice genome sequencing project: the effort to completely sequence the rice genome. Curr. Opin. Plant Biol. 3, 138–141.

Satoh, K., Doi, K., Nagata, T., 2007. Gene organization in rice revealed by full-length cDNA mapping and gene expression analysis through microarray. PLoS ONE 2, e1235.

Schatz, M.C., Maron, L.G., Stein, J.C., Wences, A.H., Gurtowski, J., Biggers, E., et al., 2014. Whole genome *de novo* assemblies of three divergent strains of rice, *Oryza sativa*, document novel gene space of *aus* and *indica*. Genome Biol. 15, 506.

Schneeberger, K., Ossowski, S., Ott, F., Klein, J.D., Wang, X., Lanz, C., et al., 2010. Reference-guided assembly of four diverse *Arabidopsis thaliana* genomes. Proc. Natl. Acad. Sci. USA 25, 10249–10254.

Seck, P.A., Diagne, A., Mohanty, S., Wopereis, M.C.S., 2012. Crops that feed the world. 7: rice. Food Secur. 4, 7–24.

Sega, G.A., 1984. A review of the genetic effects of ethyl methanesulfonate. Mutat. Res. 134, 113–142.

Sen, P., Panda, B., Bose, L.K., Misra, R.N., 2005. A partially fertile somaclone of *O. sativa* L./*O. eichingeri* A. Peter. Oryza 42, 268–270.

Sharma, S.D., 2003. Species of genus *Oryza* and their inter-relationships. In: Nanda, J.S., Sharma, S.D. (Eds.), Monograph on Genus *Oryza*. Oxford & IBH, New Delhi, pp. 73–111.

Sharma, S.D., Shastry, S.V.S., 1965. Taxonomic studies in genus *Oryza*. I. Asiatic types of *O. sativa* complex. Indian J. Genet. 25, 145–156.

Shen, Y.J., Jiang, H., Jin, J.P., Zhang, Z.B., Xi, B., He, Y.Y., et al., 2004. Development of genome-wide DNA polymorphism database for map-based cloning of rice genes. Plant Physiol. 135, 1198–1205.

Singh, N., Jayaswal, P.K., Panda, K., Mandal, P., Kumar, V., Singh, B., et al., 2015. Single-copy gene based 50 K SNP chip for genetic studies and molecular breeding in rice. Sci. Rep. 5, 11600.

Stalker, H.T., 1980. Utilization of wild species for crop improvement. Adv. Agron. 33, 111–147.

Steudel, E.G., 1855. Synopsis Plantarum Graminum. J.B. Metzler, Stuttgardie.

Sun, Q., Zhou, D.X., 2008. Rice jmjC domain-containing gene JMJ706 encodes H3K9 demethylase required for floral organ development. Proc. Natl. Acad. Sci. USA 105, 13679–13684.

Sun, J., Liu, D., Wang, J.Y., Ma, D.R., Tang, L., Gao, H., et al., 2012. The contribution of intersubspecific hybridization to the breeding of super-high-yielding japonica rice in northeast China. Theor. Appl. Genet. 125, 1149–1157.

Sun, J., Qian, Q., Ma, D.R., Xu, Z.J., Liu, D., Du, H.B., et al., 2013. Introgression and selection shaping the genome and adaptive loci of weedy rice in northern China. New Phytol. 197, 290–299.

Sweeney, M., Mc Couch, S., 2007. The complex history of the domestication of rice. Ann. Bot. 100, 951–957.

Takagi, H., Abe, A., Yoshida, K., Kosugi, S., Natsume, S., Mitsuoka, C., 2013. QTL-seq: rapid mapping of quantitative trait loci in rice by whole genome resequencing of DNA from two bulked populations. Plant J. 74, 174–183.

Tanaka, T., Antonio, B.A., Kikuchi, S., Matsumoto, T., Nagamura, Y., Numa, H., et al., 2008. The Rice Annotation Project Database (RAP-DB): 2008 update. Nucleic Acids Res. 36, D1028–D1033.

Tang, L.H., Morishima, H., 1996. Genetic characteristics and origin of weedy rice. Origin Differentiation of Chinese Cultivated Rice. China Agricultural University Press, Beijing, pp. 211–2180.

Tateoka, T., 1964a. Taxonomic studies of the genus *Oryza*. Rice Genetics and Cytogenetics. Elsevier, Amsterdam, pp. 15–21.

Tian, X.J., Zheng, J., Hu, S.N., Yu, J., 2006. The rice mitochondrial genomes and their variations. Plant Physiol. 140, 401–410.

Toki, S., Hara, N., Ono, K., Onodera, H., Tagiri, A., Oka, S., et al., 2006. Early infection of scutellum tissue with *Agrobacterium* allows high-speed transformation of rice. Plant J. 47, 969–976.

Tsai, H., Howell, T., Nitcher, R., Missirian, V., Watson, B., Ngo, K.J., et al., 2011. Discovery of rare mutations in populations: TILLING by sequencing. Plant Physiol. 156, 1257–1268.

Vaughan, D.A., 1989. The genome *Oryza* L. Current status of taxonomy. IRRI Research Paper Series 138, 1–21.

Vaughan, D.A., 1994. The Wild Relatives of Rice. IRRI, Philippines.
Vaughan, D.A., Sitch, L.A., 1991. Geneflow from jungle to farmers: wild rice genetic resources and their uses. Biol. Sci. 44, 22–28.
Vaughan, D.A., Morishima, H., Kadowaki, K., 2003. Diversity in the *Oryza* genus. Curr. Opin. Plant Mol. Biol. 6, 139–146.
Vaughan, D.A., Song, Ge., Kaga, A., Tomooka, N., 2008. Phylogeny and biogeography of the genus *Oryza*. Biotechnol. Agric. Forest. 62, 219–234.
Vogel, E.W., Natarajan, A.T., 1995. DNA damage and repair in somatic and germ cells *in vivo*. Mutat. Res. 330, 183–208.
Wang, Z., Liang, Y., Li, C., 2005. Microarray analysis of gene expression involved in anther development in rice (*Oryza sativa* L.). Plant Mol. Biol. 58, 721–737.
Wang, J., Yu, H., Xie, W., Xing, Y., Yu, S., Xu, C., et al., 2010. A global analysis of QTLs for expression variations in rice shoots at the early seedling stage. Plant J. 63, 1063–1074.
Wang, L., Wang, A., Huang, X.H., Zhao, Q., Dong, G.J., Qian, Q., et al., 2011. Mapping 49 quantitative trait loci at high resolution through sequencing-based genotyping of rice recombinant inbred lines. Theor. Appl. Genet. 7, 327–340.
Wang, L., Zheng, J., Luo, Y., Xu, T., Zhang, Q., Zhang, T., et al., 2013. Construction of a genome wide RNAi mutant library in rice. Plant Biotechnol. J. 11, 997–1005.
Wang, M., Yu, Y., Haberer, G., Marri, P.R., Fan, C., Goicoichea, J.L., et al., 2014. The genome sequence of African rice (*Oryza glaberrima*) and evidence for independence domestication. Nat. Genet. 46, 982–988.
Waters, D.L.E., Nock, C.J., Ishikawa, R., Rice, N., Henry, R.J., 2012. Chloroplast genome sequence confirms distinctness of Australian and Asian wild rice. Ecol. Evol. 2, 211–217.
Watt, G., 1891. Dictionary of Economic Products of India. VWH Allen, London.
Wing, R.A., Ammiraju, J.S., Luo, M., Kim, H., Yu, Y., Kudrna, D., et al., 2005. The Oryza Map Alignment Project: the golden path to unlocking the genetic potential of wild rice species. Plant Mol. Biol. 59, 53–62.
Wu, C., Li, X., Yuan, W., Chen, G., Kilian, A., Li, J., et al., 2003. Development of enhancer trap lines for functional analysis of the rice genome. Plant J. 35, 418–427.
Wu, J.L., Wu, C., Lei, C., Baraoidan, M., Bordeos, A., Madamba, M.R., et al., 2005. Chemical and irradiation-induced mutants of indica rice IR64 for forward and reverse genetics. Plant Mol. Biol. 59, 85–97.
Wu, C., Bordeos, A., Madamba, M.R., Baraoidan, M., Ramos, M., Wang, G.L., et al., 2008a. Rice lesion mimic mutants with enhanced resistance to diseases. Mol. Genet. Genomics 279, 605–619.
Wu, C., You, C., Li, C., Long, T., Chen, G., Byrne, M.E., et al., 2008b. RID1, encoding a Cys2/His2-type zinc finger transcription factor, acts as a master switch from vegetative to floral development in rice. Proc. Natl. Acad. Sci. USA 105, 12915–12920.
Xiong, H.B., Xu, H.Y., Xu, Q., Zhu, Q., Gan, S.X., Feng, D.D., et al., 2012. Origin and evolution of weedy rice revealed by inter-subspecific and inter-varietal hybridizations in rice. Mol. Plant Breed. 10, 131–139, (in Chinese with English abstract).
Xu, X., Liu, X., Ge, S., Jensen, J.D., Hu, F.Y., Li, X., et al., 2011. Resequencing 50 accessions of cultivated and wild rice yields markers for identifying agronomically important genes. Nat. Biotechnol. 30, 105–111.
Xu, X., Liu, X., Ge, S., Jensen, J.D., Hu, F.Y., Li, X., et al., 2012. Resequencing 50 accessions of cultivated and wild rice yields markers for identifying agronomically important genes. Nat. Biotechnol. 7 (1), 105–111.
Xue, L.J., Zhang, J.J., Xue, H.W., 2012. Genome-wide analysis of the complex transcriptional networks of rice developing seeds. PLoS ONE 7, e31081.

Yamada, K., Lim, J., Dale, J.M., 2003. Empirical analysis of transcriptional activity in the *Arabidopsis* genome. Science 302, 842–846.

Yamaki, S., Ohyanagi, H., Yamasaki, M., Eiguchi, M., Miyabayashi, T., Kubo, T., 2013. Development of INDEL markers to discriminate all genome types rapidly in the genus *Oryza*. Breed. Sci. 63, 246–254.

Yamamoto, E., Yonemaru, J.I., Yamamoto, T., Yano, M., 2012. OGRO: the overview of functionally characterized genes in rice online database. Rice 5, 26.

Yamanaka, S., Nakamura, I., Nakai, H., Sato, Y.I., 2003. Dual origin of the cultivated rice based on molecular markers of newly collected annual and perennial strains of wild rice species, *Oryza nivara* and *O. rufipogon*. Genet. Resour. Crop Evol. 50, 529–538.

Yazaki, J., Kishimoto, N., Nakamura, K., 2000. Embarking on rice functional genomics via cDNA microarray: use of 3′ UTR probes for specific gene expression analysis. DNA Res. 7, 367–370.

Yazaki, J., Shimatani, Z., Hashimoto, A., 2004. Transcriptional profiling of genes responsive to abscisic acid and gibberellin in rice: phenotyping and comparative analysis between rice and *Arabidopsis*. Physiol. Genomics 17, 87–100.

Yu, J., Wang, J., Lin, W., 2005. The genomes of *Oryza sativa*: a history of duplications. PLoS Biol. 3, e38.

Zang, L., Zou, X., Zhang, F., Yang, Z., Song, G.E., 2011. Phylogeny and species delimitation of the C-genome. J. Syst. Evol. 49 (5), 386–395.

Zhang, J., Li, C., Wu, C., Xiong, L., Chen, G., Zhang, Q., et al., 2006. RMD: a rice mutant database for functional analysis of the rice genome. Nucleic Acids Res. 34, D745–D748.

Zhang, Z.H., Deng, Y.J., Tan, J., Hu, S.N., Yu, J., Xue, Q.Z., 2007. A genome-wide microsatellite polymorphism database for the *indica* and *japonica* rice. DNA Res. 14, 37–45.

Zhang, Q.J., Zhu, T., Xia, E.H., Shi, C., Liu, W.L., Zhang, Y., et al., 2014. Rapid diversification of five Oryza AA genomes associated with rice adaptation. Proc. Natl. Acad. Sci. USA 46, E4954–E4962.

Zhao, H., Yao, W., Ouyang, Y., Yang, W., Wang, G., Lian, X., Xing, Y., Chen, L., Xie, W., 2014. RiceVarMap: a comprehensive database of rice genomic variations. Nucleic Acids Res. 43. doi: 10.1093/nar/gku894.

Zhu, T., Budworth, P., Chen, W., 2003. Transcriptional control of nutrient partitioning during rice grain filling. Plant Biotechnol. J. 1, 59–70.

Further Readings

Chen, S., Jin, W., Wang, M., Zhang, F., Zhou, J., Jia, Q., et al., 2003. Distribution and characterization of over 1000 T-DNA tags in rice genome. Plant J. 36, 105–113.

Chevalier, A., 1932. Nouvelle contribution à Pétude systematique de *Oryza*. Rev. Bot. Appl. Agric. Trop. 12, 1014–1032.

IRRI, 1985. International Rice Research: 25 Years of Partnership. IRRI, Los Baños, Philippines, p. 188.

Zeigler, R.S., 2014. Food security, climate change and genetic resources. In: Jackson, M., Ford-Lloyd, B., Parry, M. (Eds.), Plant Genetic Resources and Climate Change. CABI, Wallingford, UK, pp. 1–15.

Wheat

2

Wuletaw Tadesse, Ahmad Amri*, Francis C. Ogbonnaya[†], Miguel Sanchez-Garcia*, Quahir Sohail*, Michael Baum**
*Department of Wheat Breeding/Genetics, International Center for Agricultural Research in the Dry Areas (ICARDA), Rabat, Morocco; [†]Grains Research and Development Corporation (GRDC), Australia

2.1 Introduction

Wheat is the universal cereal of the Old World agriculture and the world's foremost consumed crop plant followed by rice and maize (FAOSTAT, 2011). It is the most widely adapted crop, growing in diverse environments spanning from sea level to regions as high as 4570 m.a.s.l. in Tibet (Percival, 1921). It grows from the Arctic Circle to the equator, but most suitably at the latitude range of 30° and 60°N and 27° and 40°S (Nuttonson, 1955). A crop of wheat is harvested somewhere in the world during every month of the year (Briggle and Curtis, 1987). Cultivated wheat is classified into two major types: (1) the hexaploid bread wheat ($2n = 6x = 42$, BBAADD) and (2) the tetraploid durum wheat ($2n = 4x = 28$, BBAA). Currently, at the global level, bread wheat accounts for 95% of all the wheat produced. Based on growth habit, wheat is classified into spring wheat and facultative/winter wheat, covering about 65 and 35% of the total global wheat production area, respectively (Braun et al., 2010; Braun and Săulescu, 2002). The flour of bread wheat is used to make French bread, Arabic bread, Chapatti, biscuits, pastry products, and the production of commercial starch and gluten. Durum wheat is specifically grown for the production of semolina for use in pasta and macaroni products. In North Africa, durum wheat is preferred for the preparation of couscous and bulgur. It is also widely used to prepare a special bread made by mixing bread and durum flours. Wheat has played a fundamental role in human civilization and improved food security at the global and regional levels. It provides about 19% of the calories and 21% of protein needs of daily human requirements at the global level (Braun et al., 2010). It is a staple food for 40% of the world's population mainly in Europe, North America, and the western and northern parts of Asia. The demand for wheat is growing fast in new wheat growing regions of the world such as eastern and southern Africa (5.8%), West and Central Africa (4.7%), and South Asia and the Pacific (4.3%). Demand is also growing in the traditional wheat growing regions of Central Asia (5.6%), Australia (2.2%), and North Africa (2.2%) (Shiferaw et al., 2013). Wheat is the most traded agricultural commodity at the global level with a trade volume of 144 million tons, with a total value of 36 billion US dollars (2010 data; Shiferaw et al., 2013). Many of the developing countries that depend on wheat as a staple crop are not self-sufficient in wheat production, and accordingly, wheat is their single most important imported commodity. Wheat also accounts for the largest share of emergency food aid (Dixon et al., 2009).

Genetic and Genomic Resources for Grain Cereals Improvement. http://dx.doi.org/10.1016/B978-0-12-802000-5.00001-0

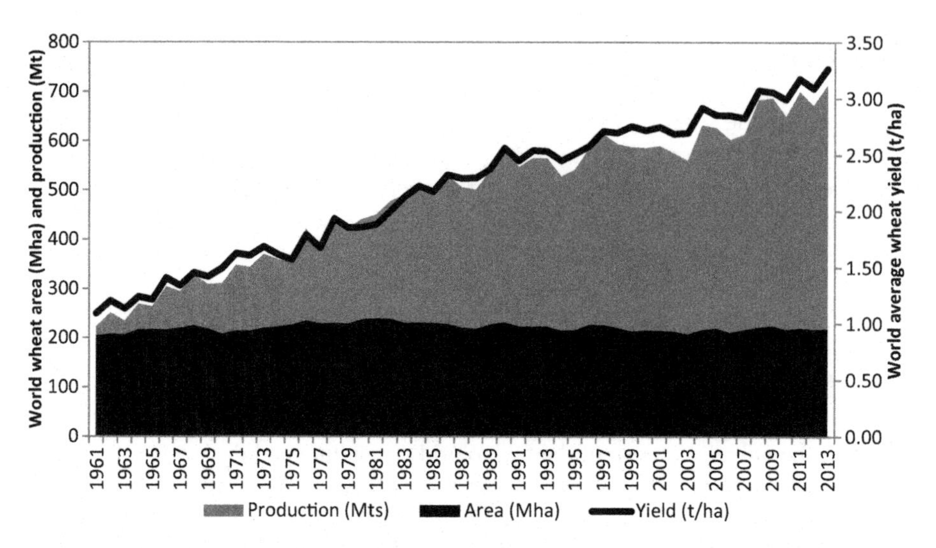

Figure 2.1 World wheat yield, production, and area from 1961 to 2013 (FAO, 2015).

In line with the increasing economic importance of wheat, governments have made significant investments in improving wheat productivity through the years. According to FAO (2015), about 732 million tons of wheat was produced on an average of 218.5 million ha with a productivity level of 3.3 t/ha in 2013, a highly significant increase from 1961, which stood at 222 million tons with a productivity level of only 1.2 t/ha (Fig. 2.1). The accelerated increase in wheat production is attributed to the adoption of the Green Revolution technology packages by farmers, in particular improved high-yielding varieties with better response to inputs (e.g., fertilizers), improved irrigation systems, and improved disease resistance and pesticides as well as better management practices, coupled with conducive policies and stronger institutions (Baum et al., 2013).

The wide cultivation and adoption of mega cultivars since the Green Revolution, however, has caused substantial genetic erosion and loss of diversity, reinforcing the genetic erosion caused by its domestication (Peng et al., 2011). Coupled with climate change, such genetic erosion has aggravated the susceptibility and vulnerability of wheat to environmental stresses, pests, and diseases (Nevo, 2009; Solh et al., 2012). Wheat genetic resources, especially wild relatives including primary, secondary, and tertiary gene pools, are important resources that need to be conserved and utilized wisely to sustain genetic gains through breeding for higher yields, better adaptation, higher nutritional end-products quality, and resistance to biotic and abiotic stresses. Though wheat genomics has lagged behind other major crops such as maize and rice, it is becoming evident that the recent progress in wheat genomics research would enable us to sequence the wheat genome, cloning and introgression of genes of interest, and thereby revolutionize the process of marker-assisted breeding. In this review, we have summarized the recent views on wheat evolution and domestication, genetic diversity and conservation, and the roles of wheat genetic resources in wheat breeding and genomics.

2.2 Evolution and origin of *Triticum*

2.2.1 Evolution

All the cultivated wheats belong to the genus *Triticum*, which in turn was divided into three major taxonomic groups (einkorn, emmer, and dinkel) by Schultz (1913). This classification was supported by the pioneering cytological study of Sakamura (1918), who found that Schultz's three wheat groups also differ in their chromosome number; the einkorns are diploids ($2n = 2x = 14$), the emmers are tetraploids ($2n = 4x = 28$), and the dinkels are hexaploids ($2n = 6x = 42$), all with the genomic basic chromosome number $x = 7$. Soon after, based on further cytogenetic analysis, Kihara (1924) designated the genome formulae for the cultivated einkorn (*Triticum monococcum* L., $2n = 2x = 14$), emmer (*Triticum turgidum* L. $2n = 4x = 28$), and dinkel (*Triticum aestivum*, $2n = 6x = 42$) as AA, AABB, and AABBDD, respectively.

The diploid einkorn wheat, *T. monococcum* var. *monococcum* ($2n = 2x = 14$, A^mA^m), was domesticated directly from its wild form, *T. monococcum* var. *aegilopoides* ($2n = 2x = 14$, A^mA^m) in the Fertile Crescent, probably in the Karacadag mountain range in southeast Turkey (Heun et al., 1997). Einkorn wheat was basically replaced by cultivated tetraploid and hexaploid wheats during the last 5000 years, and currently it is a relict crop grown for feed only in a few Mediterranean countries (Nesbitt and Samuel, 1996a). Similarly, the cultivated emmer wheat, *Triticum dicoccum* ($2n = 2x = 28$, BB A^uA^u), was domesticated from the wild emmer, *Triticum dicoccoides* ($2n = 2x = 28$, BBA^uA^u), which is an allopolyploid, arising by amphiploidy between *Triticum urartu* ($2n = 2x = 14$, A^uA^u) and the B genome ancestor, *Aegilops speltoides* ($2n = 2x = 14$, SS) (Dvorak and Akhunov, 2005; Feldman and Levy, 2005; Johnson and Dhaliwal, 1976) 300,000–500,000 years before present. *T. urartu*, though never cultivated, occurs in parts of the Fertile Crescent, and has played a significant role in the wheat evolution by contributing the A^uA^u genome to all tetraploid and hexaploid wheats (Dvorak et al., 1993). There is also another wild tetraploid wheat, *Triticum araraticum* (GGA^uA^u), which is similar to the cultivated form *Triticum timopheevii* ($2n = 4x = 28$, GGA^tA^t). *T. timopheevii*, however, has been cultivated at a very limited extent. It is believed to be domesticated from the wild emmer wheat, *T. dicoccoides* ssp. *armeniacum* (Feldman, 2001). According to Naranjo (1990) and Jiang and Gill (1994), a species-specific translocation involving chromosomes $6A^t$, 1G, and 4G distinguishes *T. timopheevii* from *T. turgidum*, which contains a translocation involving chromosomes 4A, 5B, and 7B (Devos et al., 1995). *T. dicoccoides*, wild emmer, grows naturally all over the Fertile Crescent, and it is believed that emmer wheats was first domesticated probably in southeast Turkey (Dvorak et al., 2011; Özkan et al., 2002). Emmer cultivation has declined through time and currently it is found only in limited areas in Ethiopia and Russia.

The evolution of the common wheat, *T. aestivum* ($2n = 6x = 42$, BBAADD), has been the subject of many investigations and intense discussions for several decades. As indicated in Fig. 2.1, it is now considered certain that hexaploid wheat was formed from a hybrid between the tetraploid wheat species *T. turgidum* ($2n = 4x = 28$, BBAA) and the diploid species *Aegilops tauchii* var. *strangulata* ($2n = 2x = 14$, DD) (Dvorak et al., 1998; McFadden and Sears, 1946; Riley et al., 1958). Genome analyses by Kihara (1919) and Sax (1922) on the pairing behavior of interspecific hybrids between $2x/4x$ and $4x/6x$ wheats indicated that *T. monococcum* and *T. turgidum* share one genome in common, and *T. turgidum* and *T. aestivum* have two genomes in common. However, the cytological data did not discriminate between *T. monococcum* ($2n = 2x = 14$, A^mA^m) and *T. urartu* ($2n = 2x = 14$, AA) genomes (Johnson and Dhaliwal, 1976), but the molecular evidence showed that *T. urartu* actually is the A genome donor of both tetraploid and hexaploid wheats (Dvorak et al., 1993). The other hexaploid wheat, *Triticum zhukovsky* ($2n = 6x = 42$, A^tA^t A^mA^m GG), arose from the hybridization of *T. timopheevii* ($2n = 4x = 28$, A^tA^tGG) with *T. monococcum* ($2n = 2x = 14$, A^mA^m) (Upadhya and Swaminathan, 1963).

There has been much controversy regarding the origin of the B and G genomes of polyploid wheats since the early proposal of Sarkar and Stebbins (1956) supported by Riley et al. (1958) that *Ae. speltoides* was the donor of the second genome of tetraploid wheats. Recent molecular evidence, however, is convincing that the B and G genomes of polyploid wheats were donated by *Ae. speltoides* (Daud and Gustafson, 1996; Dvorak and Zhang, 1990; Petersen et al., 2006). Furthermore, the cytoplasmic genome heterogeneity within *Ae. speltoides* indicated that it may be the maternal (cytoplasmic) donor of all polyploid wheats (Gill and Friebe, 2001; Wang et al., 1997) (Fig. 2.2).

Chromosome pairing in polyploid *Triticum* species occurs in a diploid-like fashion between homologous chromosomes and not between homoeologous (partially

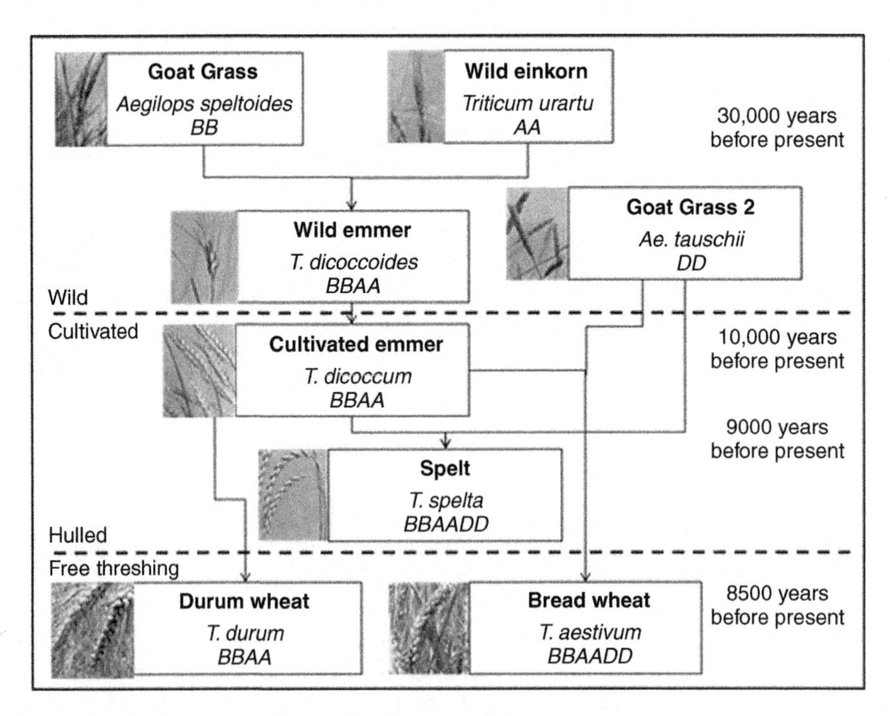

Figure 2.2 Evolutionary relationship of wheat and its ancestors.
Adapted from New Hall Mill: The Evolution of Wheat.

homologous chromosomes of the different genomes). This is due to the supressor *Ph1* (Riley and Chapman, 1958; Sears, 1976; Vega and Feldman, 1998) and *Ph2* (Dong et al., 2002; Mello-Sampayo, 1971) genes. Therefore, in plants lacking these genes, particularly the *Ph1* gene, multivalents were observed during meiosis due to pairing among the homoeologous chromosomes, resulting in partial sterility of plants, indicating the crucial role of the *Ph1* gene for diploid-like chromosome pairing and for the evolution of polyploid wheats and their domestication (Koebner and Shepherd, 1986; Riley and Chapman, 1958; Sears, 1976, 1977).

2.2.2 Origin and distribution of Triticum

The origin and evolution of a cultivated plant can be best studied following the identification of its wild progenitors and the current and past distribution of its progenitors. This may indicate the changes that led to domestication as well as the site of the initial cultivation. However, when such a wild progenitor is not found, or is extinct, understanding the complete history of that cultivated plant is greatly impaired (Feldman, 2001).

Archaeological and botanical studies of both wild and cultivated forms have indicated that the Fertile Crescent (Fig. 2.2) is the birthplace of cultivated wheats about 8,000–10,000 years ago (Gill and Friebe, 2001; Mujeeb-Kazi and Villareal, 2002). Among diploid wheats, einkorn wheat (*T. monococcum* L.) is still cultivated to a limited extent, and its wild form, *Triticum aegilopoides*, is widely distributed in the Middle East (Heun et al., 1997; Johnson, 1975).

The tetraploid hulled wheat, *T. turgidum* ssp. *dicoccum* (emmer wheat), was one of the ancient cultivated wheats. However, it is the free-threshing macaroni or durum wheat, which arose by few mutations from primitive emmer wheats that are widely cultivated in the present times (Gill and Friebe, 2001). The remains of the cultivated emmer (*T. turgidum* ssp. *dicoccum*) have been discovered at several archaeological sites in Syria dating to 7500 BC (Zohary and Hopf, 1993; Zohary, 1999). The other cultivated tetraploid wheat, *T. timopheevii* ($2n = 4x = 28$, AAGG), is of little economic importance. The wild forms of both tetraploid wheats, *T. turgidum* ssp. *dicoccoides* and *T. timopheevii* ssp. *armeniacum*, are widely distributed in the Fertile Crescent. *T. dicoccoides* is found exclusively in Lebanon, Israel, Palestine, and Syria, while *Triticum armeniacum* is dominantly found in Azerbaijan and Armenia, and yet both progenitors overlap in Turkey, northern Iraq, and possibly Iran (Feldman et al., 1995; Feldman, 2001; Gill and Friebe, 2001). The hexaploid species *T. aestivum* ($2n = 6x = 42$, BBAADD) and *T. zhukovsky* ($2n = 6x = 42$, BBAAGG) have no wild progenitors, and are only found in cultivated forms in farmers' fields by hybridization between cultivated tetraploid wheat and wild diploid species (Feldman, 2001) (Fig. 2.3).

Bread wheat arose farther northwest, away from the Fertile Crescent, in the corridor extending from Armenia in Transcaucasia to the southwest coastal areas of the Caspian Sea in Iran (Dvorak et al., 1998). In this region, *Ae. tauschii* var. *strangulata* is predominant, which evidently hybridized with cultivated emmer to produce *T. aestivum*. In a recent study, Wang et al. (2013) genotyped using single nucleotide

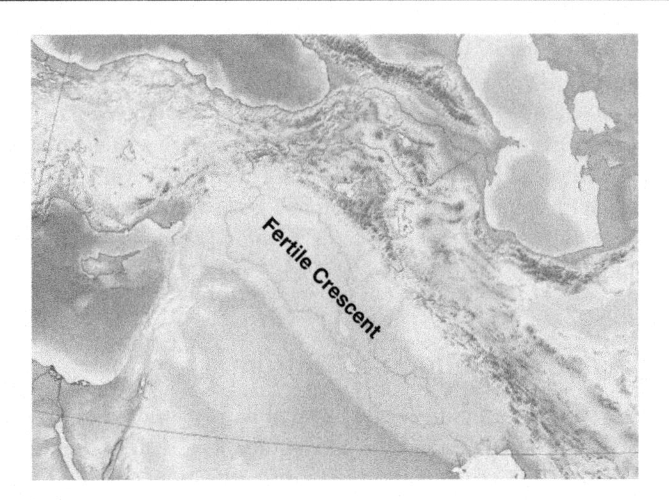

Figure 2.3 Geographic origin of *Triticum*.
Adapted from Valkoun et al. (1998).

polymorphism (SNP) markers 477 *Ae. tauschii* and wheat accessions collected from eastern Turkey to China. Their results showed that the *Aegilops* populations growing nowadays in the southern part of the Caspian Sea are the most genetically similar to the D genome of modern wheats.

There are five *T. aestivum* ($2n = 6x = 42$, BBAADD) subspecies based on spike morphology: (1) *T. aestivum* ssp. *aestivum* (*QQ cc S1S1*), (2) *T. aestivum* ssp. *compactum* (*QQ CC S1S1*), (3) *T. aestivum* ssp. *spelta* (*qq cc S1S1*), (4) *T. aestivum* ssp. *macha* (*qq CC S1S1*), and (5) *T. aestivum* ssp. *sphaerococcum* (*QQ CC s1s1*), which differ principally due to allelic variations of single major genes: *q* (the speltoid gene) and its dominant allele *Q* (which confers free-threshing grain and tough rachis) on chromosome 5A and 2D; *c* and its dominant compact-ear-producing allele *C* on chromosome 2D; *S* and its recessive spherical-grain producing *s* allele on chromosome 3D (Miller, 1987). The first bread wheats may have looked similar to *T. aestivum* ssp. *spelta* found growing in Iran from which free-threshing types were derived by mutation (McFadden and Sears, 1946). According to Ohtsuka (1998) and Yan et al. (2003), the European spelt wheats may have been derived secondarily from a hybridization involving *Triticum compactum* and emmer wheat. More recently, Matsuoka and Nasuda (2004), on the other hand, have suggested durum wheat (*Triticum durum* ssp. *durum*) as a candidate for the female progenitor (BBAA) genome of bread wheat after embryo rescue-free crossing of the durum wheat cultivar Langdon with *Ae. tauschii* line and successfully producing fertile triploid F_1 hybrids, which spontaneously (without colchicine treatment) set hexaploid F_2 seeds at average selfed seed rate of 51.5%. Currently, common wheat (*T. aestivum*, $2n = 6x = 42$, BBAADD), is the world's most widely cultivated crop grown in all temperate and in most subtropical countries with altitude levels ranging from below sea level near the Dead Sea and the Imperial Valley of California to more than 4500 m in Tibet (Stoskopf, 1985).

2.3 Wheat genetic resources and gene pools

The wheat genetic resources are composed of landraces, obsolete cultivars, wild relatives, and elite breeding lines and modern cultivars. The concept of the gene pools was proposed by Harlan and deWet (1971) and later on the basis of evolutionary distance from each other and their genomic constitution (Jiang et al., 1994), they gave the idea of the three gene pools, that is, primary, secondary, and tertiary gene pools. These gene pools are usually in relation to the cultivated species (Table 2.1). The knowledge of the ancestry cultivated wheats is important for understanding variation and genetic diversity in their primary and secondary gene pools and the potential for exploiting the valuable genes responsible for disease resistance or stress tolerance, into new varieties (Smale, 1996).

The primary gene pool of wheat is composed of wheat landraces, early domesticates, and wild species that hybridize directly with the cultivated and the diploid donors of the A and D genome to bread wheat and durum wheat. Bread wheat arose recently (6000–8000 years ago) from the hybridization of tetraploid (*T. turgidum*) and diploid, *Ae. tauschii* Coss. so these two species constitute the primary gene pool (Qi et al., 2007). The primary gene pool is often preferred due to the easiness of its crossability with wheat (Mujeeb-Kazi, 2003). The chromosomes of these species are homologous to the cultivated types and can be utilized easily by breeding methods (Feuillet et al., 2007). During the last decades many useful genes may have been lost due to crop improvement for specific environments; some of these lost genes can somehow be recovered from the primary gene pool. The primary gene pool of wheat carries a highly diverse, geographically widespread, and sexually compatible germplasm (Feuillet et al., 2007). It is important to note that only a small proportion of the existing genetic diversity of the primary gene pool for most crop species has been utilized for crop improvement (Tanksley and McCouch, 1997).

The secondary gene pool of wheat contains polyploid species that share at least one homologous genome with the cultivated types, such as polyploid *Triticum* and *Aegilops*

Table 2.1 Number of wheat genetic resource accessions held by genebanks

Triticum and related species	Number of accessions
Diploid	13,659
Tetraploid	106,398
Hexaploid	444,648
Octaploid	38
Other *Triticum* hybrid	367
Unspecified *Triticum*	167,133
Aegilops	42,026
Triticale	37,439
Total	811,708

Adapted from Knüpffer (2009).

species. These species share only one genome with wheat (Feuillet et al., 2007). Transferring genetic material is comparatively more complex, there are usually problems of hybrid seed death, female sterility of F_1 hybrids, reduced recombination (Ogbonnaya et al., 2013), and often to obtain F_1 hybrids embryo rescue is required. This gene pool includes, for instance, *T. timopheevii* (AAGG) and the diploid S-genome (similar to the B genome) species from *Aegilops* (Curtis, 2002; Qi et al., 2007).

The tertiary gene pool is composed of more distantly related diploids and polyploids with nonhomologous genomes. They have no genome constitutions of the cultivated species and transfer of genetic material is highly complex (Feuillet et al., 2007). Usually, special techniques, such as irradiation or gametocidal chromosomes, are needed for gene transfer and embryo rescue is necessary in these cases (Jiang et al., 1994; Mujeeb-Kazi and Hettel, 1995). This group includes mostly germplasm of Triticeae that are not within the primary or secondary gene pools. Most of the germplasm in this group are perennials (Feuillet et al., 2007; Mujeeb-Kazi, 2003). Although tertiary gene pool resources are highly complex to utilize, they have the potential of becoming an important means to develop new diverse wheat germplasm (Mujeeb-Kazi, 2006).

2.4 Genetic diversity and erosion from the traditional areas

Genetic diversity is the sum of genetic characteristics within any species or genus (Brown, 1983; Rao and Hodgkin, 2002). Differences within individuals of a species, whether they are at phenotypic or only at genotypic level, can contribute to improve traits of interest for crop adaptation, performance, and marketability. These variations are a result of mutation, selection, genetic drift, and gene flow processes, accumulated through time in different regions according to the environment. However, the genetic diversity can also be reduced by natural and artificial factors. Such reduction of diversity or loss of a species, landraces, cultivars, varieties, or alleles due to genetic and modernization bottlenecks is called genetic erosion. The process of extinction or genetic erosion can be due to biotic or abiotic stresses, caused by factors such as competition, predation, parasitism, and disease, or to isolation and habitat alteration due to slow geological and climatic change, natural catastrophes, or human activities. The first decrease in genetic diversity in cultivated grasses, such as bread and durum wheat, occurred with domestication. Since the early periods of wheat domestication, many traits have been altered to the interest of the domesticators and breeders. Events of the domestication syndrome started with the fixation of nonbrittle rachis and naked-grain traits in cultivated fields. Then after, in most of the main cereal species cultivated nowadays (wheat, maize, rice, and barley among others) the domestication process contributed to select in the cultivated varieties traits that have progressively evolved in a favorable direction for farmers, namely, flowering time or grain size (Buckler et al., 2001). In fact, more than 90 QTL related to domestication in wheat have been identified mostly in chromosomes 1B, 2A, 3A, and 5A (Peng et al., 2011). Such domestication and selection processes have a dramatic effect on the long-term effective size (N_e) of a population. Reductions of N_e, typically referred to as genetic bottlenecks, have usually led to decreased levels of diversity in cultivated crops relative to their wild progenitors (Buckler et al., 2001) since only a subset of the diversity in the wild form is domesticated. The extent of the bottleneck varies from species to species. For instance, it has been estimated that the genetic diversity of the DD genome of modern bread wheat cultivated varieties only account for 15% of that of *Ae. tauschii* growing in the Transcaucasia area (Dvorak et al., 1998). Similarly, Haudry et al. (2007) revealed that domestication reduced the diversity of bread and durum wheat by 69 and 84%, respectively. More recently, the analysis of the bread wheat genome using whole-genome shotgun sequencing indicated the loss of between 10,000 and 16,000 genes and several gene families in hexaploid wheat compared with the three diploid progenitors on polyploidization and domestication (Brenchley et al., 2012). At the same time, several classes of gene families with predicted roles in defense, nutritional content, energy metabolism, and growth have increased sizes in the Triticeae lineage, possibly as a result of selection during domestication (Brenchley et al., 2012). A second worldwide reduction in wheat genetic diversity with the adoption of the mega cultivars issued from the Green Revolution and the progressive decline of local landraces cultivation has been long debated. Fu and Somers (2009) have indicated an allelic reduction of 17% in the most modern varieties that affected every part of the wheat genome after a genome-wide examination of 75 Canadian wheat (*T. aestivum*) cultivars released from 1845 to 2004 by using 370 SSR markers. A reduction of diversity in the improved lines released in the 1980s as compared to the landraces was also seen in CIMMYT germplasm (Warburton et al., 2006). However, this reduction was quantified as 6% population decrease from landraces as compared to modern varieties in a recent study involving a worldwide sample of 2994 accessions of hexaploid wheat genotyped with more than 9000 SNP markers (Cavanagh et al., 2013). Peng et al. (1999) attributed the reduction in diversity in the new cultivars to the utilization of a limited number of landraces as the basis for the development of new cultivars, and a reduced diversity due to directional selection in new "key modern breeding genes" such as the dwarfing genes that led to the development the high-yielding semidwarf "Green Revolution" wheat varieties.

Nowadays, increasing awareness among breeders of the negative effects of genetic erosion together with the high mutation rates of hexaploid wheat and its buffering effects caused by polyploidy (Dubcovsky and Dvorak, 2007) indicate that the loss in cultivated wheat diversity can be reverted. In fact, Warburton et al. (2006) showed a recovery in genetic diversity among the most modern CIMMYT lines as a result of breeders' concern to prevent genetic erosion. The extent of this recovery is however not an absolute trend in all breeding programs and in spite of the recent advances some traits of interest may have been neglected (Nazco et al., 2014).

2.5 Conservation of genetic resources

Genetic resources are fundamental to sustaining global wheat production now and in the future since they embody a wide range of genetic diversity that is critical for enhancing and maintaining wheat yield potential. The adoption and cultivation of widely adapted semidwarf high-yielding and input-responsive wheat varieties of CIMMYT/ ICARDA origin has contributed in ensuring food security by increasing wheat production in developing countries with small changes in wheat cultivation area. However, the modern mega cultivars of wheat, coupled with other natural factors, such as climate change, natural catastrophes, or other human activities such as urbanization, overgrazing, and mining, have caused genetic erosion by displacing landraces and wild relatives in their regions of origin. Also, the genetic similarity of the leading cultivars results in a small diversity of disease- and pest-resistance sources. In the scenario of a major outbreak, the effects on global food security could be devastating as exemplified by the recent outbreak of stem and yellow rusts. A new stem rust race Ug99 (TTKS) was first detected in Uganda in 1999 and then spread to Kenya, Ethiopia, Yemen, Sudan, and Iran, and became a global threat to the wheat industry of the world for the very fact that it overcomes many of the known and most common stem rust resistance genes such as *Sr31*, *Sr24*, and *Sr36* (Jin and Singh, 2006). Similarly, the breakdown of yellow rust resistance genes *Yr9* in cultivars derived from "Veery" in the 1980s and *Yr27* in 2000s in major mega cultivars derived from "Attila" cross, such as PBW343 (India), Inquilab-91 (Pakistan), Kubsa (Ethiopia), and others such as Achtar in Morocco, Hidab in Algeria, and many other cultivars in the CWANA region (Solh et al., 2012), has caused significant wheat production loss. Through a coordinated international effort, many wheat varieties resistant to Ug99 and yellow rust have been released and replaced the susceptible cultivars. Given the continuous cycle of such boom and bust events, it is essential to conserve and utilize genetic resources in a sustainable manner to ensure the continuous improvement of the new wheat cultivars. There are two ways of conserving genetic resources: *in situ* and *ex situ*, which are basically complementary, not exclusive, to each other.

2.5.1 In situ *conservation*

In situ conservation, including "on-farm conservation," has been defined as "the continuous cultivation and management of a diverse set of populations by farmers in the agroecosystems where a crop has evolved" (Bellon et al., 1997), and it involves entire agroecosystems, including immediately useful species (such as cultivated crops, forages, and agroforestry species), as well as their wild and weed relatives that may be growing in nearby areas. *In situ* conservation will allow the continuation of processes of evolution and adaptation in crop plants, ensuring that new germplasm is generated over time. The continuous interaction of the local species and varieties with the changing environment and evolving pathogens enforces the appearance and selection of new positive traits for crop performance. Continuous cultivation of local and diverse landraces promotes agroecosystem stability and decreased use of chemicals in agriculture; enables to support cultural traditions; and helps mitigate the effects of pests, diseases, and other environmental stresses, all while providing new genetic material better adapted to the changing climate scenario. However, such caricature of the advantages of *in situ* conservation should not serve to impose cultural stasis and isolation of marginal communities. It is important to combine crop improvement with retention of useful diversity and to benefit from farmer and breeder expertise. There should be conducive government and international organizational policies and support to local communities and farmers who conserve landraces and wild relatives at the expense of yield and market, which they could reap from growing modern cultivars. Under harsh conditions and low-input agriculture, the cultivation of landraces is still practiced because of special quality attributes for local uses and because of their straw yield advantage.

Though improved varieties of wheat are dominating the global wheat production, there are still countries in West Asia where landraces, traditional varieties, and wild relatives are growing side by side. Durum landraces are also widely grown in North Africa and Ethiopia. Traditional emmer and einkorn cultivation persists in pockets in Italy (D'Antuono and Pavoni, 1993), Spain, Turkey, the Balkans, and India (Feldman et al., 1995). Wild emmer is most morphologically and genetically diverse in northern Israel with massive stands in the northeast and peripheral populations, which also occur along the Fertile Crescent center of diversity. However, there are few existing protected areas that are specifically targeting the *in situ* conservation of wheat wild relatives:

- Amiad reserve in Galilee, Israel, for conservation of wild *Triticum*, particularly *T. turgidum* subsp. *dicoccoides* (Anikster et al., 1997; Kaplan, 2008).
- Miyanjangal-e-Fasa, a protected area in the Fars province of Iran, for conservation of crop wild relatives including several *Aegilops* species.
- Al-Lujat biosphere reserve, established in 2008 in southern Syria, for the *in situ* conservation of wild relatives of cereals, legumes, and fruit trees.
- Erebuni nature reserve, northeast of Yerevan in Armenia, established on the recommendation of Vavilov, for *in situ* conservation of *T. urartu*, *T. monococcum* subsp. *aegilopoides, T. timopheevii* subsp. *armeniacum*, several *Aegilops* species and *Amblyopyrum muticum.*

There are several other protected areas and national parks located within the area of natural distribution of wheat wild relatives, likely to contain a number of important species. The *in situ* conservation of CWR in these areas is more or less passive and "accidental" and the extent to which these areas are contributing to the conservation of *Aegilops* and wild *Triticum* needs to be assessed.

Studies at Amrniad in Israel have shown the presence of a large diversity and coevolution of the host–pathogen interaction indicating the importance of this region for *in situ* conservation (Nevo, 1995). High diversity of *T. urartu* as well as wild emmer and einkon wheats have been also reported in the high plateau in Sweida province of Syria along the border between Lebanon and Syria, making them an ideal location for *in situ* conservation (Damania, 1994; Amri et al., 2005). Unless concerted efforts are made, these crucial resources will be highly threatened due to overgrazing, political uncertainties, land reclamation, development, and rapid urbanization. Understanding these scenarios, ICARDA has conducted a GEF-funded regional project on "promoting the conservation and sustainable use of dryland agrobiodiversity in Jordan, Lebanon, Palestine, and Syria" during 1999–2005, which allowed the development of a holistic approach for *in situ* conservation and management of landraces and wild relatives of wheat, barley, lentil, chickpea, forage legumes, and dryland fruit trees. Effective on-farm conservation of landraces requires a combination of technological, value-added, institutional, and policy options along with alternative sources of income to improve the livelihoods of the custodians of the remaining diversity. Several sites were recommended for establishing protected areas targeting the *in situ* conservation of wheat wild relatives and their management plans were also recommended (Amri et al., 2005) including

- Samta-Ajloun and around cement factory in Jordan,
- Wadi Saweid-Aarsal, Bishwet-Nabha, and Talat Sadah-Ham in Lebanon,
- Tayassir-Jenin and Wadi Sair-Hebron in Palestine, and
- Sweida in Syria.

DIVA-GIS version 7.5.0 (www.diva-gis.org) developed by Hijmans et al. (2005) was used to portray the distribution of species, calculate the species richness, and conduct complementary analysis to identify species diversity hotspots. The species richness analysis highlighted two areas ("hotspots") having the highest number of *Aegilops* species (at least 15): one in the vicinity of the border between Syria and Lebanon and the other in western Turkey (Fig. 2.4). Six areas with 12–14 *Aegilops* species are located in western Jordan, northern Iraq, western Iran, Israel/Palestine, and northern Syria/southern Turkey.

Species richness analysis identified five complementary regions of high *Aegilops* diversity: in western Syria and northern Lebanon, central Israel, and Northwest Turkey. Additional high species richness areas identified in Turkmenistan and Southern France are questionable and need further checking. Within these areas, 16 IUCN-recognized protected areas are found and these are potentially valuable sites in which to establish genetic reserves. Unfortunately, the "premier" *Aegilops* hotspots on the Syrian/Lebanese border are not coincident with any existing internationally recognized protected areas, and there is an urgent need to establish a new protected area in that region.

Reserve analysis using DIVA-GIS identified species diversity hotspots for wild *Triticum* in southern Turkey, and for *Aegilops* in the border region between Syria and Lebanon, and western Turkey. Ideally, these hotspot areas should be included in protected conservation areas, or genetic reserves, for *in situ* conservation of wild *Triticum*

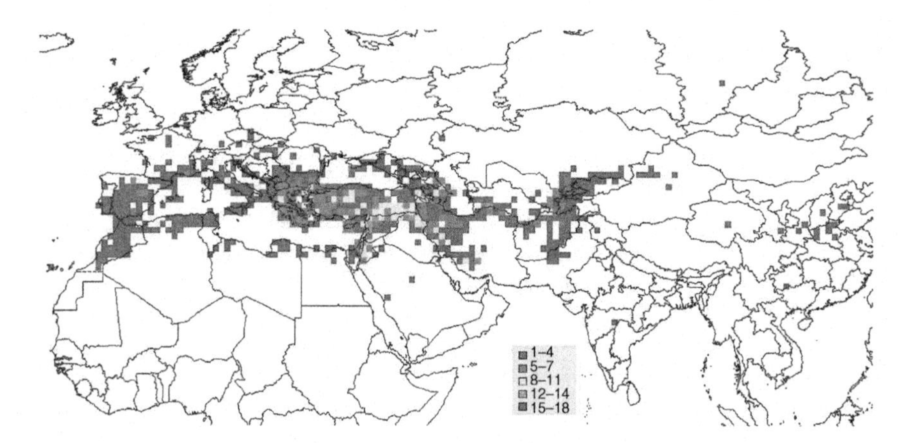

Figure 2.4 Species richness analysis for *Aegilops* species using DIVA-GIS program.

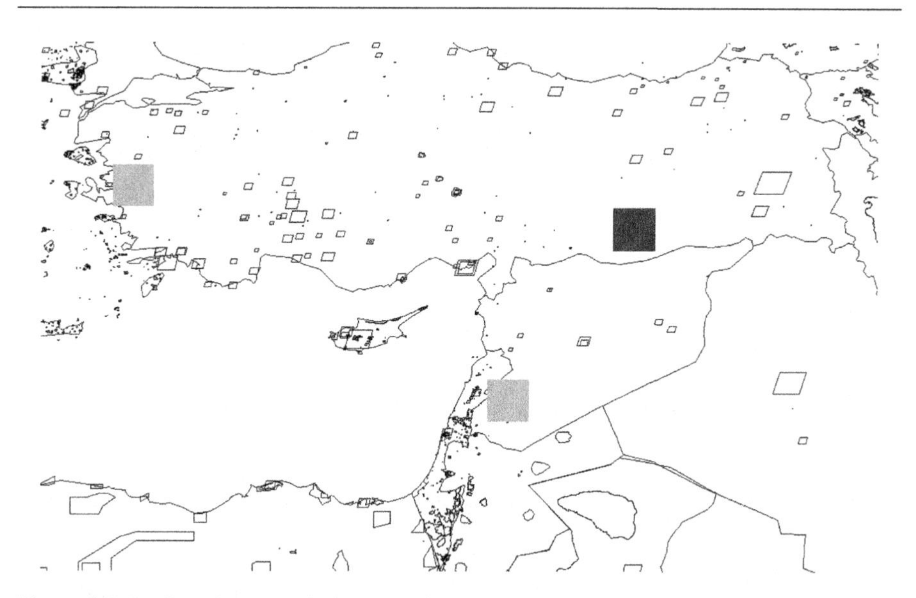

Figure 2.5 Regions for establishing genetic reserves for the conservation of *Aegilops* species (gray) and Wild *Triticum* Species (black), with the scatter of the existing protected areas.

and *Aegilops* species diversity (Fig. 2.5). None of these hotspots are at present included in existing protected areas, but the establishment of a new protected area in the valley of Qal'at Al Hosn in Syria as recommended by Maxted et al. (2008) would be an excellent first step in the *in situ* conservation of *Aegilops*.

There is a need to assess the extent to which existing protected areas are contributing to the conservation of wheat wild relatives. Preliminary observations in Jordan, Lebanon, and Syria showed that none of the recognized protected areas are specifically targeting the *in situ* conservation of *Aegilops* and wild *Triticum* species, although many of them contain some of these species. The Dana reserve in Jordan did not contain any of the wild *Triticum* species found in the adjacent region, while the protected areas in Jabal Abdulaziz and Sweida, in Syria, contained several *Aegilops* and wild *Triticum* species. In 1999, the Syrian government restricted grazing by sheep and goats in the Sweida region, and in 2007 established a biosphere reserve in Al-Lujat in Houran region and an *in situ* conservation site of 10,000 ha close to the border between Lebanon and Syria, about 30 km from Homs, to conserve wild relatives and range species.

Several other areas have been recommended as suitable for the *in situ* conservation of wheat wild relatives and other species by the GEF-funded, ICARDA-coordinated agrobiodiversity project. The establishment of these protected areas could target a number of species including wild relatives of crops, rangeland, and medicinal plants, and some neglected fruit trees. Several less-disturbed areas, in the Golan Heights, in the border areas between Lebanon and Syria, and between Turkey and Syria, were identified as being ideal for the establishment of genetic reserves for wheat wild relatives and other CWR.

For on-farm conservation of landraces, several projects were conducted in Ethiopia (Brush, 1995; Worede and Mekbib, 1992), Morocco, Palestine, Syria, Tunisia, and Yemen (ICARDA, 2014). These projects demonstrated the value of seed cleaning and treatment and participatory improvement in enhancing the yields of landraces. However, the sustainability of such activities will depend on farmers' and consumers' continued interest in these landraces and on the support and commitment of the international community and respective national governments.

2.5.2 Ex situ *conservation*

Ex situ conservation is the oldest conservation strategy targeted to preserve seeds or other propagative materials in a place separated from the environment of origin such as botanical gardens or genebanks. Practically, most of the *ex situ* conservations of wheat are in the form of seeds stored in genebanks. The underlying principle of successful seed storage is to maintain genetic integrity of accessions such as seeds with high viability for long periods. Seeds of the original sample should be stored under the best possible conditions to ensure safe long-term survival, while seeds of accessions that are frequently requested by breeders or other users should be stored in the active collection. It is important to duplicate collections in more than one location as a safety backup. Accessions available at other genebanks may require fewer duplicates than unique samples. A third safety backup can be kept in the Svalbard Global Seed Vault for added security.

The World Information and Early Warning System (WIEWS) database compiled by FAO possesses 856,167 wheat genetic resources accessions, including 25,242 of *Aegilops* and 6,015 wild *Triticum* species. In the "Crop Strategy for Wheat" conducted in 2007 by the GCDT, CGIAR centers and other partners reported around 750,000 wheat accessions held in over 80 genebanks, of which 4% were wild relatives, although no detailed information on *Aegilops* and wild *Triticum* was included (CIMMYT, 2007, http://www.croptrust.org/documents/cropstrategies/WheatStrategy.pdf).

According to FAO (2009), 95% of wheat landraces and more than 60% of wild relatives have been collected and conserved in genebanks of international centers, such as CIMMYT and ICARDA, and other national genebanks (Table 2.2). ICARDA genebank holds a total of 41,331 accessions of wheat genetic resources representing 28% of the total holdings of ICARDA genebank, placing it in the top six genebanks in the world.

Collecting missions and conservation of large number wheat germplasm accessions have been carried out at international and national genebanks as indicated earlier. Management of such accessions for characterization, regeneration, evaluation, and distribution is costly. The key questions associated with genebank management and germplasm evaluation are the following:

1. How unique are these accessions within the genebank and among genebanks at international and national levels? Crop registry was developed for wheat to identify the duplicates among all accessions held in different genebanks and the results showed that there is a lot of duplication among the main genebanks mainly for the germplasm collected by Vavilov Institute of Russia or the germplasm maintained by USDA-Genebank. Most of the wheat

Table 2.2 Germplasm collections of *Triticum*, *Aegilops*, and Triticale species

Genus	Genebank/institution	Country	Number proportion of accessions			
			Total	Wild relatives, landraces, and old cultivars (%)	Breeding lines and modern cultivars (%)	Other (%)
Triticum	Centro Internacional de Mejoramiento de Maíz y Trigo (CIMMYT)	Mexico	110,281	37	57	6
	National Small Grains Germplasm Research Facility (NSGC)	USA	57,348	61	38	<1
	Institute of Crop Germplasm Resources (ICGR-CAAS)	China	43,039	5	0	95
	National Bureau of Plant Genetic Resources (NBPGR)	India	35,889	6	10	84
	International Center for Agricultural Research in the Dry Areas (ICARDA)	Syria	34,951	80	<1	21
	National Institute of Agrobiological Sciences (NIAS)	Japan	34,652	7	31	61
	N.I. Vavilov All-Russian Scientific Research Institute of Plant Industry (VIR)	Russia	34,253	44	55	<1
	Istituto di Genetica Vegetale (IGV)	Italy	32,751	100	0	
	Leibniz Institute of Plant Genetics and Crop Plant Research (IPK)	Germany	26,842	53	44	4
	Australian Winter Cereals Collection (TAMAWC)	Australia	23,811	3	82	16
	Others		424,123	16	32	52
	Total		857,940	28	33	39
Aegilops	Lieberman Germplasm Bank (ICCI-TELAVUN)	Israel	9,146	100		<1
	International Center for Agricultural Research in the Dry Areas (ICARDA)	Syria	3,847	100		<1
	National Plant Gene Bank of Iran (NPGBI-SPII)	Iran	2,653	99		1
	National Institute of Agrobiological Sciences (NIAS)	Japan	2,433	5		95
	N.I. Vavilov All-Russian Scientific Research Institute of Plant Industry (VIR)	Russia	2,248			100

(Continued)

Table 2.2 Germplasm collections of *Triticum*, *Aegilops*, and Triticale species *(cont.)*

			Number proportion of accessions			
Genus	**Genebank/institution**	**Country**	**Total**	**Wild relatives, landraces, and old cultivars (%)**	**Breeding lines and modern cultivars (%)**	**Other (%)**
	National Small Grains Germplasm Research Facility (NSGC)	USA	2,207	100		
	Laboratory of Plants Gene Pool and Breeding (LPGPB)	Armenia	1,827	100	<1	
	Leibniz Institute of Plant Genetics and Crop Plant Research (IPK)	Germany	1,526	100		<1
	Centro Internacional de Mejoramiento de Maíz y Trigo (CIMMYT)	Mexico	1,326	99	<1	<1
	Cereal Research Centre, Agriculture and Agri-Food (WRS)	Canada	1,100	100		
	Others		13,713	77	2	21
	Total		42,026	82	1	18
Triticale	Centro Internacional de Mejoramiento de Maíz y Trigo (CIMMYT)	Mexico	17,394	<1	100	<1
	N.I. Vavilov All-Russian Scientific Research Institute of Plant Industry (VIR)	Russia	2,030			100
	National Small Grains Germplasm Research Facility (NSGC)	USA	2,009	1	99	
	Soil and Crops Research and Development Centre (SCRDC-AAFC)	Canada	2,000			100
	Institute of Plant Production N.A. V.Y. Yurjev of UAAS (IR)	Ukraine	1,748		99	1
	Institute of Genetics and Plant Breeding, University of Agriculture (LUBLIN)	Poland	1,748		96	3
	Leibniz Institute of Plant Genetics and Crop Plant Research (IPK)	Germany	1,577	2	98	<1
	Others		8,933	4	47	49
	Total		37,439	1	76	23

Adapted from FAO (2009).

genetic resources held at ICARDA are unique as they are predominately landraces, primitive wheats, *Aegilops*, and wild *Triticum*, most of which are issued from collecting missions organized jointly by ICARDA in CWANA region.

2. Are we confident that the global wheat gene pool has been adequately captured and conserved?

   The gap analysis combines the collecting locations of all available accessions with known natural range of each species to determine areas requiring further collecting. Also, future collecting missions can be based on priority levels (shown next) as outlined by Maxted et al. (2008), based on the number of accessions available. Based on this, there are 10 "high-priority" *Aegilops* species (*Aegilops bicornis, Aegilops columnaris, Aegilops juvenalis, Aegilops kotschyi, Aegilops longissima, Aegilops searsii, Aegilops sharonensis, Aegilops uniaristata, Aegilops vavilovii*, and *Aegilops ventricosa*) and two "high-priority" wild *Triticum* species (*T. timopheevii* subsp. *armeniacum* and *T. turgidum* subsp. *dicoccoides*), which have fewer than 200 accessions in total, in global collections. In addition, new collecting missions could be justified when targeting adaptive traits such as drought, heat, and salinity tolerance. The wheat genetic resources conservation strategy (CIMMYT, 2007) identified priority regions and certain "target" wild wheat relatives on which to base future collection missions, including

   a. Albania, Greece, and former Yugoslavia, which have 13 species including *Ae. uniaristata* known to possess tolerance to heavy metals.
   b. Iran, Jordan, Pakistan, and Syria, very dry areas, for heat- and drought-tolerant *Ae. searsii, Ae. tauschii*, and *Ae. vavilovii.*
   c. Iran, in the mountains near Shiraz and near Esfahan, along the Zagros mountains and in the desert and salt affected areas, all wild *Triticum* and *Aegilops* species, for drought and salt tolerance.
   d. Algeria, Cyprus, Egypt, Greece, Iran, Israel, Libya, Pakistan, Palestine, Spain, Syria, Tajikistan, Tunisia, Turkey, Turkmenistan, and Uzbekistan, targeting *Ae. bicornis, Aegilops comosa, Ae. juvenalis, Ae. kotschyi, Aegilops peregrina, Ae. sharonensis, Ae. speltoides, Ae. uniaristata,* and *Ae. vavilovii* (Maxted et al., 2008).
   e. Aegean region, western Mediterranean region (France, Portugal, Spain), North Africa, Iran, and Syria (Skovmand et al., 2000).

3. How much of the conserved germplasm has been well characterized both at phenotypic and molecular levels? Most of the conserved germplasm is characterized for major descriptors and for some agronomic traits. However, few were characterized using molecular markers techniques and CIMMYT is recently engaged in this activity using genotyping by sequencing technique and around 34,000 accessions of ICARDA will be genotyped in 2015.
4. Is the conserved germplasm accessible and useful? For the international collections access is facilitated by the use of the Standard Material Transfer Agreement (SMTA) and the same is done for some collections from national genebanks. However, accessibility to accessions in some national genebanks is still not guaranteed.

2.6 Processing to conservation

Any acquired and collected samples need to be accompanied by the required documents (Germplasm acquisition agreements or SMTA, phytosanitary certificate, and others), fumigated, checked for seed-borne pests and diseases, dried to 3–5% relative humidity (RH) in the seeds, tested for viability, then checked for duplication before assigning a unique identifier and processing to conservation. There are three types of collections: (1) active collection (medium-term conservation most often at 0–4°C and less than 20% RH) and designed for distribution; (2) base collection (long-term storage at 12°C and seeds packed in sealed and air-vacuumed aluminum foils); and (3) safety duplicate sent to a reliable genebank outside the location of active and base collection (storage conditions are similar to those of the base collection).

2.6.1 Characterization

Characterization is the description of wheat germplasm at phenotypic, biochemical, and molecular levels using standard tools and a list of descriptors specific to each species or group of species, establish their diagnostic traits and identify duplicates; classification and clustering of accessions into groups; identification of accessions with desired agronomic traits and selection of entries for more precise evaluation; development of interrelationships between or among traits and between geographic groups of cultivars; and estimating the extent of the genetic variation in the collection. The results of the wheat genome sequencing will take wheat characterization to the next level. Once implemented, wheat genome sequencing will allow curators and users to more practically characterize the accessions.

2.6.2 Documentation

Documentation is essential for genebank management to allow efficient and effective use of germplasm. Passport, characterization, and evaluation data are of little use if they are not adequately documented and incorporated into an information system that can facilitate access to data. Information plays, therefore, a significant role in biodiversity conservation and use. Accurate information about conserved materials is essential to increase its use. Computerized documentation systems enable rapid dissemination of information to users as well as to assist curators in managing the collections more efficiently. The development of a database, or an interconnected system of databases, with the capacity to manage and integrate all wheat information, including passport, characterization, and evaluation data, is key to most wheat genetic resource management activities in genebanks. The International Wheat Information System (IWIS), composed of the Wheat Pedigree Management System and the Wheat Data Management System, has enabled to assign unique wheat identifiers and genealogies, and to manage performance information and data regarding known genes. Similarly, the Genetic Resource Information Package (GRIP) has been developed using IWIS as data warehousing to collate passport information across genebanks and identify duplications (Skovmand et al., 2000). Such a strong database system would also enable to identify core subsets of collections for evaluation and identification of important traits for breeding purposes. Furthermore, such database can also guide future collection demands and priorities. GRIN Global is under development in collaboration between USDA-ARS and the Global Crop Diversity Trust for use in managing genebanks at national and international levels.

2.6.3 Multiplication and regeneration

When the seeds are not enough for distribution, a multiplication is conducted and when seed germination drops below a threshold of 85% of the initial germination, a regeneration is conducted. For both activities, there is a need to use adequate sample size, which will conserve the genetic integrity of the accessions. For some species, like *Ae. speltoides,* isolation is required to avoid outcrossing. After three cycles of using seeds from the active collection, it is advisable to refresh the active collection by multiplying seeds from the base collection. Characterization is often done during multiplication and regeneration.

2.6.4 Distribution

The genetic resources conserved in genebanks serve as a bridge linking the past and the future, and help improve crop varieties through plant breeding to meet the needs of farmers and communities, for research activities or to restore diversity lost on farm and in natural habitats. Genebanks can be more proactive in establishing links with germplasm users, breeders, researchers, farmers, and other groups to enhance the use of the germplasm. Access to genetic resources for distribution is guided by international law, specifically the Convention on Biological Diversity and the International Treaty on Plant Genetic Resources for Food and Agriculture (ITPGRFA), and is tied to sharing any benefits arising from the use of the material. Germplasm is usually distributed under a Material Transfer Agreement (MTA) that sets out the terms and conditions of access and responsibilities of recipients. In the case of crop germplasm, The ITPGRFA establishes a global system to provide farmers, plant breeders, and scientists, facilitated access to plant genetic material for research, breeding, and education using the SMTA. It also ensures that users share any benefits they derive from genetic material used in plant breeding or biotechnology with the regions from which the genetic resources were originated. The International Agricultural Research of the CGIAR signed an agreement with the Governing Body of the Treaty placing their collections in-trust with the ITPGRFA.

2.7 Role of genetic resources in wheat breeding

The ultimate goal of collection, conservation, characterization, and evaluation of germplasm is to identify accessions with traits of interest to breeders to be utilized in their breeding program in order to develop high-yielding varieties with resistance to major biotic and abiotic constraints with acceptable level of end-use quality. In this regard, the main question is how much of the genebank diversity has been utilized by breeders? Though there are few attempts to estimate the contribution of genetic resources for wheat improvement, the available reports indicate that only a limited amount (10%) of the genetic resources (landraces and wild relatives) have been utilized in crosses for prebreeding and breeding purposes (Chapman, 1986). Wheat is among the few crops where crop wild relatives are used and contribute to the varieties releases. However, more prebreeding efforts are needed. The most probable reasons for the low utilization of genetic resources by breeders are the following:

1. Most genebank materials are too wild, obsolete, and difficult to breed.
2. The characterization and evaluation data are poor and often it is done for traits that are not of interest to breeders.
3. Breeders may have traits of interest from their elite breeding materials and cultivars.
4. Genebank accession's information might not be accessible and too technical for noncurators.
5. Transferring some traits/genes from the wild relatives to modern wheat is often time-consuming and success is not always granted.
6. Even when traits are successfully transferred, there may be linkage drag associated with transferred traits.
7. Economic and labor investment is needed.
8. Nevertheless, genetic resources have played a significant role in wheat breeding in the following major areas.

2.7.1 Increasing yield potential

The first scientific breeders started selecting and crossing landraces and cultivars from different origins and genetic pools in the nineteenth century, to obtain the combinations that led to superior varieties. Especially successful cases of different gene pool uses include the Italian breeder Nazareno Strampelli, who in the first decades of the twentieth century included the Japanese variety Akakomugi in his crossing block (Salvi et al., 2013). The crosses between Italian landraces and breeding lines and Akakomugi resulted in the introgression of new alleles for phenological adjustment and suitable agronomical types – such as Ppd-D1 for photoperiod insensitivity and Rht8c for short straw – in the Italian gene pool. Varieties developed by Strampelli, like Ardito and Mentana became the backbone of most of the new varieties developed in the Mediterranean countries and some distant countries like Russia, China, and some South American countries (Dalrymple, 1986; Lupton, 1987; Yang and Smale, 1996).

A similar approach was followed by the Nobel laureate Norman E. Borlaug in the 1950s with the Norin-10/Brevor cross that introduced the Rht-B1 and Rht-D1 dwarfing alleles, further reducing plant height thus preventing lodging when the crop was subjected to increased fertilization and irrigation and led to what is commonly known as "The Green Revolution" (Borlaug, 2008; Gale and Youssefian, 1985; Worland and Snape, 2001). The dwarfing genes Rht-B1 and Rht-D1 originated from the Japanese landrace *Shiro Daruma* from which they were successfully introgressed into Norin 10 (Kihara, 1983). The introgression of these genes into modern wheat cultivars, however, is tedious and time-consuming and requires persistent efforts (Borlaug, 1988; Krull and Borlaug, 1970). These cases could be considered to be among the most successful human introgression of suitable alleles from a different gene pool that led to important increases in wheat productivity. Allele introduction from different primary gene pools is still one of the main sources of diversity and traits of interest (Reif et al., 2005). Crosses between winter and spring varieties – two gene pools that used to remain isolated from one another due to geographical and physiological barriers – have been widely used by national and international breeding centers like ICARDA and CIMMYT to obtain new high-yielding widely adapted varieties. Similarly, 1B/1R translocation has contributed to most of the mega-varieties used around the world.

2.7.2 Drought and heat tolerance

Wheat yield gains in the dry areas have been comparatively lower than in the irrigated areas and this is mainly due to the effect of drought and heat stress on wheat yield potential. The wild relatives of wheat have naturally evolved under drought and heat conditions in their original environments and can provide a number of genes selected to tolerate those stresses. They can be exploited for drought and heat tolerance through direct gene transfer or interspecific and intergeneric hybridizations.

Wheat landraces have been exploited for many abiotic and biotic stress tolerances and are the source of many important genes, such as that conferring drought tolerance (Reynolds et al., 2007). The reason behind this fact is that landraces that have been continuously grown and evolving in environments prone to these stresses have probably accumulated alleles contributing to a better adaptation to these types of environments. Wheat landraces are being tested by several institutions in the world to identify these new potential alleles promoting drought and heat adaptation. For instance, a recent study carried out in CIMMYT aiming to identify new sources of drought and heat tolerance in a collection of more than 750 landraces collected in Central and West Asia – where environments subjected to natural drought and heat events are common – showed large phenotypic differences for grain yield, aboveground biomass, and grain weight, traits associated to drought and heat tolerance (Lopes et al., 2015). This is not an isolated case; the comparison between modern varieties and landraces under near- and suboptimal conditions has led to the identification of candidates to provide traits of interest for drought and heat adaptation in several countries. In most cases wheat landraces, as compared to modern varieties, did not show higher grain yield under near- or suboptimal conditions nor higher yield stability but some candidate traits to improve adaptation to suboptimal environments have been identified (Denčić et al., 2000; Dodig et al., 2012).

On the other hand, wild relatives, synthetic hybrid wheat (SHW), synthetic derivatives (SDW), or modern lines with alien translocations have shown better tolerance compared to their modern counterparts under heat and drought stress conditions. CIMMYT identified several SHW and SDW with superior adaptation to drought conditions (Trethowan and van Ginkel, 2009). Yang et al., 2002 also identified candidates specifically adapted to heat stress although the specific adaptation to heat-stressed environments was negatively correlated to yield potential under near-optimal conditions. This suggests that specific lines with wild genetic background could be directly released in zones where heat stress is constant. A strategy to tackle down physiological traits of interest from wild relatives, identify their genetic control, assess the biochemical pathway involved, and make these resources available for breeding programs is described in a study carried out by Placido et al. (2013). Wheat lines carrying 7DL/7A translocation from *Agropyron elongatum* showed improved water stress adaptation and root biomass at deeper soil levels under drought stress (Placido et al., 2013). Two candidate genes involved in this adaptation have been identified and mapped in the translocation and the main biochemical pathways involved have been studied. Recently, the yield advantage of synthetic backcrossed derived wheat under rain-fed environments, such as northern and southern Australia, was reported to be 11 and 30% higher, respectively, compared to elite cultivars of bread wheat (Dreccer et al., 2007; Ogbonnaya et al., 2007). Similarly, Lage et al. (2008) reported that the synthetic derived wheat germplasm "Vorobey" (Croc_1/Ae. squarrosa (224)//Opata/3/Pastor) was a top yielding line in about 48 out of 52 sites globally.

2.7.3 Resistance to diseases and insects

Genetic resources have contributed to the improvement of disease resistance and insect pest resistance in wheat (Table 2.3). Most of these genes originated from wild relatives (Roelfs, 1988) or from landrace cultivars (McIntosh, 1988). The *Sr2* gene, which provided durable resistance to stem rust of wheat, was originally transferred to hexaploid wheat from Yaroslav emmer by McFadden in 1923 (Stakman and Harrar, 1957). This gene, in combination with other major and minor genes, has provided resistance to stem rust successfully in North America and in developing countries, such as Mexico, India, Pakistan, Turkey, and others, during the time of the Green Revolution and beyond. The 1RS.1BL translocation from rye carries the leaf rust resistance gene *Lr26* along with the stem rust resistance gene *Sr31* and the yellow rust resistance gene *Yr9*. Additional successful resistance genes from translocations were *Sr24* from *Thinopyrum ponticum* (1RS.1AL); *Sr38* from *Triticum ventricosaor* and *Sr36* from *T. timopheevi* (McIntosh et al., 1995). Currently, most of the effective resistance genes, such as *Sr22, Sr25, Sr26, Sr39, Sr42, Sr45*, which provide resistance against the Ug99 stem rust race, are derived from translocations of wheat wild relatives (Singh et al., 2007; Hajjar and Hodgkin, 2007). Sources of resistance for diverse diseases and insect pest, such as Septoria, tan spot, nematodes, Hessian fly, and aphids, have been reported from synthetic wheats (Villareal et al., 1992; Tadesse et al., 2007; El Bouhssini et al., 2012; Ogbonnaya et al., 2008, 2013). Additional efforts should be made to identify, introgress, and pyramid new resistance alleles in order to avoid food security threats due to resistance breakdowns.

Table 2.3 Examples of genes identified from wheat gene pools

Trait	Locus	Source	References
Disease resistance			
Leaf rust	*Lr9*	*Aegilops umbellulata*	Soliman et al. (1963)
	Lr18	*T. timopheevii*	Dyck and Samborski (1968)
	Lr19	*Thinopyrum elongatum*	Browder (1972)
	Lr23	*T. turgidum*	McIntosh and Dyck (1975)
	Lr24	*Ag. elongatum*	McIntosh et al. (1976)
	Lr32	*Ae. tauschii*	Kerber (1987)
	Lr35	*Ae. speltoides*	Kerber and Dyck (1969)
	Lr37	*Ae. ventricosa*	Bariana (1991)
	Lr47	*Ae. speltoides*	Dubcovsky et al. (1998)
	Lr57	*Ae. geniculata*	Kuraparthy et al. (2007)

Table 2.3 **Examples of genes identified from wheat gene pools *(cont.)***

Trait	Locus	Source	References
Stem rust	*Sr2*	*T. turgidum*	Ausemus et al. (1946)
	Sr5	*T. aestivum*	Ausemus et al. (1946)
	Sr22	*T. monococcum*	The (1973)
	Sr24	*Th. ponticum*	McIntosh et al. (1976)
	Sr25	*Th. ponticum*	McIntosh et al. (1976)
	Sr26	*Th. ponticum*	McIntosh et al. (1976)
	Sr38	*Ae. ventricosa*	Bariana (1991)
Stripe rust	*Yr5*	*T. spelta*	Macer (1966)
	Yr10	*T. spelta*	Macer (1975)
	Yr9	*S. cereale*	Macer (1975)
	Yr15	*T. dicoccoides*	Gerechter-Amitai et al. (1989)
	Yr17	*Ae. ventricosa*	Bariana (1991)
	Yr40	*Ae. geniculata*	Kuraparthy et al. (2007)
Powdery mildew	*Pm1*	–	Sears and Briggle (1969)
	Pm12	*Ae. speltoides*	Jia et al. (1994)
	Pm21	*Haynaldia villosa*	Qi et al. (1995)
	Pm25	*T. monococcum*	Shi et al. (1998)
Wheat streak mosaic virus	*Wsm1*	*Ag. elongatum*	Talbert et al. (1996)
Karnal bunt	QTL	*T. turgidum*	Nelson et al. (1998)
Barley yellow dwarf virus	*Bdv2*	*Thinopyrum intermedium*	Chen et al. (1998)
	Bdv3	*Th. intermedium*	Sharma et al. (1995)
Tan spot	*tsn3a,b,c*	*Ae. tauschii*	Tadesse et al. (2007)
Pest resistance			
Hessian fly	*H21*	*S. cereale*	Friebe et al. (1990)
	H23, H24	*Ae. tauschii*	Gill et al. (1986)
	H27	*Ae. ventricosa*	Delibes et al. (1997)
	H30	*Ae. triuncialis*	Martin-Sanchez et al. (2003)
	H32	*Ae. tauschii*	Sardesai et al. (2005)
Cereal cyst nematode	*Cre3, Cre4*	*Ae. tauschii*	Eastwood et al. (1991)
Abiotic stress			
Drought tolerance	QTL	*Ag. elongatum*	Placido et al. (2013)
Salinity tolerance	*Kna1*	*T. aestivum*	Dvorak and Gorham (1992)
Quality			
Preharvest sprouting	QTL	*T. aestivum*	Anderson et al. (1993)
Protein, Zn, and Fe content	QTL	*T. turgidum*	Uauy et al. (2006)

2.7.4 Resistance to salinity and micronutrients constraints

In the original wheat primary gene pool, diversity for salinity tolerance was scarce and mostly relied on a landrace called Kharcia 65 (Ogbonnaya et al., 2013). However, some genes improving wheat adaptation to saline conditions, were found to be related to the D genome donors, such as *Ae. tauschii*, which can be exploited by producing synthetic wheat lines (Gorham et al., 1987). As a result of this discovery, several primary synthetics have been tested under saline conditions finding high degrees of tolerance (Dreccer et al., 2004; Pritchard et al., 2002; Schachtman et al., 1992). These lines were able to exclude the abundance of Na^+ more efficiently and at the same time be more tolerant to the highest saline conditions of the experiment (Dreccer et al., 2004). These findings, together with boron toxicity tolerance, were found in similar synthetic lines (Dreccer et al., 2003; Emebiri and Ogbonnaya, 2015). This indicates that the wheat wild relatives can contribute to improve hexaploid wheat tolerance to soil composition constraints, increasing production, and making currently unavailable land (due to salinity and mineral deficiency) available for cultivation.

2.7.5 Improving grain quality

In most of the developing countries, wheat improvement for end-use quality is not a priority. Most of the focuses are in improving grain yield potential and resistance to diseases. However, some developing NARS are critically looking for high-quality varieties suited for the preparation of a range of end products. For farmers in the developed world, improved wheat quality provides increased market price and therefore better revenues for farmers. For instance, a 2% increase in grain protein (from 13.5% to 15.5%) of Western Red Spring wheat in Canada resulted in 2011 – an unfavorable year for wheat quality in that country – in an average premium of almost $90 per ton (Canadian Wheat Board, 2011), about 25% extra compared to the prices at the time. Besides grain protein content, increases in other micronutrients grain content, like Fe and Zn, could greatly improve the livelihood and health of people in developing countries where wheat is the main source of calories, especially for women and children (Ortiz-Monasterio et al., 2007). The introduction of the semidwarf varieties during the Green Revolution led to high yield increases. However, it also had a negative effect on Zn and Fe grain content (Monasterio and Graham, 2000). The negative correlation between grain yield and quality might have been a factor for the low priority of wheat improvement for better quality especially in the developing world (Peña et al., 2002; Sanchez-Garcia et al., 2015). Genetic resources, such as landraces, wild relatives, and synthetic wheats, have been reported as novel sources for improving wheat grain quality (Davies et al., 2006; Ogbonnaya et al., 2013; Rasheed et al., 2014). Several studies assessing the diversity across wheat gene pools for Zn and Fe grain content have been carried out (Cakmak et al., 1999; Calderini and Ortiz-Monasterio, 2003; Monasterio and Graham, 2000) and candidates with improved Zn and Fe allocation to the grain were found (Ortiz-Monasterio et al., 2007). Currently, efforts are underway at international centers and national programs to develop biofortified varieties and advanced lines with high Zn and Fe content from emmer wheat.

2.8 Strategies to enhance utilization of genetic resources

2.8.1 *Focused identification of germplasm strategy*

Random sampling and core collections were used very often to respond to seed requests by different users. The Generation Challenge Program has developed a reference subset from the core collection representing 1% of the total holding. Minicore collection was also advertised. ICARDA with its partners at Vavilov Institute in Russia, the Australian Winter Cereals Collection, and the Nordgen genebank (Nordic Region) developed the Focused Identification of Germplasm Strategy (FIGS) as the first scientific approach to agricultural genebank mining. This approach is based on the development of powerful algorithms that match plant traits with geographic and agroclimatic information about where the samples were collected (Bari et al., 2012; ICARDA, 2014; Mackay and Street, 2004). This approach was successful in finding sources of resistance to Sunn pest, Russian wheat aphid, salinity tolerance, and so on.

Development of core collection or minicore collections from the large collections of germplasm available in the genebank followed by extensive evaluation for different traits of economic importance would help to improve the utilization of genetic resources in the breeding programs. Previously, core collections were assembled based on passport and characterization data. It was reported that identification of novel sources of resistance for biotic and abiotic stresses from such core collections is often rare (Brown and Spillane, 1999; Dwivedi et al., 2007; Gepts, 2006; Pessoa-Filho et al., 2010; Polignano et al., 2001; Xu, 2010).

The FIGS is a trait-based and user-driven approach to select potentially useful germplasm for crop improvement. It searches for specific sought-after traits, using as surrogate the environment, based on the hypothesis that the germplasm is likely to reflect the selection pressures of the environment from which it was originally sampled (Mackay, 1990, 1995; Mackay and Street, 2004). FIGS uses the eco-geographical data of a reference dataset of accessions for desired traits such as resistance to either diseases or pests, and has successfully helped to identify a number of novel genes in germplasm from environmentally similar sites to those of the reference/template dataset (Bhullar et al., 2009; El-Bouhssini et al., 2009; El-Bouhssini et al. 2010; Mackay, 1995; Mackay and Street, 2004). Relationships between adaptive traits and collection site environmental parameters have also been revealed by recent studies using multivariate and multiway models such as N-PLS (multilinear partial least squares) (Endresen, 2010; Endresen et al., 2011). Modeling of stem rust resistance using geographical information system (GIS) approaches has also led to the detection of a relationship between geographical areas and incidence of resistance to stem rust (Bonman et al., 2007).

2.9 Utilization of gene introgression techniques

The different gene pools of wheat possess immense potential for providing genes of interest to improve bread and durum wheat. However, introgression of new alleles, chromosomes, or entire genomes of these relatives into common wheat had been tiresome if not impossible. With the advances in hybridization and biotechnological techniques during the first decades of the twentieth century, gene transfers from wild relative wheat has become a reality though one of the most successful introgressions of wild genomic regions into a modern wheat variety, that is, wheat/rye 1RS.1BL translocation was not the result of a directed cross made by a breeder but a natural hybridization (Zeller et al., 1973). There are two basic ways of gene introgression: (1) gene transfer through hybridization and chromosome-mediated gene transfer approaches and (2) direct gene transfer using molecular approaches.

2.9.1 Hybridization and chromosome-mediated gene transfer

The synthesis of hybrids between wheat and alien species is the basic and most important step to transfer genes from alien species into wheat. The success of this step depends on many factors. It is important to identify a highly crossable wheat variety such as Chinese Spring with *kr1*, *kr2*, *kr3*, and *kr4* genes (Luo et al., 1994). Pre- and postfertilization factors such as pre- and postpollination hormonal treatments, *in vitro* embryo culture, and cold treatment of mother spikes prior to excision of embryos and other factors do play significant roles in the synthesis of hybrids (Sharma and Ohm, 1990). After the synthesis of the F_1 hybrid, pairing between wheat and alien chromosomes is required for a successful gene transfer to happen. The *Ph1* gene in the long arm of chromosome 5B may suppress homoeologous pairing between wheat and alien chromosomes and hence restrict alien-gene transfer into wheat. There are different methods to promote chromosome pairing in such instances. Using Ph mutant of wheat, a high pairing mutation involving a small intercalary deficiency for *Ph1* has been developed from Chinese spring cultivar and designated as *ph1b*. According to Sears (1984), the level of homoeologous pairing is highest when a wheat cultivar carrying the *ph1b* mutant is used for making crosses with the donor species. For durum wheat, methods involving the *ph1b/ph2b* and *ph1c* mutants have been employed and successful gene transfers have been made (Ceoloni et al., 1996).

2.9.2 Use of 5B-deficient stocks

Intergenomic chromosome pairing and intergeneric gene transfers have been promoted in durum wheat using a durum substitution line in which chromosome 5B is substituted by chromosome 5D. The *Ph1* gene will be absent in both the substitution line and the interspecific hybrids. This method has been used to promote chromosome pairing between durum wheat and *Thinopyrum bessarabicum* and *Thinopyrum curvifolium* (Jauhar and Almouslem, 1998).

2.9.3 Use of appropriate genotypes of alien species

Another method of inducing intergenomic pairing in hybrids is by crossing wheat with alien species such as *Ae. speltoides*, *Ae. comosa,* and *Elymus sibiricus*, which inactivate the homoeologous pairing suppressor *Ph1* (Jauhar, 1975; Jauhar and Almouslem, 1998; Knott and Dvorak, 1981; Riley et al., 1968). It is important to note that for any wheat–alien exchange to be usable, it should have (1) a small exchanged segment, (2) good compensation of the alien genetic material for the wheat genetic material that it replaces, and (3) normal transmission of the introgressed genetic material to the progenies, preferably in a Mendelian fashion (Jauhar and Chibbar, 1999).

There are different approaches to produce wheat–alien translocations. The most common ones include radiation treatment, tissue culture, and spontaneous translocation, and univalent misdivision and induced homoeologous recombination (Zhang et al., 2007). Friebe et al. (1996) have made an extensive review and reported that out of the 57 spontaneous or induced wheat–alien translocations, 10 were Robertsonian, 45 were terminal with distal alien chromosome segments translocated to terminal positions on wheat chromosome arms, and only two were intercalary.

Wheat–rye translocations of 1BL.1RS and 1AL.1RS, which arose spontaneously from centromeric breakage and reunion, are perfect examples of genetically compensated Robertsonian translocations that are perhaps the most successful wheat–alien translocations used for wheat improvement (Friebe et al., 1996; Rajaram et al., 1983; Zeller and Hsam, 1983).

In the aforementioned examples, a small portion of the genome of a species was introgressed into cultivated wheat to transfer a gene or a trait of interest. In some rare cases, wheat–alien species amphiploid can directly be used for breeding or even cultivation as can be exemplified by Triticale. Triticale is a manmade crop developed by crossing either common wheat (*T. aestivum*, $2n = 6x = 42$ = AABBDDD) or durum wheat (*T. durum*, $2n = 4x = 28$ = AABB) as the female parent using rye (*Secale cereale*, $2n = 2x = 14$ = RR) as the male parent. By using the hexaploid wheat octaploids triticale ($2n = 8x = 56$ =AABBDDRR) are produced while tetraploids wheat leads to hexaploid triticale ($2n = 6x = 42$ = AABBRR). In the beginning, most of the focus was on octaploid triticale; however, hexaploid triticale showed more success. Most of the triticale breeding programs are working with hexaploid lines, due to the good vigor and stability compared to the octaploid triticale (Mergoum et al., 2009).

Another example of amphiploids is the synthetic wheats. The synthetic hexaploid wheat – also called "resynthesized synthetic wheat" ($2n = 6x = 42$, BBAADD) – is most commonly produced by crossing the *T. turgidum* spp. *durum* ($2n = 4x = 28$, BBAA) with *Ae. tauschii* ($2n = 2x = 14$, DD) (Mujeeb-Kazi and Hettel, 1995). The chromosomes in such primary synthetics are fully homologous to cultivated hexaploid wheat and therefore all the chromosomes can directly recombine with modern wheat cultivars.

Although most of the synthetic wheat lines are agronomically inferior to modern cultivars, they possess superior genes for monogenic and polygenic traits valuable for wheat improvement. Synthetic hexaploid wheat is now being used to transfer genes from wild relatives to cultivated wheat for biotic and abiotic stresses resistance and yield per se.

Synthetic hexaploid wheat has served as a bridge between wild species of wheat and the modern improved wheat varieties to transfer genes from wild relatives to cultivated wheat for biotic and abiotic stresses resistance and yield per se. Many valuable genes have been transferred into wheat using primary synthetics. For instance, the tetraploid parents have been used to transfer to wheat genes for resistance to leaf and stripe rust from durum wheat (*T. turgidum* ssp. *durum* (Desf.) Husn (Ma et al., 1995), stripe resistance from *T. turgidum* ssp. *dicoccoides* (Kema et al., 1995), and Russian wheat aphid resistance from *T. turgidum* ssp. *dicoccum* (Lage et al., 2004). Similarly, some important genes have been identified/introduced from *Ae. tauschii*, the D genome donor, such as tolerance or resistance to diseases and pests (Ma et al., 1995; May and Lagudah, 1992; Tadesse et al., 2007; Villareal et al., 1992). Besides, increases in 1000 grain weight (Cox et al., 1995), salinity tolerance (Gorham et al., 1987; Shah et al., 1987), mineral and nutrients efficiency (Cakmak et al., 1999; Genc and McDonald, 2004; Monasterio and Graham, 2000), drought tolerance (Sohail et al., 2011), bread making quality (Hsam et al., 2001; Peña et al., 1995; Tang et al., 2008), and preharvest sprouting (Imtiaz et al., 2008; Ogbonnaya et al., 2007) have been reported in synthetic wheat. About 20% of the new CIMMYT and up to 24% of ICARDA material include synthetic background (Ogbonnaya et al., 2013).

2.9.4 Direct gene transfer into wheat

Genetic engineering via direct gene transfer involves the insertion of a characterized gene(s) that has been isolated from an unrelated organism into plant cells, and the subsequent recovery of fully fertile plants with the inserted gene(s) integrated into their genome. This technology gives access to an unlimited gene pool, without the constraint of sexual compatibility. Despite its great importance, wheat was the last of the major crops to be genetically transformed (Vasil et al., 1992; Weeks et al., 1993). The relatively slow progress in the development of transgenic wheat has been attributed to, first, the nonavailability of an efficient *in vitro* regeneration system and, second, the cereals being considered to be outside the host range of *Agrobacterium*. Initial success in the production of transgenic wheat was achieved by the combination of a high-efficiency regeneration system with biolistics-mediated gene delivery. The recent successful production of *Agrobacterium*-mediated transgenic wheat (Cheng et al., 1997) may help overcome some of the detrimental effects of the biolistics-mediated production of transgenic plants (Chibbar and Kartha, 1994). Transgenic wheat has been developed to incorporate tolerance to herbicides like glyphosate by transferring the CP4 and GOX genes into the crop (Zhou et al., 2003). The barley trypsin inhibitor CMe (BTI-CMe) was introduced into wheat resulting in a significant survival reduction of the major grain storage pest, Angoumois grain moth (*Sitotroga cerealella*) (Altpeter et al., 1999). Abebe et al. (2003) reported the transfer of the mannitol-1-phosphate dehydrogenase (mtlD) coding gene to wheat, which enabled the transgenic wheat to accumulate mannitol and tolerate water stress and salinity. Similarly, transgenic wheat lines overexpressing citrate synthase or malate dehydrogenase genes have been developed to enhance aluminum tolerance (Anoop et al., 2003; Hoekenga et al., 2006; Tesfaye et al., 2001). The DREB1A gene from *Arabidopsis thaliana* was introduced to wheat conferring drought tolerance to the crop (Pellegrineschi et al., 2004). Xue et al. (2011) reported the overexpression of TaNAC69 gene to promote tolerance to dehydration and water efficiency. The beta betA encoding choline dehydrogenase from *Escherichia coli* was introduced through *Agrobacterium* into the wheat genome improving root growth and drought resistance. However, unlike many other major crops (notably maize, soybean, cotton, and canola) that now account for more than 180 million ha of commercial transgenic crop production across many countries, there is no genetically modified (GM) wheat production in any country. The GM approach would be particularly valuable for traits for which there is limited or no genetic variation within the *Triticum* species. This would include herbicide resistance, *Fusarium* resistance, novel quality traits, and technologies for creating hybrid cultivars. In addition, GM technologies hold promise for enhancing drought and heat tolerance, as well as disease and pest resistance. Major efforts are needed to break yield barriers in wheat to increase yield potential by 50% to cope with growing demand. Increasing the radiation use efficiency of wheat through modification of key enzymes (e.g., Rubisco) and biochemical pathways to increase photosynthesis, ear size, and lodging resistance are key areas for wheat research through the integration of physiological and molecular breeding methodologies to increase wheat yield potential. A further increase in yield potential could be achieved through the development of hybrid wheat systems based on native and transgenic interventions collaboratively, leveraging private sector technologies for the benefit of stakeholders in the developing world.

2.10 Utilization of genomics

Genomics, the study of an organism's entire genome, is the new field of study to enhance the use of crop genetic resources. Significant progress has been made in different areas of wheat genomics research during the last two decades particularly in the development of marker technologies such as restriction fragment length polymorphism (RFLPs), simple sequence repeat (SSRs), amplified fragment length polymorphism (AFLP), single nucleotide polymorphism (SNPs), and diversity arrays technology (DArT) markers (Gupta et al., 2008). The availability of high-throughput DNA sequencing technologies collectively named as next-generation sequencing (NGS), such as DArT array technologies, SNP platforms or genotype by sequencing (GBS), would identify accelerate marker – trait discovery, develop high-density physical and genetic maps, determine genetic diversity, under take gene mining, gene pyramiding and introgression, marker-assisted selection, and genomic selection (Gupta et al., 2008; Pérez-de-Castro et al., 2012; Varshney et al., 2005). These new marker platforms have dramatically reduced the genotyping cost, opening the door to its massive use to more accurately identify major and especially minor QTL. These new tools have also helped understand the changes that occur during wheat domestication and identifying the main QTL involved (Poland, 2015). This information has the power to increase breeder's understanding of the key processes for most polygenic traits like yield or adaptation. Practically from the genetic resource conservation point of view, genotyping of genebank accessions using the recently available NGS enables the determination of genetic diversity and clustering of accessions, avoid duplications, assemble core subsets of germplasm, and accelerate allele mining, which will in turn reduce the cost of genebank management and increase efficiency of germplasm utilization for breeding purposes. Many QTLs using biparental mapping populations have been identified in wheat as indicated by the review by Gupta et al. (1999, 2008). Genome-wide association mapping approaches using association panels have also been carried out, and significantly associated markers have been reported for many traits such as kernel size and milling quality (Breseghello and Sorrells, 2006; Rasheed et al., 2014), grain yield (Crossa et al., 2007), high-molecular-weight glutenins (Ravel et al., 2006), resistance to foliar diseases (Crossa et al., 2007; Tadesse et al., 2014), *Fusarium* head blight (FHB) resistance (Massman et al., 2011) soil-borne pathogens (Mulki et al., 2013), stem rust resistance (Zegeye et al., 2014; Jighly et al., 2015), and major insect pest resistances in wheat (Joukhadar et al., 2013). Most of the diagnostic markers reported to date have been extensively utilized in the characterization of parents and assembling of crossing blocks, and pyramiding of resistance genes. Recently, genomic selection, which basically uses breeding values (BV) determined by using genotypic and phenotypic data across all traits, has been reported to be effective especially in combining minor genes, accelerate breeding cycles, and increase its efficiency (Lado et al., 2013; Poland et al., 2012). However, genomic selection will not be a panacea and cannot replace field-based plant breeding. It is important to integrate and validate genomic selection using strong phenotyping activities across representative and key locations.

2.11 Future direction and prospects

It is expected that by the year 2050 the world population will reach 9 billion and the demand for wheat will reach more than 900 million tons. Increasing human population coupled with climate change and other production constraints, such as increasing drought/water shortage, soil degradation, reduced supply and increasing cost of fertilizers, increasing demand for biofuel, and emergence of new virulent diseases and pests, are becoming challenging issues to wheat production and genetic resources conservation. To increase wheat production and feed the ever-increasing world population while conserving the natural resource base, the following points are important to consider:

1. To tackle genetic erosion, wise and concerted efforts have been made to collect and conserve wheat genetic resources *ex situ* in national and international genebanks. However, because of the threat of climate change, war, overpopulation, and overgrazing, further collection missions need to be carried out especially in countries where wheat has originated.
2. Management of the huge collections of genetic resources available in genebanks is costly, and the utilization level to date is limited with some reports indicating it to be only around 10%. It is therefore important to design and apply efficient strategies and tools such as FIGS, cytogenetic and genomic tools, and bioinformatics in order to form core or FIGS sets and subsets to increase genetic resources utilization in the breeding programs.
3. Characterization and evaluation of the FIGS sets in key locations for key traits of interest needs to be carried out in order to identify genes for introgression in wheat prebreeding programs. Utilization of hybridization and chromosome-mediated gene transfer techniques are important to effectively transfer genes from wild relatives into wheat. In this regard, synthetic wheats have played and will play a significant role in transferring novel sources of

genes for yield potential, grain quality, and resistance/tolerance to abiotic and biotic stresses. However, it is important to develop new primary synthetics using different sources of durum and *Ae. tauschii* in order to capture the genetic diversity available from these species.

4. The availability of new molecular tools, such as GBS, would enable efficient characterization and mining novel genes and alleles, gene introgression and pyramiding through marker-assisted selection and undertake genomic selection. However, it is noteworthy to keep the right balance in investment between marker technology and field-level phenotyping. In the past, many QTLs have been identified with very little practical application in the breeding process.
5. Though there is no GM wheat currently under production anywhere in the world, it is important to invest in this direction in the future as GM wheat would be particularly valuable for traits with limited or no genetic variation within the *Triticum* species such as herbicide resistance, *Fusarium* resistance, novel quality traits, increasing yield potential through breaking the yield barrier, and technologies for creating hybrid cultivars. In addition, GM technologies hold promise for enhancing drought and heat tolerance, as well as disease and pest resistance.
6. Future investment in hybrid wheat production is believed to increase wheat production as hybrid wheats provide higher grain yield, higher thousand grain weight, more tillers, higher biomass, deeper roots, and better resistance to biotic and abiotic stresses as compared to their parents. It is anticipated that the application of biotechnological methods will enable to capture increased heterosis by direct selection of favorable alleles and development of new genetically based systems to control male sterility.
7. Establishment of efficient networking and enabling policy environments would play a major role in technology generation and promotion. In this regard, the International Wheat Improvement Network (IWIN) coordinated by CIMMYT and ICARDA has been the most successful and efficient network for making available and widespread distribution of wheat genetic resources and new wheat genotypes globally. To address global challenges in wheat science and production, the exchange of wheat genetic material and associated knowledge through existing networks and new partnerships (IWIN) will be a critically important international public good that must remain freely available to achieve impact.

References

Abebe, T., Guenzi, A.C., Martin, B., Cushman, J.C., 2003. Tolerance of mannitol-accumulating transgenic wheat to water stress and salinity. Plant Physiol. 131, 1748–1755.

Altpeter, F., Diaz, I., Mcauslane, H., Gaddour, K., Carbonero, P., Vasil, I.K., 1999. Increased insect resistance in transgenic wheat stably expressing trypsin inhibitor CMe. Mol. Breed. 5, 53–63.

Amri, A., Monzer, M., Al-Oqla, A., Atawneh, N., Shehahdeh, A., Konopka, J., 2005. Status and threats to crop wild relatives in selected natural habitats in West Asia region. Proceedings of the International Conference on: Promoting Community-driven Conservation and Sustainable Use of Dryland Agrobiodiversity on April 18–21, 2005, ICARDA, Aleppo, Syria.

Anderson, J.A., Sorrells, M.E., Tanksley, S.D., 1993. RFLP analysis of genomic regions associated with resistance to preharvest sprouting in wheat. Crop Sci. 33, 453–459.

Anikster, Y., Feldman, M., Horovitz, A., 1997. The Ammiad experiment. In: Maxted, N., Ford-Lloyd, B.V., Hawkes, J.G. (Eds.), Plant Genetic Conservation. Springer, The Netherlands, pp. 239–253.

Anoop, V.M., Basu, U., McCammon, M.T., McAlister-Henn, L., Taylor, G.J., 2003. Modulation of citrate metabolism alters aluminum tolerance in yeast and transgenic canola overexpressing a mitochondrial citrate synthase. Plant Physiol. 132, 2205–2217.

Ausemus, E.R., Harrington, J.B., Reitz, L.P., Worzella, W.W., 1946. A summary of genetic studies in hexaploid and tetraploid wheats. J. Am. Soc. Agron. 38, 1082–1099.

Bari, A., Street, K., Mackay, M., Endresen, D.T.F., De Pauw, E., Amri, A., 2012. Focused Identification of Germplasm Strategy (FIGS) detects wheat stem rust resistance linked to environmental variables. Genet. Resour. Crop. Evol. 59, 1465–1481.

Bariana, H.S., 1991. Genetic Studies on Stripe Rust Resistance in Wheat. PhD Thesis, University of Sydney.

Baum, M.W., Tadesse, W., Singh, R., Payne, T., Morgounov, A.I., Braun, H.J., 2013. Global crop improvement networks to bridge technology gaps. The Twelfth International Wheat Genetics Symposium, Yakohama, Japan.

Bellon, M.R., Pham, J.L., Jackson, M.T., 1997. Genetic conservation: a role for rice farmers. Plant Genetic Conservation. Springer, Netherlands, pp. 263–289.

Bhullar, N.K., Zhang, Z., Wicker, T., Keller, B., 2009. Wheat gene bank accessions as a source of new alleles of the powdery mildew resistance gene *Pm3*: a large scale allele mining project. BMC Plant Biol. 10, 88.

Bonman, J.M., Bockelman, H.E., Jin, Y., Hijmans, R.J., Gironella, A., 2007. Geographic distribution of stem rust resistance in wheat landraces. Crop Sci. 47, 1955–1963.

Borlaug, N.E., 1988. Challenges for global food and fiber production. J. R. Swed. Acad. Agric. Forest. Suppl. 21, 15–55.

Borlaug, N.E., 2008. Feeding a world of 10 billion people: our 21st century challenge. In: Scanes, C.G., Miranowski, J.A. (Eds.), Perspectives in World Food and Agriculture 2004. Iowa State Press, Ames, pp. 32–56.

Braun, H.J., Sãulescu, N.N., 2002. Breeding Winter and Facultative Wheat. FAO Plant Production and Protection Series. FAO, Rome.

Braun, H.J., Atlin, G., Payne, T., 2010. Multi-location testing as a tool to identify plant response to global climate change. In: Reynolds, M.P. (Ed.), Climate Change and Crop Production. CABI Publishers, Wallingford, UK, pp. 115–138.

Brenchley, R., Spannag, M., Pfeifer, M., Barker, G.L., D'Amore, R., Allen, A.M., et al., 2012. Analysis of the bread wheat genome using whole-genome shotgun sequencing. Nature 491, 705–710.

Breseghello, F., Sorrells, M.E., 2006. Association mapping of kernel size and milling quality in wheat (*Triticum aestivum* L.) cultivars. Genetics 172, 1165–1177.

Briggle, L.W., Curtis, B.C., 1987. Wheat worldwide. In: Heyne, E.G. (Ed.), Wheat and Wheat Improvement. American Society of Agronomy, Madison, WI, pp. 1–31.

Browder, L.E., 1972. Designation of two genes for resistance to *Puccinia recondita* in *Triticum aestivum*. Crop Sci. 12, 705–706.

Brown, W.L., 1983. Genetic diversity and genetic vulnerability – an appraisal. Econ. Bot. 37, 4–12.

Brown, A.H.D., Spillane, C., 1999. Implementing core collections – principles, procedures, progress, problems and promise. In: Johnson, R.C., Hodgkin, T. (Eds.), Core Collections for Today and Tomorrow. IPGRI, Rome, pp. 1–9.

Brush, S.B., 1995. *In situ* conservation of landraces in centers of crop diversity. Crop Sci. 35, 346–354.

Buckler, E.S., Thornsberry, J.M., Kresovich, S., 2001. Molecular diversity, structure and domestication of grasses. Genet. Res. 77, 213–218.

Cakmak, I., Cakmak, O., Eker, S., Ozdemir, A., Watanabe, N., Braun, H.J., 1999. Expression of high zinc efficiency of *Aegilops tauschii* and *Triticum monococcum* in synthetic hexaploid wheats. Plant Soil 215, 203–209.

Calderini, D.F., Ortiz-Monasterio, I., 2003. Are synthetic hexaploids a means of increasing grain element concentrations in wheat? Euphytica 134, 169–178.

Canadian Wheat Board, 2011. http://www1.agric.gov.ab.ca/$department/deptdocs.nsf/all/sis14235 (accessed May 2015).

Cavanagh, C.R., Shiaoman, C., Shichen, W., Huang, B.E., Stephen, S., Kiani, S., et al., 2013. Genome-wide comparative diversity uncovers multiple targets of selection for improvement in hexaploid wheat landraces and cultivars. Proc. Natl. Acad. Sci. 110 (20), 8057–8062.

Ceoloni, C., Biagetti, M., Ciaffi, M., Forte, P., Pasquini, M., 1996. Wheat chromosome engineering at the 4x level: the potential of different alien gene transfers into durum wheat. Euphytica 89, 87–97.

Chapman, C.G.D., 1986. The role of genetic resources in wheat breeding. Plant Genet. Resour. Newsl. 65, 2–5.

Chen, Q., Conner, R.L., Ahmad, F., Laroche, A., Fedak, G., Thomas, J.B., 1998. Molecular characterization of the genome composition of partial amphiploids derived from *Triticum aestivum* × *Thinopyrum ponticum* and *T. aestivum* × *Th. intermedium* as sources of resistance to wheat streak mosaic virus and its vector, *Aceria tosichella*. Theor. Appl. Genet. 97, 1–8.

Cheng, M., Fry, J.E., Pang, S., Zhou, H., Hironaka, C.M., Duncan, D.R., et al., 1997. Genetic transformation of wheat mediated by *Agrobacterium tumefaciens*. Plant Physiol. 115, 971–980.

Chibbar, R.N., Kartha, K.K., 1994. Transformation of plant cells by bombardment with microprojectiles. In: Shargool, P.D., Ngo, T.T. (Eds.), Biotechnological Applications of Plant Cultures. CRC Press, Boca Raton, FL, pp. 37–60.

CIMMYT, 2007. Global strategy for the *ex situ* conservation with enhanced access to wheat, rye and triticale genetic resources. International Maize and Wheat Improvement Center (CIMMYT), Mexico. http://www.croptrust.org/documents/web/Wheat-Strategy-FINAL-20Sep07.pdf (accessed January 2008).

Cox, T.S., Sears, R.G., Bequette, R.K., Martin, T.J., 1995. Germplasm enhancement in winter wheat × *Triticum tauschii* backcross populations. Crop Sci. 35, 913–919.

Crossa, J., Burgueno, J., Dreisigacker, S., Vargas, M., Herrera-Foessel, S.A., Lillemo, M., et al., 2007. Association analysis of historical bread wheat germplasm using additive genetic covariance of relatives and population structure. Genetics 177, 1889–1913.

Curtis, B.C., 2002. Wheat in the world. In: Curtis, B.C., Rajaram, S., Macpherson, H.G. (Eds.), Bread Wheat Improvement and Production. FAO Plant Production and Protection Series. FAO, Rome, pp. 1–19.

D'Antuono, L.F., Pavoni, A., 1993. Phenology and grain growth of *Triticum dicoccum* and *T. monococcum* from Italy. In: Damania, A.B. (Ed.), Biodiversity and Wheat Improvement. Wiley-Sayce, London, UK, pp. 273–286.

Dalrymple, D.G., 1986. Development and Spread of High-Yielding Wheat Varieties in Developing Countries. Bureau for Science and Technology, U.S. Agency for International Development, Washington, DC.

Damania, A.B., 1994. *In situ* conservation of biodiversity of wild progenitors of cereal crops in the Near East. Biodiver. Lett., 56–60.

Daud, H.M., Gustafson, J.P., 1996. Molecular evidence for *Triticum speltoides* as a B-genome progenitor of wheat (*Triticum aestivum*). Genome 39, 543–548.

Davies, J., Berzonsky, W.A., Leach, G.D., 2006. A comparison of marker-assisted and phenotypic selection for high grain protein content in spring wheat. Euphytica 152, 117–134.

Delibes, A., Del Moral, J., Martin-Sanchez, J.A., Mejias, A., Gallego, M., Casado, D., et al., 1997. Hessian fly-resistance gene transferred from chromosome 4Mv of *Aegilops ventricosa* to *Triticum aestivum*. Theor. Appl. Genet. 94, 858–864.

Denčić, S., Kastori, R., Kobiljski, B., Duggan, B., 2000. Evaluation of grain yield and its components in wheat cultivars and landraces under near optimal and drought conditions. Euphytica 113, 43–52.

Devos, K.M., Dubcovsky, J., Dvorak, J., Chinoy, C.N., Gale, M.D., 1995. Structural evolution of wheat chromosomes 4A, 5A, and 7B and its impact on recombination. Theor. Appl. Genet. 91, 282–288.

Dixon, J., Braun, H.J., Crouch, J., 2009. Transitioning wheat research to serve the future needs of the developing world. In: Dixon, J., Braun, H.J., Kosina, P. (Eds.), Wheat Facts and Futures. CIMMYT, Mexico, DF, pp. 1–19.

Dodig, D., Zorić, M., Kandić, V., Perović, D., Šurlan-Momirović, G., 2012. Comparison of responses to drought stress of 100 wheat accessions and landraces to identify opportunities for improving wheat drought resistance. Plant Breed. 131, 369–379.

Dong, C., Whitford, R., Langridge, P., 2002. A DNA mismatch repair gene links to the *Ph2* locus in wheat. Genome 45, 116–124.

Dreccer, M.F., Ogbonnaya, F.C., Borgognone, G., Wilson, J., 2003. Boron tolerance is present in primary synthetic wheats. In: Pogna, N. (Ed.), Proceedings of Tenth International Wheat Genetics Symposium, Paestum.

Dreccer, M.F., Ogbonnaya, F.C., Borgognone, G., 2004. Sodium exclusion in primary synthetic wheats. Proceedings of the Eleventh Wheat Breeding Assembly, pp. 118–121.

Dreccer, M.F., Borgognone, M.G., Ogbonnaya, F.C., Trethowan, R.M., Winter, B., 2007. CIMMYT-selected derived synthetic bread wheats for rainfed environments: yield evaluation in Mexico and Australia. Field Crops Res. 100, 218–228.

Dubcovsky, J., Dvorak, J., 2007. Genome plasticity a key factor in the success of polyploid wheat under domestication. Science 316, 1862–1866.

Dubcovsky, J., Lukaszewski, A.J., Echaide, M., Antonelli, E.F., Porter, D.R., 1998. Molecular characterization of two *Triticum speltoides* interstitial translocations carrying leaf rust and greenbug resistance genes. Crop Sci. 38, 1655–1660.

Dvorak, J., Akhunov, E.D., 2005. Tempos of gene locus deletions and duplications and their relationship to recombination rate during diploid and polyploid evolution in the *Aegilops–Triticum* alliance. Genetics 171, 323–332.

Dvorak, J., Gorham, J., 1992. Methodology of gene transfer by homoeologous recombination into *Triticum turgidum*: transfer of K^+/Na^+ discrimination from *Triticum aestivum*. Genome 35, 639–646.

Dvorak, J., Zhang, H.B., 1990. Variation in repeated nucleotide sequences sheds light on the phylogeny of the wheat B and G genomes. Proc. Natl. Acad. Sci. 87, 9640–9644.

Dvorak, J., di Terlizzi, P., Zhang, H.B., Resta, P., 1993. The evolution of polyploid wheats: identification of the A genome donor species. Genome 36, 21–31.

Dvorak, J., Luo, M.C., Yang, Z.L., Zhang, H.B., 1998. The structure of the *Aegilops tauschii* gene pool and the evolution of hexaploid wheat. Theor. Appl. Genet. 97, 657–670.

Dvorak, J., Luo, M.C., Akhunov, E.D., 2011. NI Vavilov's theory of centres of diversity in the light of current understanding of wheat diversity, domestication and evolution. Czech. J. Genet. Plant Breed. 47, S20–S27.

Dwivedi, S., Puppala, N., Upadhyaya, H.D., Manivannan, N., Singh, S., 2007. Development and Evaluation of Valencia Core Collection for the US Peanut Germplasm. Western Society of Crop Science Annual Meeting, June, Las Cruces, New Mexico, pp. 18–19.

Dyck, P.L., Samborski, D.J., 1968. Genetics of resistance to leaf rust in the common wheat varieties Webster, Loros, Brevit, Carina, Malakof and Centenario. Can. J. Genet. Cytol. 10, 7–17.

Eastwood, R.F., Lagudah, E.S., Appels, R., Hannah, M., Kollmorgen, J.F., 1991. *Triticum tauschii*: a novel source of resistance to cereal cyst nematode (*Heterodera avenae*). Crop Pasture Sci. 42, 69–77.

El Bouhssini, M., Ogbonnaya, F.C., Chen, M., Lhaloui, S., Rihawi, F., Dabbous, A., 2012. Sources of resistance in primary synthetic hexaploid wheat (*Triticum aestivum* L.) to insect pests: Hessian fly, Russian wheat aphid and Sunn pest in the Fertile Crescent. Genet. Resour. Crop Evol. 60, 621–627.

El-Bouhssini, M., Street, K., Joubi, A., Ibrahim, Z., Rihawi, F., 2009. Sources of wheat resistance to Sunn pest, *Eurygaster integriceps* Puton, in Syria. Genet. Resour. Crop Evol. 56, 1065–1069.

El-Bouhssini, M., Street, K., Amri, A., Mackay, M., Ogbonnaya, F.C., Omran, A., et al., 2010. Sources of resistance in bread wheat to Russian wheat aphid (*Diuraphis noxia*) in Syria identified using the focused identification of germplasm strategy (FIGS). Plant Breed. 130, 96–97.

Emebiri, L.C., Ogbonnaya, F.C., 2015. Exploring the synthetic hexaploid wheat for novel sources of tolerance to excess boron. Mol. Breed. 35, 1–10.

Endresen, D.T.F., 2010. Predictive association between trait data and ecogeographic data for Nordic barley landraces. Crop Sci. 50, 2418–2430.

Endresen, D.T.F., Street, K., Mackay, M., Bari, A., De Pauw, E., 2011. Predictive association between biotic stress traits and ecogeographic data for wheat and barley landraces. Crop Sci. 51, 2036–2055.

FAO, 2009. Draft Second Report on the State of the World's Plant Genetic Resources for Food and Agriculture. Commission on Genetic Resources for Food and Agriculture. CGRFA-12/09/Inf.7 Rev.1, FAO, Rome.

FAO, 2011. FAOSTAT. FAO, Rome, Italy. http://faostat.fao.org (accessed 11.03.2015).

FAO, 2015. FAOSTAT.FAO, Rome, Italy. http://faostat.fao.org (accessed 11.03.2015).

Feldman, M., 2001. Origin of cultivated wheat. In: Bonjean, A.P., Angus, W.J. (Eds.), The World Wheat Book: A History of Wheat Breeding. Lavoisier Publishing, Paris, France.

Feldman, M., Levy, A.A., 2005. Allopolyploidy – a shaping force in the evolution of wheat genomes. Cytogenet. Genome Res. 109, 250–258.

Feldman, M., Lupton, F.G.H., Miller, T.E., 1995. Wheats. In: Smart, J., Simonds, N.W. (Eds.), Evolution of Crop Plants. Longman Group Ltd, London, UK, pp. 184–192.

Feuillet, C., Langridge, P., Waugh, R., 2007. Cereal breeding takes a walk on the wild side. Trends Genet. 24, 24–32.

Friebe, B., Hatchett, J.H., Sears, R.G., Gill, B.S., 1990. Transfer of Hessian fly resistance from "Chaupon" rye to hexaploid wheat via a 2BS/2RL wheat-rye chromosome translocation. Theor. Appl. Genet. 79, 385–389.

Friebe, B., Jiang, J., Raupp, W.J., McIntosh, R.A., Gill, B.S., 1996. Characterization of wheat-alien translocations conferring resistance to diseases and pests: current status. Euphytica 91, 59–87.

Fu, Y.B., Somers, D.J., 2009. Genome-wide reduction of genetic diversity in wheat breeding. Crop Sci. 49, 161–168.

Gale, M.D., Youssefian, S., 1985. Dwarfing genes in wheat. In: Russell, G.E. (Ed.), Progress in Plant Breeding. Butterworths, London, UK, pp. 1–35.

Genc, Y., McDonald, G.K., 2004. The potential of synthetic hexaploid wheats to improve zinc efficiency in modern bread wheat. Plant Soil 262, 23–32.

Gepts, P., 2006. Plant genetic resources conservation and utilization. Crop Sci. 46, 2278–2292.

Gerechter-Amitai, Z.K., van Silfhout, C.H., Grama, A., Kleitman, F., 1989. *Yr15* – a new gene for resistance to *Puccinia striiformis* in *Triticum dicoccoides* sel. G-25. Euphytica 43, 187–190.

Gill, B.S., Friebe, B., 2001. Cytogenetics, phylogeny and evolution of cultivated wheats. In: Bonjean, A.P., Angus, W.J. (Eds.), The World Wheat Book: A History of Wheat Breeding. Lavoiser Publishing, Paris, France, pp. 71–88.

Gill, B.S., Hatchett, J.H., Cox, T.S., Raupp, W.J., Sears, R.G., Martin, T.J., 1986. Registration of KS85WGRC01 Hessian fly-resistant hard red winter wheat germplasm. Crop Sci. 26, 1266–1267.

Gorham, J., Hardy, C., Wyn Jones, R.G., Joppa, L.R., Law, C.N., 1987. Chromosomal location of a K/Na discrimination character in the D genome of wheat. Theor. Appl. Genet. 74, 584–588.

Gupta, P.K., Varshney, R.K., Sharma, P.C., Ramesh, B., 1999. Molecular markers and their applications in wheat breeding. Plant Breed. 118, 369–390.

Gupta, P.K., Mir, R.R., Mohan, A., Kumar, J., 2008. Wheat genomics: present status and future prospects. Int. J. Plant Genomics 2008, 896451_1–_1896451.

Hajjar, R., Hodgkin, T., 2007. The use of wild relatives in crop improvement: a survey of developments over the last 20 years. Euphytica 156, 1–13.

Harlan, J.R., deWet, J.M.J., 1971. Toward a rational classification of cultivated plants. Taxonomy 20, 509–517.

Haudry, A., Cenci, A., Ravel, C., Bataillon, T., Brunel, D., Poncet, C., et al., 2007. Grinding up wheat: a massive loss of nucleotide diversity since domestication. Mol. Biol. Evol. 24, 1506–1517.

Heun, M., Schafer-Pregl, R., Klawan, D., Castagna, R., Accerbi, M., Borghi, B., et al., 1997. Site of Einkorn wheat domestication identified by DNA fingerprinting. Science 278, 1312–1314.

Hijmans, R.J., Guarino, L., Jarvis, A., O'Brien, R., Mathur, P., Bussink, C., et al., 2005. DIVA-GIS version 5.2 Manual. Available from: www.diva-gis.org.

Hoekenga, O.A., Maron, L.G., Pineros, M.A., Cancado, G.M.A., Shaff, J., Kobayashi, Y., et al., 2006. *AtALMT1*, which encodes a malate transporter, is identified as one of several genes critical for aluminum tolerance in Arabidopsis. Proc. Natl. Acad. Sci. USA 103, 9738–9743.

Hsam, S.L.K., Kieffer, R., Zeller, F.J., 2001. Significance of *Aegilops tauschii* glutenin genes on bread making properties of wheat. Cereal Chem. 78, 521–525.

ICARDA, 2014. A new approach to mining agricultural gene banks – to speed the pace of research innovation for food security. Research to Action 3. Available from: www.icarda.org.

Imtiaz, M., Ogbonnaya, F.C., Oman, J., van Ginkel, M., 2008. Characterization of quantitative trait loci controlling genetic variation for preharvest sprouting in synthetic backcross-derived wheat lines. Genetics 178, 1725–1736.

Jauhar, P.P., 1975. Genetic control of diploid-like meiosis in hexaploid tall fescue. Nature 254, 595–597.

Jauhar, P.P., Almouslem, A.B., 1998. Production and meiotic analyses of intergeneric hybrids between durum wheat and *Thinopyrum* species. In: Jaradat, A.A. (Ed.), Proceedings of the Third International Triticeae Symposium. Scientific Publishers, New Hampshire, pp. 119–126.

Jauhar, P.P., Chibbar, R.N., 1999. Chromosome-mediated and direct gene transfers in wheat. Genome 42, 570–583.

Jia, J., Miller, T.E., Reader, S.M., Gale, M.D., 1994. RFLP tagging of a gene *Pm12* for powdery mildew resistance in wheat (*Triticum aestivum* L.). Sci. China Chem. 37, 531.

Jiang, J., Gill, B.S., 1994. Different species-specific chromosome translocations in *Triticum timopheevii* and *T. turgidum diphyletic* origin of polyploid wheats. Chromosome Res. 2, 59–64.

Jiang, J., Friebe, B., Gill, B.S., 1994. Recent advances in alien gene transfer in wheat. Euphytica 73, 199–212.

Jighly, A., Oyiga, B.C., Makdis, F., Nazari, K., Youssef, O., Tadesse, W., et al., 2015. Genome-wide DArT and SNP scan for QTL associated with resistance to stripe rust (*Puccinia striiformis* f. sp. *tritici*) in elite ICARDA wheat (*Triticum aestivum* L.) germplasm. Theor. Appl. Genet. 128, 1277–1295.

Jin, Y., Singh, R.P., 2006. Resistance in US wheat to recent eastern African isolates of *Puccinia graminis* f. sp. *tritici* with virulence to resistance gene *Sr31*. Plant Dis. 90, 476–480.

Johnson, B.L., 1975. Identification of the apparent B genome donor of wheat. Can. J. Genet. Cytol. 17, 21–39.

Johnson, B.L., Dhaliwal, H.S., 1976. Reproductive isolation of *Triticum boeoticum* and *Triticum urartu* and the origin of the tetraploid wheat. Am. J. Bot. 63, 1088–1094.

Joukhadar, R., El-Bouhssini, M., Jighly, A., Ogbonnaya, F.C., 2013. Genome-wide association mapping for five major pest resistances in wheat. Mol. Breed. 32, 943–960.

Kaplan, D., 2008. A designated nature reserve for *in situ* conservation of wild emmer wheat (*Triticum dicoccoides* (Körn.) Aaronsohn) in northern Israel. In: Maxted, N., Ford-Lloyd, B.V., Kell, S.P., Iriondo, J.M., Dulloo, M.E. (Eds.), Crop Wild Relative Conservation and Use. CAB International, Wallingford, UK, pp. 389–393.

Kema, G.H.J., Lange, W., van Silfhout, C.H., 1995. Differential suppression of stripe rust resistance in synthetic wheat hexaploids derived from *Triticum turgidum* subsp. dicoccoides and *Aegilops squarrosa*. Phytopathology 85, 425–429.

Kerber, E.R., 1987. Resistance to leaf rust in hexaploid wheat, *Lr32*, a third gene derived from *Triticum tauschii*. Crop Sci. 27, 204–206.

Kerber, E.R., Dyck, P.L., 1969. Inheritance in hexaploid wheat of leaf rust resistance and other characters derived from *Aegilops squarrosa*. Can. J. Genet. Cytol. 11, 639–647.

Kihara, H., 1919. Uber zytologische Studien bei einigen Getreidearten. I. Spezies-Bastarde des Weizens Und Weizenmggen-Bastarde. Bot. Mag. (Tokyo) 33, 17–38.

Kihara, H., 1924. Cytologische und genetische Studien bei wichtigen Getreidearten mit besonderer Rucksicht auf das Verhalten der Chromosomen und die Sterilität in den Bastarden, vol. 8. Memoirs of the College of Science, Kyoto Imperial University, pp. 1–200.

Kihara, H., 1983. Origin and history of 'Daruma', a parental variety of Norin 10. In: Sakamoto, S. (Ed.), Proceedings of the Sixth International Wheat Genetics Symposium, Plant Germplasm Institute, University of Kyoto, China Agricultural Scientech Press, Kyoto, Japan, Beijing.

Knott, D.R., Dvorak, J., 1981. Agronomic and quality characteristics of wheat lines with leaf rust resistance derived from *Triticum speltoides*. Can. J. Genet. Cytol. 23, 475–480.

Knüpffer, H., 2009. Triticeae genetic resources in *ex situ* genebank collections. In: Feuillet, C., Muehlbauer, G.J. (Eds.), Genetics and Genomics of the Triticeae. Springer, US, pp. 31–79.

Koebner, R.M.D., Shepherd, K.W., 1986. Controlled introgression to wheat of genes from rye chromosome arm 1RS by induction of allosyndesis. Theor. Appl. Genet. 73, 197–208.

Krull, C.F., Borlaug, N.E., 1970. The utilization of collections in plant breeding and production. In: Frankel, O., Benett, E. (Eds.), Genetic Resources in Plants – Their Exploration and Conservation. Blackwell Scientific Publications, Oxford, pp. 427–439.

Kuraparthy, V., Chhuneja, P., Dhaliwal, H.S., Kaur, S., Bowden, R.L., Gill, B.S., 2007. Characterization and mapping of cryptic alien introgression from *Aegilops geniculata* with new leaf rust and stripe rust resistance genes *Lr57* and *Yr40* in wheat. Theor. Appl. Genet. 114, 1379–1389.

Lado, B., Matus, I., Rodríguez, A., Inostroza, L., Poland, J., Belzile, F., et al., 2013. Increased genomic prediction accuracy in wheat breeding through spatial adjustment of field trial data. G3 Genes Genomes Genet. 3, 2105–2114.

Lage, J., Skovmand, B., Andersen, S.B., 2004. Field evaluation of emmer wheat-derived synthetic hexaploid wheat for resistance to Russian wheat aphid (Homoptera: Aphididae). J. Econ. Entomol. 97, 1065–1070.

Lage, J., Trethowan, R.M., Hernandez, E., 2008. Identification of site similarities in western and central Asia using CIMMYT international wheat yield data. Plant Breed. 127, 350–354.

Lopes, M.S., El-Basyoni, I., Baenziger, P.S., Singh, S., Royo, C., Ozbek, K., et al., 2015. Exploiting genetic diversity from landraces in wheat breeding for adaptation to climate change. J. Exp. Bot. 66, 3477–3486.

Luo, M.C., Yen, C., Yang, J.L., 1994. Crossability percentages of bread wheat [*Triticum aestivum*] land races from Hunan and Hubei provinces, China with rye [*Secale cereale*]. Wheat Inform. Ser. 78, 34

Lupton, F.G.H., 1987. History of wheat breeding. In: Lupton, F.G.H. (Ed.), Wheat Breeding: its Scientific Basis. Chapman and Hall, London, UK.

Ma, H., Singh, R.P., Mujeeb-Kazi, A., 1995. Resistance to stripe rust in *Triticum turgidum*, *T. tauschii* and their synthetic hexaploids. Euphytica 82, 117–124.

Macer, R.C.F., 1966. The formal and monosomic genetic analysis of stripe rust (*Puccinia striiformis*) resistance in wheat. In: MacKey, J. (Ed.), Proceedings of the Second International Wheat Genetics Symposium, vol. 2. Lund, Sweden 1963. Hereditas Supplement, pp. 127–142.

Macer, R.C.F., 1975. Plant pathology in a changing world. Trans. Br. Mycol. Soc. 65, 351–374.

Mackay, M.C., 1990. Strategic planning for effective evaluation of plant germplasm. In: Srivastava, J.P., Damania, A.B. (Eds.), Wheat Genetic Resources: Meeting Diverse Needs. Wiley, Chichester, pp. 21–25.

Mackay, M.C., 1995. One core collection or many? In: Hodgkin, T., Brown, A.H.D., Van Hintum, T.J.L., Morales, A.A.V. (Eds.), Core Collections of Plant Genetic Resources. Wiley, Chichester, pp. 199–210.

Mackay, M.C., Street, K., 2004. Focused identification of germplasm strategy – FIGS. In: Black, C.K., Panozzo, J.K., Rebetzke, G.J. (Eds.), Proceedings of the 54th Australian cereal chemistry conference and the 11th wheat breeders' assembly, Royal Australian Chemical Institute, Melbourne, pp. 138–141.

Martin-Sanchez, J.A., Gomez-Colmenarejo, M., Del Moral, J., Sin, E., Montes, M.J., Gonzalez-Belinchon, C., et al., 2003. A new Hessian fly resistance gene (*H30*) transferred from the wild grass *Aegilops triuncialis* to hexaploid wheat. Theor. Appl. Genet. 106, 1248–1255.

Massman, J., Cooper, B., Horsley, R., Neate, S., Dill-Macky, R., Chao, S., et al., 2011. Genome-wide association mapping of *Fusarium* head blight resistance in contemporary barley breeding germplasm. Mol. Breed. 27, 439–454.

Matsuoka, Y., Nasuda, S., 2004. Durum wheat as a candidate for the unknown female progenitor of bread wheat: an empirical study with a highly fertile F1 hybrid with *Aegilops tauschii* Coss. Theor. Appl. Genet. 109, 1710–1717.

Maxted, N., White, K., Valkoun, J., Konopka, J., Hargreaves, S., 2008. Towards a conservation strategy for *Aegilops* species. Plant Genet. Resour. 6, 126–141.

May, C.E., Lagudah, E.S., 1992. Inheritance in hexaploid wheat of *Septoria tritici* blotch resistance and other characteristics derived from *Triticum tauschii*. Aust. J. Agric. Res. 43, 433–442.

McFadden, E.S., Sears, E.R., 1946. The origin of *Triticum spelta* and its free-threshing hexaploid relatives. Heredity 37, 81–89.

McIntosh, R.A., 1988. The role of specific genes in breeding for durable stem rust resistance in wheat and triticale. Breeding Strategies for Resistance to the Rusts of Wheat. CIMMYT, Mexico, DF, pp. 1–9.

McIntosh, R.A., Dyck, P.L., 1975. Cytogenetical studies in wheat. VII Gene *Lr23* for reaction to *Puccinia recondita* in Gabo and related cultivars. Aust. J. Biol. Sci. 28, 201–212.

McIntosh, R.A., Dyck, P.L., Green, G.J., 1976. Inheritance of leaf rust and stem rust resistances in wheat cultivars Agent and Agatha. Aust. J. Agric. Res. 28, 37–45.

McIntosh, R.A., Wellings, C.R., Park, R.F., 1995. Wheat Rusts: An Atlas of Resistance Genes. CSIRO Publications, East Melbourne.

Mello-Sampayo, T., 1971. Genetic regulation of meiotic chromosome pairing by chromosome 3D of *Triticum aestivum*. Nat. New Biol. 230, 22–23.

Mergoum, M., Singh, P.K., Pena, R.J., Lozano-del Río, A.J., Cooper, K.V., Salmon, D.F., et al., 2009. Triticale: a "new" crop with old challenges. In: Carena, M.C. (Ed.), Cereals. Springer, US, pp. 267–287.

Miller, T.E., 1987. Systematics and evolution. In: Lupton, F.G.H. (Ed.), Wheat Breeding and its Scientific Bases. Chapman and Hall, London, UK.

Monasterio, J.I., Graham, R.D., 2000. Breeding for trace minerals in wheat. Food Nutr. Bull. 21, 393–396.

Mujeeb-Kazi, A., 2006. Utilization of genetic resources for bread wheat improvement. Genet. Resour. Chromosome Eng. Crop Improv. 2, 61–97.

Mujeeb-Kazi, A., Hettel, G.P, 1995. Utilizing Wild Grass Biodiversity in Wheat Improvement: 15 Years of Wide Cross Research at CIMMYT. CIMMYT Research Report No. 2. CIMMYT, Mexico, DF.

Mujeeb-Kazi, A., Villareal, R.L., 2002. Wheat. In: Chopra, V.L., Prakash, S. (Eds.), Evolution and Adaptation of Cereal Crops. Science Publishers, Inc, Enfield.

Mujeeb-Kazi, A., 2003. Wheat improvement facilitated by novel genetic diversity and *in vitro* technology. Plant Cell Tissue Organ Cult 13, 179–210.

Mulki, M.A., Jighly, A., Ye, G., Emeribi, L.C., Moody, D., Ansari, O., et al., 2013. Association mapping for soilborne pathogen resistance in synthetic hexaploid wheat. Mol. Breed. 31, 299–311.

Naranjo, T., 1990. Chromosome structure of durum wheat. Theor. Appl. Genet. 79, 397–400.

Nazco, R., Peña, R.J., Ammar, K., Villegas, D., Crossa, J., Moragues, M., et al., 2014. Variability in glutenin subunit composition of Mediterranean durum wheat germplasm and its relationship with gluten strength. J. Agric. Sci. (Cambridge) 152, 379–393.

Nelson, J.C., Autrique, J.E., Fuentes-Dávila, G., Sorrells, M.E., 1998. Chromosomal location of genes for resistance to Karnal bunt in wheat. Crop Sci. 38, 231–236.

Nesbitt, M., Samuel, D., 1996. From staple crop to extinction? The archaeology and history of the hulled wheats. In: Padulosi, D., Hammer, K., Heller, J. (Eds.), Hulled Wheats. Proceedings of the First International Work-shop on Hulled Wheats, Castelvecchio Pascoli, Italy, pp. 41–100.

Nevo, E., 1995. Asian, African and European biota meet at 'Evolution Canyon' Israel: local tests of global biodiversity and genetic diversity patterns. Proc. R. Soc. Lond. Ser. B 262, 149–155.

Nevo, E., 2009. Ecological genomics of natural plant populations: the Israeli perspective. Methods Mol. Biol. 513, 321–344.

Nuttonson, M.Y., 1955. Wheat-Climatic Relationships and the Use of Phenology in Ascertaining the Thermal and Photothermal Requirements of Wheat. American Institute of Crop Ecology, Washington, DC.

Ogbonnaya, F.C., Ye, G., Trethowan, R., Dreccer, F., Lush, D., Shepperd, J., et al., 2007. Yield of synthetic backcross-derived lines in rainfed environments of Australia. Euphytica 157, 321–336.

Ogbonnaya, F.C., Imtiaz, M., Bariana, H.S., McLean, M., Shankar, M.M., Hollaway, G.J., et al., 2008. Mining synthetic hexaploids for multiple disease resistance to improve bread wheat. Crop Pasture Sci. 59, 421–431.

Ogbonnaya, F.C., Abdalla, O., Mujeeb-Kazi, A., Kazi, A.G., Gosnian, N., Lagudah, E.S., 2013. Synthetic hexaploids: harnessing species of the primary gene pool for wheat improvement. Plant Breed. Rev. 37, 35–122.

Ohtsuka, I., 1998. Origin of the central European spelt wheat. In: Slinkard, A.E. (Ed.), Proceedings of the Ninth International Wheat Genetics Symposium, University of Saskatchewan, University Extension Press, Canada, pp. 303–305.

Ortiz-Monasterio, J.I., Palacios-Rojas, N., Meng, E., Pixley, K., Trethowan, R., Peña, R.J., 2007. Enhancing the mineral and vitamin content of wheat and maize through plant breeding. J. Cereal Sci. 46, 293–307.

Özkan, H., Brandolini, A., Schäfer-Pregl, R., Salamini, F., 2002. AFLP analysis of a collection of tetraploid wheats indicates the origin of emmer and hard wheat domestication in southeast Turkey. Mol. Biol. Evol. 19, 1797–1801.

Pellegrineschi, A., Reynolds, M., Pacheco, M., Brito, R.M., Almeraya, R., Yamaguchi-shinozaki, K., et al., 2004. Stress-induced expression in wheat of the *Arabidopsis thaliana* DREB1A gene delays water stress symptoms under greenhouse conditions. Genome 47, 493–500.

Peña, R.J., Zarco-Hernandez, J., Mujeeb-Kazi, A., 1995. Glutenin subunit compositions and bread-making quality characteristics of synthetic hexaploid wheats derived from *Triticum turgidum* × *Triticum tauschii* (Coss.) Schmal crosses. J. Cereal Sci. 21, 15–23.

Peña, R.J., Trethowan, R., Pfeiffer, W.H., Ginkel, M.V., 2002. Quality (end-use) improvement in wheat: compositional, genetic, and environmental factors. J. Crop. Prod. 5, 1–37.

Peng, J.R., Richards, D.E., Hartley, N.M., Murphy, G.P., Devos, K.M., Flintham, J.E., et al., 1999. "Green revolution" genes encode mutant gibberellin response modulators. Nature 400, 256–261.

Peng, J.H., Sun, D., Nevo, E., 2011. Domestication evolution, genetics and genomics in wheat. Mol. Breed. 28, 281–301.

Percival, J., 1921. The Wheat Plant. A Monograph. E.P. Dutton & Company, New York.

Pérez-de-Castro, A.M., Vilanova, S., Cañizares, J., Pascual, L., Blanca, J.M., Díez, M.J., et al., 2012. Application of genomic tools in plant breeding. Curr. Genomics 13, 179.

Pessoa-Filho, M., Rangel, P.H., Ferreira, M.E., 2010. Extracting samples of high diversity from thematic collections of large gene banks using a genetic-distance based approach. BMC Plant Biol. 10, 127.

Petersen, G., Seberg, O., Yde, M., Berthelsen, K., 2006. Phylogenetic relationships of *Triticum* and *Aegilops* and evidence for the origin of the A, B, and D genomes of common wheat (*Triticum aestivum*). Mol. Phylogenet. Evol. 39, 70–82.

Placido, D.F., Campbell, M.T., Folsom, J.J., Cui, X., Kruger, G.R., Baenziger, P.S., et al., 2013. Introgression of novel traits from a wild wheat relative improves drought adaptation in wheat. Plant Physiol. 161, 1806–1819.

Poland, J., 2015. Breeding-assisted genomics. Curr. Opin. Plant Biol. 24, 119–124.

Poland, J., Endelman, J., Dawson, J., Rutkoski, J., Wu, S., Manes, Y., et al., 2012. Genomic selection in wheat breeding using genotyping-by-sequencing. Plant Genome 5, 103–113.

Polignano, G.B., Uggenti, P., Scippa, G., 2001. Diversity analysis and core collection formation in Bari faba bean germplasm. Plant Genet. Resour. Newsl. 125, 33–38.

Pritchard, D.J., Hollington, P.A., Davies, W.P., Gorham, J.L., Diaz de Leon, F., Mujeeb-Kazi, A., 2002. K+/Na+ discrimination in synthetic hexaploid wheat lines: transfer of the trait for K+/Na+ discrimination from *Aegilops tauschii* into a *Triticum turgidum* background. Cereal Res. Commun. 30, 261–267.

Qi, L., Chen, P., Li, D., Zhou, B., Zhang, S., Sheng, B., et al., 1995. The gene *Pm21* – a new source for resistance to wheat powdery mildew. Acta Agron. Sin. 21, 257–262.

Qi, L., Friebe, B., Zhang, P., Gill, B.S., 2007. Homoeologous recombination, chromosome engineering and crop improvement. Chromosome Res. 15, 3–19.

Rajaram, S., Mann, C.E., Ortiz-Ferrara, G., Mujeeb-Kazi, A., 1983. Adaptation, stability and high yield potential of certain 1B/1R CIMMYT wheats. In: Sakamoto, S. (Ed), Proceedings of the Sixth International Wheat Genetics Symposium. Plant Germplasm Institute, Faculty of Agriculture, Kyoto University, Kyoto.

Rao, V.R., Hodgkin, T., 2002. Genetic diversity and conservation and utilization of plant genetic resources. Plant Cell Tissue Organ Cult. 68, 1–19.

Rasheed, A., Xia, X., Yan, Y., Appels, R., Mahmood, T., He, Z., 2014. Wheat seed storage proteins: advances in molecular genetics, diversity and breeding applications. J. Cereal Sci. 60, 11–24.

Ravel, C., Praud, S., Murigneux, A., Linossier, L., Dardevet, M., Balfourier, F., et al., 2006. Identification of *Glu-B1-1* as a candidate gene for the quantity of high-molecular-weight glutenin in bread wheat (*Triticum aestivum* L.) by means of an association study. Theor. Appl. Genet. 112, 738–743.

Reif, J.C., Zhang, P., Dreisigacker, S., Warburton, M.L., van Ginkel, M., Hoisington, D., et al., 2005. Wheat genetic diversity trends during domestication and breeding. Theor. Appl. Genet. 110, 859–864.

Reynolds, M.P., Dreccer, F., Trethowan, R., 2007. Drought adaptive traits derived from wheat wild relatives and landraces. J. Exp. Bot. 58, 177–186.

Riley, R., Chapman, V., 1958. Genetic control of the cytologically diploid behaviour of hexaploid wheat. Nature 182, 713–715.

Riley, R., Unrau, J., Chapman, V., 1958. Evidence on the origin of the B genome of wheat. Heredity 49, 91–98.

Riley, R., Chapman, V., Johnson, R., 1968. The incorporation of alien disease resistance in wheat by genetic interference with the regulation of meiotic chromosome synapsis. Genet. Res. 12, 199–219.

Roelfs, A.P., 1988. Resistance to leaf and stem rust in wheat. In: Simmonds, N.W., Rajaram, S. (Eds.), Breeding Strategies for Resistance to the Rusts of Wheat. CIMMYT, Mexico, DF.

Sakamura, T., 1918. Kurze Mitteilung über die Chromoso-menzahlen und die Verwandtschaftsverhältnisse der Triticum-Arten. Bot. Mag. (Tokyo) 32, 151–154.

Salvi, S., Porfiri, O., Ceccarelli, S., 2013. Nazareno Strampelli, the 'Prophet' of the Green Revolution. J. Agric. Sci. (Cambridge) 151, 1–5.

Sanchez-Garcia, M., Álvaro, F., Peremarti, A., Martín-Sánchez, J.A., Royo, C., 2015. Changes in bread-making quality attributes of bread wheat varieties cultivated in Spain during the 20th century. Eur. J. Agron. 63, 79–88.

Sardesai, N., Nemacheck, J.A., Subramanyam, S., Williams, C.E., 2005. Identification and mapping of H32, a new wheat gene conferring resistance to Hessian fly. Theor. Appl. Genet. 111, 1167–1173.

Sarkar, P., Stebbins, G.L., 1956. Morphological evidence concerning the origin of the B genome in wheat. Am. J. Bot. 43, 297–304.

Sax, K., 1922. Sterility in wheat hybrids. II. Chromosome behavior in partially sterile hybrids. Genetics 7, 513.

Schachtman, D.P., Lagudah, E.S., Munns, R., 1992. The expression of salt tolerance from *Triticum tauschii* in hexaploid wheat. Theor. Appl. Genet. 84, 714–719.

Schultz, A., 1913. Die Geschichte der kultivierten Getreide. Nebert, Halle.

Sears, E.R., 1976. Genetic control of chromosome pairing in wheat. Annu. Rev. Genet. 10, 31–51.

Sears, E.R., 1977. Induced mutant with homoeologous pairing in common wheat. Can. J. Genet. Cytol. 19, 585–593.

Sears, E.R., 1984. Mutations in wheat that raise the level of meiotic chromosome pairing. In: Gustafson, J.P. (Ed.), Gene Manipulation in Plant Improvement. Springer, US, pp. 295–300.

Sears, E.R., Briggle, L.W., 1969. Mapping the gene *Pml* for resistance to *Erysiphe graminis* f. sp. *tritici* on chromosome 7A of wheat. Crop Sci. 9, 96–97.

Shah, S.H., Gorham, J., Forster, B.P., Wyn Jones, G.R., 1987. Salt tolerance in Triticeae – the contribution of the D genome to cation selectivity in hexaploid wheat. J. Exp. Bot. 38, 254–269.

Sharma, H.C., Ohm, H.W., 1990. Crossability and embryo rescue enhancement in wide crosses between wheat and three *Agropyron* species. Euphytica 49, 209–214.

Sharma, H., Ohm, H., Goulart, L., Lister, R., Appels, R., Benlhabib, O., 1995. Introgression and characterization of barley yellow dwarf virus resistance from *Thinopyrum intermedium* into wheat. Genome 38, 406–413.

Shi, A.N., Leath, S., Murphy, J.P., 1998. A major gene for powdery mildew resistance transferred to common wheat from wild einkorn wheat. Phytopathology 88, 144–147.

Shiferaw, B., Smale, M., Braun, H.J., Duveiller, E., Reynolds, M., Muricho, G., 2013. Crops that feed the world 10. Past successes and future challenges to the role played by wheat in global food security. Food Secur. 5, 291–317.

Singh, R.P., Huerta-Espino, J., Sharma, R., Joshi, A.K., Trethowan, R., 2007. High yielding spring bread wheat germplasm for global irrigated and rainfed production systems. Euphytica 157, 351–363.

Skovmand, B., Mackay, M.C., Sanchez, H., van Niekerk, H., He, Z., Flores, M., et al., 2000. GRIP II: genetic resources package for Triticum and related species. In: Skovmand, B., Mackay, M.C., Lopez, C., McNab, A. (Eds.), Tools for the New Millennium. CIMMYT, Mexico, DF.

Smale, M., 1996. Understanding Global Trends in the Use of Wheat Diversity and International Flows of Wheat Genetic Resources. Economics Working Paper 96-02. CIMMYT, Mexico, DF.

Sohail, Q., Inoue, T., Tanaka, H., Eltayeb, A.E., Matsuoka, Y., Tsujimoto, H., 2011. Applicability of *Aegilops tauschii* drought tolerance traits to breeding of hexaploid wheat. Breed. Sci. 61, 347–357.

Solh, M., Nazari, K., Tadesse, W., Wellings, C.R., 2012. The growing threat of stripe rust worldwide. Borlaug Global Rust Initiative (BGRI) Conference, Beijing.

Soliman, A.S., Heyne, E.G., Johnston, C.O., 1963. Resistance to leaf rust in wheat derived from Chinese *Aegilops umbellulata* translocation lines. Crop Sci. 3, 254–256.

Stakman, E.C., Harrar, J.G., 1957. Principles of Plant Pathology. Ronald Press, New York.

Stoskopf, N.C., 1985. Cereal Grain Crops. Reston Publishing Company, Inc, Reston, Virginia.

Tadesse, W., Schmolke, M., Mohler, V., Wenzel, G., Hsam, S.L.K., Zeller, F.J., 2007. Molecular mapping of resistance genes to tan spot (*Pyrenophora tritici-repentis* race 1) in synthetic wheat lines. Theor. Appl. Genet. 114, 855–862.

Tadesse, W., Ogbonnaya, F.C., Jighly, A., Nazari, K., Rajaram, S., Baum, M., 2014. Association mapping of resistance to yellow rust in winter wheat cultivars and elite genotypes. Crop Sci. 54, 607–616.

Talbert, L.E., Bruckner, P.L., Smith, L.Y., Sears, R., Martin, T.J., 1996. Development of PCR markers linked to resistance to wheat streak mosaic virus in wheat. Theor. Appl. Genet. 93, 463–467.

Tang, Y.I., Yang, W.Y., Tian, J.C., Li, J., Chen, F., 2008. Effect of HMW-GS 6 + 8 and 1.5 + 10 from synthetic hexaploid wheat on wheat quality traits. Agric. Sci. China 7, 1161–1171.

Tanksley, S.D., McCouch, S.R., 1997. Seed banks and molecular maps: unlocking genetic potential from the wild. Science 277, 1063–1066.

Tesfaye, M., Temple, S.J., Allan, D.L., Vance, C.P., Samac, D.A., 2001. Overexpression of malate dehydrogenase in transgenic alfalfa enhances organic acid synthesis and confers tolerance to aluminum. Plant Physiol. 127, 1836–1844.

The, T.T., 1973. Transference of Resistance to Stem Rust from *Triticum monococcum* L. to Hexaploid Wheat. PhD Thesis, The University of Sydney.

Trethowan, R.M., van Ginkel, M., 2009. Synthetic wheat – an emerging genetic resource. In: Carver, B.F. (Ed.), Wheat: Science and Trade. Wiley-Blackwell, Ames.

Uauy, C., Distelfeld, A., Fahima, T., Blechl, A., Dubcovsky, J., 2006. A NAC gene regulating senescence improves grain protein, zinc, and iron content in wheat. Science 314, 1298–1301.

Upadhya, M.D., Swaminathan, M.S., 1963. Genome analysis in *Triticum zhukovskyi*, a new hexaploid wheat. Chromosoma 14, 589–600.

Valkoun, J., Waines, J.G., Konopka, J., 1998. Current geographical distribution and habitat of wild wheats and barley. In: Damania, A.B., Valkoun, J., Willcox, G., Qualset, C.O. (Eds.), The Origins of Agriculture and Crop Domestication. ICARDA, Aleppo, Syria, pp. 293–299.

Varshney, R.K., Graner, A., Sorrells, M.E., 2005. Genic microsatellite markers in plants: features and applications. Trends Biotechnol. 23, 48–55.

Vasil, V., Castillo, A.M., Fromm, M.E., Vasil, I.K., 1992. Herbicide resistant fertile transgenic wheat plants obtained by microprojectile bombardment of regenerable embryogenic callus. Nat. Biotechnol. 10, 667–674.

Vega, J.M., Feldman, M., 1998. Effect of the pairing gene *Ph1* on centromere misdivision in common wheat. Genetics 148, 1285–1294.

Villareal, R.L., Singh, R.P., Mujeeb-Kazi, A., 1992. Expression of resistance to *Puccinia recondita* f.sp. *tritici* in synthetic hexaploid wheats. Vortr Pflanzenzuchtg 24, 253–255.

Wang, G.Z., Miyashita, N.T., Tsunewaki, K., 1997. Plasmon analyses of *Triticum* (wheat) and *Aegilops*: PCR–single-strand conformational polymorphism (PCR-SSCP) analyses of organellar DNAs. Proc. Natl. Acad. Sci. 94, 14570–14577.

Wang, J., Luo, M.C., Chen, Z., You, F.M., Wei, Y., Zheng, Y., et al., 2013. *Aegilops tauschii* single nucleotide polymorphisms shed light on the origins of wheat D-genome genetic diversity and pinpoint the geographic origin of hexaploid wheat. New Phytol. 198 (3), 925–937.

Warburton, M.L., Crossa, J., Franco, J., Kazi, M., Trethowan, R., Rajaram, S., et al., 2006. Bringing wild relatives back into the family: recovering genetic diversity in CIMMYT improved wheat germplasm. Euphytica 149, 289–301.

Weeks, J.T., Anderson, O.D., Blechl, A.E., 1993. Rapid production of multiple independent lines of fertile transgenic wheat. Plant Physiol. 102, 1077–1084.

Worede, M., Mekbib H., 1992. Local/indigenous knowledge and agricultural research in Ethiopia, In: Boef, W. (Ed.), Local Knowledge and Agricultural Research: Report on WAU-ENDA-Zimbabwe-CGN-Grain Seminar. LUW, Brondesbury Park, London.

Worland, A.J., Snape, J.W., 2001. Genetic basis of worldwide wheat varietal improvement. In: Bonjean, A.P., Angus, W.J. (Eds.), The World Wheat Book: A History of Wheat Breeding. Lavoisier Publishing, Paris, France.

Xu, Y., 2010. Plant genetic resources: management, evaluation and enhancement. In: Xu, Y. (Ed.), Molecular Plant Breeding. CAB International, Wallingford, UK, pp. 151–194.

Xue, G.-P., Way, H.M., Richardson, T., Drenth, J., Joyce, P.A., McIntyre, C.L., 2011. Overexpression of TaNAC69 leads to enhanced transcript levels of stress up-regulated genes and dehydration tolerance in bread wheat. Mol. Plant 4, 697–712.

Yan, Y., Hsam, S.L.K., Yu, J.Z., Jiang, Y., Ohtsuka, I., Zeller, F.J., 2003. HMW and LMW glutenin alleles among putative tetraploid and hexaploid European spelt wheat (*Triticum spelta* L.) progenitors. Theor. Appl. Genet. 107, 1321–1330.

Yang, N., Smale, M., 1996. Indicators of Wheat Genetic Diversity and Germplasm Use in the People's Republic of China. CIMMYT, Mexico, DF, NRG Paper 96-04.

Yang, J., Sears, R.G., Gill, B.S., Paulsen, G.M., 2002. Quantitative and molecular characterization of heat tolerance in hexaploid wheat. Euphytica 126, 275–282.

Zegeye, H., Rasheed, A., Makdis, F., Badebo, A., Ogbonnaya, F.C., 2014. Genome-wide association mapping for seedling and adult plant resistance to stripe rust in synthetic hexaploid wheat. PLoS ONE 9, e105593.

Zeller, F.J., Hsam, S.L., 1983. Broadening the genetic variability of cultivated wheat by utilizing rye chromatin. In: Sakamoto, S. (Ed), Proceedings of the Sixth International Wheat Genetics Symposium, Plant Germ-Plasm Institute, Faculty of Agriculture, Kyoto University, Kyoto.

Zeller, F.J., Sears, E.R., Sears, L.M.S., 1973. 1B/1R wheat-rye chromosome substitutions and translocations. In: Sears, E.R., Sears, L.M.S. (Eds.), Proceedings of the Fourth International Wheat Genetics Symposium, Agricultural Experiment Station, College of Agriculture, University of Missouri, USA, pp. 209–221.

Zhang, P., Friebe, B., Gill, B., Park, R.F., 2007. Cytogenetics in the age of molecular genetics. Crop Pasture Sci. 58, 498–506.

Zhou, H., Berg, J.D., Biest, N.A., Blank, S.E., Buehler, R.E., Chay, C.A., et al., 2003. Development of roundup ready wheat through transgenic approach. Crop Sci. 43, 1072–1075.

Zohary, D., 1999. Monophyletic vs. polyphyletic origin of the crops on which agriculture was founded in the Near East. Genet. Resour. Crop Evol. 46, 133–142.

Zohary, D., Hopf, M., 1993. Domestication of Plants in the Old World, second ed. Clarendon Press, Oxford.

Barley

3

Lakshmi Kant, Shephalika Amrapali*[†]*, Banisetti Kalyana Babu***

*Crop Improvement Division, Indian Council of Agricultural Research, Vivekananda Pravatiya Krishi Anusandhan Sansthan (Vivekanada Institute for Hill Agriculture), Almora, Uttarakhand, India; [†]Indian Council of Agricultural Research, Directorate of Floriculture Research, College of Agriculture Campus, Shivaji Nagar, Pune, India; **Indian Council of Agricultural Research, Indian Institute of Oil Palm Research (IIOPR), Pedavegi, Andhra Pradesh, India

3.1 Introduction

Barley (*Hordeum vulgare* L.), a member of the grass family, is one of the eight founder crops (einkorn wheat, emmer wheat, barley, lentil, pea, chick pea, bitter vetch, and flax). Barley is one of the first cultivated grains. It is considered as one of the most important cereals worldwide owing to its multiple uses as human food, animal feed, and substrate for malting. Barley is predominantly considered as a food crop in many parts of the globe like the semiarid regions of Africa (Morocco, Algeria, Libya, and Tunisia), Middle East (Saudi Arabia, Iran, Iraq, and Syria), highlands of Nepal, Ethiopia, and Tibet, Andean countries of South America (Peru and Chile), and in some Asian countries (China and North Korea) and the Himalayas. The husk of barley protects the coleoptile (acrospire) during germination process, aids in filtration, provides firm texture of grain, and its amylase activity makes it the most preferred cereal for malt recovery. Malt can be utilized for brewing, distillation (preparation of beer and other hard liquor such as whisky and brandy), baby foods, confectionaries, cocoa-malt drinks, and medicinal syrups. It is also considered as a functional food and used in many bakery products and recipes owing to its higher dietary fiber and lower low-density lipoprotein (LDL) content, and is rich in tocols including tocopherols and tocotrienols (known to reduce serum LDL cholesterol through their antioxidant action). A decoction of barley in water called "barley water" is valued medicinally and used for the treatment of inflammation. In India, its utilization as a food crop (mainly hull-less type) is restricted to tribal areas of hills and plains. The barley products like "*Sattu*" (in summers because of its cooling effects on the human body) and missi roti (for its better nutritional quality) have been traditionally used in India (Verma et al., 2012). Besides these important uses, barley is considered as a model experimental crop because of its short life span and morphologic, physiologic, and genetic characteristics. The archeologic evidences indicate that barley was once more important than wheat in agriculture-based civilizations and was mainly grown for human consumption. But now it stands fourth in world cereal production after wheat, rice, and maize, and is mainly used as animal feed (around 85% of global production) (Schulte et al., 2009).

Genetic and Genomic Resources for Grain Cereals Improvement. http://dx.doi.org/10.1016/B978-0-12-802000-5.00003-4

3.2 Origin

There are different views about the origin of barley and two schools of its origin, namely, monophyletic and diphyletic, exist among scientists. The proponents of monophyletic origin of barley believe that barley was first domesticated in the area of the Middle East known as the "Fertile Crescent" about 10,000 years ago from its two-rowed wild progenitor *H. vulgare* L. ssp. *spontaneum* (Zohary and Hopf, 1993; Harlan and Zohary, 1966). This was supported by the fact that the wild progenitor of barley is still colonizing its primary habitats in the Fertile Crescent (Harlan and Zohary 1966; Nevo, 1992) and in some areas also occupies an array of secondary habitats, such as open Mediterranean marquis, abandoned fields, and roadsides. However, similar marginal habitats have been colonized in the Aegean region, South Eastern Iran, and central Asia, including Afghanistan, and the Himalayan region (Zohary and Hopf, 1993). Indeed, the Himalayas, Ethiopia, and Morocco have occasionally been considered centers of barley domestication (Aberg, 1938; Bekele, 1983; Molina-Cano et al., 1987). The proponents of diphyletic origin of barley (Freisleben, 1940; Takahashi, 1955) suggest that the six-rowed cultivated barley in the Oriental region is derived from ssp. *agriocrithon*, and the two-rowed cultivated barley in Southwest Asia originated from ssp. *spontaneum* (wild barley). Badr et al. (2000) demonstrated that barley was first brought into culture in the "Fertile Crescent," and that the Himalayas are a diversification region of domesticated barley. However, in a recent study, it was proposed that the domestication of the cultivated form started in the Indus Valley, at the site of Mehrgarh (Pakistan) during 7000 BC (Morrell and Clegg, 2007).

The theory of single domestication has now been abandoned in the light of current research on domestication and domestication syndromes using molecular analysis of closely linked marker of the loci controlling domestication traits. Now it is believed that barley was domesticated more than once, at least twice, over the period of time at two different locations. However, it is still believed that barley originated from a two-rowed wild progenitor, that is, ssp. *spontaneum* and the ssp. *agriocrithon* indeed is a result of secondary mutation or hybridization. It is supported by the fact that a brittle rachis type, six-rowed barley can be created within ssp. *vulgare* by hybridization or mutation (Lundqvist et al., 1997).

According to the most commonly followed system, barley (*Hordeum*) belongs to the tribe Triticeae, family Poaceae (Gramineae), and the tribe Triticeae comprises of around 350 species out of which *Hordeum* has about 31 species including wild and cultivated ones, all with a basic chromosome number of $x = 7$. These have been divided in two units, that is, *Hordeum sensu stricto,* which includes only two species *H. vulgare* and *Hordeum bulbosum,* and *critesion,* which comprises the remaining species. In *H. vulgare* the wild forms were put under subspecies *spontaneum* and the cultivated forms are *H. vulgare* ssp. *vulgare*. Cultivated barley, *H. vulgare* L., and its wild progenitor *H. vulgare* ssp. *spontaneum* (C. Koch.) are diploid species with $2n = 2x = 14$ chromosomes. The cultivated ssp. *vulgare* is further subdivided into var. *distichon* (two rows) and var. *hexastichon* (six rows) barley (Verma et al., 2012).

3.3 Domestication syndrome

Barley is one of the first known domesticated crop as a result of high selection pressure either conscious or unconscious, that is, as a result of human selection or environmental selection. Variants for different traits developed because of mutation and natural hybridization followed by selection pressure created by environmental conditions in different geographical regions and human management practices including planting and harvesting techniques resulted in development of present day cultivated barley. This further created possibilities for larger number of combinations. The gradual accumulation of favorable traits resulted in domestication of barley and the traits which resulted in domestication are called "domestication syndrome." The three key traits, which were involved in domestication of barley are – nonbrittle rachis, six-rowed spike, and naked caryopsis (Salamini et al., 2002). There are other traits such as reduced seed dormancy, reduced vernalization requirement, and photoperiod insensitivity associated with the transition of wild to cultivated barley and also responsible for the spread of cultivated barley to places outside its actual place of origin.

3.3.1 Nonbrittle rachis

Hordeum spike is characterized by the presence of typical wedge-shaped spikelets (von Bothmer et al., 1995). Wild and cultivated barley differ in disarticulation scars on their rachis. In wild barley, disarticulation scars are smooth, whereas, in cultivated barley it forms rough dehiscence scars. Smooth scars help in seed dispersal by shattering of seed at maturity. Seed shattering in barley has two forms – brittle/weak rachis – and is a characteristic of its wild relatives. These types of rachis make harvesting of mature spikes difficult and result in grain loss. Change of brittle/weak rachis to nonbrittle rachis resulted in efficient harvest without any grain loss. This trait gradually accumulated in the population as they stayed longer on the plant even after maturation and were harvested more frequently. Most probably this is the most important trait for domestication.

Development of nonbrittle rachis is a result of mutation, which resulted in loss of gene function. This trait is controlled by recessive gene *btr1/btr2* located on chromosome 3HS. Phylogenetic studies using closely linked marker to *btr1/btr2* genes determined that cultivated barley consists of two geographical types: western and eastern types (Komatsuda and Tanno, 2004; Azhaguvel and Komatsuda, 2007). This supports the two independent domestication of barley as proposed by Takahashi (1955).

3.3.2 Six-rowed spike

Generally, human selection is directed toward increased yield. The appearance of the six-rowed variant is one of the most evident selection in this direction. Six-rowed barley produces three times as many seeds per spikes as two-rowed barley and is a change of great agronomic importance.

The proponents of diphyletic origin of barley suggest that six-rowed cultivated barley was derived from six-rowed wild barley known as *Hordeum agricrithon* found in Western China. However, proponents of the single evolutionary line suggest that *H. vulgare* is originally a two-rowed spontaneum type with a brittle rachis. The wild six-rowed form at times designated as *H. agricrithon* and collected from Tibet was considered as the wild progenitor of the present-day six-rowed barley by the proponents of diphyletic origin. However, with advances in research it was found that it is possible to create six-rowed brittle rachis type within the ssp. *vulgare* (Bothmer et al., 1991) and the appearance of *H. agricrithon* may be a result of secondary mutation or hybridization. The presence of sporadic six-rowed element among the two-rowed materials in the archeologic evidences dated at about 9000 BC from Ali Kosh (Helbaek, 1959) also suggests that the six-rowed character in barley was derived from two-rowed barley during domestication and is in favor of barley scientists.

The development of the six-rowed type is a result of mutation followed by loss of gene function. At least five independent loci are involved in this trait. However, *vrs1,* a recessive gene located on chromosome 2HL, is uniformly present in all six-rowed type. Lundqvist et al. (1997) developed a six-rowed type by inducing mutation at this loci in two-rowed type. This supported the single evolutionary line of barley. But the study of closely linked marker to *vrs1* loci indicated that barley was domesticated more than once in history (Tanno et al., 2002). Another dominant allele, which is responsible for the development of six-rowed barley, was *int-c-h,* which promotes development of anther and seed set in lateral spikelet.

3.3.3 Naked caryopsis

Generally, barley has covered caryopsis in which the hull is glued to the pericarp epidermis at maturity and is difficult to thresh. The free threshing or easy threshing type of barley, popularly called naked barley, is also available and preferred for human consumption. Naked barley has worldwide distribution; however, it is most widely distributed in East Asian countries, namely, China, Korea, and Japan. It is very common in Tibet and the northern part of Nepal, India, and Pakistan. Because of this, Vavilov (1926) considered southern Asia as the center of origin of naked barley. However, naked barley was also reported to be grown in Turkey and in parts of Northern Europe.

This trait is controlled by a single recessive gene "*nud*" located on chromosome 7HL. The appearance of naked barley is a result of a single gene mutation, which resulted in damaged gene function. The molecular analysis of linked marker to "*nud*" gene supports a single evolutionary line of barley (Choo et al., 2001).

3.3.4 Reduced dormancy

Dormancy is defined as the temporary inability of a viable seed to germinate under favorable environmental conditions (Simpson, 1990). Dormancy helps the seed to survive adverse environmental conditions and at the same time helps in seed dispersal (Snape et al., 2001). High level of dormancy after harvest is undesirable when rapid seed germination is required on planting. Also, as in the case of malt house rapid, uniform germination is required upon imbibition of water. Therefore, a high level of dormancy is economically undesirable. At the same time, selection against dormancy will result in preharvest sprouting, which is again not a desirable trait. Therefore, a balance has to be maintained between both.

Dormancy in barley is a quantitative trait controlled by many quantitative trait loci (QTL)s. The major QTLs are SD1 and SD2 located on chromosome 5H. SD1 is located near centromere and SD2 is located on the long arm near the distal end. SD2 controls moderate dormancy and is desirable for breeding purpose (Gao et al., 2003).

3.3.5 Reduced vernalization

Vernalization is the requirement for a period of low temperature for transition of a plant from vegetative to reproductive phase. Vernalization requirement is characteristic of winter growth habit present in almost all wild relatives except a few strains. Development of spring growth habit and reduced requirement of vernalization is one of the key developments that are responsible for the spread of barley to places far from its actual place of origin.

Three genes, *sgh1, Sgh2,* and *Sgh3,* located on 4H, 5H, and 7H chromosomes, respectively, are present in spring barley and its allelic forms, Sgh1, sgh2, and sgh3, are required for winter growth habit. The only one genotype containing Sgh1, sgh2, sgh3 allele is responsible for winter growth. Dominant mutation at *sgh2* resulted in spring type (Takahashi et al., 1963, 1968). The gradation in vernalization requirement observed in barley is due to multiple alleles of Sgh2 locus.

3.3.6 Photoperiod insensitivity

Photoperiod in association with vernalization requirement is important to regulate flowering time in barley. In nature, plants evolved to ensure that the flowering occurs when there is greater chance of pollination, seed development, and seed dispersal. However, to ensure the spread of barley throughout the world, it is important to modify flowering time by human selection. Wild barley is mainly a quantitative long-day (LD) species and heading time increases with increase in day length (Boyd et al., 2003). This trait in wild barley is controlled by many genes including the photoperiod response gene. Spring barley accumulated LD insensitive mutants and resulted in the expansion of barley to other areas (Laurie et al., 1995).

3.4 Distribution

Barley has a wide distribution mostly in the temperate areas of the world and occupies a wide range of ecologic niche. Presently, barley is cultivated from sea level to an altitude of up to 4000–5000 m in the Himalayas and in the Andes. The spread of barley from its original place of evolution to other parts of the world is most probably by long-distance migration by birds or wind. The migration followed by speciation gave rise to many native species in different parts of the world.

The main barley-growing countries in the world are Russia, Canada, and Germany. In Tibet, Nepal, Ethiopia, and the Andes, farmers cultivate barley on the mountain slopes at elevations higher than other cereals. In the dry regions of North Africa, Middle East, Afghanistan, Pakistan, Eritrea, and Yemen, with little irrigation, barley is often the only suitable cereal. Developing countries account for about 18% of global production and 25% of the harvested area of barley.

In India, barley is cultivated in 0.65 million ha on the plains (Uttar Pradesh, Punjab, Rajasthan, Haryana, Madhya Pradesh, and Bihar) and on the hills up to an elevation of around 4000 m (Himachal Pradesh, Uttarakhand, and Jammu and Kashmir). In the Himalayas, the six-rowed covered types are common. Two-rowed types, both covered and naked, are grown only to a limited extent. Cultivation of the six-rowed naked forms is confined to the higher Himalayan ranges. In these areas, barley is generally a crop of marginal, low-input, drought-stressed environments. The landraces grown in these areas are favored by farmers for their quality, both as grain and straw. However, the area under the naked barley landraces, which were very popular and widely grown in higher Himalayan ranges three to four decades ago, has declined considerably.

3.4.1 Taxonomy

The genus *Hordeum* is an ancient monophyletic group, which belongs to one of the most economically important plant group, that is, tribe Triticeae in the grass family Poaceae splitting from *Triticum* some 13 million years ago. The genus is characterized by the presence of a triplet – a three-flowered spikelet at each rachis node. The lateral spikelets are stalked in wild species, whereas these are sessile in the cultivated and wild form of cultivated barley.

Efforts have been made for the taxonomic delimitation of *Hordeum*. The pioneering work in this area was done by a Russian taxonomist, Nevski, in 1941 (Nevski, 1941). The system of classification of barley is artificial and just a means of classifying morphologically similar forms. However, the availability of cytologic, genomic, and molecular data related to domestication syndromes have shed light in this direction and given a different and more complex story.

The C-banding pattern and analysis based on fluorescent *in situ* hybridization (FISH) and genomic *in situ* hybridization (GISH) pattern are partly species-specific and can be used for the determination of phylogeny and taxonomic classes. The analysis of meiotic pairing, intraspecific hybrids, molecular studies of nucleus and plastid suggested the presence of four basic haplomes. However, the studies on genome relationship, molecular data, and morphologic data are not consistent with each other and no system of classification can be developed using all this information. Love (1982, 1984), on the basis of genome relationship, proposed a taxonomy of *Hordeum* and was adopted by Dewey (1984).

At present, the genus *Hordeum* altogether contains 31 species and many intraspecific units, which are basically diploid, tetraploid, and hexaploid with a basic chromosome number of 7. These possess annual and perennial life forms (Table 3.1). Majority are perennials whereas the annual form includes summer and winter annuals. The reproduction behavior varies from outbreeding to inbreeding. Inbreeding predominates annual forms with few exceptions. The two species *Hordeum murinum* and *Hordeum intercedens* are more or less obligate inbreeders with cleistogamous flowers. The

Table 3.1 Species concept of the genus *Hordeum*

S. No.	Species	Subspecies		$2n$ ($2x/4x/6x$)	Genome	Life form	Distribution
I	Section. Vulgare						
1	*H. vulgare* L.	2	*vulgare* *spontaneum*	14	H	Annual	Worldwide as cultivated form, wild form is distributed from Greece to Afghanistan
2	*H. bulbosum* L.			14,28	H	Perennial	Mediterranean, east to Afghanistan
3	*H. murinum* L.	3	*murinum* *glaucum* *leporinum*	14,28,42	Xu	Annual	Worldwide as weed: Europe, Mediterranean to Afghanistan
II	Section. Anisolepis						
4	*Hordeum pusillum* Nuttall			14	I	Annual	USA, North Mexico, and South Canada
5	*H. intercedens* Nevski			14	I	Annual	South west California and North Mexico
6	*Hordeum euclaston* Steudel			14	I	Annual	Central Argentina, Uruguay, and South Brazil
7	*Hordeum flexuosum* Steudel			14	I	Annual/ Perennial	Argentina and Uruguay
8	*Hordeum muticum* Presl			14	I	Perennial	Western South America
9	*H. chilense* Roemer and Schultes			14	I	Perennial	Central Chile and West Argentina
10	*Hordeum cordobense* Bothmer, Jacobsen and Nicora			14	I	Perennial	Argentina
11	*Hordeum stenostachys* Godron			14	I	Perennial	Argentina, Uruguay, and Southern Brazil

(Continued)

Table 3.1 Species concept of the genus *Hordeum* (cont.)

S. No.	Species	Subspecies		2*n* (2*x*/4*x*/6*x*)	Genome	Life form	Distribution
III	Section. Critesion						
12	*Hordeum pubiflorum* Hooker	2	*pubiflorum* *breviaristatum*	14	I	Perennial	West Argentina, Chile, Bolivia, and Peru
13	*Hordeum comosum* Presl			14	I	Perennial	West Argentina and Chile
14	*Hordeum jubatum* L.			28	I	Perennial	Western North America to East Russia
15	*Hordeum arizonicum* Covas			42	I	Annual/ Perennial	South USA to North Mexico
16	*Hordeum procerum* Nevski			42	I	Perennial	Central Argentina
17	*Hordeum lechleri* (Steudel) Schenck			42	I	Perennial	Chile and Argentina
IV	Section Stenostachys						
18	*H. marinum* Hudson	2	*marinum* *gussneanum*	14,28	Xa	Annual	Mediterranean region to Afghanistan
19	*Hordeum secalinum* Schreber			28	I	Perennial	West Europe and North Africa
20	*Hordeum capense* Thunberg			28	I	Perennial	South Africa
21	*Hordeum bogdanii* Wilensky			14	I	Perennial	Central Asia
22	*Hordeum roshevitzii* Bowden			14	I	Perennial	Central Asia
23	*Hordeum brevisubulatum* (Trinius) Link	5	*brevisubulatum* *nevskianum* *turkestanicum* *violaceum* *Iranicum*	14,28,42	I	Perennial	West turkey to North East China
24	*Hordeum brachyantherum* Nevski	2	*brachyantherum* *californicum*	14,28,42	I	Perennial	Western North America To Kamchatka

25	*Hordeum depressum* (Scribner and Smith) Rydberg			28	I	Annual	West USA
26	*Hordeum guatemalense* Bothmer, Jacobsen and Jorgensen			28	I	Perennial	North Guatemala
27	*Hordeum erectifolium* Bothmer, Jacobsen and Jorgensen			14	I	Perennial	Central Argentina
28	*Hordeum tetraploidum* Covas			28	I	Perennial	South Argentina
29	*Hordeum fuegianum* Bothmer, Jacobsen and Jorgensen			28	I	Perennial	South Argentina and South Chile
30	*Hordeum parodii Covas*			42	I	Perennial	South Argentina and South Chile
31	*Hordeum patagonicum* (Haumann) Covas	5	*patagonicum* *setifolium* *santacrucense* *mustersii* *magellanicum*	14	I	Perennial	South Argentina and South Chile

cultivated barley shows inbreeding with ~1% out-crossing. Two species, *H. bulbosum* and *Hordeum brevisubulatum,* are self-incompatible and obligate outbreeders.

The *Hordeum* species are biologically and morphologically distinct with strict isolation barrier in the form of hybrid sterility with some exceptions. Within species biologic and morphologic pattern of variation is observed, which resulted in their division into subspecies. Transition forms are observed where the area of distribution overlaps indicating the possibility of natural hybridization and gene flow between species. The considerable variation in the morphology of *Hordeum* species is the result of morphologic plasticity and *H. murinum* is a good example of extreme plasticity.

The cultivated species *H. vulgare* is divided into two subspecies, that is, ssp. *vulgare,* the cultivated form, and ssp. *spontaneum,* the wild progenitor. This is the most variable species within the genus with many easily observable variations in characteristics, namely,

1. Tough/brittle rachis
2. Two-/four-/six-rowed spike
3. Normal/deficient lateral spikelets
4. Normal/multiflowered spikelets
5. Normal-/long-awned glumes
6. Long-/short-haired rachilla
7. Short/long hairs on rachis edge
8. Normal/hooded lemma
9. Covered/naked kernel
10. Scabrid/hairy lemma.

3.4.2 Gene pool

Harlan and De Wet (1971) classified the crop genetic diversity in terms of primary, secondary, and tertiary gene pool, which has less taxonomic importance but is useful in terms of utilization of diversity in crop improvement. All the species in the genus *Hordeum* were earlier considered to be related and all together constitute genetic resources for breeding purpose, though sterility barrier was present. However, owing to the advances in research, it was claimed that these are more distantly related and comprised different group of species. These species on the basis of their relation to cultivated barley, ease of crossability, and gene transfer are divided into different gene pools (Fig. 3.1).

3.4.2.1 Primary gene pool

This comprises of closely related species with no or very weak biologic barrier for gene transfer. The primary gene pool of barley consists of landraces, breeding lines, genetic stocks, and so on. The progenitor of cultivated barley, *H. vulgare* ssp. *spontaneum,* also belongs to this gene pool. Several traits of interest, particularly disease resistance genes, are available in this gene pool.

3.4.2.2 Secondary gene pool

The secondary gene pool of barley comprises the wild and weedy forms, that is, *H. bulbosum,* including perennial diploid and tetraploid species that have comparatively good

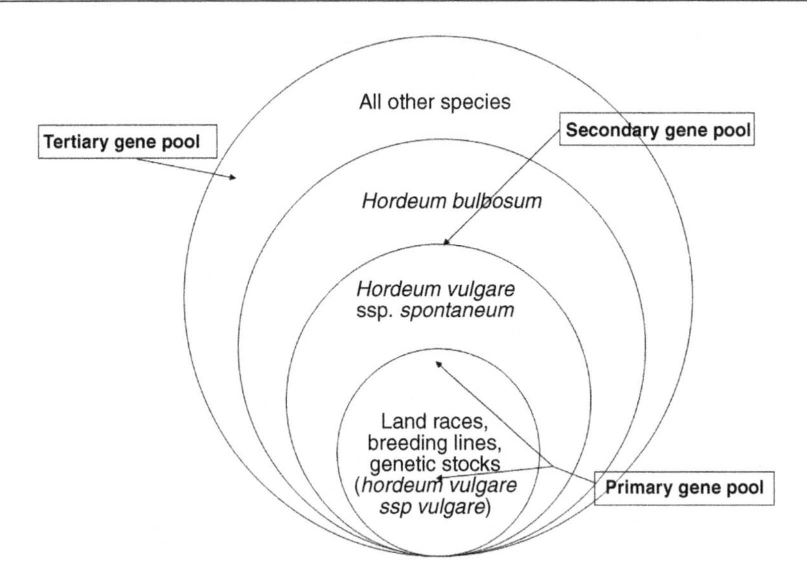

Figure 3.1 Gene pool concepts in barley.

crossability with cultivated species but in which certain sterility factors are operating like selective elimination of *H. bulbosum* chromosomes from the hybrid or sterility despite high chromosomes pairing. The gene transfer is possible with some difficulty. It is the closest relative of cultivated barley apart from ssp. *spontaneum*. Several traits from *H. bulbosum* are of interest like resistance to powdery mildew and β-amylase activity.

3.4.2.3 Tertiary gene pool

All the other species of barley belong to the tertiary gene pool, where embryo rescue is a must for hybrid development. These species possess genome other than cultivated barley; therefore, crossing of cultivated barley is difficult because of substantial genomic differences. Due to this bottleneck, so far the germplasm from this gene pool has not contributed to the progress of barley breeding. However, with the development of new techniques, genetic content of these taxa may become available for transfer in the future.

3.5 Erosion of genetic diversity from the traditional areas

Genetic diversity is the basis of any crop improvement program. But in the hands of modern agriculture the diversity is continuously eroding. For the first time the work by distinguished geneticist H.V. Harlan in 1930s cautioned about the loss of barley diversity in the hands of modern agriculture (Harlan, 1931). One of the major factors resulting in the decrease of crop diversity is genetic erosion. Genetic erosion as defined by Guarino (2003) is "the permanent reduction in richness or evenness of common localized alleles or the loss of combination of alleles over time in a defined area."

The definition suggests that the loss of locally adapted allele both in terms of number as well as frequency is a key event in genetic erosion. Genetic erosion is caused by anthropogenic and/or natural changes, including replacement of local varieties, land clearing, over exploitation, population pressures, environmental degradation, changing agricultural systems, over grazing, pests and diseases, weeds, inappropriate legislation, and policies (FAO, 2010).

The landraces are a highly heterogeneous group of populations and may serve as a major source of diversity. These landraces are continuously being replaced by new varieties. With the realization of eroding genetic diversity and importance of landraces as a diverse source of important traits scientists started evaluating genetic erosion among cereal landraces (Martos et al., 2005; Rocha et al., 2008; Ruiz et al., 2002). A study using 19 allozyme loci by Brown and Munday (1982) on wild barley, landraces, and composite crosses revealed the following ranking in terms of allozyme diversity: wild barley > landraces > composite crosses. The direct comparison of wild barley from Israel and Iranian landrace population of cultivated barley showed that landraces possessed about 73% of genetic diversity as observed in the case of wild barley. In a study, both wild and cultivated barleys of the eastern Mediterranean region appeared to be rich sources of genetic diversity in comparison with the global diversity represented by 998 random USDA accessions. However, there is a huge gap in the knowledge regarding the origin and structure of cereal landraces, which allow only theoretical anticipation of genetic erosion.

During domestication, crops initially suffer reduction in diversity for certain genes due to selection pressure and dispersal bottleneck. However, crops having high naturally occurring gene flow between domesticated and wild relatives observed a gradual increase in the diversity for the genes that were not selected during domestication. Farmers select and use the introgressed types (Jarvis and Hodgkin, 1999). As a consequence, the diversity found in landraces often increased after the initial bottleneck, sometimes to near similar levels as found in the wild species. This has been observed in the case of eastern Mediterranean barley landraces (Jana and Pietrzak, 1988).

At times reduction in regional diversity makes the total gene pool more vulnerable to loss and extinctions and might lead to reduction in evenness and richness globally. However, a decrease in diversity in a specific region does not necessarily result in genetic erosion at a larger geographical scale. In Australian wheat, no change in diversity (using coefficients of parentage) was observed at the national level, although in some states a narrowing of the genetic base was observed (Brennan and Fox, 1998). A similar study on barley diversity in the Nordic and Baltic countries, observed a decrease in the allelic diversity in some of the countries studied, whereas overall diversity level was maintained (Kolodinska et al., 2004).

3.6 Germplasm evaluation and maintenance

In recent years, the use of barley as a source of raw material for the malting and brewing industries has increased. To meet the requirement of different types of products, diverse germplasm for the development of product-specific varieties is needed. The rapid spread of improved crop varieties throughout the world has replaced many of the genetic resources essential for their continued improvement and could pose serious threat to barley production worldwide unless proper care is taken to diversify the parental lines in breeding programs.

Sustainable utilization of genetic diversity provides a strong foundation to a successful breeding program. Therefore, collection and maintenance of genetic diversity has become a priority area in breeding programs across the world. The essential requirement for utilization of diverse genetic resources for future breeding programs is the thorough knowledge of its potential value. Therefore, it is important to give high priority to the evaluation of germplasm for identifying donors for different traits. Knowledge of the basic agromorphologic traits that are strongly associated with each other is equally important to define the plant architecture needed for a specific production condition and also in analyzing germplasm collections of different categories. This helps in the development of a core subset of germplasm in order to keep the maximum diversity within a minimum number of accessions, which further facilitate maintenance as well as utilization of diversity.

In India, the National Bureau of Plant Genetic Resources, New Delhi is maintaining the base collection of barley germplasm, and the Indian Institute of Wheat and Barley Research, Karnal in association with other national active germplasm sites is involved in maintaining the active collection and is also coordinating the breeding program on barley. The active collection comprising indigenous as well as exotic germplasm are being evaluated and trait-specific germplasm identified are used in barley improvement. Indian barley breeding program with the help of a network of national and international institutions has put major emphasis on incorporation of genetic diversity in the last four decades. This has helped in developing and releasing over 115 varieties for different agroclimatic conditions as feed, food, and malt barley varieties under irrigated and rain-fed conditions with diversity in morphologic, agronomic, and biotic and abiotic stresses (Verma et al., 2012).

3.7 Conservation of genetic resources

Conservation of genetic resources is important in fulfilling the present and future needs of any crop improvement program. Genetic resources can be conserved in its native place under field condition, called *in situ* conservation, or outside its original habitat in seed genebanks or field genebanks, called *ex situ* conservation. These two methods of conservation together constitute a complementary system of germplasm conservation.

3.7.1 Ex situ *conservation*

Ex situ conservation mainly in the form of seeds in genebanks is the most significant and widespread means of conserving genetic resources. The genetic resources conserved in the genebanks throughout the world comprise about 7.9 million accessions. Nearly 45% of the world collection accounts for cereal accessions. Barley accessions in the genebanks account for the second-largest collection among cereal accessions after wheat. The Food and Agriculture Organization (FAO, 1996) in its first report on The State of World's Plant Genetic Resources for Food and Agriculture, estimated that about 0.48 million barley accessions exist in *ex situ* germplasm collections such as genebanks, breeders', and researchers' collections. FAO reported a decrease in global barley holdings as observed from 1996 to 2010, that is, 0.48–0.47 million, respectively, in the second report on The State of World's Plant Genetic Resources for Food and Agriculture (FAO, 2010) (Table 3.2).

Knupffer (2009), after surveying available online information resources, documented a total of 466,531 accessions of barley, of which 63,511 accessions were identified only at the genus level, whereas 390,097 accessions were identified at the species level. The material is maintained in 199 holding institutions worldwide.

Landraces and breeding lines form the major part of any germplasm holding. A total of 470,470 accessions of barley germplasm were reported by FAO (2010), of which 23% are classified as landraces maintained in 204 genebanks worldwide. Pinheiro et al. (2013) presented a review of the landrace genetic resources in worldwide genebanks. A survey of the global cereal holdings (Bettencourt and Konopka, 1990) identified a total of 283,138 accessions of barley maintained in 94 germplasm collections in 47 countries; 32,316 of these accessions (11.4%) were identified as landraces.

At the global level, the International Centre for Agricultural Research in Dry Areas (ICARDA), is maintaining the world collection of barley and has representations from all the barley growing regions. A total of 26,679 barley accessions are conserved in the ICARDA Genebank. Other major international institutes that hold barley germplasm of more than 20,000 and are actively involved in germplasm conservation are Plant Genetic Resources of Canada (PGRC), Canada, National Small Grain Collection (NSGC), USA, Recursos Geneticos eBiotecnologia (CENARGEN), Brazil, National Institute of Agrobiological Science (NIAS), Japan and Leibniz Institute of Plant Genetics and Crop Plant Research (IPK), Germany. In India, the National Bureau of Plant Genetic Resources (NBPGR), New Delhi is maintaining around 9000 accessions of barley (Table 3.2).

3.7.2 In situ *conservation*

In situ conservation is complementary to *ex situ* conservation. It is a dynamic mode of germplasm conservation as compared to the static nature of *ex situ* conservation. It allows the continuous evolution of barley by allowing natural selection to act upon it. These days *in situ* conservation has attracted much attention and efforts are being made to conserve the genetic resources under its native environment. It is important for conservation of species that are difficult to conserve under *ex situ* conditions, especially crop wild relatives (CWR). With the development of new biotechnologic methods, crop wild relatives are becoming increasingly important in crop genetic improvement programs. It has been estimated that there are 50,000–60,000 CWR species worldwide, of these 700 are of highest priority, as these species comprise primary and secondary gene pools of the world's most important food crops and barley is one of them.

In situ conservation often takes place in protected areas or habitats as opposed to *ex situ* conservation. The second report on The State of World Plant Genetic Resources

Table 3.2 **Barley germplasm holding of major genebanks across the world (FAO, 2010)**

		Genebank			Accessions					
					Percentage					
S. No.	Country	Institute code	Acronym	Number	Total	Wild species	Landraces	Breeding lines	Advanced cultivar	Others
1	Canada	CAN004	PGRC	40,031	9	12	41	27	13	7
2	USA	USA029	NSGC	29,874	6	7	56	23	15	
3	Brazil	BRA003	CENARGEN	29,227	6					100
4	Syria	SYR002	ICARDA	26,679	6	7	67		<1	25
5	Japan	JPN003	NIAS	23,471	5	<1	6	15		79
6	Germany	DEU146	IPK	22,093	5	6	56	12	24	2
7	China	CHN001	ICGR-CAAS	18,617	4					100
8	Korea	KOR011	RDAGB-GRD	17,660	4		25	10	<1	64
9	Russia	RUS001	VIR	16,791	4		25			75
10	Ethiopia	ETH085	IBC	16,388	4		94			6
11	Mexico	MEX002	CIMMYT	15,473	3	<1	3	77	11	9
12	Sweden	SWE054	NORDGEN	14,109	3	5	5	84	4	2
13	Great Britain	GBR011	IPSR	10,838	2		17	30	23	29
14	India	IND001	NBPGR	9,161	2	11	3	13	2	71
15	Australia	AUS091	SPB-UWA	9,031	2					100
16	Iran	IRN029	NPGBI-SPII	7,816	2					100
17	Israel	ISR003	ICCI-TELAVUN	6,658	1	100	<1			<1
18	Poland	POL003	IHAR	6,184	1		2	94	2	2
19	Bulgaria	BGR001	IPGR	6,171	1	<1	<1	4	7	88
			Others	140,259	30					
			Total	466,531	100					

for Food and Agriculture, 2010, has indicated an increase in the number of protected areas. The Erebuni reserve has been established in Armenia to conserve populations of *Hordeum spontaneum, H. bulbosum,* and *Hordeum glaucum* along with cereal wild relatives. Research in West Asia has found significant CWR diversity in cultivated areas especially at the margins of fields and along roadsides. Rare CWR of barley along with wheat, lentil, pea, and faba bean have been reported in the modern apple orchard of Jabal Sweida in the Syrian Arab Republic. In order to protect CWR, the Syrian Arab Republic in 2007 established a protected area at Alujat and has banned grazing of wild ruminants in the Sweida region. Besides these, the priority locations for conservation have been identified for *Hordeum* species in America. Chile is the high-priority location for *Hordeum chilense* identified as one of the high priority CWRs. For wild species of *Hordeum*, namely, *H. vulgare* ssp. *spontaneum* and *H. bulbosum,* the highest priority location for conservation has been identified in the Near East.

3.7.3 Utilization of wild relatives

Barley genetic resources with special reference to wild barley are a rich source of diversity (Brown et al., 1978; Nevo et al., 1979, 1986), which has not been explored and utilized to the fullest. *H. vulgare* ssp. *spontaneum* and *H. bulbosum,* which belong to primary and secondary gene pools, respectively, are the rich and largely unexploited sources of unique disease-resistance genes, yield attributes, and quality genes.

3.7.3.1 H. vulgare *ssp.* spontaneum

H. vulgare ssp. *spontaneum* is the progenitor of cultivated barley and belongs to the primary gene pool. The wild barley is a rich source of unique alleles for resistance, yield components, and nutritive quality for straw purpose, which owing to the advances in research, is being successfully utilized for breeding purposes. This has also been explored for developing an efficient molecular marker system (Bothmer and Komatsuda, 2011). It has been extensively used in barley improvement programs at ICARDA to develop drought-tolerant lines. The lines derived from the crosses between ssp. *spontaneum* with ssp. *vulgare* are higher yielding and taller under drought conditions than any material based on cultivated barley germplasm. These lines also retain spike extrusion under severe drought conditions.

This has also been used to establish permanent pasture for grazing in certain areas because of its winter annual nature (El-Shatnawi et al., 2004). Grando et al. (2005) developed a marker system for nutritive quality of straw, which is being improved in cultivated barley by introgression of genes from ssp. *spontaneum.*

QTLs for yield and yield components have also been identified in ssp. *spontaneum* (Von Korff et al., 2004b). Recombinant chromosome substitution lines (RCSL) have been developed to introduce chromosomal segments in cultivated barley for traits including yield components, malting quality, and domestication syndromes (Matus et al., 2003; Hori et al., 2005). *H. vulgare* ssp. *spontaneum* is also cross-compatible with wheat and can be used to introduce agronomically important traits (Taketa et al., 2008).

3.7.3.2 Hordeum bulbosum

H. bulbosum is the closest relative of cultivated barley after ssp. *spontaneum* and belongs to the secondary gene pool. It has several traits of agronomic interest and is an important source of resistance to disease and insect pest. *H. bulbosum* as a source of resistance to disease and insect pest has successfully been explored to transfer genes for resistance to diseases caused by soil-borne viruses (Ruge et al., 2003; Ruge-Wehling et al., 2006), scald (Singh et al., 2004; Pickering et al., 2006), leaf rust, and mosaic (Walther et al., 2000).

The other important phenomenon associated with *H. bulbosum* is selective elimination of chromosome, which has been widely exploited in barley and wheat breeding for the development of double haploid (DH).

3.8 Limitation in germplasm use

The CWR of barley are a source of many agronomically important traits. The use of these species as donors for breeding purposes is restricted mainly because of the diploid nature of barley, which makes it more sensitive to genetic imbalances. This is the reason why barley wild relatives cannot be exploited the way the wild relatives have been used in wheat breeding, which has a buffering effect to genetic imbalances because of polyploidy. However, with the advances in breeding techniques and utilization of biotechnologic tools, it became easier to introgress genes from wild to cultivated barley successfully. However, there is still a vast source of diversity in the form of distantly related wild species of barley, which has not been exploited yet due to isolation barrier.

3.9 Genomic resources

Barley is a self-pollinating diploid crop with $2n = 2x = 14$ chromosomes. Its genome has been estimated to comprise approximately 5.3×10^9 bp (Bennett and Smith, 1976). Like other cereals, its genome consists of a complex mixture of unique and repeated nucleotide sequences (Flavell, 1980). Recent progress related to the generation of genetic maps, QTL mapping, and association-mapping studies have been reviewed in the following sections. Moreover, available data on comparative mapping between rice and barley and available literature on marker-assisted selection (MAS) have also been summarized for the efficient use of molecular biotechnologic tools for basic and applied genome research.

3.9.1 Genetic maps

In barley, the first genetic map for chromosome 6H appeared in 1988 and a partial map of the whole genome by using RFLPs, morphologic, isozyme, and PCR markers was published in 1990 (Kleinhofs et al., 1988; Shin et al., 1990). The first comprehensive molecular marker maps were developed by three different workers using different F_1-derived doubled haploid populations, namely, Igri/Franka (Graner et al., 1991), Proctor/Nudinka (Heun et al., 1991), and Steptoe/Morex (Kleinhofs et al., 1993). So far, more than 3000 genetic markers have been mapped in barley. However, this large number may not be difficult for integration of the individual maps. Only a limited number of maps contain sufficient common markers to merge their information in consensus maps as they were constructed by several groups (Qi et al., 1996). Based on the data of the most comprehensive maps, the genetic length of the barley genome can be estimated to be between 1050 centimorgan (cM) and 1400 cM. Several factors may account for the differences in map size. The following table depicts the various studies involved in the generation of molecular maps in barley (Table 3.3).

3.9.2 QTL mapping

A QTL is a chromosomal region containing a gene(s) that contributes to a quantitative trait. In QTL mapping experiments the genetic basis of quantitative traits is dissected into their single components. Many traits of agricultural importance are quantitative, that is, polygenic in nature. The objective of genetic mapping is to identify simply inherited markers in close proximity to genetic factors affecting quantitative traits. The QTL mapping has largely been accomplished using linkage mapping of QTL and association mapping (AM) approaches. Both the approaches had their own advantages and disadvantages.

Table 3.3 List of some comprehensive genetic maps available for barley

Map	Population	Parents used for making crosses	References
RFLP	Double haploids	Proctor/Nudinka	Heun et al. (1991)
	DH populations	Igri/Franka, *H. vulgare* ssp. Vada/*H. vulgare* ssp. *spontaneum* line 1B-87	Graner et al. (1991)
	DH	Steptoe/Morex	Kleinhofs et al. (1993)
	DH	Harrington/TR306	Kasha and Kleinhofs (1994)
	F_2 population	Ko A/Mokusekko	Miyazaki et al. (2000)
	F_1 full-sib families	*H. bulbosum*; PB1/PB11	Salvo-Garrido et al. (2001)
SSR	DH	*H. vulgare* var Lina/*H. spontaneum* Canada Park	Ramsay et al. (2000)
AFLP	DH	Proctor/Nudinka	Becker et al. (1995)
	Recombinant inbred lines	L94/Vada	Qi et al. (1998)
	DH	Proctor/Nudinka	Castiglioni et al. (1998)

An updated version of these maps can be found at GrainGenes website (http://wheat.pw.usda.gov/).

3.9.2.1 Mapping for yield and other related traits

Crop yield is an integrative trait that reflects plant vigor and physiologic efficiency, which is influenced by several environmental and genetic factors. Therefore, it seems that barley yield and its related agronomic traits are affected by many genes and further influenced by several environmental factors. Some QTLs apparently have large effects on yield, which are controlled by one or few major genes or QTLs. Lot of efforts have been made to identify the QTL regions influencing the yield and its related traits. Initially, Hayes et al. (1993) found that yield-related QTLs were often coinciding with QTL of plant height and lodging. An RFLP map constructed from 99 doubled haploid lines of a cross between two spring barley varieties (Blenheim Kym) was used to localize QTL controlling grain yield and yield components by marker regression and single-marker analysis (Bezant et al., 1997). Eight QTLs were detected for thousand-grain weight and five for ear grain number. A total of 11 QTLs were detected for plant grain weight over 2 years and 14 for ear grain weight over 3 years. Seven QTLs were detected for plot yield. Reducing plant height has played an important role in improving crop yields. The success of a breeding program relies on the source of dwarfing genes. For a dwarfing or semidwarfing gene to be successfully used in a breeding program, the gene should have minimal negative effects on yield. Wang et al. (2014) developed 182 doubled haploid lines, generated from a cross between TX9425 and Naso Nijo, and were grown in six different environments. A QTL for plant height was identified on 7H. This QTL showed no significant effects on other agronomic traits and yield components and consistently expressed in the six environments, which may be considered as very effective QTL. Ren et al. (2013) performed QTL analysis for 10 agronomic traits and 1 quality trait using a DH population of 122 lines derived from the cross of Huaai 11/Huadamai 6. Composite interval mapping (CIM) procedures detected a total of 17 QTLs, which were mapped onto five chromosomes. The QTL on chromosome 7H has an effect on grain weight per plant, grain protein content, and main spike length, which accounts for 35.11, 45.74, and 54.88% of phenotypic variation, respectively. Xue et al. (2010) identified QTLs for yield and yield components of barley under different growth conditions. Water logging is a major abiotic stress limiting barley yield and its stability in areas with excessive rainfall. Identification of genomic regions influencing the response of yield and its components to water logging stress will enhance our understanding of the genetics of water logging tolerance and the development of more tolerant barley cultivars for water logging resistance. A total of 31 QTLs were identified for the measured characters influencing the water logging from two experiments with two growth environments. The phenotypic variation explained by individual QTLs ranged from 4.74% to 55.34%. Several major QTLs determining kernel weight (KW), grains per spike (GS), spikes per plant (SP), spike length (SL), and grain yield (GY) were detected on the chromosome 2H. Spring radiation frost is a major abiotic stress in southern Australia, reducing yield potential and grain quality of barley by damaging sensitive reproductive organs in the latter stages of development.

3.9.2.2 Mapping for abiotic stress-tolerant genes

The world population continues to increase rapidly and is expected to reach 9 billion by 2050 (FAO, 2010). Agriculture will have to increase its crop productivity by

70–110% in 2050 to feed that world (Tester and Langridge, 2010; Tilman et al., 2011). Climate change associates with increased abiotic stress factors, such as water scarcity, elevated temperature, flooding, and salinity, have major impacts on crop yields. There is an increased demand for new stress-tolerant crop varieties. In the case of barley, several studies were conducted to identify the QTLs linked to several factors influencing the salt tolerance. Biparental (traditional) mapping based on a single segregating population derived from two homozygous parental genotypes has been the common approach for genetic dissection of salt tolerance in barley (Mano and Takeda, 1997; Ellis et al., 2002; Xue et al., 2009; Witzel et al., 2009). Yoshiro and Takeda (1997) mapped QTL for salt tolerance at germination and the seedling stage in barley. The QTLs for salt tolerance at the seedling stage were located on chromosomes 2(2H), 5(1H), 6(6H), and 7(5H) in the DH lines of Steptoe/Morex, and on chromosome 7(5H) in the DH lines of Harrington/TR 306. Drought is one of the most severe stresses endangering crop yields worldwide. Drought is one of the main factors limiting yield worldwide. Due to climate change, extreme weather events are predicted to occur more frequently and an altered pattern of drought occurrence is expected. Honsdorf et al. (2014) used high-throughput phenotyping to detect drought tolerance QTL in wild barley introgression lines. The use of a nondestructive high-throughput phenotyping platform was implemented to map QTL controlling vegetative drought stress responses in barley. In total, 44 QTLs for 11 out of 14 investigated traits were mapped, which for example controlled growth rate and water use efficiency. Reinheimer et al. (2004) identified QTL on chromosome 2HL for frost-induced floret sterility in two different populations at the same genomic location.

3.9.2.3 Mapping for biotic stress-tolerant genes

The QTLs linked to disease- or pest-resistant genes is a major task in barley crop improvement. The disease resistance governed by single genes or multiple genes can be effectively identified using molecular marker techniques through traditional biparental linkage mapping studies. Several major diseases have been reported in barley and QTLs for some of them are identified and discussed later in the chapter. The major difficulty in finding the effective QTL is their percentage of phenotypic variance and influence by environmental factors. Hence, experiments conducted at multiple locations with large number of populations will give better results. Valè et al. (2005) studied genetic mapping of major genes and QTLs for leaf stripe disease has revealed resistance loci on chromosomes 1 (7H), 2 (2H), and 3 (3H). Leaf stripe is a common disease in barley characterized by a cold climate during the sowing season. Mapping studies have led to the identification of the disease-resistance loci to seed-borne diseases on several barley chromosomes. Net blotch (*Pyrenophora teres* f. *Teres)* is one of the most devastating diseases causing significant losses in barley yield and quality. Adawy et al. (2013) constructed a genetic linkage map and QTL analysis of net blotch resistance in barley. Single-point analysis was used to identify the genomic regions associated with net blotch resistance in barley. A total of 14 QTLs with a significance ranging from 0.01% to 5% were identified on four linkage groups (2, 4, 5, and 6). The most significant QTL was found on chromosome 6H. A subset of near-isogenic lines (NILs) were developed for fine-mapping a locus controlling leaf rust (Marcel et al., 2007). Close et al. (2009) identified a significant number of genic SNPs from ESTs and sequenced PCR amplicons and used them to develop two Illumina barley oligo pool assays (BOPA1 and BOPA2), each enabling the simultaneous genotyping of 1536 SNPs. Inga et al. (2011) used 73 introgression lines (S42ILs) originating from a cross between the spring barley cultivar Scarlett (*H. vulgare* ssp. *vulgare*) and the wild barley accession ISR42-8 (*H. vulgare* ssp. *spontaneum*), which were subjected to high-resolution genotyping with an Illumina 1536-SNP array. The *thresh-1* gene was fine-mapped within a 4.3 cM interval that was predicted to contain candidate genes involved in regulation of plant cell wall composition. Septoria speckled leaf blotch (SSLB), caused by *Septoria passerinii*, has emerged as one of the most important foliar diseases of barley. A sequence-characterized amplified region (SCAR) marker (E-ACT/M-CAA-170a) was developed that cosegregated with not only *Rsp2* in the Foster/CIho 4780 population but also resistance gene *Rsp3* in the Foster/CIho 10644 population. This result indicates that *Rsp3* is closely linked to *Rsp2* on the short arm of chromosome 5(1H). The utility of SCAR marker E-ACT/M-CAA-170a for selecting *Rsp2* in two different breeding populations was validated. Von Korff et al. (2004a) developed a BC_2DH population from a cross between the German spring barley cultivar Scarlett (Hv) and the Israeli wild barley accession ISR42-8. Likewise, several genes responsible for seed-borne diseases in barley were mapped using different molecular markers (Table 3.4).

3.9.3 Association mapping

Over the past decade, MAS has become a standard tool in breeding programs. As a first step to MAS, marker loci that are linked to QTL need to be identified. The objective of genetic mapping is to identify simply inherited markers in close proximity to genetic factors affecting quantitative traits (QTL). The QTL mapping has largely been accomplished using linkage mapping of QTL. However, the focus is now trending toward the use of AM, which was initially applied in human disease genetics. The advantages of AM include the potential of high resolution in localizing a QTL controlling a trait of interest to identify a single polymorphism within a gene that is responsible for phenotypic differences. It also involves searching for genotype–phenotype correlations among unrelated individuals and high resolution is accounted by the historical recombination accumulated in natural populations and collections of landraces, breeding materials,

Table 3.4 Mapped genes for resistance to seed-borne diseases

Disease	Gene	Chromosome	References
Spot blotch	*Rcs5*	7HS	Steffenson et al. (1996)
Spot blotch	*Rcs6*	1HS	Bilgic et al. (2004)
Loose smut	*Un8*	1HL	Eckstein et al. (2002)
Leaf stripe	*Rdg1a*	2HL	Thomsen et al. (1997)
Leaf stripe	*Rdg2a*	7HS	Tacconi et al. (2001)
Barley stripe mosaic	*Rsm*	7H	Edwards and Steffenson (1996)

and varieties. Rostoks et al. (2006) used high-throughput SNP genotyping assays and demonstrated that the linkage disequilibrium (LD) present in elite germplasm, after accounting for population structure, can be effectively exploited to map traits by using whole-genome association scans with several hundreds to thousands of markers. Wang et al. (2012) concentrated on comparing various statistical approaches for AM in barley. They showed the superiority of mixed model methodology for genome-wide association (GWA) analysis to assess marker-trait association for even complex traits in barley.

Recently, Mohamed et al. (2014) used AM approaches for shoot traits related to drought tolerance in barley in a structured population grown under well-watered and drought stress conditions. A total of 56 QTLs have been identified for the shoot traits, and located on the whole barley genome, and these QTLs had main and/or interaction effects on improving or reducing the traits of interest under well-watered and drought stress conditions. A total of 8, 16, 8, 8, 9, 5, and 5 QTLs were detected for wilting score (WS), number of tillers/plant (TILS), shoot fresh weight/plant (SFW), shoot dry weight/plant (SDW), relative water content (RWC), proline content (PC), and osmotic potential (OP), respectively. Frost tolerance is a key trait with economic and agronomic importance in barley because it is a major component of winter hardiness, and, therefore, limits the geographical distribution. Visioni et al. (2013) studied genome-wide AM of frost tolerance in barley. The GWAS analyses identified 12 and 7 positive SNP associations at Foradada and Fiorenzuola, respectively, using Eigenstrat, and 6 and 4, respectively, using kinship matrices. Haplotype analysis revealed that most of the significant SNP loci were fixed in the winter or facultative types, whereas, they were freely segregating within the unadapted spring barley gene pool. A new major low-temperature tolerance QTL, designated Fr-H3, has recently been discovered on barley chromosome 1H in the "NB3437f" (facultative)/"OR71" (facultative) and the "NB713" (winter)/"OR71" barley populations (Fisk et al., 2013) but the gene underlying Fr-H3 has yet to be identified. Other loci with minor effects on freezing tolerance at the vegetative stage have been mapped on barley chromosomes 1HL, 4HS, and 4HL in the "Dicktoo" (facultative)/"Morex" (spring) barley mapping population together with Fr-H1 (Skinner et al., 2006). GWA study was done with 816 markers comprised of 710 diversity array technology (DArT), 61 single nucleotide polymorphism (SNP), and 45 microsatellite or simple sequence repeat (SSR) markers (Varshney et al., 2012). Although a few QTLs were identified that differed between the dry and wet site, these QTLs explained low phenotypic variation and could not unequivocally be related to drought tolerance when compared to earlier linkage-mapping-based QTL analysis studies.

Pauli (2014) conducted AM of agronomic QTLs in US spring barley breeding germplasm to identify marker-trait associations for yield, heading date, plant height, test weight, kernel plumpness, and protein content. They identified 41 significant marker-trait associations, of which 31 had been previously reported as QTL using biparental mapping techniques, whereas, 10 novel marker-trait associations were identified. The results of this work showed that genes with major effects were still segregating in US barley germplasm and demonstrated the utility of GWAS in barley breeding populations. GWA analysis of yield, yield components, developmental, physiologic, and anatomic traits was conducted for a barley germplasm collection consisting of

185 cultivated (*H. vulgare* L.) and 38 wild (*H. spontaneum* L.) genotypes, originating from 30 countries of four continents. Kraakman et al. (2004) identified QTLs for yield and yield stability traits in a collection of modern spring barley cultivars using AFLP markers. They used linkage disequilibrium mapping of yield and yield stability in modern spring barley cultivars. Associations between markers and complex quantitative traits were investigated in a collection of 146 modern two-row spring barley cultivars, representing the current commercial germplasm in Europe. Using 236 AFLP-markers, associations between markers were found for markers at a distance of 10 cM using 32 phenotypes in the inbreeding crop barley. Cockram et al. (2010) reported GWA mapping of 15 morphologic traits across nearly 500 cultivars genotyped with 1536 SNPs. They observed high levels of linkage disequilibrium within and between chromosomes. Despite this, GWA analysis readily detected common alleles of high penetrance. Wang et al. (2012) used GWAs for 32 morphologic and 10 agronomic traits in a collection of 615 barley cultivars genotyped by genome-wide polymorphisms from a recently developed barley oligonucleotide pool assay. The present study reports significant associations for 16 morphologic and 9 agronomic traits and demonstrated the power and feasibility of applying GWAS to explore complex traits in highly structured plant samples. Inostroza et al. (2008) identified QTL for grain yield and plant height from 80 recombinant chromosome substitution lines (RCSLs) of barley in six environments with contrasting available moisture profiles. The association analysis revealed 21 chromosomal regions that were highly correlated with differences in grain yield, plant height, and/or yield adaptability.

3.9.4 Comparative mapping and synteny

Comparative mapping provides access to the model genome of rice, which gives insight to identify the syntenic regions in the close crops. An obvious strategy emerging from the concept of syntenous relationships was the vast amount of genomic information and resources available for rice, which can be transferred to the barley genome. Together with the barley ESTs, the availability of the complete sequence of rice chromosomes (Goff et al., 2002) facilitated computational approaches to identify syntenic regions between rice and barley at high resolution. There were some studies initiated for studying the synteny existing between barley with other crops (Table 3.5). However, there were some exceptions where no synteny was observed for some genes between rice and barley. No rice ortholog of the barley stem rust resistance gene *Rpg1* exists in rice although the gene order surrounding the locus was highly conserved between the two species (Kilian et al., 1997; Han et al., 1999). Similarly, in a study on comparative mapping of the barley *ppd-h1* (photoperiod response) on barley chromosome 2HS and its orthologous region *Hd2* (heading date) on rice chromosome 7L, disruption of colinearity was observed (Dunford et al., 2002; Griffiths et al., 2003).

3.9.5 Marker-assisted selection

Barley, like wheat crop, is a self-pollinating species and in terms of the breeding system and the economic structure of its market it resembles wheat. However, MAS

Table 3.5 Syntenic relationship between barley and other cereals, studied on the basis of molecular markers

Cereal	Species	References
Barley	Wheat	Hohmann et al. (1995); Hernandez et al. (2001)
Barley	Rye	Wang et al. (1992)
Barley	Rice	Kilian et al. (1995); Kilian et al. (1997); Smilde et al. (2001)
Barley	Wheat, rye	Devos and Gale (1993); Börner et al. (1998)
Barley	Oat, maize	Yu et al. (1996)
Barley	Wheat, rice	Dunford et al. (1995); Kato et al. (2001)

seems to have progressed further than in wheat, which was probably due to the simpler, diploid genome. The main focus of marker selection in barley is breeding for barley yellow mosaic virus resistance and rust resistance. In contrast to wheat, barley varieties have been released that are based on MAS. In the USA, the variety "Tango," carrying two QTL for adult resistance to stripe rust, was released in 2000 (Hayes et al., 2003), claiming to be the first commercially released barley variety using MAS. However, "Tango" yields less than its recurrent parent and is, therefore, primarily seen as a genetically characterized source of resistance to barley stripe rust rather than a variety of its own. As a result of the South Australian Barley Improvement Program the malting variety "Sloop" was improved with cereal cyst nematode resistance introgressed from the variety "Chebec" and released in 2002 as "SloopSA" (Barr et al., 2000; Eglinton et al., 2006). Jefferies et al. (2008) did marker-assisted backcross introgression of the *Yd2* gene conferring resistance to barley yellow dwarf virus. The YLM, a codominant polymerase chain reaction (PCR) marker linked to *Yd2*, could substantially improve the precision and efficiency of barley yellow dwarf virus (BYDV) resistance breeding. The *Yd2* gene was introgressed into a BYDV-susceptible background through two cycles of marker-assisted backcrossing. The YLM marker was shown to be effective in the introgression of *Yd2*. Lines carrying the YLM allele associated with resistance produced significantly fewer leaf symptoms and showed a reduction in yield loss when infected with BYDV. Along with these discussions there were some more reports available on the marker-assisted introgression and pyramiding of genes for important traits in barley, which are given in Table 3.6.

3.9.6 Genome sequencing

Steered by Nils Stein, the International Barley Genome Sequencing Consortium (IBSC) constituted in 2006 executed the whole genome sequencing (WGS) using the next-generation sequencing (NGS) strategies and the sequence information was released on November 29, 2012 (IBGSC, 2012). The WGS data unveiled that ~84% of the barley genome encompasses mobile elements and other repeat structures. The barley genome has a higher percentage of long terminal repeat (LTR) – *Gypsy* belongs to the retrotransposon super family, and moreover, consisted of about 15 million nonredundant single-nucleotide polymorphisms. The results showed a higher synteny

Table 3.6 **Important marker-assisted backcross breeding and gene pyramiding**

Trait	Marker	Objective	References
Biotic stress	RFLP	Resistance to barley yellow mosaic virus	Okada et al. (2003)
Biotic stress	RFLP, SSR, STS	Stripe rust resistance gene *Rspx* and QTLs	Castro et al. (2003)
Biotic stress	RAPD, SSR, STS	Barley yellow mosaic virus complex (BaMMV, BaYMV, BaYMV-2)	Werner et al. (2005)
Abiotic stress	SSR	Boron toxicity	Emebiri et al. (2009)
Quality traits	RFLP	Malt parameters	Schmierer et al. (2004)
Biotic stress	SSR	Spot form of net blotch	Eglinton et al. (2006)

of barley toward *Arabidopsis thaliana*, with an estimate of 30,400 genes within its genome. The transcriptome analysis showed that about 36–55% of barley genes were differentially regulated, demonstrating the intrinsic dynamics of gene expression. The release of WGS data will impeccably accelerate the aforementioned research. The WGS data will be essential for the advancement of true genomics-informed breeding strategies and for deciphering the full potential of natural genetic variation toward the improvement of agronomic traits.

3.10 Future perspectives

The DNA markers linked with different traits provide a tool to identify genotypes with desirable gene combinations via MAS. This chapter presents a comprehensive survey of the literature on genetic maps constructed in barley using different markers, QTL mapping studies for important abiotic, biotic stress-resistant genes, yield, and important agromorphologic traits, and MAS. The book chapter also provides a scope on the sequence of the barley genome and its implications for barley improvement. It appears that several important agronomic traits, including yield and disease resistance against most important diseases, have been mapped with molecular markers in barley. However, only few markers have been or are being used in practical breeding with limitations such as tight linkage of markers. Very precise studies need to be conducted at multilocational traits for identification of effective QTLs of important agronomic traits. A large number of ESTs are found in barley. Hence, genes can be searched in the EST database, and with the help of this database perhaps DNA-chips could be developed to conduct MAS. The comparative synteny existing between barley with rice and wheat may be exploited for precise identification of genes, molecular evolution, and their further applications for barley improvement through marker-assisted breeding and genetic engineering approaches.

References

Aberg, E., 1938. *Hordeum agriocrithon* nova sp., a wild six-rowed barley. Ann. R. Agric. Coll. Sweden 6, 159–216.

Adawy, S.S., Ayman, A.D., Abdel-Hadi, I.S., Shafik, D.E., Shafik, I.E., Mahmoud, M.S., 2013. Construction of genetic linkage map and QTL analysis of net blotch resistance in barley. Int. J. Adv. Biotechnol. Res. 4 (3), 348–363.

Azhaguvel, P., Komatsuda, T., 2007. A phylogenetic analysis based on nucleotide sequence of a marker linked to the brittle rachis locus indicates a diphyletic origin of barley. Ann. Bot. (Lond.) 100, 1009–1015.

Badr, A., Müller, K., Schäfer-Pregl, R., El-Rabey, H., Effgen, S., Ibrahim, H.H., et al., 2000. On the origin and domestication history of barley (*Hordeum vulgare*). Mol. Biol. Evol. 17, 499–510.

Barr, A.R., Jefferies, S.P., Warner, P., Moody, D.B., Chalmers, K.J., Langridge, P., 2000. Marker-assisted selection in theory and practice. Proceedings of the Eighth International Barley Genetics Symposium, vol. I., Adelaide, Australia, pp. 167–178.

Becker, J., Vos, P., Kuiper, M., Salamini, F., Heun, M., 1995. Combined mapping of AFLP and RFLP markers in barley. Mol. Genet. Genomics 249, 65–73.

Bekele, E., 1983. A differential rate of regional distribution of barley flavonoid patterns in Ethiopia, and a view on the centre of origin of barley. Hereditas 98, 269–280.

Bennett, M.D., Smith, L.B., 1976. Nuclear DNA amounts in angiosperms. Philosophical genomes. Theor. Appl. Genet. 102, 980–985.

Bettencourt, E., Konopka, J., 1990. Directory of Crop Germplasm Collections. Cereals: *Avena*, *Hordeum*, Millets, *Oryza*, *Secale*, *Sorghum*, *Triticum*, *Zea* and Pseudo Cereals. International Board for Plant Genetic Resources, Rome.

Bezant, J., Laurie, D., Pratchett, N., Chojecki, J., Kearsey, M., 1997. Mapping QTL controlling yield and yield components in a spring barley (*Hordeum vulgare* L.) cross using marker regression. Mol. Breed. 3, 29–38.

Bilgic, H., Steffenson, B.J., Hayes, P., 2004. Mapping QTL for resistance to pathotypes of *Cochliobolus sativus* in barley. In: Braunova, M. (Ed.), Proceedings of the Ninth International Barley Genetics Symposium, pp. 764–767.

Börner, A., Korzun, V., Worland, A.J., 1998. Comparative genetic mapping of loci affecting plant height and development in cereals. Euphytica 100, 245–248.

Brennan, J.P., Fox, P.N., 1998. Impact of CIMMYT varieties on the genetic diversity of wheat in Australia, 1973–1993. Aust. J. Agric. Res. 49 (2), 175–178.

Brown, A.H.D., Munday, J., 1982. Population-genetic structure and optimal sampling of land races of barley from Iran. Genetica 58, 85–96.

Brown, A.H.D., Nevo, E., Zohary, D., Dagan, O., 1978. Genetic variations in natural populations of wild barley *(Hordeum spontaneum)*. Genetica 49, 97–108.

Boyd, W.J.R., Li, C.D., Grime, C.R., Cakir, M., Potipibool, S., Kaveeta, L., 2003. Conventional and molecular genetic analysis of factors contributing to variation in the timing of heading among spring barley (*Hordeum vulgare* L.) genotypes grown over a mild winter growing season. Aust. J. Agric. Res. 54, 1277–1301.

Castiglioni, P., Pozzi, C., Heun, M., Terzi, V., Muller, K.J., Rohde, W., et al., 1998. An AFLP-based procedure for the efficient mapping of mutations and DNA probes in barley. Genetics 149, 2039–2056.

Castro, A.J., Capettini, F., Corey, A.E., Filichkina, T., Hayes, P.M., Kleinhofs, A., et al., 2003. Mapping and pyramiding of qualitative and quantitative resistance to stripe rust in barley. Theor. Appl. Genet. 107, 922–930.

Choo, T.M., Ho, K.M., Martin, R.A., 2001. Genetic analysis of a hulless, covered cross of barley using doubled-haploid lines. Crop Sci. 41, 1021–1026.

Close, T.J., Prasanna, R.B., Stefano, L., Yonghui, W., Nils, R., Luke, R., et al., 2009. Development and implementation of high-throughput SNP genotyping in barley. BMC Genomics 10, 582.

Cockram, J., White, J., Diana, L.Z., Smith, D., Comadran, J., Macaulay, M., 2010. Genome-wide association mapping to candidate polymorphism resolution in the unsequenced barley genome. Proc. Natl. Acad. Sci. 107 (50), 21611–21616.

Devos, K.M., Gale, M.D., 1993. Extended genetic maps of the homoeologous group-3 chromosomes of wheat, rye and barley. Theor. Appl. Genet. 85, 649–652.

Dewey, D.R., 1984. The genomic system of classification as a guide to intergeneric hybridization in the perennial Triticeae. In: Gustafson, J.P. (Ed.), Gene Manipulation in Plant Improvement. Plenum Publishing Corporation, New York, pp. 209–279.

Dunford, R.P., Yano, M., Kurata, N., Sasaki, T., Huestis, G., Rocheford, T., et al., 2002. Comparative mapping of the barley Ppd-H1 photoperiod response gene region, which lies close to a junction between two rice linkage segments. Genetics 161, 825–834.

Dunford, R.P., Kurata, N., Laurie, D.A., Money, T.A., Minobe, Y., Moore, G., 1995. Conservation of fine-scale DNA marker order in the genomes of rice and the Triticeae. Nucleic Acids Res. 23, 2724–2728.

Eckstein, P.E., Krasichynska, N., Voth, D., Duncan, S., Rossnagel, B.G., Scoles, G.J., 2002. Development of PCR-based markers for a gene (*Un8*) conferring true loose smut resistance in barley. Can. J. Plant Pathol. 24, 46–53.

Edwards, M.C., Steffenson, B.J., 1996. Genetics and mapping of barley stripe mosaic virus resistance in barley. Phytopathology 86, 184–187.

Eglinton, J. K., Coventry, S. J., Chalmers, K., 2006. Breeding outcomes from molecular genetics – breeding for success: diversity in action, Proceedings of the Thirteeth Australasian Plant Breeding Conference, Christchurch, New Zealand, 18–21 April, pp. 743–749.

El - Shatnawi, M.K.J., Al–Qurran, L.Z., Ereifej, K.I., Saoub, H.M., 2004. Management optimization of dual-purpose barley (*Hordeum spontaneum* C. Koch) for forage and seed yield. J. Range Manage. 57, 197–202.

Ellis, R.P., Forster, B.P., Gordon, D.C., Handley, L.L., Keith, R.P., Lawrence, P., et al., 2002. Phenotype/genotype associations for yield and salt tolerance in a barley mapping population segregating for two dwarfing genes. J. Exp. Bot. 53, 1163–1176.

Emebiri, L., Michael, P., Moody, D., 2009. Enhanced tolerance to boron toxicity in two-rowed barley by marker-assisted introgression of favourable alleles derived from Sahara 3771. Plant Soil 314, 77–85.

FAO, 1996. FAO State of the World's Plant Genetic Resources for Food and Agriculture, Rome.

FAO, 2010. The Second Report on State of the World's Plant Genetic Resources for Food and Agriculture, Rome.

Fisk, S.P., Cuesta-Marcos, A., Cistue, L., Russell, J., Smith, K.P., Baenziger, S., et al., 2013. FR-H3: a new QTL to assist in the development of fall-sown barley with superior low temperature tolerance. Theor. Appl. Genet. 126, 335–347.

Flavell, R., 1980. The molecular characterization and organization of plant chromosomal DNA sequences. Annu. Rev. Plant Physiol. 31, 569–596.

Freisleben, R., 1940. Die phylogenetische Bedeutung asiatischer Gersten. Der Zuchter 12, 257–272.

Gao, W., Clancy, J., Han, F., Prada, D., Kleinhofs, A., Ullrich, S.E., 2003. Molecular dissection of a dormancy QTL region near the chromosome 7 (5H) L telomere in barley. Theor. Appl. Genet. 107, 552–559.

Goff, S.A., Ricke, D., Lan, T.-H., Presting, G., Wang, R., Dunn, M., et al., 2002. A draft sequence of the rice genome (*Oryza sativa* L. ssp.*japonica*). Science 296, 92–100.

Grando, S., Baum, M., Ceccarelli, S., Godchild, A., JabyEl-Haramein, F., Jahoor, A., et al., 2005. QTLs for straw quality characteristics identified in recombinant inbred lines of a *Hordeum vulgare* × *H. spontaneum* cross in a Mediterranean environment. Theor. Appl. Genet. 110, 688–695.

Graner, A., Jahoor, A., Schondelmaier, J., Siedler, H., Pillen, K., Fischbeck, G., et al., 1991. Construction of an RFLP map of barley. Theor. Appl. Genet. 83, 250–256.

Griffiths, S., Dunford, R.P., Coupland, G., Laurie, D.A., 2003. The evolution of CONSTANS-like gene families in barley, rice and *Arabidopsis*. Plant Physiol. 13, 1855–1867.

Guarino, L., 2003. Approaches to measuring genetic erosion. PGR Documentation and Information in Europe – towards a sustainable and user-oriented information infrastructure. EPGRIS Final Conference combined with a meeting of the ECP/GR Information and Documentation Network, Prague, Czech Republic, pp. 11–13.

Han, F., Kilian, A., Chen, J.P., Kudrna, D., Steffenson, B., Yamamoto, K., Matsumoto, T., Sasaki, T., Kleinhofs, A., 1999. Sequence analysis of a rice BAC covering the syntenous barley *Rpg1* region. Genome 42, 1071–1076.

Han, F., Ullrich, S.E., Clancy, J.A., Romagosa, I., 1999. Inheritance and fine mapping of a major barley seed dormancy QTL. Plant Sci. 143, 113–118.

Harlan, H.V., 1931. The origin of *Hooded* barley. J. Hered. 22, 265–272.

Harlan, J.R., De Wet, J.M.J., 1971. Towards a rational classification of cultivated plants. Taxonomy 20, 509–517.

Harlan, J.R., Zohary, D., 1966. Distribution of wild wheat and barley. Science 153, 1074–1080.

Hayes, P.M., Corey, A.E., Mundt, C., Toojinda, T., Vivar, H., 2003. Registration of 'Tango' Barley. Crop Sci. 43, 729–731.

Hayes, P.M., Liu, B.H., Knapp, S.J., Chen, F., Jones, B., Blake, T., et al., 1993. Quantitative trait locus effects and environmental interaction in a sample of North American barley germplasm. Theor. Appl. Genet. 87, 392–401.

Helbaek, H., 1959. Domestication of food plants in old world. Science 130, 365–372.

Hernandez, P., Dorado, G., Prieto, P., Gimenez, M.J., Ramirezm, M.C., Laurie, D.A., et al., 2001. A core genetic map of *Hordeum chilense* and comparisons with maps of barley (*Hordeum vulgare*) and wheat (*Triticum aestivum*). Theor. Appl. Genet. 102, 1259–1264.

Heun, M., Kennedy, A.E., Anderson, J.A., Lapitan, N.L.V., Sorrells, M.E., Tanksley, S.D., 1991. Construction of a restriction fragment length polymorphism map for barley (*Hordeum vulgare*). Genome 34, 437–447.

Hohmann, U., Graner, A., Endo, T.R., Gill, B.S., Herrmann, R.G., 1995. Comparison of wheat physical maps with barley linkage maps for group 7 chromosomes. Theor. Appl. Genet. 91, 618–626.

Honsdorf, N., March, T.J., Berger, B., Tester, M., Pillen, K., 2014. High-throughput phenotyping to detect drought tolerance QTL in wild barley introgression lines. PLoS ONE 9 (5), e97047.

Hori, K., Sato, K., Nankaku, N., Takeda, K., 2005. QTL analysis in recombinant chromosome substitution lines and doubled haploid lines derived from across between *Hordeum vulgare* ssp. *vulgare* and *Hordeum vulgare* ssp. *spontaneum*. Mol. Breed. 16, 295–331.

Inga, S., Timothy, J.M., Thomas, B., Robbie, W., Klaus, P., 2011. High-resolution genotyping of wild barley introgression lines and fine-mapping of the threshability locus thresh-1 using the Illumina GoldenGate assay. Genes Genomes Genet. 1, 187–196.

Inostroza, L., Del Pozo, A., Matus, I., Castillo, D., Hayes, P., Machado, S., 2008. Association mapping of plant height, yield, and yield stability in recombinant chromosome substitution lines (RCSLs) using *Hordeum vulgare* subsp. *spontaneum* as a source of donor alleles in a *Hordeum vulgare* subsp. *vulgare* background. Mol. Breed. 23 (3), 365–376.

International Barley Genome Sequencing Consortium, 2012. A physical, genetic and functional sequence assembly of the barley genome. Nature 491, 711–716.

Jana, S., Pietrzak, L.N., 1988. Comparative assessment of genetic diversity in wild and primitive cultivated barley in a center of diversity. Genetics 119, 981–990.

Jarvis, D.I., Hodgkin, T., 1999. Wild relatives and crop cultivars: detecting natural introgression and farmer selection of new genetic combinations in agro-ecosystems. Mol. Ecol. 8, 159–173.

Jefferies, S.P., King, B.J., Barr, A.R., Warner, P., Logue, S.J., Langridge, P., 2008. Marker-assisted backcross introgression of the *Yd2* gene conferring resistance to barley yellow dwarf virus in barley. Plant Breed. 122 (1), 52–56.

Kasha, K.J., Kleinhofs, A., 1994. The North American Barley Genome Mapping Project. Mapping of the barley cross Harrington × TR306. Barley Genet. Newsl. 23, 65–69.

Kato, K., Nakamura, W., Tabiki, T., Miura, H., Sawada, S., 2001. Detection of loci controlling seed dormancy on group 4 chromosomes of wheat and comparative mapping with rice and barley. Trans. R. Soc. Lond. Biol. Sci. 274, 227–274.

Kilian, A., Chen, J., Han, F., Steffenson, B., Kleinhofs, A., 1997. Towards map-based cloning of the barley stem rust resistance genes *Rpgl* and *rpg4* using rice as an intergenomic cloning vehicle. Plant Mol. Biol. 35, 187–195.

Kilian, A., Kudrna, D.A., Kleinhofs, A., Yano, M., Kurata, N., Steffenson, B., et al., 1995. Rice–barley synteny and its application to saturation mapping of the barley *Rpg1* region. Nucleic Acids Res. 23, 2729–2733.

Kleinhofs, A., Chao, S., Sharp, P.J., 1988. Mapping of nitrate reductase genes in barley and wheat. In: Miller, T.E., R. Koebner, R.M.D. (Eds.) Proceedings of the Seventh International Wheat Genetics Symposium, vol. I. Institute of Plant Science Research, Cambridge, pp. 541–546.

Kleinhofs, A., Kilian, A., Saghai-Maroof, M.A., Biyashev, R.M., Hayes, P.M., Chen, F.Q., et al., 1993. A molecular, isozyme and morphological map of barley (*Hordeum vulgare*) genome. Theor. Appl. Genet. 86, 705–712.

Knupffer, H., 2009. Triticeae genetic resources in *ex situ* genebank collections. I. In: Feuillet, C., Muehlbauer, G.J. (Eds.), Genetics and Genomics of the Triticeae, Plant Genetics and Genomics: Crops and Models 7. Springer, New York.

Kolodinska, B.A., von Bothmer, R., Dayeg, C., Rashal, I., Tuvesson, S., Weibull, J., 2004. Inter simple sequence repeat analysis of genetic diversity and relationships in cultivated barley of Nordic and Baltic region. Hereditas 14, 186–192.

Komatsuda, T., Tanno, K., 2004. Comparative high resolution map of the six-rowed spike locus 1 (vrs1) in several populations of barley *Hordeum vulgare* L. Hereditas 141 (1), 68–73.

Kraakman, A.T.W., Niks, R.E., Van den Berg, P., Stam, P., Van Eeuwijk, F.A., 2004. Linkage disequilibrium mapping of yield and yield stability in modern spring barley cultivars. Genetics 168 (1), 435–446.

Laurie, D.A., Pratchett, N., Bezant, J.H., Snape, J.W., 1995. RFLP mapping of five major genes and eight quantitative trait loci controlling flowering time in a winter × spring barley (*Hordeum vulgare* L.) cross. Genome 38, 575–585.

Love, A., 1982. Generic evolution of the wheat grass. Biol. Abstr. 101, 199–212.

Love, A., 1984. Conspectus of the Triticeae. Feddes Repert. 95, 425–521.

Lundqvist, U., Franckowiak, J.D., Konishi, T., 1997. New and revised descriptions of barley genes. Barley Genet. Newsl. 26, 22–516.

Mano, Y., Takeda, K., 1997. Mapping quantitative trait loci for salt tolerance at germination and the seedling stage in barley (*Hordeum vulgare* L.). Euphytica 94, 263–272.

Marcel, T.C., Varshney, R.K., Barbieri, M., Jafary, H., de Kock, M.J.D., Graner, A., et al., 2007. A high density consensus map of barley to compare the distribution of QTLs for partial resistance to *Puccinia hordei* and of defence gene homologues. Theor. Appl. Genet. 114, 487–500.

Martos, V., Royo, C., Rharrabti, Y., Garcia del Moral, L.F., 2005. Using AFLPs to determine phylogenetic relationships and genetic erosion in durum wheat cultivars released in Italy and Spain throughout the 20th century. Field Crops Res. 91, 107–116.

Matus, I., Corey, A., Filichkin, T., Hayes, P.M., Vales, M.I., Kling, J., et al., 2003. Development and characterization of recombinant chromosome substitution lines (RCSLs) using *Hordeum vulgare* subsp. *spontaneum* as a source of donor alleles in a *Hordeum vulgare* subsp. *vulgare* background. Genome 46, 1010–1023.

Miyazaki, C., Osanai, E., Saeki, K., Hirota, N., Ito, K., Ukai, Y., et al., 2000. Construction of a barley RFLP linkage map using an F2 population derived from a cross between Ko A and Mokusekko 3. Barley Genet. Newsl. 30, 41–43.

Mohamed, N.E.M., Said, A.A., Naz, A.A., Bauer, A., Mathew, B., Reinders, A., Anne, R., Jens, L., 2014. Association mapping for shoot traits related to drought tolerance in barley. Int. J. Agric. Innov. Res. 3 (1), 1473–2319.

Molina-Cano, L., FraMon, P., Salcedo, O., Aragoncilo, C., Roca De Togores, F., Garcia-Olmedo, F., 1987. Morocco as a possible domestication centre for barley: biochemical and agro-morphological evidences. Theor. Appl. Genet. 73, 531–536.

Morrell, P.L., Clegg, M.T., 2007. Genetic evidence for a second domestication of barley (*Hordeum vulgare*) east of the Fertile Crescent. Proc. Natl. Acad. Sci. USA 104, 3289–3294.

Nevski, S.A., 1941. Beitragezurkenntnis der wildwachsendengersten in zusammenhangmit der Frage den Ursprung von *Hordeum vulgare* L. und *Hordeum distichon* L. zuklaren (Versucheiner Monographie der Gattung *Hordeum* Trudy). Botanski Inst. Akad. Nauk. SSSR Ser. 15, 255–264.

Nevo, E., Beiles, A., Zohary, D., 1986. Genetic resources of wild barley in the Near East: structure, evolution and application in breeding. Biol. J. Linnean Soc. 27, 255–380.

Nevo, E., Brown, A.H.D., Zohary, D., 1979. Genetic diversity in the wild progenitor of barley in Israel. Experimentia 35, 1027–1029.

Nevo, E., 1992. Origin, evolution, population genetics and resources for breeding of wild barley, *Hordeum spontaneum*, in the Fertile Crescent. In: Shewry, P.R. (Ed.), Barley: Genetics, Biochemistry, Molecular Biology and Biotechnology. C.A.B. International, UK, pp. 19–43.

Okada, Y., Kanatani, R., Arai, S., Asakura, T., Ito, K., 2003. Production of a novel virus-resistant barley line introgression to the rym1 locus with high malting quality using DNA marker assisted selection. J. Inst. Brewing 109, 99–102.

Pauli, W.D., 2014. Application of Genomic Assisted Breeding for Improvement of Barley Cultivars. PhD Thesis, Montana State University, Bozeman, Montana.

Pickering, R., Ruge-Wehling, B., Johnston, P.A., Schweizer, G., Ackermann, P., Wehling, P., 2006. The transfer of a gene conferring resistance to scald (*Rhynchosporium secalis*) from *Hordeum bulbosum* into *H. vulgare* chromosome 4HS. Plant Breed. 125 (6), 576–579.

Pinheiro, M.A.A., Bebeli, P.J., Bettencourt, E., Dias, S., Dos Santos, T.M.M., Slaski, K.J., 2013. Cereal landraces genetic resources in worldwide genebanks. A review.

Qi, X., Stam, P., Lindhout, P., 1996. Comparison and integration of four barley genetic maps. Genome 39, 379–394.

Qi, X., Stam, P., Lindhout, P., 1998. Use of locus-specific AFLP markers to construct a high-density molecular map in barley. Theor. Appl. Genet. 96, 376–384.

Ramsay, L., Macaulay, M., Ivanissevich, D.S., MacLean, K., Cardle, L., Fuller, J., et al., 2000. A simple sequence repeat-based linkage map of barley. Genetics 156, 1997–2005.

Reinheimer, J.L., Barr, A.R., Eglinton, J.K., 2004. QTL mapping of chromosomal regions conferring reproductive frost tolerance in barley (*Hordeum vulgare* L.). Theor. Appl. Genet. 109, 1267–1274.

Ren, X., Dongfa, S., Genlou, S., Chengdao, L., Wubei, D., 2013. Molecular detection of QTL for agronomic and quality traits in a doubled haploid barley population. Aust. J. Crop Sci. 7 (6), 878–886.

Rocha, F., Bettencourt, E., Gaspar, C., 2008. Genetic erosion assessment through the re-collecting of crop germplasm. Counties of Arcode Valdevez, Melgaço, Montalegre, Ponte da Barca and Terras de Bouro (Portugal). Plant Genet. Resour. Newsl. 154, 6–13.

Rostoks, N., Ramsay, L., MacKenzie, K., Cardle, L., Bhat, P.R., Roose, M.L., et al., 2006. Recent history of artificial outcrossing facilitates whole-genome association mapping in elite inbred crop varieties. Proc. Natl. Acad. Sci. USA 103, 18656–18661.

Ruge, B., Linz, A., Pickering, G., Greif, P., Wehling, P., 2003. Mapping of $Rym14^{Hb}$, a gene introgressed from *Hordeum bulbosum* and conferring resistance to BaMMV and BaYMV in barley. Theor. Appl. Genet. 107, 965–971.

Ruge-Wehling, B., Linz, A., Habekuß, A., Wehling, P., 2006. Mapping of $Rym16^{Hb}$, the second soil-borne virus-resistance gene introgressed from *Hordeum bulbosum*. Theor. Appl. Genet. 113, 867–873.

Ruiz, M., Rodriguez-Quizano, M., Metakovsky, E.V., Vazquez, J.F., Carrillo, J.M., 2002. Polymorphism, variation and genetic identity of Spanish common wheat germplasm based on gliadins alleles. Field Crops Res. 79, 185–196.

Salamini, F., Ozkan, H., Brandolini, A., Schafer-Pregl, R., Martin, W., 2002. Genetics and geography of wild cereal domestication in the Near East. Nat. Rev. Genet. 3, 429–441.

Salvo-Garrido, H., Laurie, D.A., Jaffe, B., Snape, J.W., 2001. An RFLP map of diploid *Hordeum bulbosum* L. and comparison with maps of barley (*H. vulgare* L.) and wheat (*Triticum aestivum* L.). Theor. Appl. Genet. 103, 869–880.

Schmierer, D.A., Kandemir, N., Kudrna, D.A., Jones, B.L., Ullrich, S.E., Kleinhofs, A., 2004. Molecular marker-assisted selection for enhanced yield in malting barley. Mol. Breed. 14, 463–473.

Shin, J.S., Corpuz, L., Chao, S., Blake, T.K., 1990. A partial map of the barley genome. Genome 33, 803–808.

Simpson, G., 1990. Seed Dormancy in Grasses. Cambridge University Press, UK.

Singh, A.K., Rossnagel, B.G., Scoles, G.J., Pickering, R.A., 2004. Identification of a qualitatively inherited source of *Hordeum bulbosum* derived scald resistance from barley line 926K2/11/1/5/1. Can. J. Plant Sci. 84, 935–938.

Skinner, J.S., Szucs, P., von Zitzewitz, J., Marquez-Cedillo, L., Filichkin, T., Stockinger, E.J., et al., 2006. Mapping of barley homologs to genes that regulate low temperature tolerance in *Arabidopsis*. Theor. Appl. Genet. 112, 832–842.

Smilde, D.W., Haluskova, J., Sasaki, T., Graner, A., 2001. New evidence for the synteny of rice chromosome 1 and barley chromosome 3H from rice expressed sequence tags. Genome 44, 361–367.

Snape, J.W., Sarma, R., Quarrie, S.A., Fish, L., Galiba, G., Sutka, J., 2001. Mapping gene for flowering time and frost tolerance in cereals using precise genetic stocks. Euphytica 120, 309–315.

Steffenson, B.J., Hayes, P.M., Kleinhofs, A., 1996. Genetics of seedling and adult plant resistance to net blotch (*Pyrenophora teres* f. *teres*) and spot blotch (*Cochliobolus sativus*) in barley. Theor. Appl. Genet. 92, 552–558.

Schulte, D., Close, T.J., Graner, A., Langridge, P., Matsumoto, T., Muehlbauer, G., Sato, K., Schulman, A.H., Waugh, R., Wise, R.P., Stein, N., 2009. The international barley sequencing consortium – at the threshold of efficient access to the barley genome. Plant Physiol. 149, 142–147.

Tacconi, G., Cattivelli, L., Faccini, N., Pecchioni, N., Stanca, A.M., Valè, G., 2001. Identification and mapping of a new leaf stripe resistance gene in barley (*Hordeum vulgare* L.). Theor. Appl. Genet. 102, 1286–1291.

Takahashi, R., 1955. The origin and evolution of cultivated barley. In: Demerec, M. (Ed.), Advances in Genetics. Academic Press, New York, pp. 227–266.

Takahashi, R., Hayashi, S., Yasuda, S., Hiura, U., 1963. Characteristics of the wild and cultivated barleys from Afghanistan and its neighbouring regions. Ber. Ohara Inst. Landwirtsch. Biol. Okayama Univ. 12, 1–23.

Takahashi, R.I., Hayashi, T., Hiura, U., Yasuda, S., 1968. A study of cultivated barley from Nepal, Himalayas and North India with special reference to their phylogenetic differentiation. Ber. Ohara Inst. Landwinsch. Biol. Okayama Univ. 14, 85–122.

Taketa, S., Amano, S., Tsujino, Y., Sato, T., Saisho, D., Kakeda, K., Nomura, M., et al., 2008. Barley grain with adhering hulls is controlled by an ERF family transcription factor gene regulating a lipid biosynthesis pathway. Proc. Natl. Acad. Sci. USA 105 (10), 4062–4067.

Tanno, K., Taketa, S., Takeda, K., Komatsuda, T., 2002. A DNA marker closely linked to the vrs1 locus (row-type gene) indicates multiple origins of six-rowed cultivated barley (*Hordeum vulgare* L.). Theor. Appl. Genet. 104, 54–60.

Tester, M., Langridge, P., 2010. Breeding technologies to increase crop production in a changing world. Science 327, 818–822.

Thomsen, S.B., Jensen, H.P., Jensen, J., Skou, J.P., Jorgensen, J.H., 1997. Localization of a resistance gene and identification of sources of resistance to barley leaf stripe. Plant Breed. 116, 455–459.

Tilman, D., Balzer, C., Hill, J., Befort, B.L., 2011. Global food demand and the sustainable intensification of agriculture. Proc. Natl. Acad. Sci. USA 108, 20260–20264.

Valè, G., Nicola, P., Davide, B., Gianni, T., 2005. Genetic basis of barley resistance to the leaf stripe agent *Pyrenophora graminea*. In: Tuberosa, R., Phillips, R.L., Gale, M. (Eds.), Proceedings of the International Congress "In the Wake of the Double Helix: From the Green Revolution to the Gene Revolution", May 27–31, 2003, Bologna, Italy, pp. 267–278.

Varshney, R.K., Paulob, M.J., Grandod, S., van Eeuwijkb, F.A., Keizerb, L.C.P., Guod, P., et al., 2012. Genome wide association analyses for drought tolerance related traits in barley (*Hordeum vulgare* L.). Field Crops Res. 126, 171–180.

Vavilov, N.I., 1926. Studies on the origin of cultivated plants. Bull. Appl. Bot. Genet. Plant Breed. 16, 1–248.

Verma, R.P.S., Kumar, V., Sarkar, B., Kharub, A.S., Kumar, D., Selakumar, R., et al., 2012. Barley cultivars released in India: name, parentages, origins and adaptations. Directorate of Wheat Research, Karnal, Haryana. Res. Bull. 29, 26.

Visioni, A., Tondelli, A., Francia, E., Pswarayi, A., Malosetti, M., Russell, J., et al., 2013. Genome-wide association mapping of frost tolerance in barley (*Hordeum vulgare* L.). BMC Genomics 14, 424.

von Bothmer, R., Jacobsen, N., Baden, C., Jørgensen, R.B., Linde-Laursen, I., 1995. An ecogeographical study of the genus Hordeum, second ed. Systematic and Ecogeographic Studies on Crop Genepools 7. International Plant Genetic Resources Institute, Rome.

von Bothmer, R., Jacobsen, N., Jørgensen, R.B., Linde-Laursen, 1991. An Ecogeographical Study of the Genus *Hordeum*. Systematic and Ecogeographic Studies on Crop Genepools 7. International Board for Crop Genetic Resources, Rome.

von Bothmer, R., Komatsuda, T., 2011. Barley origin and related species. In: Ullrich, S.E. (Ed.), Barley: Production, Improvement & Uses. Blackwell Publishing Ltd, Oxford, UK, pp. 14–68.

Von Korff, M.H., Wang, H., Le'on, J., Pillen, K., 2004a. Development of candidate introgression lines using an exotic barley accession (*Hordeum vulgare* ssp. *spontaneum*) as donor. Theor. Appl. Genet. 109, 1736–1745.

Von Korff, M.H., Wang, H., Le'on, J., Pillen, K., 2004. Detection of QTL for agronomic traits in an advanced backcross population with introgressions from wild barley (*Hordeum vulgare* ssp. spontaneum). Proceedings of the Seventeenth EUCARPIA General Congress, Tulln, Austria, September 8–11, 2004, pp. 207–211.

Walther, U., Rapke, H., Proeseler, G., Szigat, G., 2000. *Hordeum bulbosum* – a new source of disease resistance-transfer of disease resistance to leaf rust and mosaic viruses from *H. bulbosum* into winter barley. Plant Breed. 199, 215–218.

Wang, J., Yang, J., Jia, Q., Zhu, J., Shang, Y., 2014. A new QTL for plant height in barley (*Hordeum vulgare* L.) showing no negative effects on grain yield. PLoS ONE 9 (2), e90144.

Wang, M., Jiang, N., Jia, T., Leach, L., Cockram, J., Waugh, R., et al., 2012. Genome-wide association mapping of agronomic and morphologic traits in highly structured populations of barley cultivars. Theor. Appl. Genet. 124, 233–246.

Wang, M.L., Atkinson, M.D., Chinoy, C.N., Devos, K.M., Gale, M.D., 1992. Comparative RFLP-based genetic maps of barley chromosome-5 (1H) and rye chromosome-1 R. Theor. Appl. Genet. 84, 339–344.

Werner, K., Friedt, W., Ordon, F., 2005. Strategies for pyramiding resistance genes against the barley yellow mosaic virus complex (BaMMV, BaYMV, BaYMV-2). Mol. Breed. 16, 45–55.

Witzel, K., Weidner, A., Surabhi, G.K., Borner, A., Mock, H.P., 2009. Salt stress-induced alterations in the root proteome of barley genotypes with contrasting response towards salinity. J. Exp. Bot. 60, 3545–3557.

Xue, D.W., Huang, Y.Z., Zhang, X.Q., Wei, K., Westcott, S., Li, C.D., et al., 2009. Identification of QTLs associated with salinity tolerance at late growth stage in barley. Euphytica 169, 187–196.

Xue, D.W., Zhou, M.X., Zhang, X.Q., Chen, S., Wie, K., Zeng, F.R., et al., 2010. Identification of QTLs for yield and yield components of barley under different growth conditions. J. Zhejiang Univ. Sci. B 11 (3), 169–176.

Yoshiro, M., Takeda, K., 1997. Mapping quantitative trait loci for salt tolerance at germination and the seedling stage in barley (*Hordeum vulgare* L.). Euphytica 94, 263–272.

Yu, G.X., Bush, A.L., Wise, R.P., 1996. Comparative mapping of homoeologous group 1 regions and genes for resistance to obligate biotrophs in *Avena*, *Hordeum*, and *Zea mays*. Genome 39, 155–164.

Zohary, D., Hopf, M., 1993. Domestication of Plants in Old World. The Origin and Spread of Cultivated Plants in West Asia, Europe and the Nile Valley. Clarendon Press, Oxford.

Oat

4

*Maja Boczkowska**, *Wiesław Podyma*[†,**], *Bogusław Łapiński*[‡]
*Department of Functional Genomics, Plant Breeding and Acclimatization Institute (IHAR), National Research Institute, Radzików, Poland; [†]Organic Farming Section, Plant Breeding and Acclimatization Institute (IHAR), National Research Institute, Radzików, Poland; **Laboratory of Gene Bank, Polish Academy of Sciences Botanical Garden, Center for Biological Diversity Conservation in Powsin, Warsaw, Poland; [‡]National Centre for Plant Genetic Resources, Plant Breeding and Acclimatization Institute (IHAR), National Research Institute, Radzików, Poland

4.1 Introduction

Oat is a minor cereal used primarily for animal feed, human food, and industry purposes. It is the seventh most economically important cereal after corn, rice, wheat, barley, sorghum, and millet. It is cultivated all over the world, but majority of its production is located in the Northern Hemisphere between the latitudes of 40° and 60°N in North America, Europe, and Asia. The main producers are Russia, Canada, and Poland. Over the last decade, the average annual production was around 23 million tons per year and this is half of the production in the 1960s (FAOSTAT, 2013). This occurred due to the increase of mechanization and production of more profitable cereals, decrease of horse population, and higher specialization in the agriculture sector. Oat is better adapted to cool, moist climate and acidic soils than other cereals, but it is sensitive to water deficit and heat during seed formation and maturity (Murphy and Hoffman, 1992). The highest yields were recorded in European Nordic countries, those in North America are ~30% lower, while in Africa the average yield does not exceed 30% of the European level (FAOSTAT, 2013). Such large differences are caused by many factors; the most important are climate and advancement in agriculture, including breeding of crops. Common oat (*Avena sativa* L.) is the cultivated species, but in some regions three other species are also cultivated, that is, red oat (*Avena byzantina* C. Koch.), bristle oat (*Avena strigosa* Schreb.), and Ethiopian oat (*Avena abyssinica* Hochst.).

Oats were domesticated much later than wheat and barley, although all of them come from the Middle East. The cultivation began no earlier than 4000 years ago (Zohary and Hopf, 1988). Archeologic sites indicated that oats initially appeared as a weed in wheat and barley fields (Hansen and Renfrew, 1978; Hillman et al., 1989; Hopf, 1969; Renfrew, 1969). Domestication of oats probably occurred outside the center of origin. Although the exact time and place remain unknown, it is believed that during the Neolithic revolution, it had been accidentally introduced further to the north of Europe as a weed in cereal seeds. Due to better adaptation to the cooler and more humid climate, oat gained advantage over wheat and barley (Murphy and Hoffman, 1992). It is evident that the domestication of all four species of oats was independent (Harlan, 1977; Zohary, 1971). The earliest archeologic findings of domesticated forms of oats

Genetic and Genomic Resources for Grain Cereals Improvement. http://dx.doi.org/10.1016/B978-0-12-802000-5.00004-6

in central Europe have been dated to the beginning of the modern era (Zohary and Hopf, 1988). In a similar time, oat cultivation appeared in China (Baum, 1977). In medieval Europe oat served both as feed and food and was cultivated usually on poor soils unsuitable for cultivation of other cereals (Moore-Colyer, 1995). Oat with other cereals was introduced into the Americas in the sixteenth and seventeenth centuries. As the first one, red oat, was brought by Spaniards to the southern parts of North America to South America; common oat came later with the English and German settlers (Murphy and Hoffman, 1992). The English settlers introduced oat to Australia as well (Holland, 1997). Oats are grown for grain, pasture, forage, and bedding or as a rotation crop depending on the regions in the world. Traditionally, it has served as horse feed, but it is used for calves and young stock, and poultry and sheep feeding as well (Andersson and Börjesdotter, 2011; Marshall et al., 2013). Only 10% of oats produced is used for human consumption (Andersson and Börjesdotter, 2011). Nowadays, oat grain and its derivatives belong to the group of trendy, healthy, functional food. People, especially belonging to developed countries, are eating not only porridge but also oat pasta, bread, rolls, cookies, cakes, and so on (Andersson and Börjesdotter, 2011). This results from the content of β-glucan, which improves the cardiovascular system by reducing blood cholesterol level (Ryan et al., 2007). Oat is a good source of valuable proteins, lipids, vitamins, minerals, and antioxidants (Peterson, 2001). The unique composition of oat grain caused an increase in interest from the food, pharmaceutical, and cosmetic industries (Marshall et al., 2013). In some regions of the world, oat is used primarily as forage. It grows mainly in cold areas with very short vegetation period (Suttie and Reynolds, 2004). Recent studies on oats have focused on diverse topics such as nutrition value in human diet, special dietary needs of livestock and race horses, role in sustainable agriculture, and using grain as a biofuel. In this chapter, the genetic and genomic resources of oat are reviewed mainly for origin, domestication, diversity, and taxonomy. Also addressed are its conservation, characterization, evaluation, and maintenance status of germplasm and its use in crop improvement.

4.2 Origin, distribution, and diversity

4.2.1 Taxonomy

The genus *Avena* L. belongs to tribe Aveneae of the subfamily Pooideae in the family Poaceae. It includes a group of diploid, tetraploid, and hexaploid species in which all but one are annual and self-pollinated. The species classification method and their names vary depending on the taxonomist; since the number of species distinguished varies between 7 (Ladizinsky and Zohary, 1971) and 30 (Baum, 1977). The first attempt of classification of this genus was made by (Linnaeus, 1753, 1762) and since then, many other taxonomists published systematics of genus *Avena*. The most detailed morphologic classification was prepared by Malzev (1930). According to him, the genus *Avena* contains two subsections, *Aristulatae* Malz. and *Denticulatae* Malz. The first one is divided into three sections, namely, seria *inaequaliglumis* Malz. containing two species, that is, *Avena clauda* Dur. and *Avena pilosa* M. B. seria *Stipitae* Malz. containing *Avena longiglumis* Dur. and *Avena ventricosa* Balan., and seria *Eubarbatae* Malz. containing only one species, *A. strigosa* Schreb. In the second subsection, two species are included, that is, *Avena fatua* L. and *Avena sterilis* L. Both of these species, as well as *A. strigosa* and *A. ventricosa*, are divided into subspecies. In total, the genus *Avena* contains 22 subspecies. For more details please see Table 4.1. The second very important and also often citied classification was published by Baum (1977). Based on 27 characteristics such as type of lodicules and ploidy level, he separated 7 sections and located in them as many as 30 species (Table 4.1). In stark contrast to this classification is the one conceived 6 years earlier by Ladizinsky and Zohary (1971), in which only seven so-called "biological species" were sectioned, that is, *A. clauda* Dur., *A. ventricosa* Bal., *A. longiglumis* Dur., *A. strigosa* Schreb., *Avena magna* Murphy et Terrell, *Avena murphyi* Ladiz., and *A. sativa* L. Currently according to the new scientific finding Ladizinsky (2012) recognizes 13 biological species. The classifications of Malzev, Baum, and Ladizinsky and Zohary are listed in Table 4.1. The most recent *Avena* taxonomy, which is used for purposes of this chapter, was prepared by Loskutov (2008). In it, based on detailed morphologic traits, 26 species divided into two subgenus *Avenastrum* (C. Koch) Losk. and *Avena* were identified (Table 4.2). The main taxonomic features are related to the structure of generative organs. The complete descriptors to identify the species of the genus *Avena* was published by Loskutov and Rines (2011).

4.2.2 Diversity

Morphologic differentiation within the genus as well as within the species is quite significant. Plants with the lowest height (30 cm) were observed in *A. sterilis* and the tallest in *Avena barbata* (210 cm). Out of 26 species reported, 15 are characterized by erect type of flowering stem, 8 have a geniculate one, and in 3 both types are present. Differentiation is also observed in the way in which juvenile plants are growing. In 10 species it is prostrate, in 1 erect, and in the remaining 15 it is variable. In *A. sativa*, *A. strigosa*, and *A. longiglumis* occur as uni- and equilateral type of panicle, while the other species have either an equilateral (12 species) or unilateral (11 species) one. Number of florets varies from two, for instance in *A. sativa*, *A. abyssinica*, *A. strigosa*, up to eight in *Avena macrostachya*. The main difference between domesticated and wild species is in florets separation from the glumes. The cultivated species have no disarticulating florets and the wild ones have all or only lower florets disarticulating at the mature plant. For more details please see Table 4.3. Significant morphologic variation also occurs within species, for example, in *A. sativa* 31 botanical varieties have been described, in *A. byzantina* 14, and in *A. strigosa* and *A. abyssinica*, 7 and 6 varieties, respectively. The complete key to identifying the species of the *Avena* genus is included in the review of Loskutov and Rines (2011).

Variation within this genus relates to the ploidy level as well. Species form a polyploid series of di-, tetra-, and hexaploids. The basic number of chromosomes is always equal to seven. Four basic genome types, that is, A, B, C, and D, are distinguished. Based on numerous studies it is considered that A and C genomes are the most initial ones (reviewed by Loskutov and Rines (2011)). Additionally, several variants of the A genome were identified while the C-genome underwent only minor changes (reviewed by Loskutov and Rines (2011)). On the basis of chloroplast genome DNA sequences it has been suggested that an A-genome diploid species might be a maternal

Table 4.1 Summary of three different taxonomy of genus *Avena*

Malzev (1930)		Baum (1977)		Ladizinsky and Zohary (1971)
Species	**Subspecies**	**Section**	**Species**	**Biological species**
A. clauda Dur.		*Avenatrichon* (Holub) Baum	*A. macrostachya* Bal. ex Coss. et Dur.	*A. clauda* Dur.
A. pilosa M.B.				*A. ventricosa* Bal.
A. longiglumis Dur.		*Ventricosa* Baum	*A. clauda* Dur.	*A. longiglumis* Dur.
			Avena eriantha Dur.	*A. strigosa* Schreb.
A. ventricosa Balan.	*ventricosa* (Balan.) Malz.		*A. ventricosa* Bal. ex Coss.	*A. magna* Murphy et Terrell
				A. murphyi Ladiz.
	bruhnsiana (Grun.) Malz.			*A. sativa* L.
A. strigosa Schreb.	*strigosa* (Schreb.) Thell.	*Agraria* Baum	*Avena brevis* Roth	
	hirtula (Lagas.) Malz.		*Avena hispanica* Ard.	
	barbata (Pott) Thell.		*A. nuda* L.	
	wiestii (Steud.) Thell.		*A. strigosa* Schreb.	
		Tenuicarpa Baum	*A. agadiriana* Baum et Fed.	
			A. atlantica Baum	
	vaviloviana Malz.		*A. barbata* Pott ex Link	
	abyssinica (Hochst.) Thell.		*A. canariensis* Baum, Rajh. et Samp.	
A. fatua L.	*septentrionalis* Malz.		*A. damascena* Rajh. et Baum	
	nodipilosa Malz.		*A. hirtula* Lag.	
	meridionalis Malz.		*A. longiglumis* Dur.	
	macrantha (Hack.) Malz.		*Avena lusitanica* Baum	
	fatua (L.) Thell.		*Avena matritensis* Baum	
	sativa (L.) Thell.		*A. prostrata* Ladiz.	
	cultiformis Malz.		*A. wiestii* Steud.	
	praegravis (Kraus.) Malz.	*Ethiopica* Baum	*A. abyssinica* Hochst.	

A. sterilis L.	*ludoviciana* (Dur.) Gill. et Magn.		*A. vaviloviana* (Malz.) Mordv.	
	pseudo-sativa Thell.			
	trichophylla (C. K. et Hausskn.) Malz.	*Pachycarpa* Baum	*A. maroccana* Gdgr.	
	nodipubescens Malz.		*A. murphyi* Ladiz.	
	macrocarpa (Monch.) Briq.	*Avena* Baum	*Avena atherantha* Presl.	
	byzantina (C. K.) Thell.		*A. fatua* L.	
			Avena hybrida Peterm.	
			A. occidentalis Dur.	
			A. sativa L.	
			A. sterilis L.	
			Avena trichophylla C. Koch	

Table 4.2 Taxonomy of genus *Avena* based on Loskutov and Rines (2011) with information about genome type, ploidy level, and distribution

Subgenus	Section	Species	Genome	2*n*	Distribution
Avenastrum Koch		*A. macrostachya* Bal. ex Coss. et Dur.*,**	C	28	Algeria Atlas Mountains
Avena L.	*Aristulatae* (Malz.)	*A. clauda* Dur.[†]	C_p	14	Bulgaria, Greece, Turkey, Iran, Iraq, Uzbekistan, Azerbaijan, Jordan, Israel, Lebanon, Syria, Algeria, Morocco
		A. pilosa M.B.[†]	C_p	14	Spain, Greece, Bulgaria, Ukraine, Russia, Iran, Turkey, Iraq, Uzbekistan, Syria, Jordan, Israel
		A. prostrata Ladiz.	A_p	14	Spain, Morocco
		A. damascena Raj. et Baum[†]	A_d	14	Syria, Morocco
		A. longiglumis Dur.	A_l	14	Spain, Portugal, Greece, Italy, Syria, Libya, Morocco, Algeria, Israel, Jordan
		A. wiestii Steud.	A_s	14	Spain, Azerbaijan, Turkey, Iraq, Iran, Syria, Jordan, Israel, Algeria, Egypt, northern Sahara, Arabic Peninsula
		A. atlantica Baum*	A_s	14	Morocco
		A. strigosa Schreb.[‡]	A_s	14	Europe
		A. hirtula Lagas.	A_s	14	Spain, Portugal, France, Italy, Greece, Algeria, Morocco, Tunisia, Israel, Turkey, Syria, Jordan
		A. barbata Pott[†]	AB	28	Mediterranean basin, European Atlantic coast, Asia Minor, Himalayas, Ethiopia, Brazil, Japan, Australia
		A. abyssinica Hochst.[‡]	AB	28	Ethiopia, Saudi Arabia, Algeria
		A. vaviloviana (Malz.) Mordv.*	AB	28	Ethiopia, Eritrea, Yemen

	Avenae (L.)	*A. ventricosa* Balan.	C_v	14	Cyprus, Algeria, Iraq
		A. bruhnsiana Grun.*	C_v	14	Azerbaijan
		A. canariensis Baum*	A_c	14	Canary Islands
		A. agadiriana Baum et Fed.*	AB?	28	Morocco
		A. magna Mur. et Terr.*	AC	28	Morocco
		A. murphyi Ladiz.	AC	28	Spain, Morocco
		A. insularis Ladiz.	CD?	28	Sicily, Tunisia
		A. fatua L.†	ACD	42	All over the world
		A. sterilis L.	ACD	42	Spain, Portugal, Italy, Switzerland, France, Iraq, Turkey, Ukraine, Northern Africa, Ethiopia, Japan, South Korea,
		A. byzantina C. Koch‡	ACD	42	Spain, Portugal, North Africa, Brazil, Australia
		A. occidentalis Dur.	ACD	42	Canary Islands, Portugal, Egypt, Ethiopia, Azores, Madeira, Algeria
		A. ludoviciana Dur.	ACD	42	Europe, Ukraine, Russia, Azerbaijan, Central and south-western Asia, Iran, Asia Minor, Afghanistan, Northern Africa, Mediterranean basin, Australia, New Zealand
		A. sativa L.‡	ACD	42	All over the world

* Endemic.
** Perennial.
†Weeds.
‡ Cultivated.

Table 4.3 Morphologic description of oat species (Loskutov and Rines, 2011; Clayton et al., 2006)

Oat species	Life cycle	Juvenile growth	Flowering stem	Height (cm)	Pannicum	Florets	Glumes	Lemma tips	Florets dis-articulation at maturity
A. clauda	Annual	Prostrate–semi-prostrate	Erect	60–100	Unilateral	3–5	Very unequal	Biaristulate–bisubulate	All
A. pilosa	Annual	Prostrate–semi-prostrate	Erect	55–85	Unilateral	2–3	Very unequal	Biaristulate–bisubulate	Only lower
A. ventricosa	Annual	Prostrate	Erect	65	Unilateral	2	Nearly unequal	Bisubulate	Only lower
A. bruhnsiana	Annual	Prostrate	Erect	70–110	Unilateral	2	Nearly uneqal	Bisubulate	Only lower
A. longiglumis	Annual	Semierect–prostrate	Erect	50–180	equilateral/unilat-eral	2–3	Nearly unequal	Biaristulate	All
A. damascena	Annual	Prostrate	Geniculate	70–80	Equilateral	3	Equal	Biaristulate	All
A. prostrata	Annual	Prostrate	Geniculate	50–60	Unilateral	2–3	Equal	Biaristulate	All
A. canariensis	Annual	Prostrate	Erect	50-75	Unilateral	2–3	Equal	Bidentate	Only lower
A. wiestii	Annual	Semierect–prostrate	Erect	75–140	Equilateral	2	Nearly unequal	Biaristulate	All
A. hirtula	Annual	Semierect–prostrate	Erect	70–150	Equilateral	2–3	Nearly unequal	Biaristulate	All
A. atlantica	Annual	Prostrate	Geniculate	95	Equilateral	2–3	Nearly uneqal	Biaristulate	Only lower
A. barbata	Annual	Prostrate–erect	Erect	65–210	Equilateral	2–4	Nearly unequal	Biaristulate	All
A. vaviloviana	Annual	Prostrate–erect	Erect	80–110	Unilateral	2–3	Equal	Biaristulate	All

A. agadiriana	Annual	Prostrate	Geniculate	60	Unilateral	2	Nearly unequal	Bidentate	Only lower
A. magna	Annual	Prostrate	Geniculate	65–100	Unilateral	3–4	Nearly unequal	Bidentate	Only lower
A. murphyi	Annual	Prostrate	Geniculate	70–80	Equilateral	2–6	Equal	Bidentate	Only lower
A. insularis	Annual	Prostrate	Geniculate	60	Unilateral	3–5	Equal	Bisubulate/ shortly biaristate	Only lower
A. macro-stachya	Peren-nial	Semipros-trate	Erect	100	Equilateral	6–8	Very unequal	Bisubulate	All
A. sterilis	Annual	Erect–prostrate	Erect	30–145	Equilateral	3–5	Equal	Bidentate	Only lower
A. ludoviciana	Annual	Erect–prostrate	Erect	40–150	Equilateral	3	Equal	Bidentate	Only lower
A. fatua	Annual	Erect–prostrate	Erect	40–150	Equilateral	2–3	Equal	Bidentate	All
A. occidentalis	Annual	Semierect–prostrate	Geniculate	45–100	Equilateral	3–4	Equal	Bisubulate	All
A. strigosa	Annual	Semipros-trate–erect	Erect/ge-niculate	75–125	equilateral/ unilat-eral	2	Equal	Biaristulate	Nonshatter-ing
A. abyssinica	Annual	Erect	Erect	50–90	Unilateral	2	Equal	Biaristulate	Nonshatter-ing
A. byzantina	Annual	Erect–prostrate	Erect/ge-niculate	60–150	Equilateral	3–4	Equal	Biaristulate	Nonshatter-ing
A. sativa	Annual	Erect–prostrate	Erect/ge-niculate	40–180	Equilateral/ unilat-eral	2	Equal	Biaristulate	Nonshatter-ing

donor of germplasm for all polyploid species (Yan et al., 2014). The same studies indicated that the A genome in the AB-tetraploid species originated from species with the As variant of the genome. The other polyploid species have diverse origins of the A genome (Nikoloudakis and Katsiotis, 2008; Li et al., 2009; Peng et al., 2010; Yan et al., 2014). The B and D genomes were not identified in any of the diploid species. They were most likely formed by evolution from the A genome. The origin of the C genome, suspected of being the male parent in the evolution of the polyploid species, is not entirely clear (Yan et al., 2014). It is considered that the tetraploid species could evolve through duplication of chromosome number or by spontaneous hybridization of two closely related species. Among tetraploids, species with CC, AB, and AC genomes are identified. *A. macrostachya* (CC) is considered as the most primitive tetraploid (Loskutov and Rines, 2011). All hexaploid species have similarly composed karyotype, that is, ACD, which was formed by hybridization of tetraploid species carrying A, C, and D genomes (Loskutov, 2008). The diagram of phylogenetic relationship in *Avena* genus is shown in Fig. 4.1. For more details please read Loskutov and Rines (2011).

4.2.3 Origin and distribution

The western Mediterranean region is considered to be a center of diversity for the genus *Avena* (Baum and Fedak, 1985) and wild *Avena* species mainly occur there and in the Middle East. Only three of the biological species of oats *Avena insularis*, *Avena canariensis*, and *A. macrostachya* have not been recorded in Morocco. It is also interesting that *Avena maroccana*, *Avena agadiriana*, and *Avena atlantica* have not been found

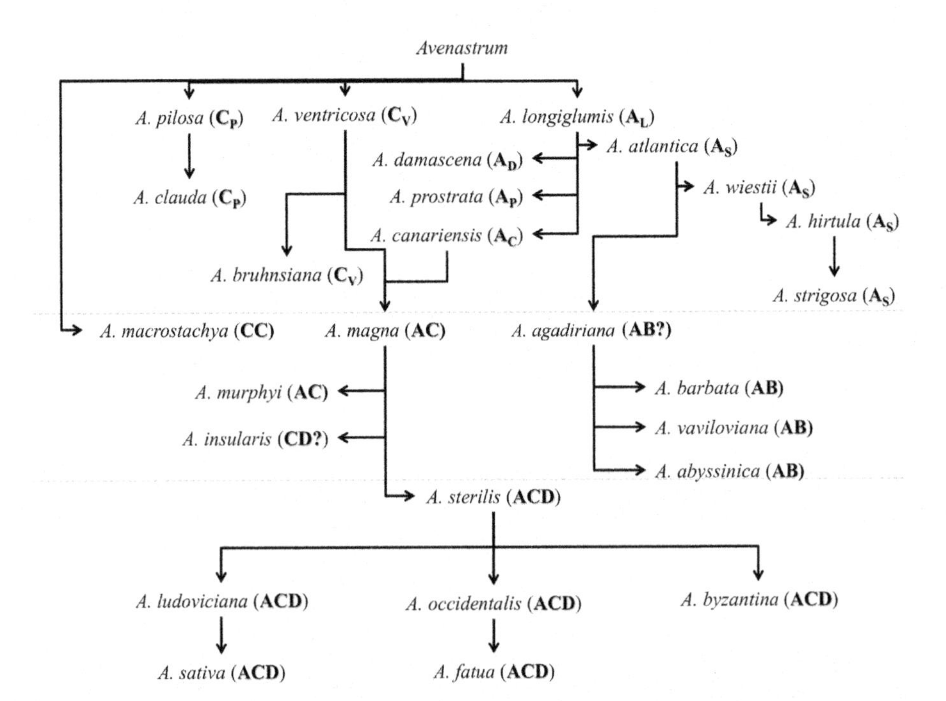

Figure 4.1 Diagram of phylogenetic relationship in *Avena* genus based on Loskutov (2008).

outside Morocco (Leggett et al., 1992). The secondary center of formation of *Avena* species and origin of cultivated oat (*A. sativa*) is situated within the Asia Minor center of crop origin. The analysis of intraspecific diversity of landraces helped to identify centers of morphogenesis for all cultivated oat species. The center of origin and diversity for the diploid species (*A. strigosa*) is Spain with Portugal, for the naked forms (*A. sativa* subsp. *nudisativa* (Husnot.) Rod. et Sold.) is Great Britain, for the tetraploid species (*A. abyssinica*) is Ethiopia, for the hexaploid species (*A. byzantina*) it is Algeria and Morocco, for the hulled forms of *A. sativa* it is Iran, Georgia, and Russia, and for its unhulled forms, it is Mongolia and China (Loskutov, 2008). The occurrence of the largest diversity of polyploids in the eastern part of Anterior Asia, where soil and climate conditions are harder than in the western Mediterranean areas, was confirmed by Vavilov (1926) with his statement about greater hardiness of this group of species, as compared with the diploid ones. Allopolyploid species promoted the development of extremely differentiated ecotypes, which played an important role in the evolution. Proceeding from the center of origin toward the southwestern Asiatic center, smaller-seeded and more adaptive hexaploid forms of wild species began to occur (Loskutov, 2007, 2008). As shown in Table 4.2, among 26 oat species 4 were domesticated and are cultivated and 5 are treated as weeds. Common oat (*A. sativa*), which is the most economically important species, is spread all over the world. Red oat (*A. byzantina*) is grown in the south of Europe mainly in Spain and Portugal, but also in the north of Africa, Southwest Asia, South America, and Australia. Two other species are cultivated regionally, that is, *A. strigosa* (cultivars in Europe and Brazil, landraces on European islands) and *A. abyssinica* (in Ethiopia). Except for common oat only one species occurs around the world, that is, *A. fatua*, which is considered as a noxious weed. Seven of the wild species of genus *Avena* are endemic and are the most threatened by genetic erosion or even extinction in their natural habitats.

4.3 Erosion of genetic diversity from the traditional areas

The brief descriptions of the habitats, distributions, and threats of *Avena* species in the Mediterranean, Near East, and Anterior Asia areas are based on observations made by participants of expeditions undertaken during the last 25 years (Guarino et al., 1991; Katsiotis and Ladizinsky, 2011; Ladizinsky, 2012; Leggett et al., 1992) and the previous ones.

A. ventricosa is native to Algeria, Cyprus, Iraq, Saudi Arabia, and Azerbaijan, in Europe is present only on the Island of Cyprus. In Cyprus during a survey in 2009 *A. ventricosa* was recorded between Kampos and Nicosia, near the Cypriot Agricultural Institute (ARI), Athalass Farm (National Park), Macheras Mt, and Salt Lake (Alykes) near Larnaka airport (Katsiotis and Ladizinsky, 2011). This is a rare and scattered species, with probably a relic nature, which is found only in undisturbed habitats.

A. pilosa and *A. clauda* grow on steppe mountain passes of the western Transcaucasia and more rarely, in the semidesert and desert zones and mountain regions of Central Asia (Uzbekistan). Both species are found in Asia Minor, Iran, Turkish and Iraqi Kurdistan, Lebanon, and Syria. *A. pilosa* and *A. clauda* form large populations in Turkish and Iraqi Kurdistan and Iran but small and scattered populations in the Mediterranean region (Baum et al., 1972). In Turkey, the populations were found in the regions Chardak and Cheilapinar and on the coast of the Aegean Sea. *A. clauda* is an endangered species in

Israel according to International Union for the Conservation of Nature and Natural Resources (IUCN) status. It occurs in three geographic regions of Lower Galilee, the hills of Judah, and Samaria Desert. The species is protected in a nature reserve Um Zuka in Samaria. Currently, it occurs at four stands, while it is extinct in three additional places (Shmida et al., 2011). A small population was found in Jordanian highlands and in Algeria near the towns of Oran and Batna. In Europe, *A. clauda* was recorded in Bulgaria and Greece near Athens, Attica, Macedonia, and Thrace as well as on the Crete Island. *A. pilosa* also occurs in various parts of Spain. Populations of *A. clauda* and *A. pilosa* were found in Morocco in the foothills of the Middle Atlas Mountains near the town of Azrou. *A. clauda* contaminates wheat and barley and irrigated alfalfa fields. Together with *A. barbata* and *A. pilosa*, it grows along the roads and near various buildings. *A. clauda* is mainly a segetal and ruderal plant. *A. pilosa* is abundant in the maquis among oaks and pistachio trees, in the cenoses of abandoned pastures, on limestone slopes, and in the narrow cracks on mountain slopes (Shelukhina et al., 2008).

A. longiglumis is sporadically distributed on the Iberian Peninsula, North Africa, and Israel, and according to observations done by Leggett et al. (1992) is quite common in Morocco. It consists of two ecotypes, one a coastal type of robust, tall plant with large drop panicles; the other a desert type of shorter, slender plant with small panicles. Both ecotypes are specialized for sandy soils. In Morocco, occurrence of most populations is recorded in eucalyptus plantations where grazing is restricted (Leggett et al., 1992).

A. atlantica is an endemic species of Morocco (El Oualidi et al., 2012). The species was collected during expeditions in 1985–1988 by Baum and Fedak (1985) at the Atlantic coast from Essaouira to Tiznit. Distribution area of the species reaches northwestern slopes of the Atlas Mountains to a height of 1000 m above sea level (Loskutov, 2007). Ladizinsky (2012) examined the *A. atlantica* in its natural habitat in southwestern Morocco in an area with annual rainfall of about 200 mm. Even in a relatively rainy year it is rather rare and forms small and disjunctive populations. The *A. atlantica* was restricted mainly to sandy soil but some populations were also occasionally found on brown soil.

For several years *Avena damascena* was known from a single location in Syria, about 60 km from Damascus in the Syrian Desert (Rajhathy and Baum, 1972). During a collection trip to Morocco, Mike Leggett and Per Hagberg collected seeds from several populations on the east-facing slopes of the Atlas range. The altitude of the collection sites varied from 700 m to 1650 m, and the soil type was variable but basically sandy loam and dry (Leggett et al., 1992). Later, following hybridization with *A. damascena* from Syria, these proved to be of the same species (Leggett, 1992). This was an exciting discovery because it indicated that *A. damascena* occurs at the two extremities of the Mediterranean basin, and raises questions about the origin and evolution of this oat species (Ladizinsky, 2012). In the opinion of Leggett et al. (1992) the species occurs quite widely in Morocco, and it would not appear to be under any immediate extinction threat as long as the goat populations do not increase.

Avena prostrata was identified by Ladizinsky (1971). Morphologically it is very similar to other species such as *Avena hirtula*, *Avena wiestii*, or *A. barbata*, and usually grows in mixed stands with them in restricted areas of southeast Spain (Murcia and Almeria Provinces). It also occurs at a number of locations in Morocco (Ladizinsky, 2012; Leggett et al., 1992). Because of the mountainous terrain and poor quality soils on which this species grows in Spain, it seems unlikely that Spanish populations are under any immediate threat of extinction. However, in the area of Morocco where *A. prostrata* was collected (rich red-brown soil), there is a very real threat due to the encroachment of improved farming practice (Leggett et al., 1992). The European Cooperative Programme for Plant Genomic Resources (ECPGR) considers that *A. prostrata* should possibly be a Red List species based on a small number of described populations and because the primary habitat where it grows has been progressively taken over by the glasshouse industry (Leggett, 1992).

Avena canariensis, described by Baum et al. (1973), is an endemic species to the Canary Islands, especially in Lanzarote and Fuerteventura. This species is listed as "vulnerable" (IUCN, 1994) in the "Lista Roja 2008 de la Flora Vascular Española" (Moreno et al., 1998). Some subpopulations occur in protected areas, on Fuerteventura and Lanzarote (Table 4.4). *A. canariensis* is assessed as "least concern" as it is fairly common and locally abundant and its population is stable, despite being affected by

Table 4.4 Locations of *Avena* species covered by different categories of protection

Species	Country/region	Protection site	Type of protection
A. canariensis	Canary Islands	Fuerteventura	Biosphere Reserve
		Lanzarote	Biosphere Reserve
A. clauda	Samaria	Um Zuka	Nature reserve
A. hirtula	Spain	La Breña y Marismas de Barbate	Natura 2000
		Doñana,	Natura 2000
A. longiglumis	Spain	La Breña y Marismas de Barbate	Natura 2000
		Doñana,	Natura 2000
A. macrostachya	Algeria	Djurdjura	National Park
A. murphyi	Spain	Los Alcornocales	Natura 2000
		Estrecho	Natura 2000
A. strigosa	Scotland and Ireland	South Uist Machair and Lochs; Ness and Barvas, Lewis; Aird and Borve, Benbecula; Eoligarry, Barra; Tiree; North Uist Machair; Rinns of Islay.	Natura 2000*
A. ventricosa	Cyprus	Athalassa	National Forest Park
		Alykes Larnakas	Natura 2000
		Alykos Potamos – Agios Sozomenos	Natura 2000

* Arable crops grown on the machair are based on indigenous varieties of cereals (among other *A. strigosa*), usually grown as a mixed cereal crop to be fed to cattle.

grazing pressure at some localities (Lista Roja, 2008). Its distribution, outside Islands, requires clarification. Izquierdo et al. (Lista de especies silvestres de Canarias, 2004), Ladizinsky (2012), and others specify it as endemic to the Canary Islands, while Valdés and Scholz; with contributions from Valdés et al. (2009), recorded it as native to the Canary Islands and Morocco (Guerra and Betancort, 2013).

Avena agadiriana grows along the Atlantic Coast of Morocco. This species is distributed from Casablanca to Tiznit. Its distribution area is divided into two parts by the Haut-Atlas mountains (altitude 4165 m). The northern part has very heavy clay soil while the southern part is very dry and has sandy soil. *A. agadiriana* is reported as an endemic species for North Africa and Morocco (El Oualidi et al., 2012).

Avena magna (synonym *A. maroccana* Gdgr.) is reported as endemic species for Morocco (El Oualidi et al., 2012). The first samples of the species *A. magna* were collected in 1964 on the Moroccan coast; later numerous populations were found to the south of Rabat at an altitude of 1000–1300 m above sea level, to the southeast of Casablanca at an altitude of 500 m above sea level, and to the northwest of Fes on the slopes of Atlas Mountains at an altitude to 600 m above sea level. It is a typical weedy species forming massive stands together with *A. sterilis* in cereal fields on heavy alluvial soil to which its large spikelet is well adapted. However, unlike *A. sterilis*, *A. magna* less frequently weeds cultivated crops. As the populations of this species were found in a region of rich alluvial soils intensely used in farming, this may lead to eradication or displacement of this oat species by a more aggressive member of this community, *A. sterilis* (Shelukhina et al., 2007). Leggett et al. (1992) noted that in both the major areas of Morocco where *A. magna* were collected, agriculture is becoming more and more intensive. In the collection area, this species was found only on the edges of cultivation, never growing in competition with the associated crops. Advanced soil plant management will be a real danger for the existence of this species, especially since much of the primary habitat has already been disappearing owing to overgrazing, changes in farming practice, or both (Leggett et al., 1992).

A. murphyi has been found in the areas in the south of Spain (between Tarifa and Vejer de la Frontera), where it grows in a typically Mediterranean climate on rich alluvial soils in undamaged association with the hexaploid oat species *A. sterilis*, which is morphologically similar. In addition, sampling of this species has been done in southwestern Spain (coastal zone), where a number of *A. murphyi* specimens were found. Recently, only several *A. murphyi* plants were found in southern Spain in the Cadiz province, confirming the assumption about a drastic decrease in the population of this species or its partial disappearance from the territory of Spain (Shelukhina et al., 2007). New sites with small *A. murphyi* populations were located close to cities of Atlanterra and Barbate during a field survey conducted in 2010. Another site where this species is present is north of Alcala de los Gazules (Katsiotis and Ladizinsky, 2011). It was discovered in not numerous forms in the northern part of Morocco near Tangier. The Moroccan population of *A. murphyi* inhabits rich alluvial soils with intensive agricultural activities, which may result in eradication of this species or its displacement by a more aggressive member of this community, the hexaploid *A. sterilis*, which intensely weeds the cultivated crops (Leggett et al., 1992). In order to protect this species, the Andalusian Government included *A. murphyi* in the Red List. It is also included in the Red List of Spanish Vascular Flora in the vulnerable category (Garcia et al., 2007) and as endangered species on the European Red List of Vascular Plants (Bilz et al., 2011).

A. insularis was first discovered in Sicily in 1996. The four populations of *A. insularis* were detected in a restricted area between Gela and Butera, in southern Sicily (Ladizinsky, 1998), where they grew in undisturbed cenoses on the hills at an altitude of 50–150 m above sea level on alluvial clay soils with sand–clay and conglomerate–stony subsoil. Later the populations of this species were found in Tunisia (Shelukhina et al., 2007). *A. insularis* plants were found by Ladizinsky (2012) around Bargou always on uncultivated land, sometimes along narrow strips between wheat fields and dry water courses, where they were protected from grazing. It appears that grazing represents the most serious threat to *A. insularis* in Tunisia (Ladizinsky, 2012). The uncultivated sites colonized by this species on Sicily had been planted with eucalyptus trees in some places. As a result, *A. sterilis* had invaded the habitat and mixed stands of the two oats were not uncommon (Ladizinsky, 2012). *A. insularis* is classified as endangered species on the European Red List of Vascular Plants (Bilz et al., 2011).

A. macrostachya has only been recorded from the higher regions of two mountain chains in northeastern Algeria, the Aures, and the Djurdjura mountains. It is included in the list of endemic species of Algeria (El Oualidi et al., 2012). *A. macrostachya* was collected from all the areas of the Aures and Bellezma mountains and was also encountered further east, in the Djebel Chenntgouma area just west of Kenchala by Guarino et al. (1991).

A. macrostachya is not particularly common in the Aures area. In contrast, large populations do occur in the Bellezma. In Djurdjura the species is common throughout the *Festuca* sp. grassland areas above approximately 1500 m, which is a result of degradation of the cedar woodland. The distribution of the species seemed to be much more continuous than in the Aures and the Bellezma. The situation in the Djurdjura is more secure, as the species is more common and abundant and the area is to some extent protected, although cattle are allowed in the park and grazing pressure is severe in places. The national park in Djurdjura was created in September 1925 under the French colonial government of the time, and following the independence of Algeria it was reestablished as a national park to protect this unique ecosystem. Once established *A. macrostachya* can produce very large, vigorous tussocks and is probably difficult to dislodge completely. However, it is apparently much eaten by livestock, and in some heavily grazed areas the only flowering culms are found inside unpalatable or spiny shrubs (Guarino et al., 1991).

The main threats of wild *Avena* species are destruction of their natural habitats, grazing, and changing of agricultural practices, in some cases afforestation. In the case of endemic species with narrow or disjunctive distribution the climate changes may have an influence on its existence.

The more complicated situation is in the case of semicultivated and cultivated species, where diversity could be lost very quickly, by abandonment of cultivation, if survival through a soil seed bank is not ensured, so the crop has to be replanted annually.

Among the species present in the Red List of Plants of Ethiopia and Eritrea there are *A. abyssinica* and *Avena vaviloviana* (Vivero et al., 2006). These two related species are endemic to Ethiopia, Eritrea, and Yemen where they are distributed everywhere and grow as mixtures with the main crop. *A. vaviloviana* usually occurs as a weed in crops, field margins, prefers cultivated fertile soils, but also has meadow forms. It was recorded on the Ethiopian plateau at an altitude 2200–2800 m primarily as segetal plant in the fields of wheat and barley. *A. abyssinica*, in the north of Ethiopia, in the more humid areas, displaced other crops, such as barley, and became a cultivated plant. In the south, it is a segetal weed of emmer and barley (Loskutov, 2007). Ladizinsky confirmed that *A. abyssinica* is very widespread as a weed in Ethiopia. It is not threatened at all and probably no need is seen for the conservation of this species (Germeier, 2008). However, it should be considered that changes of agricultural practice and use of new crops or varieties can affect the existence of these species as it has a place with landraces of common oats.

A. strigosa bristle oat has a Mediterranean origin and has been used in European countries since the Bronze Age for cereal or fodder (Kropač, 1981; Kubiak, 2009). Historically, bristle oat is believed to have been the primary oat crop grown in Europe, particularly on poor or marginal soils (Kuszewska and Korniak, 2009). Its use in Europe declined dramatically with the introduction of the more productive common oat (*A. sativa*). Small populations of bristle oat survive in areas where common oats do not produce seed without significant inputs (Podyma et al., 2005; Scholten et al., 2009).

The area of origin and distribution of *A. strigosa* (sensu stricto) embraces most of West and Central Europe and, at least historically, also ranged into the Nordic countries. Most of the grown oats in Great Britain and Ireland until the seventeenth century was *A. strigosa* (Hunter, 1924). Findlay (1956) reported that bristle oat was grown commonly in Scotland, not least due to its tolerance to acid soils, and is still being cultivated in some of the West Scottish isles (Podyma et al., 2005; Scholten et al., 2009). *A. strigosa* remained popular particularly in some mountainous regions. In the Karpaty, bristle oat was cultivated until 1980 as feed crop for horses and pigs (Podyma, 1993). However, several authors have regarded bristle oat as a disappearing species. Kropač (1981) concluded that *A. strigosa* was on the verge of extinction in former Czechoslovakia (Frey, 1989, 1991a) and followed this conclusion by providing information from Poland and other parts of Europe. Podyma (1993) noted that this synanthropic species, although historically important as a crop and survived as a weed in Poland, nowadays seems to be disappearing owing to the changing agricultural practices. Korniak (1997), however, reported new findings from northeastern Poland and argued that in some areas the species was again increasing. Meanwhile, the status of *A. strigosa* in other parts of Northern Europe has also rapidly deteriorated (Weibull et al., 2001).

New populations of *A. strigosa* were discovered in the eastern part of Lithuania and more specifically within the Aukštaitija national park. As the localities are all situated within villages where extensive agriculture is practiced, there is reasonably good hope that the populations will prevail. Earlier known populations in the southern parts of the country (Alytus district), however, were seemingly extinct after the landowner had recently changed his seed. This shows that *A. strigosa* is extremely sensitive to changing agricultural practices, and does not appear to survive through a seed reserve but has to be replanted annually together with the crop. Additionally, unfavorable agroeconomic policies together with controversies regarding land properties left large areas under fallow (Weibull et al., 2001).

A major change in the characteristics of the common oat cultivars took place as a consequence of the transition from cultivation of traditional genetically and morphologically heterogeneous landraces to more uniform "line" cultivars (Grau Nersting et al., 2006). As also documented by Ahokas and Manninen (2000), a large variation is revealed in the landraces, and a subsequent loss of diversity during the transition to purified cultivars. Grau Nersting et al. (2006) studied genetic diversity in microsatellites and agronomic characters in Nordic oat cultivars (*A. sativa*) from the twentieth century, ranging from landraces to new cultivars. A clear development in agronomic characters has taken place in this period. But when comparing cultivars released after 1940 to the landraces, the loss of diversity revealed for the agronomic characters was also indicated by the molecular data. The study showed that breeding of Nordic oat has clearly resulted in a decline in number of alleles from landraces to recent cultivars; a tendency was also reported previously for Canadian oat by Fu et al. (2003).

Changes of diversity of weedy forms of hexaploid oats should be also noted. In the IUCN Red List of Threatened Species (IUCN, 2014) is documented the history of *Avena volgensis* (Vavilov) Nevski syn. *A. fatua* subspecies *nodopilosa* var. *subglabra* subvar. *speltiformis* Vavilov ex Malzev, which is a specialized weed of a local landrace of *Triticum dicoccum*. The landrace has been replaced with commercial varieties since the 1950s and the population of this weed has therefore declined dramatically. It has probably declined by 80%. It is suspected that the taxon faces a high risk of extinction (Smekalova and Maslovky, 2013).

4.4 Status of germplasm resources conservation

In general conservation of genetic resources is based on two prevalent methods: *ex situ* and *in situ*. The first one consists of maintaining genetic resources outside of their natural habitat or cultivation area. The idea of the second one is to preserve genetic resources in their natural ecosystems, habitats or in the case of domesticates or cultivated species, in the surroundings where they have developed their distinctive properties. Protection of *Avena* genetic diversity, as well as other cultivated species, is based on the *ex situ* conservation. Germplasm is stored as seed sample in collections of gene banks worldwide. According to the FAO (FAOwiews), the world's *Avena* collection consists of approximately 131,000 accessions stored by 125 institutions in 63 countries. This is the eighth most numerous collection of crop germplasm, after wheat, rice, barley, maize, bean, sorghum, and soybean. In total, only 14 countries held more than 80% of the world's genetic resources of oat (Fig. 4.2). The largest collections are held in Canada (~40,000), the United States (~22,000), and Russia (~12,000). Significant collections, that is, containing several thousands of accessions, are stored in Germany, Kenya, China, Australia, and the United Kingdom. Collections, gathered by the particular institutions, differ in terms of accession number, species composition, origin and methods of sample classification, evaluation, and storage. Among preserved accessions 56% represents cultivated gene pool and 24% are identified as

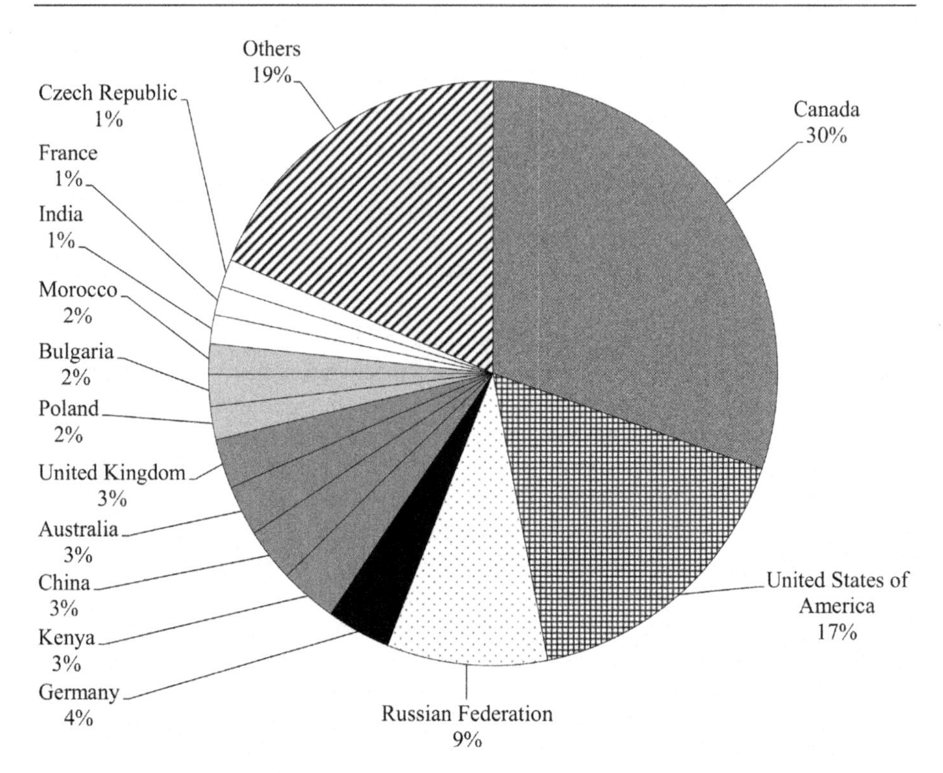

Figure 4.2 Participation in world *Avena* collection by countries.

wild related species. As much as 20% of the accessions of the *Avena* world collection have not been assigned to the species. In Canada and Russia cultivated species and numerous wild species are stored, while in the United States emphasis was placed on *A. sativa* and its wild relatives from the primary gene pool, that is, *A. sterilis* and *A. fatua*. Under the terms of the uniqueness a very interesting collection is the one stored in Morocco. The Moroccan gene bank holds most accessions of *A. damascena*, *A. atlantica*, *A. agadiriana*, and *A. murphyi* (Table 4.5). Regeneration of the collection is a very urgent concern for this collection (Germeier, 2008). Wild species, which are very poorly represented in other gene banks, form its core. An interesting collection, from the botanical point of view, is in possession of the United Kingdom. It contains 27 species from the genus *Avena*.

4.4.1 Conservation of cultivated gene pool

About 75,000 accessions of cultivated oat species, that is, *A. sativa*, *A. byzantina*, *A. strigosa*, and *A. abyssinica*, are preserved. Genetic resources of common oat (*A. sativa*) represent 95% of accessions of all cultivated gene pool. The largest collections are held in Canada (17%), the United States (15%), and Russia (12%). The remaining 56% of the samples are stored in the various sizes of collections in 57 countries throughout the world. From a global point of view, 28% of the accessions came from

Table 4.5 Summary of species structure of accessions in 14 of the biggest collections

Species	Number of accessions	Canada	USA	Russia	Germany	Kenya	China	Australia	United Kingdom	Poland	Bulgaria	Morocco	India	France	Czech	Others
A. abyssinica	690	254	241	53	8	0	0	0	67	1	5	0	0	0	1	60
A. agadiriana	257	14	0	3	0	0	0	0	6	0	0	234	0	0	0	0
A. atlantica	40	15	0	2	0	0	0	0	10	1	0	12	0	0	0	0
A. barbata	2,984	2,097	619	83	40	0	0	2	16	3	7	15	0	0	0	102
A. bruhnsiana	2	0	0	2	0	0	0	0	0	0	0	0	0	0	0	0
A. byzantina	2,015	0	0	910	417	0	0	0	158	75	42	3	0	0	90	320
A. canariensis	78	45	0	7	7	0	0	0	19	0	0	0	0	0	0	0
A. clauda	133	91	0	13	0	0	0	0	3	0	1	17	0	0	0	8
A. damascena	271	3	0	4	0	0	0	0	0	9	0	255	0	0	0	0
A. fatua	2,593	642	1,322	219	96	0	0	0	74	20	19	2	1	0	1	197
A. hirtula	143	50	0	10	0	0	0	0	13	6	0	56	0	0	0	8
A. insularis	11	3	0	1	0	0	0	0	0	7	0	0	0	0	0	0
A. longiglumis	212	45	7	11	0	0	0	0	8	2	1	118	1	0	0	19
A. ludoviciana	440	0	0	433	0	0	0	0	2	0	0	0	0	0	1	4
A. macrostachya	18	1	0	1	0	0	0	0	1	10	0	0	0	0	0	5
A. magna	40	0	0	14	0	0	23	0	1	0	0	0	0	0	0	2
A. murphyi	97	2	1	0	2	0	0	0	14	0	0	75	0	0	0	3
A. occidentalis	72	62	0	7	0	0	0	0	3	0	0	0	0	0	0	0
A. pilosa	21	0	0	12	0	0	0	0	3	0	0	0	0	0	0	6
A. prostrata	42	0	0	2	0	0	0	0	7	0	0	33	0	0	0	0

(Continued)

Table 4.5 **Summary of species structure of accessions in 14 of the biggest collections** ***(cont.)***

Species	Number of accessions	Canada	USA	Russia	Germany	Kenya	China	Australia	United Kingdom	Poland	Bulgaria	Morocco	India	France	Czech	Others
A. sativa	70,845	11,955	10,933	8,766	2,270	4,239	3,849	249	2,901	1,967	2,148	26	2,082	1,410	1,901	16,149
A. sterilis	23,182	11,496	8,234	1,045	65	0	0	5	103	37	6	385	23	0	4	1,779
A. strigosa	845	155	120	191	45	0	0	5	80	70	12	0	1	0	0	166
A. vavilo-viana	240	135	43	45	0	0	0	0	15	0	1	0	0	0	0	1
A. ventricosa	11	4	0	2	0	0	0	0	3	0	0	2	0	0	0	0
A. wiestii	82	46	4	17	5	0	0	0	0	0	0	2	0	0	0	8
*A. brevis**	101	39	22	0	0	0	0	0	11	0	8	0	0	0	0	21
*A. eriantha**	175	133	4	0	0	0	0	0	2	0	1	35	0	0	0	0
*A. hispanica**	16	16	0	0	0	0	0	0	0	0	0	0	0	0	0	0
*A. hybrida**	27	23	1	0	1	0	0	0	0	0	1	0	0	0	0	1
*A. lusitanica**	30	30	0	0	0	0	0	0	0	0	0	0	0	0	0	0
*A. maroc-cana**	988	35	2	0	16	0	0	0	15	0	0	916	0	0	0	4
*A. nuda**	65	13	8	0	4	0	1,719	4	3	1	16	0	0	0	0	16
*A. tricho-phylla**	1	0	0	0	0	0	0	0	1	0	0	0	0	0	0	0
Unknown	24,566	12,374	276	4	1,823	9		3,416	86	198	43	10	2	683	13	5,629
Total	131,333	39,778	21,837	11,857	4,799	4,248	3,872	3,681	3,625	2,407	2,311	2,197	2,110	2,093	2,11	24,507
		30%	17%	9%	4%	3%	3%	3%	3%	2%	2%	2%	2%	2%	2%	19%

* taxons not recognized by Loskutov (2008).

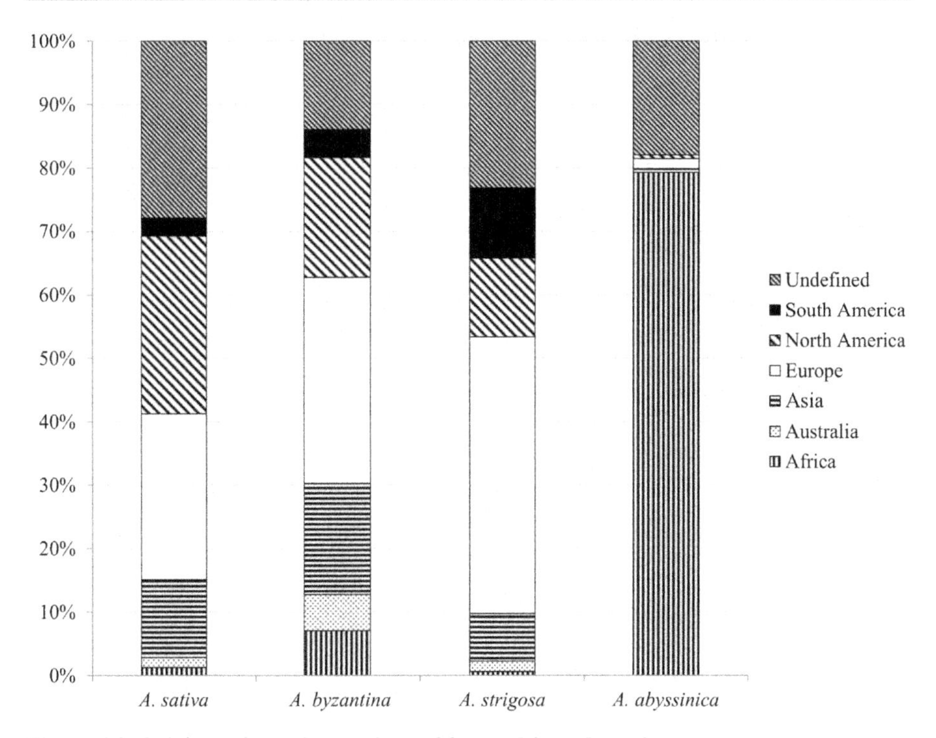

Figure 4.3 Origin regions of accessions of four cultivated species.

North America, 26% from Europe, 12% from Asia, 3% from South America, 2% from Australia and Oceania, and 1% from Africa (Fig. 4.3). The accessions originated from nearly 100 countries and 23% of them came from the United States, 5% each from Russia and Canada, and 39% from other countries. For 28% of the accessions information about the country of origin is missing in the passport data (Fig. 4.4). In terms of the status of the sample 19% are advanced cultivars as well as landraces, breeder's lines represent 17% of the collection, and 42% of accessions do not have a specified status (Fig. 4.5).

The global resources gathered about 2000 samples of red oat (*A. byzantina*), which is approximately 3% of the cultivated gene pool. It should be taken into account that this species by many gene banks is classified as *A. sativa* spp. *byzantina*. Majority of accessions are stored in Russian (45%) and German (21%) collections and the remaining 34% germplasm of this species is kept in 17 countries. Of the collected samples, 87% have known origin and were derived mainly from Europe (34%) and North America (19%). Based on information from passport data, 19% of accessions came from the United States but in the national collection this species is included as *A. sativa* as described earlier. Valuable in this collection is that almost half of the accessions have the status of traditional cultivar or landrace. Unfortunately, every third of the preserved samples have no designated status at all. In the gene banks located in 25 countries, 849 accessions of bristle oat (*A. strigosa*) are stored and they represent

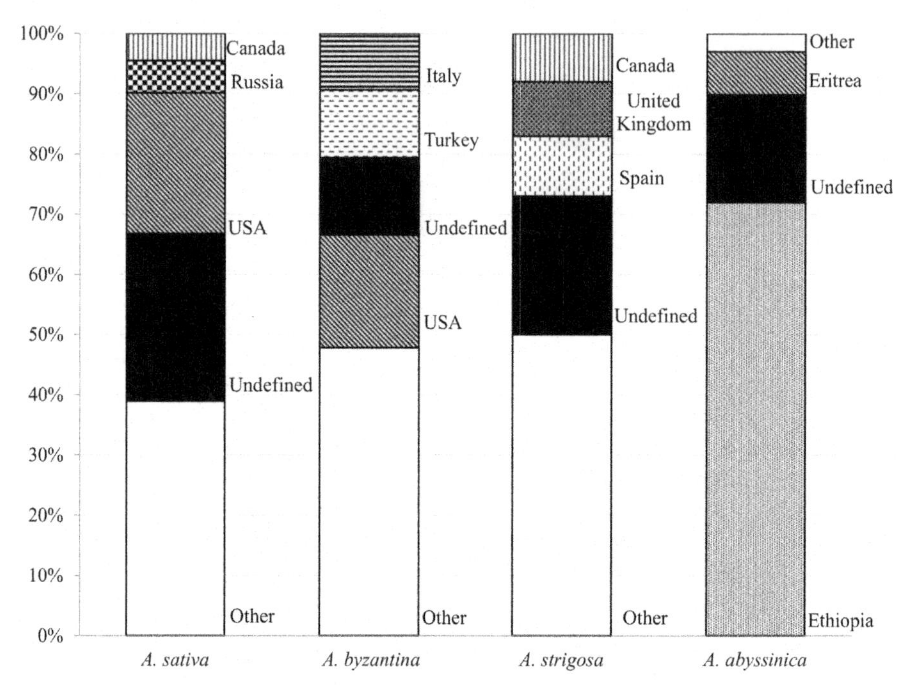

Figure 4.4 Country of origin of four cultivated species.

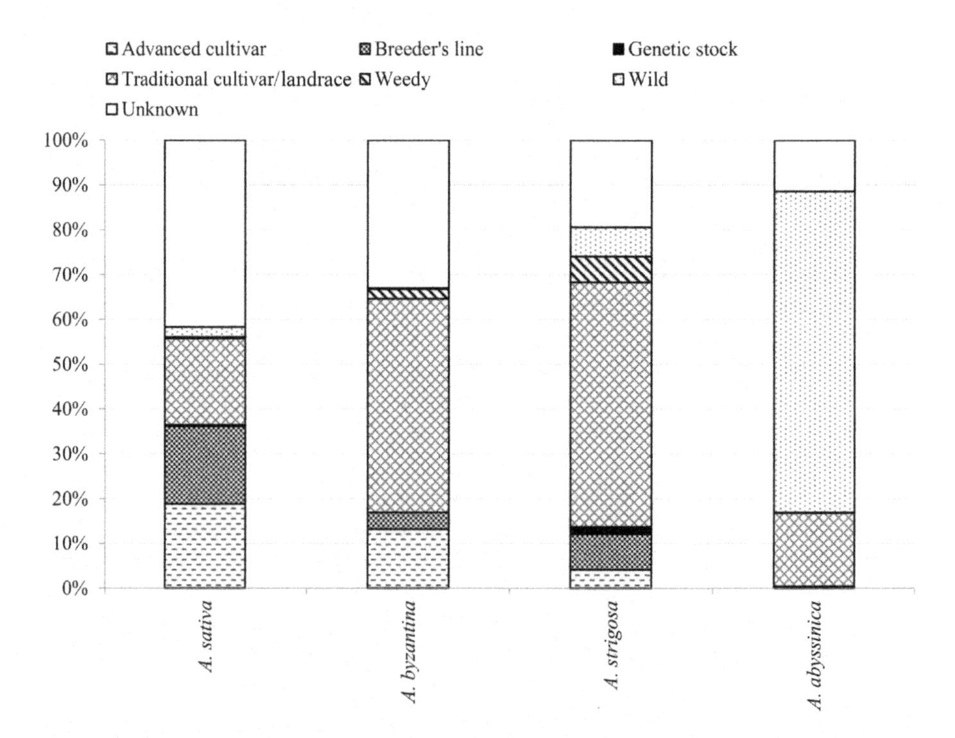

Figure 4.5 The structure of accessions according to sample status.

merely 1% of germplasm of the cultivated *Avena* species. The most numerous collections are held in Russia, Canada, and the United States (23, 18, and 14%, respectively). Approximately half of the accessions originated from Europe and only slightly less than a quarter from both Americas (Fig. 4.3). A large number of specimens came from Spain (10%), United Kingdom (9%), and Canada (8%). For 23% of the samples, information about the origin is missing. The traditional cultivar or landrace status has been determined for 50% of accessions. Of the collection, 7% has been annotated as wild ones and 6% as weeds. Ethiopian oat (*A. abyssinica*) is the last of the cultivated species in the genus *Avena.* It has long been used in Ethiopia and is well adapted to local conditions such as high altitude. Around 700 samples of this species are held in 15 countries. Majority of them (72%) are preserved in the United States and Canada germplasm collections (37 and 35%, respectively). They originated mainly from Ethiopia and neighboring Eritrea, while 18% have an undefined place of origin. More than 70% of collected specimens are wild populations, 16% have a status of traditional cultivar or landrace, and 11% have no defined status.

4.4.2 Conservation of wild gene pool

As a result of progressive devastation of the environment, protection of crop-related wild species (CWR) became a necessity. Although these species do not affect directly the food security of mankind, they constitute a source of desirable traits such as resistance to biotic and abiotic stresses or phytonutrients. Slightly more than 24% of the accessions in the world's *Avena* collection are classified as wild species. Unfortunately, a direct comparison of the wild species composition of germplasm collections among gene banks is hindered by the use of different taxonomic classifications within the genus *Avena* by their curators. A list of species described by Baum (1977) and Loskutov (2007), the two classifications used by the major holders, with the number of accessions in the most important collections is included in Table 4.5. Among wild species the most numerous are hexaploid species collections, which were assigned into the primary gene pool (Leggett and Thomas, 1995). Four species from this group, that is, *A. sterilis*, *A. fatua*, *Avena ludoviciana*, and *Avena occidentalis* represent 46% of all collected CWR. Out of them the most numerous is the *A. sterilis* collection. It is held by 24 countries, but 92% of all accessions are stored in only three of them, that is, Canada (50%), United States (36%), and Israel (6%). Germplasm samples come from 51 countries and 66% of them have the Israel origin. Only 1% of the accessions have unknown country of origin.

A. fatua is widespread throughout the world and is listed among the most noxious weeds. In the world's genetic resources it is represented by approximately 2500 of seed samples, which are located in the collections deployed in 25 countries. Three of the largest store a total of 84% of *A. fatua* accessions, and they are in the United States (51%), Canada (25%), and Russia (8%). The samples come from 59 countries, with nearly half of them originating from the United States. The other two species of this gene pool, namely, *A. ludoviciana* and *A. occidentalis*, are much less frequent. The first one is represented by 440 accessions while the second one by only 72 samples. Their genetic resources are stored primarily in Russia and Canada, respectively.

A. ludoviciana germplasm samples originated from 27 different countries and *A. occidentalis* from only two, that is, Spain and Morocco.

In the genus *Avena* seven tetraploid species were defined: *A. agadiriana, A. barbata, A. insularis, A. macrostachya, A. magna, A. murphyi,* and *A. vaviloviana*. Among wild tetraploids *A. barbata* has the most numerous collections, which is 70% stored in the Canadian germplasm resources. *A. agadiriana* and *A. vaviloviana* are represented by about 250 samples each. Less than a hundred samples are preserved of *A. magna* and *A. murphyi*. While for *A. insularis* and *A. macrostachya* only about a dozen of accessions have been collected. A very unique collection of *A. macrostachya* is stored in a Polish repository. There is also a very high probability that *A. insularis* occurs in *A. sterilis* collections due to the considerable morphologic similarity of these two species (Germeier, 2008). Among this group, there are two endangered species, that is, *A. insularis* and *A. murphyi* (Bilz et al., 2011).

The diploid species are the most numerous group of species in the genus *Avena*; however, it is also the least represented in germplasm collection. Together, 12 species constitute no more than one hundredth of a fraction of all genetic resources of oat. The most numerous collections of wild diploid species have *A. damascena* and *A. longiglumis*. Almost all of the accessions of these two species are stored in Morocco. Only 13 accessions were collected jointly for species *Avena bruhnsiana* and *A. ventricosa*, whose names are considered to be synonymous. For more details please see Table 4.5.

4.4.3 In situ *conservation*

Most conservation activities today focus on managing genetic material *ex situ*, that is, maintaining genetic resources away from their original growing sites. This is the situation for all seed material kept in gene bank freezers or in field collections or *in vitro* conserved of vegetatively propagated species such as apples, onions, and rhubarb to mention a few.

The Convention on Biological Diversity (CBD, 2013), however, specifically points out the complementarity between *ex situ* and *in situ* approaches, and stresses the need for *in situ* programs to be developed. CBD (2013) stresses that the two conservation strategies should be practiced as complementary approaches to conservation, each providing a safety backup for the other. The goal of applying the two conservation strategies is ultimately to serve the present needs of plant breeders on one hand, and the need to maintain genetic resources that are always in tune with the environment to deal with future unpredictable changes on the other hand (Veteläinen et al., 2009).

The primary objective of *in situ* conservation is to conserve genetic material and the processes that give rise to diversity. Common approaches for *in situ* conservation are genetic reserve conservation and on-farm conservation. The genetic reserve conservation is defined as management, monitoring of genetic diversity of natural populations of CWR in specific areas for the long-term preservation. On-farm conservation is focused on cultivated species and in particular on landraces and traditional cultivars and consists of agrobiodiversity preservation in a dynamic agroecosystem that is self-supporting and favoring evolutionary processes.

CWR are usually conserved within ecosystems that have to be maintained as part of an overall environmental management strategy. The most of taxa are likely to be conserved as components of ecosystems, which are being protected for other reasons (Table 4.4). This strategy tends to lead to greater sustainability and efficacy of conservation actions. However, this is not a necessary prerequisite and *in situ* conservation of CWR may also be necessary outside protected areas. For some extremely important wild relatives, such as *Triticum dicoccoides*, it has been considered worthwhile to set up specific reserves to ensure their maintenance. In 1984, a study of the natural dynamics of wild emmer populations was launched in Israel, Amiad in eastern Galilee. It was to serve as a precursor to genetic reserve conservation of selected wild cereal populations in their native ecosystems (Anikster et al., 1997).

Decisions on where to concentrate efforts will therefore be based on general criteria (for example, those developed by organizations such as the IUCN) that are relevant to all species, together with specific criteria concerning potential usefulness with respect to present or future crop production (Maxted et al., 1997).

In Europe, on the basis of *inter alia* of the Habitats Directive (EC, 1992), the Natura 2000 network has been established. Its overall objective (EC, 1992) is to maintain and restore, at a favorable conservation status, natural habitats and species of wild fauna and flora of community interest. It is necessary to establish a coherent European ecological network of special areas of conservation (called Natura 2000). This network is composed of sites hosting the natural habitat types listed in Annex I and habitats of the species listed in Annex II. No single species of *Avena* has been included in Annex I but in some types of endangered habitats (Appendix II) oats are natural components of cenosis, for example, machair (Table 4.4).

Selecting taxa for *in situ* conservation is complex and many factors need to be taken into account when choosing which species to conserve and where and how to employ resources. In the case of wild species, considerations include the closeness of their relationship to the cultivated species and the ease with which they can be used as donors of useful traits. The presence of known useful variation also is important, the status and distribution of the species and the degree to which they are threatened in particular habitats (Maxted et al., 1997).

The issue of *in situ* protection of genetic resources in genus *Avena* has been specifically targeted in the framework of the European AEGRO project (Frese et al., 2013). It is aimed at the creation of conservation strategies for CWR and landraces, and to transfer them into practice. For the four wild oat species, locations in protected areas suited for the establishment of genetic reserves have been recommended. They are located in Spain (*A. murphyi*, *A. longiglumis*, and *A. hirtula*) and Cyprus (*A. ventricosa*) (Table 4.4).

As the Convention on Biological Diversity and the International Treaty (IT) reinforce the national sovereignty on biological diversity, centers of diversity need primary consideration in developing a global conservation strategy, also with respect to complementary actions of *ex situ* and *in situ* conservation. For wild oat species, the western Mediterranean region is undoubtedly the center of diversity.

For some of the wild *Avena* species, studies on the distribution and abundance, *in situ* conservation projects, and further collection of the restricted species are still needed. INRA Morocco stresses the need for a conservation strategy to safeguard the endangered species by *in situ* conservation. In Morocco further collecting trips should be undertaken for all species and especially those that are reported to be under the threat of genetic erosion.

Despite the fact that Spain is rich in wild oat species and they were poorly represented in *ex situ* collections, only some projects have been funded to collect and establish a representative collection of this material (Garcia et al., 2007). Molecular analysis of the Spanish endangered species, that is, *A. canariensis*, *A. murphyi*, and *A. prostrata*, showed the presence of significant differentiation among distinct populations within each tested species. In order to reach this genetic structure, gene flow among *A. canariensis* must be restricted and/or there is natural selection favoring particular genotypes in different sites. It is stated that the populations appeared to be adapted to distinct microhabitats, especially soil conditions. For *A. murphyi* high level of variability is observed, caused by limited gene flow as the main agent for the population differentiation (Garcia et al., 2007). Thus, a loss of populations implies a loss of genetic diversity. The results can help to define which populations would be of most interest in order to preserve *in situ* the maximum genetic variability in these species (Garcia et al., 2007).

4.4.4 On-farm *conservation*

Lately, on-farm conservation has been advocated to complement *ex situ* conservation. However, for a long time on-farm conservation of crop landraces was considered as impractical and inappropriate especially in developed countries. However, this is a concern for developing countries where rural societies maintain agricultural biodiversity because it is essential for their survival. They select and breed new varieties for the same reason. There is no useful distinction for them between conservation and development. By its very nature, on-farm conservation is dynamic because the varieties that farmers manage continue to evolve in response to natural and human selection. It is believed that in this way crop populations retain adaptive potential for the future (Maxted et al., 1997).

In addition to the fact that landraces are evidently an important genetic resource used in crop enhancement programs, there are two more strong arguments for their on-farm conservation. First, landraces that are still managed by people today are a vivid cultural heritage and also contribute to the identity of people in the regions they are living in. The typical traditional products belonging to their own regional culture are a value in itself. Second, the conservation of landraces on-farm contributes to the flexibility and security of agricultural production by increasing the number of independently managed breeding populations.

Contrary to CWR growing in their natural habitat, landraces would not continue to exist without active maintenance by the people interested in their use. Conservation actions have to consider not only the conservation per se but also the promotion of the economic use of landraces, preferable at the local or regional level. People reproducing and cultivating landraces would like to receive similar revenue when compared to cultivating mainstream crops and use of commercial seeds.

The EU member states have already adapted their agricultural policies with the objective of conservation of agricultural biodiversity. First, the European Commission released a number of directives ("the conservation variety directives") providing for certain derogations for marketing of seed of landraces and varieties, in order to facilitate the trade of seeds and other propagating materials of landraces and varieties that have been traditionally grown in particular localities and regions (EC, 1999, 2008). The rules have been implemented in national seed laws. In the Common catalog of varieties of agricultural plant species, 11 conservatory varieties of common oat have been registered, including one unhulled form (EC, 2014a).

Second, payments to farmers growing landraces are foreseen in the European Agricultural Fund for Rural Development (EAFRD) in order to facilitate continuous cultivation of landraces, the development of products based on these landraces, and to promote the expansion of market niches. Both these measures, the conservation variety directives and the EAFRD, facilitate the development of specific landrace products and thus support the maintenance of landraces in regional agricultural production (Maxted et al., 1997). European countries, in a coordinated manner in the framework of ECPGR, initiated the development of inventories of landraces (Veteläinen et al., 2009).

Efforts to revive bristle oat cultivation in Europe are ongoing (Podyma, 1993; Scholten et al., 2009) often with use of the materials maintained in gene banks.

The Scottish Landrace Protection Scheme was launched by the Science and Advice for Scottish Agriculture (SASA) in August 2006 to provide a safety net for the continued use of landraces by storing seed produced by each grower each year. In the event of a poor harvest, a grower can request some of the seeds already deposited and stored at SASA. Survey work undertaken for the compilation of the UK National Inventory of Plant Genetic Resources identified landraces that were still being grown and used in agriculture in Scotland, among them populations of common and bristle oats (Green, 2008).

The Nordic Gene Bank (Alnarp, Sweden) initiated a project to look for *A. strigosa* at its historic localities in Denmark and in Lithuania (Weibull et al., 2001). As result four new populations were discovered in the eastern part of Lithuania and more specifically within the Aukštaitija national park. The area of Švencionys in east Lithuania represents a suitable site for carrying out detailed studies on on-farm conservation. The historically and geographically isolated bristle oat populations could serve as an interesting object of genetic studies (Weibull et al., 2001).

The Council Directive 98/95/EC (EC, 1999) opened the possibility of establishing specific "conditions under which seed may be marketed in relation to the conservation *in situ* and the sustainable use of plant genetic resources (PGR)." The Parliament of Finland included the idea in the Seed Trade Act of 2000 (FAOLEX, 2000) by allowing the seed of landraces to be marketed uncertified in order to conserve genetic diversity. As a result of the "Landrace Project" on-farm inventory project, the first European support system for *in situ* (on-farm) cultivation of landraces and old cultivars was developed in Finland. The support has been paid as a special subsidiary within the EU agri-environmental scheme (Veteläinen et al., 2009).

The aim of the subsidiary system is to enhance the continuity of cultivation of landraces and old cultivars by offering annual financial support to a farmer, who is voluntarily involved in conservation. Many European countries, following the example of

Finland, included into the agri-environment-climate scheme of subsidies, the package on conservation of genetic resources with support for cultivation and seed production of landraces and old cultivars, taking the opportunity given by Regulation No. 1305/2013 on the support for rural development by the EAFRD (EC, 2013) and strengthen the "agriculture at the margin" (Weibull et al., 2001). But for many regions, on-farm conservation is not synonymous with "ancient" agricultural techniques, rather the reverse is true. There the principles of sustainable, organic agriculture and sometimes quality testing and certification of each productive step often drive production. Also there, for example, Italy, about a third of landraces still extant are used in large-scale production and are cultivated using high-input agronomic techniques or under highly skilled modern (sometimes organic) agricultural techniques (Veteläinen et al., 2009). The spread of organic farming has improved the prospects for preserving landrace crops, since local varieties can be suited better for organic farming methods than ordinary farming. There is also good atmosphere for encouraging the development of regional products.

4.4.5 Access to genetic resources

The Convention on Biological Diversity (CBD, 2011) seeks to create a holistic legal regime for the genetic, species, and ecosystem levels of biodiversity. The basic element of the regime is affirmation of the sovereign rights of states over their genetic resources. The convention also creates an obligation to endeavor conditions to facilitate access to genetic resources by other parties. Therefore, under FAO auspices, and in harmony with the CBD, the International Treaty on Plant Genetic Resources for Food and Agriculture (ITPGRFA) (FAO, 2009) was adopted on November 3, 2001. The Treaty establishes a global system to provide farmers, plant breeders, and scientists with access to PGRFA, and ensures that recipients share benefits they derive from the use of these genetic materials. The Parties have agreed to forget their individual rights to negotiate separate access and benefit sharing terms and to insist on giving their prior informed consent on a bilateral basis.

The main mechanism of the treaty is a multilateral system (MLS) that seeks to facilitate access to a negotiated list of PGR, annexed to the treaty (Annex I), as well as the fair and equitable sharing of benefits arising from their use. The genus *Avena* is listed in Annex I of the ITPGRFA (FAO, 2009). Genetic resources listed on the MLS are to be circulated with facilitated access for research, breeding, and training for food and agriculture. Parties are encouraged to place germplasm in the MLS in exchange for benefit sharing in areas of information exchange, technology transfer, and capacity building. Facilitated access is accorded through a standard material agreement.

New plant varieties can be protected by "effective" *sui generis* systems or patents. The International Convention for the Protection of New Varieties of Plants (UPOV, 1994b) establishes the International Union for the Protection of New Varieties of Plants (UPOV), which creates plant breeder's rights, is one possible *sui generis* system that would appear to meet the requirements of the TRIPS Agreement (WTO, 1994). The 1991 revised act provides some exceptions, for example, breeder exception. Protected varieties can be used for experimental purposes and for the purpose of breeding other varieties. Another form of IPRs are patents. Patents are exclusive rights granted to inventors that prevent others from making, using, selling or importing the patented invention (WTO, 1994). The criteria for granting patents are novelty, inventiveness, and industrial applicability. The American solutions provide possibility of patenting varieties (e.g., oat variety ROMAR-07 US 8697956 B1) but in accordance with European patent rights a living organism cannot be the subject.

The Nagoya Protocol on Access to Genetic Resources and the Fair and Equitable Sharing of Benefits Arising from their Utilization (ABS) to the Convention on Biological Diversity (CBD, 2011) adopted on October 29, 2010 is a supplementary agreement to the Convention on Biological Diversity. The Nagoya Protocol applies to genetic resources that are covered by the CBD, and to the benefits arising from their utilization. The Nagoya Protocol does not affect the rules implemented by the ITPGRFA. The European Union countries implemented the Protocol by Regulation (EU) No 511/2014 (EC, 2014b).

4.5 Germplasm evaluation and maintenance

Describing plants is one of the most important ways that PGR users can contribute to germplasm utilization and conservation efforts. Descriptors in the characterization category are observations about plant characteristics that can be used for diagnostic purposes to describe the plants of an accession and differentiate them from those belonging to another accession. Therefore, data gathered during characterization are used for distinguishing accessions. They provide information on the type of plants that are in a collection, and information potentially useful in crop development. It should be noted that the intraspecific concepts in taxonomy have been extensively elaborated and used down to the botanical variety level by "Eastern Europeans School of Genetic Resources." Theoretically, 576 character state combinations (morphologic groups) can be described for *A. sativa* based on 3 phenologic and 34 morphologic characters (Germeier, 2008). Evaluation descriptors are of great interest to plant breeders and are useful in crop improvement. They include descriptors such as yield, agronomic and other economically important traits, biochemical traits (content of specific chemical compounds, dry matter content, etc.), and reaction to biotic and abiotic stresses.

Over the years, different guidelines for PGR documentation have been developed by Bioversity and its predecessors, UPOV, COMECON, USDAGRIN, and others. In addition, several national programs have developed descriptors for crops of national interest. Within the PGR community, a descriptor is defined as an attribute, characteristic, or measurable trait that is observed in an accession of a gene bank.

The original aim of descriptor lists was to provide a minimum number of characteristics to describe a particular crop and allow communicating with other institutions. Lack of compatibility in documentation systems for describing PGR seriously hampered data exchange between collections. With the integration of collections at the national level into multicrop collections, it became evident that common descriptors needed to be more consistent across crops. As a result, Bioversity and FAO, with substantial contributions from European countries through the ECPGR Network and Consultative Group on International Agricultural Research centers through the

Systemwide Information Network for Genetic Resources system, developed a subset of passport descriptors: the FAO/IPGRI List of Multi-crop Passport Descriptors (MCPD) (FAO/IPGRI, 2001). The MCPD list is a reference tool that provides international standards to facilitate germplasm passport information exchange across crops. Passport data are basic data that describe and identify the particular material. They normally include the accession identifier number assigned in the collection, species, subspecies and other taxonomic descriptor, the varietal of local name, the biological status, for example, cultivated or wild, the origin country, geographical location, and date of collection and the identity of the collections.

Bioversity crop descriptor lists are compatible with the descriptors used for the FAO World Information and Early Warning System (WIEWS) (FAOwiews) on PGR, with CG Centers and with European countries through the EURISCO Catalogue. Pinheiro de Carvalho et al. (2013) studied oat landraces genetic resources in worldwide gene banks on the basis of WIEWS (FAOwiews) data, and pointed out the difficulties in comparing data from different collections. He stated that different systems of resource documentation, as well as differences in taxonomic systems and the limitations of the WIEWS system, prevented comparative surveys of landrace collections from different gene banks. Crop-specific information systems, for example, the European *Avena* database, are able to deal with parallel multiple taxonomic systems (Germeier, 2008).

The descriptor list for oat (*Avena* spp.) developed in the initial stage of the process is based upon a list of descriptors selected by the Oat Working Group of the ECPGR networks (IBPGR, 1985). For documentation purposes the MCPD list has been implemented. The technical guidelines developed by the UPOV have been developed specifically for testing the distinctness, uniformity, and stability (DUS) of new cultivars of crops. DUS traits are central to the breeder's work since they are necessary to obtain legal protection for a bred variety. UPOV descriptor lists are constructed with the thoroughness of legal documents. Requirements for the minimum amount of seed, number of vegetation periods, minimum number of plants, and maximum number of aberrant plants are defined. Precise rules for scoring are given, along with example varieties for each trait and level of manifestation. Many countries have adopted the UPOV guidelines for identifying and registering new plant varieties. Oats guidelines contain 24 descriptors (UPOV, 1994a). In 1977, the member countries of the Council for Mutual Economic Aid (COMECON) joined forces to develop descriptor lists for crops of primary economic importance. By 1990, 48 bilingual (Russian/English) descriptor lists had been published. The passport category of these descriptor lists contained 13 fields. In addition to descriptive data, the characterization category (six fields) contained detailed geographical information on the location of collections in COMECON countries. The characterization descriptors included data on morphology, biology, disease and pest resistance, chemical composition, economic utilization, and other descriptors; botanical keys were also included. At the country level the descriptor lists were used directly or were adapted to the national needs. The characterization and evaluation of germplasm were routine activities in the collections. In Bulgaria, from 1986 to 2002, over 2000 accessions of oats from the collection were evaluated for characteristics that would be of use to breeding programs (Antonova, 2005).

In VIR all accessions of oats were analyzed by their diverse morphologic and agricultural traits at the fields of VIR Experimental Stations network from the 1920s up to now. The research was based on the field guidelines and the COMECON international descriptors of *Avena* L. (COMECON, 1984). All this information is maintained in field registers. The US National Plant Germplasm System (NPGS) has developed descriptor lists for many major food plants. Descriptor lists allow NPGS curators to enter plant trait data into the Genetic Resources Information Network (GRIN). GRIN contains information on 21,371 oats accessions (USDA, 2014). In this database data on the following categories can be found: chemical (β-glucans, lipid and protein content), cytology (chromosome number), disease (reaction to: barley yellow dwarf virus (BYDV), crown rust, smut, and stem rust), growth (growth habit and plant height), insect (reaction to cereal leaf beetle and greenbug), morphology (awn frequency, awn type, hull, kernel per spikelet, lemma color, lodging, panicle density, panicle type, shattering, spikelets per panicle, straw breakage, and straw color), phenology (days to anthesis), and production (1000 kernels weight, bundle weight, test weight, and yield).

Also, the Plant Genetic Resources of Canada (PGRC) in consultation with Canadian and international oat breeders, a descriptor list using phenologic, morphologic, and agronomic characters was generated according to international gene bank standards (Diederichsen et al., 2001). In the framework of PGRC are maintained 27,790 accessions of oats (PGRC). The data structure in this database is quite similar to that described for the United States. Several categories have been specified: agronomic (bushel, drought tolerance, grain yield, maturity, oat type, sensitivity to daylight, and sprouting tendency), chemical (percent of oil and lipids, β-glucan, and protein), chemical resistance (specify fungicide and herbicide), cytology (chromosome number), disease (BYDV, crown rust, covered and loose smut, *Helminthosporium Avenae*, powdery mildew, septoria leaf, septoria steam, smut reaction, *Puccinia graminis*, and stem rust reaction), disease reaction (anthracnose, bacterial stripe blight, blast, *Fusarium* blight, gray speck, halo blight, *Helminthosporium* leaf blotch, and *Scoleotrichum* leaf blotch), insect (cereal leaf beetle resistance and reaction, greenbug reaction), morphology (awn type, awn frequency, awn insertion position, attitude of panicle, awn color, floret scar hairiness, floret scar shape, fluorescence of seeds, glume length, glume relative length, juvenile growth habit, hull characteristic, leaf blade margin hairiness, leaf length and width, lemma color, hairiness, and length; lemma tip lateral teeth, lemma tip shape, ligula, lemma awns, number of tillers, panicle density, length, and type; panicles per row, seeds per spikelet, shattering, spikelet disarticulation, stem node hairiness, straw breakage, color and lodging, spikelet separation and disarticulation, tip-awn lemma ratio, and tip-awn length), phenology (date of sowing, days to anthesis, days to maturity, and days to emergence), production (bundle weight, groat percent, kernel weight, test weight, weight per volume, and yield), quality (hull percent), and stress (winter survival). Obviously, only part of the accessions of these two collections has been evaluated, and among them only part of the listed characteristics have been evaluated.

In 2000, a project to establish a European Plant Genetic Resources Information Infra-Structure (EPGRIS) had been started. The aim of the project was to establish an infrastructure for information on PGR maintained *ex situ* in Europe by supporting the creation and providing technical support to national PGR inventories and creating a European PGR search catalog with passport data on *ex situ* collections maintained in Europe. The final output of this project was the European Search Catalogue – EURISCO.

The central search catalog (EURISCO) was created and maintained until 2014 by Bioversity International. Since 2014, EURISCO has been maintained and advanced by the Leibniz Institute of Plant Genetics and Crop Plant Research (EURISCO, 2014). EURISCO is based on a European network of *ex situ* national inventories (NIs). Currently, EURISCO comprises passport data of about 1.1 million samples, representing 5,929 genera and 39,630 species from 43 countries, among them 24,852 oats accessions. Database development plans assume the inclusion of characterization and evaluation data.

A more specific work is the European *Avena* Database (EADB), which is an inventory of *ex situ* accessions maintained in an international, decentralized network of gene banks (EADB). This information system allows the search for passport, characterization, and evaluation data on gene bank accessions. The EADB established in 1984 is now managed by the JKI – Federal Research Institute for Cultivated Plants with passport data of 32,910 accessions representing collections from 26 European contributors and nearly 170,000 characterization and evaluation points for 3,134 accessions (Germeier, 2008).

The AVEQ project, *Avena* genetic resources for quality in human consumption (AGRI GEN RES 061), evaluated and characterized *Avena* genetic resources (600 accessions) for traits relevant for the quality of oats in human nutrition (contents of protein, fat, minerals, dietary fiber – especially ß-glucan – antioxidants, and phenolic compounds), analyzed resistance to *Fusarium* infection and mycotoxin contamination, and tested cultivars to cold tolerance (Germeier et al., 2011). All results are available in the EADB. Screening of oats collections is carried out for national breeding programs for selected traits or conditions: grain and fodder production adapted to the Mediterranean conditions (Iannucci et al., 2011), *Fusarium* head blight (Gagkaeva et al., 2012), and morphologic traits (Dumlupinar et al., 2012). Data from the evaluation of oat accessions can also be obtained from websites of GrainGenes (GrainGenes). Molecular studies are used more and more often to describe the genetic resources. In the molecular approach the following techniques could be distinguished: molecular markers, single nucleotide polymorphism (SNPs), genotyping by sequencing, Kompetetive Allele Specific PCR (KASP^{TM}), and so on. Obtained in this manner data are used to calculate the genetic distance, diversity, and mapping or genotyping samples for the presence of the desired trait. The molecular approach will be discussed in detail in one of the further sections of this chapter.

As we wrote earlier, 125 gene banks participated in oat germplasm conservation. Institutions, located in 63 countries, which store oat seeds, are affiliated with governments, universities, and international organizations. At the beginning, it must be emphasized that in the vast majority of oat genetic resources are safeguarded in long-term vaults. The predominant form of germplasm protection is a long-term storage, although storage standards for each institution are different. In most of them accessions are maintained at a temperature below −10°C. The only exception is the Polish collection, which is kept at 0°C but with controlled, low moisture content, that is,

5–6%. Out of 10 of the largest collections only one does not store the accessions in conditions compatible with long-term storage conditions. Such a situation takes place in the Russian collection, which is kept in the N.I. Vavilov Research Institute of Plant Industry. Most of these oat accessions are stored in the medium-term conditions. A relatively common practice is to maintain additional active collection under medium- or short-term storage conditions. Particular institutions also differ in terms of testing seeds viability and regeneration procedures. For example, in Canada PGR accessions viability is tested randomly and if less than 85% of viability is found in a regeneration group, the rejuvenation procedure is initiated. During regenerations morphologic and agronomic traits are observed and compared to historic notes. In contrast, in Poland and the United States, seeds viability is tested every 10 or 5 years, respectively. The common for all of collections is field regeneration, but the conditions strongly depend on the climate, soil type, fertility, and presence or absence of irrigation facilities. Another important thing in the successful germplasm preservation is establishing management systems and written protocols for acquisition, viability testing, regeneration, characterization, maintenance and storage, safety-duplicates, documentation, phytosanitary control and certifications, and distribution. In most of the oat collections such procedures were developed and are being implemented. Further details about maintenance conditions and procedures can be found in the global strategy for the *ex situ* conservation of oats (*Avena* spp.) complied by Germeier (2008).

Regeneration and evaluation of any species depends on its reproductive system. All oats except one (*A. macrostachya*) are self-pollinated species, thus they do not require mechanical or spatial isolation during field procedures. Unfortunately, wild oats have all or only lower florets disarticulating in the mature plant; therefore, panicles require putting into isolators during maturation.

4.6 Use of germplasm in crop improvement

Sharing of stored germplasm for research and breeding is among the main tasks of gene banks, while the main directions of contemporary breeding are in increasing crop productivity and improving quality traits. So, for every crop there is a point where germplasm preservation and breeding meet their tasks. This happens in the moment when the crop gene pool has to be expanded. As shown in many studies, the gene pool of common oat is quite narrow (Boczkowska and Tarczyk, 2013; Diederichsen, 2007; Fu et al., 2005); therefore, breeders and researchers have started looking for particular traits and encoding the genes in gene bank collections. Many potential donors of genes useful in breeding have been identified among approximately 30,000 accessions of oat CWR species. In Table 4.6, the presence of particular traits interesting for breeding are checked in all wild species.

4.6.1 Yield

The major objective of oat breeding is grain yield improvement. The main source of germplasm for yield increase is within the cultivated species. However, the wild ones

Table 4.6 Sources of important traits for breeding

	Yield			Disease resistance						Pests resistance				Abiotic stresses resistance		Other		Grain quality			
	Number of spike-lets/higher density of latter	Grain size	Pro-ductive tiller-ing	Crown rust	Stem rust	Pow-dery mil-dew	BYDV	Smut	Sep-toria	Nem-atodes	Fruit fly	Aphids	Cereal leaf beetle	Water defi-cient	Winter hardi-ness	Low height	Lodg-ing resis-tance	Pro-tein	Fat	β-Glu-cans	Starch
A. abyssinica	+			+	+	+		+		+									+		+
A. agadiriana						+						+				+			+	+	
A. atlantica						+			+									+	+	+	
A. barbata	+			+	+	+	+	+	+	+	+	+						+	+	+	
A. bruhnsiana				+		+															
A. byzantina				+	+	+	+		+									+	+	+	
A. canariensis				+	+		+		+	+	+					+		+	+		
A. clauda				+	+	+	+					+				+		+	+		
A. damascena			+	+	+	+	+		+							+		+	+	+	
A. fatua	+	+	+	+	+	+	+	+	+	+	+	+	+	+	+	+	+	+	+	+	+
A. hirtula	+		+	+	+	+	+		+		+					+		+	+	+	
A. insularis		+		+	+																
A. longiglumis				+	+	+	+		+			+						+	+	+	
A. ludoviciana		+		+	+	+	+		+	+	+	+		+	+		+	+	+	+	+
A. macro-stachya				+	+	+	+	+	+			+			+						
A. magna		+	+	+	+		+			+	+	+				+	+	+	+	+	
A. murphyi		+	+	+		+	+		+	+		+						+	+	+	
A. occidentalis				+	+	+	+					+		+		+	+	+	+	+	
A. pilosa				+	+	+					+	+				+		+	+		
A. prostrata	+		+	+		+						+				+					
A. sativa				+	+	+	+		+			+				+		+	+	+	
A. sterilis	+	+		+	+	+	+	+	+	+	+	+		+	+	+	+	+	+	+	+
A. strigosa				+	+	+	+	+	+	+								+	+	+	+
A. vaviloviana	+			+		+	+	+	+	+	+								+		
A. ventricosa				+	+	+	+					+				+		+			
A. wiestii	+		+	+		+		+	+	+		+		+				+	+		

show frequently high level of particular yield components. High number of panicles per area unit is a direct derivative of productive tillering. This trait has been identified in some forms of diploid and tetraploid species, such as, *A. prostrata*, *A. damascena*, *A. wiestii*, *A. hirtula*, *A. murphyi*, *A. magna*, *A. barbata*, and *A. sterilis* and also in hexaploid *A. ludoviciana* (Kanan and Jaradat, 1996; Ladizinsky, 1988; Loskutov, 1998a; Mal, 1987). High number of grains per panicle have been observed in some forms of *A. prostrata*, *A. hirtula*, *A. wiestii*, *A. vaviloviana*, *A. abyssinica*, *A. barbata*, *A. fatua*, and *A. sterilis* (Kanan and Jaradat, 1996; Loskutov, 2008; Mal, 1987; Rezai and Frey, 1988, 1990; Thomas and Griffiths, 1985; Trofimovskaya et al., 1976). Large grain size has been identified among species from primary and secondary gene pools, such as, *A. fatua*, *A. ludoviciana*, *A. sterilis*, *A. magna*, *A. murphyi*, and *A. insularis* (Ladizinsky, 1988, 1998; Loskutov, 1998a, 2008; Martens et al., 1980; Rezai and Frey, 1989, 1990).

4.6.2 Lodging

Breeding for high yield must be supported by selection of other traits that increase the yield stability. Lodging resistance is one of the most important factors in the production of cereals. It is connected with straw stiffness and plant height. Intraspecific variation of straw length based on the Dw genes is not much useful because of negative side-effects on plant morphology and physiology. The landraces are usually tall and not useful in this respect. Wild *Avena* species are highly variable in terms of height and are a potential source of new variation. Among diploid and tetraploid species short straw has been noticed for *A. prostrata*, *A. canariensis*, *A. agadiriana*, *A. pilosa*, *A. ventricosa*, *A. clauda*, *A. damascena*, *A. magna*, and *A. hirtula* (Loskutov, 2008). Some semishort straw forms have been also found among hexaploids *A. fatua*, *A. ludoviciana*, and *A. sterilis* (Loskutov, 2008). Other morphologic traits that could be used for improving lodging resistance have been found among accessions of *A. occidentalis* (erect juvenile growth), *A. magna* (stiff straw), and *A. wiestii* (adaptation of root system).

4.6.3 Disease resistance

Oat breeding is oriented on the development of disease-resistant cultivars. Resistance to the most important oat pathogens is controlled genetically and sources of resistance genes are available in conserved germplasm. The main oat diseases that could be limited by host resistance are rust, BYDV, powdery mildew, and soil-borne oat mosaic. Landraces of *A. sativa* and *A. byzantina* that were introduced in North America from other regions of the world were identified as carriers of resistance genes to various pathogens (Coffman, 1961).

Puccinia coronata Cda. f. sp. *Avena* e Fraser et Led., which causes the crown rust disease, is regarded as the most formidable oat pathogen, which leads to yield reduction, lower test weight, and increased lodging. It occurs in almost all regions of oat production, but is most serious in humid areas. Current sources of resistance to this pathogen become useless because of virulence pattern shift (Chong and Zegeye, 2004;

Leonard et al., 2004). Therefore, new sources of resistance are extremely needed. Among wild oat species resistance to this pathogen is quite common. In almost all of them resistant forms have been identified. More than 100 crown rust resistance genes (*Pc*) have been described (Cabral et al., 2014). Many of them were derived from *A. sterilis* accessions (Leonard et al., 2004). Also, *A. strigosa* and *A. barbata* were identified as a rich source of the resistance genes. Within *A. barbata* very resistant and strongly susceptible forms can be found (Loskutov and Rines, 2011). Stem rust (*Puccinia graminis* Pers. F. sp. *Avenae* Eriks & Henn.) is capable of causing similar damage as crown rust. Wild *Avena* species remain a large untapped reservoir of resistance to this pathogen. Out of 17 oat stem rust resistance (*PG*) genes only few have been utilized in oat breeding programs (Fetch and Fetch, 2011).

Another pathogen spread worldwide is *Ustilago* spp., which causes smut disease. Among ~55,000 of entries of the USDA world oat collection 10% were identified as either immune or highly resistant to composite smut populations (Nielsen, 1977). Substantial levels of resistance have been found also in wild oat accessions. It is considered that *A. strigosa*, *A. wiestii*, *A. barbata*, *A. vaviloviana*, *A. fatua*, *A. sterilis*, and *A. abyssinica* are resistant to smut (Forsberg and Shands, 1989; Frey, 1991b; Harder et al., 1992; Leggett, 1992; Nielsen, 1978, 1993; Rodionova et al., 1994). However, among *A. strigosa* and *A. barbata* two radically different levels of resistance have been identified (Vavilov, 1965, 1992).

Powdery mildew (*Blumeria graminis* DC. f. sp. *Avena* e Em. Marchal.) is one of the pathogenic fungi most difficult to control by means of crop management (e.g., rotation). It is observed in all oat cultivation regions. The new races of this fungus overcome cultivar resistance increasingly. So, new sources of effective resistance are necessary. Among wild species new sources were identified in *A. magna*, *A. murphyi*, and *A. sterilis* (Okoń et al., 2014). Many other wild oat species have been proven to possess resistance to the powdery mildew disease.

Nowadays the most harmful disease of oat is the BYDV, which is transmitted by aphids and caused by *Hordeum virus nanescens* Rademacer et Schwarz. Multiple genes conferring varying degrees of resistance to this pathogen have been found in cultivated and wild oats such as *A. fatua* an *A. occidentalis* (Comeau, 1982; Endo and Brown, 1964; Rines et al., 1980).

4.6.4 Pest resistance

Various pests cause damage to oats at each stage of the life cycle from seedling until the grain. However, the presence of pests is not synonymous with a decrease in the yield. A moderate-size population of certain pests in cereals has a stimulating effect on the tillering and ultimately increases yield. The economic injury levels indicator points to density of a particular pest population when economic damage will occur. Among many pest species are only a few causing a major threat for yield such as aphids and cereal cyst nematode. Resistance to pests is not so common among wild *Avena* species, but still sufficient sources can be found. Cereal cyst nematode (*Heterodera avenae* e Woll.) resistance has been found amid such species as *A. canariensis*, *A. wiestii*, *A. strigosa*, *A. barbata*, *A. vaviloviana*, *A. abyssinica*, *A. magna*, and *A. murphyi*

(Harder et al., 1992; Leggett, 1992). Aphids (*Rhopalosiphum padi* L. and *Sitobion avenae* Fabricius.) can inflict double damage, that is, by themselves and as a vector of BYDV. Resistant forms have been identified among such species as *A. clauda, A. pilosa, A. ventricosa, A. longiglumis, A. prostrata, A. wiestii, A. barbata, A. macrostachya, A. occidentalis*, and *A. fatua* (Leggett, 1992; Loskutov, 2008; Weibull, 1986).

4.6.5 Abiotic stress resistance

Besides biotic stress-resistant forms, breeders are looking for donors of abiotic stresses resistance. A certain group of wild species is resistant to drought, heat, and salinity stress, that is, *A. wiestii, A. ludoviciana*, and *A. sterilis* (reviewed by Loskutov and Rines (2011)). High tolerance to aluminum is present in the A genome species while the C genome diploids and tetraploids have low resistance (Loskutov, 2008). For temperate zone breeders the most important abiotic stress is cold and connected with its winter hardiness. It is defined as the ability of a plant to tolerate the numerous and complex stresses during winter. Therefore, winter hardiness is a genetically complex trait strongly influenced by genotype–environment interactions. Usually, the winter killing is mainly caused by direct freezing injury (Levitt, 1972). The resistance is common among weedy hexaploid oats (Leggett, 1992). *A. byzantina* seems to be a good source of genetic diversity for winter hardiness, which has not been heavily exploited. *A. macrostachya*, the only perennial oat species, is characterized by increased winter hardiness (Loskutov, 2008).

4.6.6 Grain quality

The next important direction of oat breeding is improvement of grain quality, which depends on the composition of particular chemical compounds. The majority of oat grain is used as animal feed, but recently its importance in human diet has been constantly increasing. Oat, like the other cereals, is a valuable source of carbohydrates. However, it also contains well-balanced proteins, several essential vitamins, fatty acids, and soluble fiber.

4.6.6.1 Protein

The nutritional value of oat proteins is specified by three features: protein concentration, balance of amino acids, and digestibility. Among staple cereals, oat has the highest protein content, but still it is significantly lower than in animal products and legumes. Some wild oat species are noted for high (above 20%) protein content and these are diploids *A. atlantica, A. canariensis, A. clauda, A. damascena, A. hirtula, A. longiglumis, A. pilosa, A. strigosa, A. ventricosa*, and tetraploids *Avena agadiriana, A. barbata, A. magna*, and *A. murphyi* (Butler-Stoney and Valentine, 1991; Harder et al., 1992; Hoppe and Hoppe, 1991; Ladizinsky, 1988; Ladizinsky and Fainstein, 1977; Leggett, 1992; Loskutov, 2008; Miller et al., 1993; Trofimovskaya et al., 1976; Welch et al., 2000). Two hexaploid weedy species, *A. fatua* and *A. sterilis*, are considered as good partners in breeding for high protein yield (Briggle et al., 1975; Frey, 1991b; Leggett, 1992; Loskutov, 2008; Thompson, 1966;

Trofimovskaya et al., 1976; Welch et al., 2000; Zillinsky and Murphy, 1967). Some accessions of other hexaploid *Avena* species have also been identified as containing a large amount of protein, that is, *A. ludoviciana* and *A. occidentalis* (Miller et al., 1993; Trofimovskaya et al., 1976). Protein quality is the result of the relative balance of amino acids and digestibility. Cereals in general are considered to be limited in some amino acids such as lysine, methionine, threonine, isoleucine, and tryptophan. According to the FAO standards for food lysine content in oat protein is the highest among cereals. According to Loskutov (2008) wild hexaploid oats demonstrated the percentage of lysine comparable to cultivated forms. *A. barbata* was identified as having the highest levels of methionine, valine, tyrosine, and isoleucine. At the same time it had the lowest content of leucine and contains less glutamic acid and proline than cultivated species (Loskutov, 2008).

4.6.6.2 Lipids

The concentration of lipids is highly variable among oat cultivars (Brown and Craddock, 1972). Triglycerides are the major fraction of total lipids and the remaining are phospholipids, glycolipids, free fatty acids, partial glycerides, sterols, and sterol esters (Youngs, 1986). Lipids composition of oat grain is beneficial because of the high percentage of unsaturated fatty acids. The high content of linoleic acid, an essential fatty acid for human nutrition, occurs in oat grain. Unfortunately, breeding to improve concentration and/or composition of lipids is difficult. Cultivars with high lipid content usually have worse ratio of oleic to palmitic and linoleic acids as compared to those with lower lipids concentration (de la Roche et al., 1977; Frey et al., 1975; Welch, 1975). The genes controlling lipid content in cultivated and wild species have been identified as nonallelic (Frey, 1991b; Thro and Frey, 1985). The high content of lipids has been identified in majority of accessions of diploid species such as *A. atlantica*, *A. clauda*, *A. canariensis*, *A. damascena*, *A. hirtula*, *A. longiglumis*, and *A. wiestii* (Leonova et al., 2008; Loskutov, 2008; Welch et al., 2000). Among tetraploids *A. agadiriana*, *A. barbata A. magna*, *A. murphyi*, and *A. vaviloviana* high-lipids forms have been also observed (Ladizinsky, 1988; Leonova et al., 2008; Loskutov, 2008; Welch et al., 2000). All five hexaploid of oat wild species have forms with high lipid content (Frey, 1991b, 1994; Leggett, 1992; Leonova et al., 2008; Loskutov, 2008; Thro, 1982; Trofimovskaya et al., 1976; Welch et al., 2000). By recurrent selection of *A. sativa* and *A. sterilis* forms with increased lipid content up to 18% was obtained (Branson and Frey, 1989a, 1989b; Frey and Holland, 1999; Schipper and Frey, 1991).

4.6.6.3 Nonstructural carbohydrates

Starch is the major constituent of oat grain and it is the main determinant of grain digestibility. In general, starch and protein contents are inversely correlated in oat. The same as in wheat, oat starch has an intermediate fraction between amylopectin and amylose, which is predominantly linear (Paton, 1986). Wild hexaploid species *A. sterilis*, *A. ludoviciana*, and *A. fatua* have similar content of amylose to the cultivated ones (Rodionova et al., 1994).

4.6.6.4 Fiber

Dietary fiber is defined as plant carbohydrates and lignin that are indigestible for humans. It includes insoluble in water cellulose, hemicellulose, and lignin, and other hydrophilic components. The soluble fraction is high relative to other cereals and contains high concentration of β-glucans [(1→3),(1→4)- β-D-glucans]. Only barley is reported to have higher content of β-glucans, but a higher proportion of oat β-glucans are soluble (Aaman and Graham, 1987). Based on data from wild oat species, there is a wide range of β-glucans concentration among wild species and in intra- and inter-species substantial differences were observed (Welch et al., 2000). An increased content has been found in diploid *A. atlantica*, *A. damascena*, *A. hirtula*, *A. longiglumis*, and *A. strigosa*. Among tetraploid species the ranges of β-glucans content were far narrower and the highest values were found in *A. agadiriana*, *A. barbata*, *A. magna*, and *A. murphyi* (Howarth et al., 2000; Legget, 1996; Miller et al., 1993; Welch et al., 2000). All hexaploid wild species contain more β-glucans than the tetraploids (Cho and White, 1993; Frey, 1991b; Miller et al., 1993; Welch et al., 1991).

4.7 Limitations in germplasm use

Modern oat breeding no longer relies only on cultivars, pure lines, and other breeding materials that were collected or/and derived by particular breeders. The importance and potential of oat germplasm is well understood and using them during breeding programs became a routine. However, full utilization of genetic resources is not and probably will never be possible. An employment of gene banks resources entails a whole series of restriction and limitations. Some of them partially result from the absence or incompleteness of information about stored accessions. As we mentioned earlier, a part of *Avena* accessions has not been taxonomically identified. Quite a lot of accessions suffer from basic passport data deficiency and the less information is available about the individual accessions, the less the interest in them by the farmers is. From the breeders' point of view, even if the accession has incomplete basic passport data, but was evaluated in detail, it could be potentially useful in developing new cultivars. Therefore, the lack of evaluation data is the next limiting factor in germplasm use. If the breeder decides to obtain from the gene bank such an accession, he/she has no guarantee at all that it has the desired characteristics. Furthermore, the verification process of such an accession is expensive and time-consuming. Another limiting factor in the use of oat germplasm is insufficient information about the level of genetic diversity. This issue will be discussed in detail later in this chapter. Breeders, during the search for new genes encoding particular qualitative or quantitative traits, are increasingly turning to oat CWR; however, their potential is limited by crossing barriers (Table 4.7). Also, this issue will be presented in greater detail in the subsequent part of this chapter. Access to certain genetic resources is also forbidden by various legal retrenchments, which were already mentioned earlier. The last of the limiting factors of germplasm usage is associated with maintenance and regeneration of collected accessions. Due to the high cost of maintaining large collections and their field regeneration, gene banks

Table 4.7 Gene pools of *A. sativa* and crossability potential between species (Loskutov, 2001)

	Cultivated				**I Gene pool**				**II Gene pool**		**III Gene pool**															
Avena	***A. abyssinica***	***A. byzantina***	***A. sativa***	***A. strigosa***	***A. fatua***	***A. ludoviciana***	***A. occidentalis***	***A. sterilis***	***A. magna***	***A. murphyi***	***A. agadiriana***	***A. barbata***	***A. insularis***	***A. macrostachya***	***A. vaviloviana***	***A. atlantica***	***A. bruhnsiana***	***A. canariensis***	***A. clauda***	***A. damascena***	***A. hirtula***	***A. longiglumis***	***A. pilosa***	***A. prostrata***	***A. ventricosa***	***A. wiestii***
A. abyssinica	x	+/−	+/−	−	− −	− −	− −	− −	− −			+			+						− −	− −	− − −		− − −	
A. byzantina		x	+	− −	+	+	+	+	+/−			−			− −						− −	+/−				
A. sativa			x	− −	+	+	+	+	+/−	+/−		−	+/−	−	− −	−	+/−	+/−			+/−	+/−	+/−	− −	+/−	
A. strigosa				x	− −	− −	− −	− −	+/−	− −		+/−	+/−	− − −	− −	+	− − −	−	− − −	− − −	+	− −	− − −	− −	− − −	+

+, Fertile progeny; −, partial sterile; − −, strong sterility; − − −, noncrossing.

are limiting the number of shared seeds. In the case of wild oat species, seed regeneration is additionally extremely difficult and as an effect of that, some part of the accessions are temporarily or permanently unavailable for ordering.

4.8 Germplasm enhancement through wide crosses

Related wild and cultivated species are an inexhaustible source of genetic variation for breeding and wide crossing has been verified as a much successful way of crop improvement. Alien genes for qualitative characters frequently show good expression when transferred to the recipient species. Different genetic determination of quantitative traits in distant relatives enables transgression effects and breaking negative correlations between desired characters. Usefulness of alien variation is frequently restricted by difficult access related to interspecific isolation barriers and by the risk of incompatibility of donor species genes in the interaction with the genetic background of a recipient species, which may decrease expression of a desired character or bring negative side effects. Due to this and to other limitations described in the previous chapter, wide crossing is used in breeding mainly in cases when the intraspecific variation is insufficient.

The alien variation has been widely used in oat improvement, particularly in North America and the United Kingdom in the second half of the last century. Predominantly, the recipient forms for the transferred genes belonged in two hexaploid species: *A. sativa* and *A. byzantina.* Their gene pools were mixed in some modern cultivars, which were not classified as wide crossing products, because of the lack of reproductive barriers and easy recombination. In practice, both species are considered as different subunits of *A. sativa*; the same is assumed for the following text in this section. The work on wide crossing in oat delivered not only new germplasm lines and cultivars. Pioneer experimentation, mainly in the United States, on the incorporation of alien variation for quantitative traits resulted in the development of complex breeding strategies of general usefulness. The reviews of Frey (1986) and Holland (1997) are helpful in searching for information on the achievements of the last century. Alien introgressions are reported mainly for disease resistance genes, which are usually well expressed independently on a different genetic background. Abundance of possible sources is reported for all wild species and all diseases (see Section 4.6 and Table 4.6). Wide crossing of oat has contributed significantly to building up the resistance to crown rust (*P. coronata*) – one of the most destructive disease of common oat on the world scale. Genes for complete (race-specific, vertical) or partial (horizontal) resistance have been transferred to *A. sativa* mainly from the most closely related hexaploid *A. sterilis* (Frey, 1982, 1991b; McDaniel, 1974; McKenzie et al., 1984; McMullen and Patterson, 1992; Ohm et al., 1995; Simons et al., 1987; Stuthman, 1995). Other resistance genes came from hexaploid *A. fatua* (Šebesta and Kühn, 1990), tetraploid *A. magna* (Rothman, 1984), and diploids *A. strigosa* (Aung et al., 1996; Forsberg et al., 1991; Reinbergs, 1983; Rines et al., 2007; Sharma and Forsberg, 1977) and *A. longiglumis* (Rooney et al., 1994). The *A. strigosa* resistance was transferred also to the tetraploid hybrids with *A. abyssinica* (Zillinsky et al., 1959). The work on the transfer of crown rust resistance from exotic sources led to the registration of numerous cultivars and germplasm lines. In some cultivars percentages of *A. sterilis* or *A. fatua* genes were particularly high (even 25% and 50%), which indicates a relatively small evolutionary distance between these species and the cultivated oat. Totally, among the known genes for crown rust resistance (approaching 100 *Pc*-genes) nearly half have been transferred to the cultivated oats from *A. sterilis* (Gnanesh et al., 2014). Durability of their effects, until new virulent races of the pathogen evolved, was generally low and did not exceed several years. Relatively stable effects of the resistance were recorded in multiline cultivars (composed of sublines with different resistance genes) (Frey, 1982, 1992) and cultivars with cumulated genes for partial resistance (Castell and Stuthman, 2002). The most effective single gene for stable crown rust resistance (*Pc94*) comes from *A. strigosa* (Aung et al., 1996), which is presently the second important source of the resistance (21 *Pc*-genes) (Gnanesh et al., 2014).

In parallel, an effort has been put on the improvement of resistance to stem rust (*P. graminis* Pers. f.sp. *avenae* Eriks. and Henn.). Successful transfers were reported from *A. sterilis* (McKenzie et al., 1986; McMullen and Patterson, 1992; Rothman, 1984), tetraploid *A. barbata* (Brown, 1985), and diploids *A. strigosa* (Rothman, 1984) and *A. longiglumis* (Rothman, 1986).

Resistance to powdery mildew (*B. graminis* D.C. f.sp. *avenae* Em Marchall) was introduced to breeding materials with the use of germplasm from *A. sterilis* (Hayes and Jones, 1966), *A. barbata* (Aung et al., 1977; Thomas et al., 1980b), and diploids *A. pilosa* (Hoppe and Kummer, 1991; Sebesta et al., 1986), *A. hirtula* (Thomas, 1968), *A. ventricosa* (Thomas, 1970), and *A. prostrata* (Griffiths, 1984; Morikawa, 1995). In more recent studies of Herrmann (Yu and Herrmann, 2006) the resistance was transferred from the tetraploid perennial oat *A. macrostachya*.

Relatively little has been done in the application of wide crossing for improvement of biotic stress toleration when quantitative type of variation is involved, which requires more effort than simple backcrossing sufficient for majority of the rust or mildew resistance genes. BYDV is such an important pathogen, when considering remarkable losses caused by the disease. Transfer of partial resistance from *A. sterilis* to the hybrids with *A. sativa* has been reported by Landry et al. (1984). Much remains to do with less popular but locally important oat diseases from the genera *Drechslera*, *Septoria*, *Ustilago*, and *Myrothecium*. No reports are available also on wide crossing application in decreasing contamination of grain with mycotoxins produced by the fungi *Fusarium* and *Alternaria*. Insect pests in oats are easy to control chemically and less significant than diseases. Nematodes are more problematic. Wide crossing was reported for the cereal cyst nematode resistance (*H. avenae* Woll.); the genes have been transferred from *A. sterilis* (Marschall and Shaner, 1992; Mattsson, 1988).

In relation to the achievements of alien transfer on the area of biotic stress tolerance, those concerning abiotic stress in oats are scarce. Drought resistance, which is the most and increasingly critical factor in oat cultivation, remains a challenge for the future. Rich genetic resources are easily accessible in *A. sterilis* and *A. fatua*. These widely adapted weeds occur even in desert oases and on roadsides in dry areas. The tetraploid and diploid relatives of common oat grow widely in water-deficient habitats in large areas of northern Africa and Central Asia (Baum, 1977).

Almost all wild species contain ecotypes showing high tolerance to drought, extreme temperatures, soil salinity, and acidity (Section 4.6, Table 4.6, and Loskutov and Rines, 2011). Selection of early maturing forms is a way of terminal drought escape (and frequently disease escape). Earliness, shattering resistance, and large seed size were combined in some cultivars selected from hybrids with *A. fatua* by Suneson (1967). Acceleration of growth in lines derived from a similar cross was reported by Takeda and Frey (1976).

The rich gene pool of *A. sterilis* was useful in the improvement of winter hardiness, as reported by Bacon (1991) for the cultivar "Ozark." Several other wild species contain cold-resistant forms, especially those of winter or semiwinter type (Section 4.6, Table 4.6). Among oats, the perennial species *A. macrostachya* shows the highest resistance to winter killing (Baum and Rajhathy, 1976) and was selected as a source of variation in the studies in North Carolina University (Jia et al., 2006; Santos et al., 2002). They reported transgressive segregation in resistance to low temperatures and, additionally, to the soil-borne mosaic virus. A progress in winter hardiness was also reported in Polish winter oat created with participation of *A. macrostachya* (Łapiński et al., 2013). The "Dormoat" lines from Canada are another option for the areas where green plants do not survive severe winters. They emerge in spring after autumn sowing and lying dormant over winter. The trait was transferred from *A. fatua* (Burrows, 1986).

Among agronomically important traits improved with the use of *A. sterilis* genetic variation are herbicide resistance (Barr and Tasker, 1992) and increases of growth rate, biomass quantity, forage quality (Cox and Frey, 1984; Gupta et al., 1987), and undefined factors increasing grain yield (Frey, 1992). Novel genes from *A. fatua* enabled the combination of high forage yield with good grain yield in the cultivar "Mesa" (Thompson, 1967). New transfers of plant dwarfness from *A. fatua* were reported by Morikawa (1988) and Morikawa et al. (2007).

Improvement of grain chemical composition was an important target of wide crossing experiments in oat. Increases in groat protein and oil content are classical examples of successful application of alien variation in germplasm development for quantitative traits. Transgression effects for grain protein percentage in the *A. sativa* × *A. sterilis* hybrids have been frequently reported in literature, for example, by Ohm and Patterson (1973), Iwig and Ohm (1978), Cox and Frey (1984), and Kuenzel and Frey (1985). The positive result of Thomas et al. (1980b) is based on a more distant cross *A. sativa* × *A. maroccana*. Tetraploid oats *A. murphyi* and *A. magna* (=*A. maroccana*) were donors of high protein content in the studies of Hagberg and Mattson (1986). For this character, almost all wild species are superior to cultivars. The main obstacle is that the protein content increases at the cost of decrease in yield. Breaking this negative correlation, which contributed to raising protein yield per area, is a success of Frey and coworkers from Iowa State University who took advantage of recurrent selection within a set of carefully preselected forms of *A. sterilis*, *A. sativa*, and their hybrids. Lack of the adverse correlation was the primary preselection criterion, before the protein and yield scores (McFerso and Frey, 1991; Moser and Frey, 1994). The recorded progress in protein quantity did not worsen the amino acid composition.

An interesting option in germplasm improvement through wide crossing was tried by Ladizinsky (1995). In his studies the wild tetraploid oats *A. magna* and *A. murphyi* with large and protein-rich seed and adaptation to hot and dry environments were the recipients for "domestication" genes transferred from *A. sativa*. The resulting tetraploid lines were morphologically similar to the cultivated parent and showed higher protein content. The experiment indicated a low number of genes differentiating between wild and cultivated forms.

The recurrent selection method was applied at Iowa State University also for raising oil content in groat (Frey, 1992; Frey and Holland, 1999). The initial set of lines was preselected after two backcrosses of *A. sativa* × *A. sterilis* hybrids to *A. sativa* (Branson and Frey, 1989b). After nine cross-selection cycles the oil content was nearly doubled and the top result of 18.1% in the best line was approximately 1/3 better than the best values reported for the whole *Avena* genus. The oil yield per area increased by 27.7% with parallel improvement in proportion of unsaturated fatty acids and content of antioxidants (Schipper and Frey, 1991). However, resistances to lodging and diseases dropped significantly and grain yield was decreased by nearly 30% (Frey and Holland, 1999).

The same breeding method and variation source of *A. sterilis* proved to be successful in differentiating β-glucan content in groat (Cervantes-Martinez et al., 2001).

Seed size of hexaploid oat has been augmented in the progeny of crosses with *A. magna* (= *A. maroccana*), in combination with high protein content (Thomas et al., 1980b) or with earliness and high groat percentage (Hagberg, 1988). *A. fatua* proved to be also a successful source (Stevens and Brinkman, 1986). As reported by Allen and Kaufmann (1979) and Burrows (1986) the small seed of the diploid parent *A. strigosa* was not an obstacle in the improvement of kernel size of common oat. Enlarged grain and broken negative correlation between yield and seed size was also reported for spring naked hexaploid oat derived from a cross of *A. sativa* with *A. macrostachya* (Łapiński et al., 2014). Some octoploid husked derivatives from the same crossing program reached thousand kernels weight (TKW) values above 60 g, without increase of husk content, in spite of a very small seed of the wild *A. macrostachya* parent (TKW ~11 g) (Łapiński et al., 2013).

4.8.1 Germplasm for target traits

The alien genetic variation used successfully in widening the gene pool of cultivated oat represents only a small proportion of resources stored in gene banks and a minute part in relation to the resources existing *in situ*. Some collections of oat, in Canada, the United States, Russia, and Morocco, contain thousands of accessions of the relative species (see Section 4.4.2 and Table 4.5). Therefore, special strategies have been tried in order to cope with the proper choice of donor genotypes from large numbers of available accessions. Expression level of a target trait is obviously important. In gene banks the body of information on resistances to various environmental factors and on grain chemical composition is constantly growing. Based on the published results and on access to the gene bank databases (Section 4.5), proper selection of parental forms for a wide crossing experiment is presently much easier than before. However, usefulness of the accessions as donors of target traits for transfer is only partially predictable, as not only the expressed trait level but also evolutionary distance and genetic compatibility with the recipient species are important. The core collections containing maximal genetic diversity in a restricted number of accessions should be much helpful in the restriction of area of search. Creation of core collections of wild oats, based on the ~2000 accessions of the VIR collection from Sankt Petersburg, Russia, was announced by Loskutov (1998b). Quickly growing information on variation pattern and evolutionary relationships between bank accessions, based on DNA or isozyme biochemistry (see Section 4.9.1), serve additional valuable hints. Murphy and Phillips (1993) recommend three general strategies for the molecular search of new individual sources of variation, verified during *A. sterilis* selection for the common oat improvement. The first one addresses attention to the presumable center of origin of the species (for *A. sterilis* – Turkey), where genetic variation is maximal. The second hint recommends selection of an equal proportion of accessions from major clusters distinguished after multivariate analysis of molecular (and other) research data. The third strategy relies on choice of accessions with isozyme (or DNA) banding patterns absent in the recipient species.

The previously mentioned criteria restrict the area of search, but are never sufficient for individual genotypes before they are tried experimentally. Interaction of an introgressed gene with the genetic background of a recipient form is unpredictable. Therefore, working on broad genetic basis, that is, choice of multiple sources and multiple backgrounds, is the only remedy to avoid the undesired incompatibility effects.

4.8.2 Breaking the interspecific isolation barriers

Reproductive isolation between species works as a mechanism for maintenance of separation and individuality of different gene pools that are incompatible and/or adapted to different environments. Among the related species such isolation is usually not complete, which enables some level of gene exchange and taking benefit of those genes, which are or may easily become compatible to the other species' genetic background.

Ploidy level and differences in genomic composition are principal factors influencing effectiveness of crossing and fertility of the F_1 wide hybrids. Practically, no difficulties are to be expected at hybridization of hexaploid cultivated oats *A. sativa* and *A. byzantina* with their close relatives *A. sterilis*, *A. fatua*, *A. occidentalis*, and *A. ludoviciana*, which show the same ploidy level and common genomic formula (AACCDD). Occasional occurrence of univalents in meiosis of hybrids within this group (McMullen et al., 1982) does not restrict the interspecific gene flow. Hybrids of *A. sativa* with *A. sterilis* and *A. fatua* occur even spontaneously (Andersson and Carmen de Vincente, 2010) and the alien variation is easily assimilated into cultivated forms. Some high-yielding cultivars released in the second half of the last century contained even 50% of germplasm from *A. sterilis*, for example, "Ozark" (Bacon, 1991) or from *A. fatua*, for example, "Mesa" (Thompson, 1967). According to Leggett and Thomas (1995) these species are classified as the primary gene pool for *A. sativa*. For the tetraploid cultivated oat *A. abyssinica* the similar most closely related group is formed by *A. barbata*, *A. vaviloviana* (spontaneous hybrids reported by Baum (1977)), and

A. agadiriana (all with common genome formula AABB). A corresponding primary gene pool for diploid cultivated oat *A. strigosa* would involve *A. atlantica*, *A. hirtula*, and *A. wiestii* (carriers of the As genome).

For *A. sativa*, species of oat with 4*x* ploidy level but the same homologous genomes have been classified by Leggett and Thomas as the secondary gene pool. Crossability of common oat with these species is relatively high but sterility occurs in pentaploid F_1 hybrids (Table 4.7) due to incomplete chromosome pairing in meiosis. In spite of this, fertility is usually sufficient to perform successful backcrossing to a parental species (usually *A. sativa*). The secondary gene pool includes tetraploids *A. maroccana* and *A. murphyi* sharing the A and C genomes with *A. sativa*. The recently discovered tetraploid *A. insularis* is another candidate to the group as its genomic formula is close to CCDD and hybrids with *A. sativa* are also partially fertile (Ladizinsky, 1999).

Very low levels of crossability and F_1 fertility are reported for crosses of common oat with diploid species carrying variants of the C genome (*A. pilosa*, *A. ventricosa*, *Avena bruhnsiana*) or the A genome (*A. longiglumis*, *A. canariensis*, and *A. hirtula*) (Table 4.7). All diploids, together with the AABB tetraploids (*A. abyssinica*, *A. barbata*, *A. agadiriana*, *A. vaviloviana*) and tetraploid *A. macrostachya*, have been classified by Leggett and Thomas to the tertiary gene pool. Larger differences in ploidy level, lack of homology between different genomes' chromosomes, and lack of coadaptation between genes from *A. sativa* and species of this group cause strong reproductive isolation, which cannot be overcome without special procedures such as embryo rescue and doubling of chromosome number (usually with colchicine). More detailed data on crossability and sterility between all *Avena* species are available in Table 4.7 and in the review of Loskutov (2001).

For the highest chance of seed set in the interploidy crosses of oats, Rajhathy and Thomas (1974) recommend use of the lower ploidy species as a pistillate parent. However, the opposite cross direction may work better in some crosses (e.g., *A. sativa* × *A. macrostachya* (Łapiński et al., 2013)). Reciprocal crossing is worthy of trial because of the possible cytoplasmic effects, which are to be expected even in less distant crosses of oat, as enhanced yield of the lines with *A. sterilis* cytoplasm reported by Beavis and Frey (1987) or different level of disease resistance (Simons et al., 1985).

Application of growth regulators after pollination, embryo rescue, colchicine treatment, and intense vegetative propagation of highly sterile hybrids are powerful procedures that greatly increase chances for alien gene transfer. However, for the most recalcitrant interploidy crosses the best solution may be chromosome doubling in the lower ploidy parent or production of an artificial bridging alloploid (Sadanaga and Simons, 1960). Increasing of ploidy level (and genetic diversity) through a combination of a preliminary wide cross and chromosome doubling proved to be much useful in oat. At higher ploidy levels, the presence of common genomes in F_1 hybrids exerts a buffering effect on disturbances caused by the lack of meiotic pairing or incompatible gene action. Therefore, it is recommended, based on a variety of diploid and polyploid species with recognized genomic composition, to produce first a proper artificial alloploid (optimally hexaploid or octoploid) with not more than one critical genome different than in the acceptor species. Next, the alloploid should be used as a cross partner with the target species. This way proved to be successful in the transfer of numerous genes from the tertiary gene pool, for example, for powdery mildew resistance from *A. pilosa* (Sebesta et al., 1986). On the other hand, the use of a bridge species third component may complicate restitution of fully fertile and agronomically acceptable lines (Rines et al., 2007).

Difficulties at alien transfer are often not restricted to crossability and F_1 sterility. More serious problems may appear later, as a restricted recombination between native and alien chromosomes, which disturbs separation of target genes from unadapted or undesired neighbor genes from the same chromosome. In this context, the line Cw57 of *A. longiglumis* is a particularly valuable cross partner for preliminary crosses. It causes increase of homeologous pairing and intrachromosomal recombination frequency (Thomas et al., 1980a), acting similarly to the deficiency of *Ph1* locus used in wheat chromosome engineering.

Transfer of mildew resistance genes from *A. barbata* (tertiary gene pool) to *A. sativa* is an example of the classical procedure. It started from colchicine-facilitated production of *sativa* + *barbata* alloploid (10*x*), followed by backcrossing to *A. sativa* and selfing, aimed at the production of a disomic addition line. Next, irradiation of this line caused fragmentation of the alien chromosome, which led to the desired translocation (Aung et al., 1977). In another methodical version the same addition line was crossed to an alloploid (8*x*) carrying the *A. longiglumis* "CW57" genome (added to *A. sativa* cv. "Pendek") in order to promote intergenomic crossing over (Thomas et al., 1980a). Another irradiation stimulating the transfer of stem rust resistance from *A. barbata* was described by Brown (1985). The "CW57" system was used in the *Pc94* crown rust resistance gene transfer from *A. strigosa* (Aung et al., 1996).

Intense vegetative propagation (up to thousands of plants) of colchicine-treated highly sterile F_1 hybrids, followed by extensive spontaneous pollination with mixture of *A. sativa* lines on a specially prepared field, was a successful strategy to overcome the strong sterility barrier in the Polish *A. sativa* × *A. macrostachya* crosses (on average, one germinable seed was formed per 210 panicles of F_1 generation). In one of the F_1 hybrids, the procedure facilitated occurrence of very rare reduced functional gametes from a semirandom assortment in the disturbed meiotic divisions. Among 57 germinable seeds obtained nearly half gave rise to semisterile plants with chromosome numbers between 40 and 49 (the rest were of alloploid type with ~70 or 56 chromosomes). The variation among lines derived from this material was sufficient to select hexaploid forms with large grain and winter hardiness superior to the lines used as standards (Łapiński et al., 2013).

The ability to participate in interspecific hybridization is not equally distributed in the populations and frequently is restricted to a small proportion of cross-compatible genotypes. Therefore, much may depend on the number and diversity of parental forms selected for a difficult wide cross. Independently on different ploidy and different genomic composition the individual allelic differences remain an essential factor for crossability, F_1 sterility, and genetic compatibility with the other parent. There is no evidence on the possible links between an individual crossability or sterility and usefulness of the resulting hybrids. Anyway, in case of a highly difficult cross, the better breeding strategy is hybridization between possibly high number of genetically differentiated parents instead of crossing a pair of genotypes.

Experimentation with gynogenetic production of doubled haploid lines of oat contributed to the extension of the tertiary gene pool beyond the *Avena* genus. Matzk

(1996) succeeded in crossing oats with maize (*Zea mays* L.) and pearl millet (*Pennisetum americanum* L.). More recently, Kynast et al. (2004) reported on the production of a whole set of oat lines carrying disomic additions of each of 10 maize chromosomes. The advantages of this material have not been recognized yet. Surely, more intergeneric cross combinations will appear in the future.

Progress in the recombinant DNA technology is expected to omit any reproductive isolation barriers and its role will be growing. However, especially for the quantitative traits, the theoretically unlimited area of choice among genetic resources is remarkably restricted by cognitive and technological possibilities of science. Incomprehensive variation in wild gene pools and their evolving character create dynamic systems with thousands of variables and astronomic number of solutions. Only a minute part of these solutions have been realized. Therefore, no decline is expected for the usefulness of classical search for new variation through wide crossing.

4.8.3 Further improvement of oat interspecific derivatives

Interspecific recombinants of early generations seldom show satisfactory usefulness for agriculture. Broken coadaptation of genes decreases vigor and fertility. Some adverse correlations may appear, usually related to linkage with undesired "wild" characters. The level of expression of the desired transferred genes may be insufficient. In oat, suppression effects have been reported, for example, for the introgressed crown rust genes (Wilson and McMullen, 1997). All these require additional effort and special breeding strategies. The simplest cases are related to improvement of qualitatively inherited characters. Transfer of a single biotic stress resistance gene through a series of backcrosses to an elite cultivar is a typical case; it is usually sufficient for good expression of the character. Linkage drag may remain a problem, because a chance for appropriate crossing over is frequently much decreased by insufficient homology between the related chromosomal fragments. Application of irradiation (Sharma and Forsberg, 1977) or chemicals causing chromosome fragmentation is a possible way to break adverse linkages, but the random changes in chromosomal structure frequently bring new problems with gene dosage related to deficiency and/or duplication of chromosome fragments. The previously mentioned use of *A. longiglumis* CW57 in a wide crossing program, as an additional preliminary cross partner, facilitates homoeologous chromosome pairing and interspecific crossing over, which is less destructive to karyotype than random heterologous translocations caused by irradiation (Thomas et al., 1980a, 1980b).

Recently, introgression and stacking of alien genes has been facilitated with growing number of molecular markers tagging the most useful genes of interest. The markers enable detection of genes in early growth stages, independently on their expression. The molecular markers have been developed for crown rust resistance alien genes *Pc94*, *Pc91*, and *Pc68* (Chen et al., 2006, 2007; Gnanesh et al., 2013). The markers for stem rust resistance genes *Pg3* and *Pg9* from *A. sterilis* were reported by Orr et al. (1999). The SCAR markers have been produced for the gene of mildew resistance transferred from *A. macrostachya* (Yu and Herrmann, 2006).

All the attempts may fail when genetic incompatibility concerns the transferred gene itself (or a strong linkage block of coadapted genes). If the genetic background of the recipient cultivar is inconvenient, the breeding of a new background is recommended. Replacement of a backcross recurrent parent with a different cultivar(s) may deliver variation, which could be sufficient for compensation of the side effects. Otherwise, more variation and several cross- selection cycles may be necessary.

In the case of a genetically complex character, the difficulties mentioned earlier may overlap with those caused by incompleteness of a coadapted multilocus gene system, which should be transferred as a whole in order to work properly. A special breeding strategy is required, directed on accumulation of the multiple introgressions occurring independently in different chromosomes and different hybrid lines. The moderately intense recurrent selection is the most appropriate method for such multiple transfers, as it creates necessary isolation from local breeding populations and generates intense recombination. Simmonds proposed for the procedure a separate term "incorporation" instead of "introgression" reserved for the transfer of one or a few genes without necessity of profound rebuilding of the genetic structure (Simmonds, 1993). The method of recurrent selection within an isolated gene pool of oat wide hybrids has proven its value in the experiments carried out at Iowa State University on the increase of oil, protein, and β-glucan content in groat. Recent progress in saturation of oat genomes with molecular markers, based mainly on the diversity array technology (Tinker et al., 2009) (see also Section 4.9.2), delivers new tools for controlling and shortening the incorporation process. As yet, no reports have been published on the application of the marker-assisted recurrent selection (Johnson et al., 2004) or genomic selection (Meuwissen et al., 2001) at work on the improvement of wide hybrids.

Substitution or addition of a whole genome, instead of chromosomes or their fragments, is the more easy option to be considered; however, with care of time necessary to establish a new intergenomic coadaptation, which may require a few decades. Triticale breeding history may deliver useful hints for similar manipulations in oat. Addition of the diploid rye genome to bread wheat (hexaploid) or durum wheat (tetraploid) was not sufficient to produce competitive cultivars. Decrease of yield, poor toleration of low temperatures, cytogenetic instability, and grain shrivelling in the octoploid (8*x*) and primary hexaploid (6*x*) triticales reduced their usefulness to components for production of the more valuable secondary hexaploid triticale. However, predictions based on results from different taxa may be deceptive. Unlike the wheat–rye crosses, the addition of 14 chromosomes of *A. macrostachya* to the *A. sativa* complement immediately improved winter hardiness and exerted no negative effect on grain plumpness. Nonuniform ripening and ~30% decrease of yield potential (in relation to *A. sativa* cultivars) of the resulted octoploids remain problematic (Łapiński et al., 2013); however, the distance to cultivar standards is generally smaller than between primary triticales and their wheat components. As yet, none of the reported octoploid primary synthetics of oats were recommended by their authors as prospective new man-made species for agriculture. Since, as yet, no experimental results support the idea of increasing ploidy level in oats; however, it does not allow to be definitely rejected.

In triticale, rather the whole genome substitution, not the addition, has proven to be the appropriate solution. Replacement of the bread wheat D genome with the R genome of rye formed the modern triticale karyotype. The substitution proceeded after crossing of primary hexaploids to 8*x* triticale or to 6*x* wheat (Pissarev, 1963; Zillinsky and

Borlaug, 1971). In the following period of breeding yield potential and grain plumpness have been much improved, while the expression of major adaptive target traits from rye (drought, salt, soil acidity toleration) has been maintained. In the CIMMYT triticale breeding program, yield potential of secondary triticale grew from 2.4 t/ha in 1968 to more than 10 t/ha in 2000, exceeding that of bread wheat (Hede, 2001). There are no similar attempts reported on the improvement of adaptation in oat. The small number of genes responsible for domestication, specified by Ladizinsky (1995), should encourage similar experimentation with the whole genome substitution in the oat genus.

4.9 Integration of genomic and genetic resources in crop improvement

Thorough knowledge of the range and structure of genetic diversity is essential to the effective exploitation of collected oat germplasm for crop improvement. Precise evaluation of genetic diversity plays a significant role in the characterization of breeding materials and it allows predicting the appropriate parental forms to crosses. The heterosis effect in the progeny is inversely proportional to the genetic similarity of parental forms (Cox and Murphy, 1990). Although numerous attempts to describe the genetic diversity of a particular collection have been made the existing gene pools of cultivated oats are in fact lacking of comprehensive characterization, especially on the molecular level. Knowledge of the genetic diversity of wild species is even weaker. The studies on CWR focused mainly on the variation among species and only a few of them analyzed differentiation within species. Knowledge of the diversity of wild species gene pools is priceless in terms of their *in situ* and *ex situ* conservation.

4.9.1 Wild species diversity

A significant imbalance exists in the amount of data on the diversity of particular wild species. The most comprehensive information is available about the species most interesting for breeding or for those considered as noxious weeds. *A. sterilis* has been the most often analyzed. One thousand five *A. sterilis* accessions from 23 countries preserved in the United States collection were analyzed with isozymes (Phillips et al., 1993). Based on these data, Turkey was identified as the center of diversity of this germplasm collection. Turkish and Lebanese accessions showed the highest polymorphism, and the highest probability of containing unique genotypes was found in populations from Iran, Turkey, Iraq, and Lebanon. One hundred seventy-three accessions from the same gene bank, which had originated from eight countries of Africa and Southwest Asia, were also analyzed by restriction fragment length polymorphism (RFLP) (Goffreda et al., 1992). The highest variation was identified among accessions originating from Iran and the lowest among those collected in Ethiopia. These results also indicated large genetic divergence of the Iran–Iraq accessions from the other regional collections surveyed. Once again 24 accessions gathered in the United States collection were subjected to comparative analysis by random amplified polymorphic DNA (RAPD) and isozymes technique. Zhou et al. (1999) evaluated the similarity of *A. sterilis* genotypes with diverse origins by RAPD too. Fu et al. (2007) assessed the genetic diversity of 369 *A. sterilis* accessions, from the 26 countries, preserved in the Canadian collection using the SSR system. ISSR markers analysis was conducted by Paczos-Grzęda et al. (2009) in order to determine the genetic diversity of 27 *A. sterilis* genotypes used in crossing with *A. sativa*. Value of genetic similarity indices testify to the high diversity of the designated set of accessions.

A. fatua is one of the most invasive weeds in temperate climate cereal crops, so diversity studies are focused on natural populations structure or herbicides resistance (Andrews et al., 1998; Cavan et al., 1998; Li et al., 2007; Mengistu et al., 2005). *A. barbata* is another quite often tested wild oat species analyzed in terms of natural populations' genetic variability (Garcia et al., 1989; Guma et al., 2006; Kahle et al., 1980). Little information on the diversity within other oat species is available (Francisco-Ortega et al., 2000; García et al., 1991).

4.9.2 Cultivated species diversity

The genetic diversity level of *A. sativa*, the main cultivated species, has been widely analyzed. As well as in other crops its narrowing in modern cultivars is apparent. This is due to the widespread use in breeding programs closely related and well-acclimatized prebreeding materials and recent cultivars. These universal statements have been confirmed by many studies on common oat germplasm (Boczkowska et al., 2014, 2015; Boczkowska and Tarczyk, 2013; Fu et al., 2003; Leisova et al., 2007). The morphologic analysis of 10,105 hexaploid *Avena* accessions preserved in the PGRC indicated that 90% of them belong to only 13 morphologic groups out of 118 detected in total (Diederichsen, 2007). The most diverse were found to be groups of accessions originating from countries with temperate climates and intensive oat breeding programs. In this study another cultivated oat *A. byzantina* was also analyzed. The highest morphologic diversity of this species was identified in accessions originating from West Asia while the United States could be considered a secondary center of diversity. Based on amplified fragment length polymorphism (AFLP) analysis of 114 oat cultivars of worldwide origin, Achleitner et al. (2008) postulated that in general, genetic diversity included in European cultivars is significantly lower than those in North and South America. Molecular analysis based on AFLP markers, which was performed for 670 accessions of the PGRC oat core collection, demonstrated that the majority (89.9%) of variance resided within country of origin and only 6.2% of the differences existed within major geographic regions (Fu et al., 2005). The within-country diversity was the highest in Russia and the United States while the gene pool of Mediterranean accessions was the most distinctive. In general, the genetic diversity of *A. byzantina* accessions was higher than *A. sativa* ones. For sure, the United States and Canada oat germplasm collections have the best diversity description. They were also described by several other methods such as pedigree analysis (Souza and Sorrells, 1989), phenotypic data (Souza and Sorrells, 1991), and isozymes (Murphy and Phillips, 1993; Phillips et al., 1993). Results of Fu et al. (2003) and Leisova et al. (2007) indicated that breeding programs have had a significant impact on the decrease of common oat genetic diversity. Both of them also suggested the need to expand the gene pool of *A. sativa*. However, later, Fu et al. (2005) pointed out that the genetic variability of oat landraces is not greater than that within the breeding materials. Boczkowska and Tarczyk (2013) considered the evident impact of local climatic conditions on the genetic distinctiveness of Polish landraces.

Nowadays, it is virtually impossible to refer unambiguously genetic differentiation to the place of accession origin. Based on such results, the issue of the necessity of oat cultivar gene pool expansion is absolutely indisputable, whereas the source of lacking diversity remains questionable. On the one hand, preserved wild oat relatives are the largest reservoir of unused variation; on the other hand, its mobilization is difficult as discussed earlier. Among preserved accessions about 19% are traditional cultivars and landraces. These materials are properly adapted to eco-geographic conditions in the places of their origin. They also are potential rich sources of valuable genetic variation that had been lost in modern breeding (Boczkowska and Tarczyk, 2013; Grau Nersting et al. 2006). Unfortunately, the opinions of their higher level of variation with respect to modern oat cultivars are divergent (Boczkowska et al., 2015; Fu et al., 2005). However, among gene bank accessions another interesting source of variability, that is, very old/historical cultivars could be found. Their diversity level could be comparable to the landraces' one (Boczkowska et al., 2014).

4.9.3 Genetic linkage mapping

Genetic maps provide tools to understanding the genetic background of phenotypic traits and transfer selection from phenotype to genotype. They also allow identifying and cloning the specific gene. The first two molecular maps of oat were obtained in the beginning of the 1990s and were constructed in crosses among closely related diploid species, that is, *A. atlantica* × *A. hirtula* (O'Donoughue et al., 1992) and *A. strigosa* × *A. wiestii* (Rayapati et al., 1994). The initial idea of constructing these two maps based on wild diploid species instead of cultivated hexaploid oat was to simplify the alleles of homeological loci identification and to analyze a lower number of linkage groups. Both of the maps were based on RFLP markers. At the beginning of the next decade two more maps of *A. strigosa* × *A. wiestii* were constructed (Kremer et al., 2001; Yu and Wise, 2000). Unfortunately, for various reasons, these three maps have not been integrated. In the meantime, the first map of hexaploid oats was constructed. It was based on a population of recombinant inbred lines of two genetically very distant cultivars, that is, *A. byzantina* cv. Kanota × *A. sativa* cv. Ogle (O'Donoughue et al., 1995). Initially this map contained 561, mainly RFLPs, loci. Because it is considered as a basic map of hexaploid-domesticated oat it has been saturated with other types of markers (Becher, 2007; Cheng et al., 2002; Irigoyen et al., 2004; Orr and Molnar, 2007, 2008; Pal et al., 2002; Tinker et al., 2009; Wight et al., 2003). Since then, a lot of genetic maps of oat have been created (reviewed by Rines et al. (2006). But still the Kanota × Ogle remains the most saturated with molecular marker oats map and is widely used as a reference in creating next ones. Several genes were successfully localized on this map (reviewed by Rines et al. (2006)). The first physically anchored consensus map of hexaploid oat was developed in 2013 based on SNP assay (Oliver et al., 2013). It included 985 SNPs and 68 previously published markers, and 21 linkage groups with a total map distance of 1838.8 cM were resolved.

4.9.4 Molecular breeding

Along with the development of molecular techniques and increasing easiness of access to them, classical breeding theory has changed from quantitative to molecular genetics with emphasis on quantitative trait loci (QTL) identification and marker-assisted selection (MAS). The major condition that must be met to carry out successful MAS is the availability of molecular markers linked to QTLs. Unfortunately, in oat many gene-tagging studies were performed on inconvenient for MAS RFLP, RAPD, and AFLP markers. Some of them have been successfully converted into the sequenced characterized amplified region markers (details reviewed by Rines et al. (2006)). Despite numerous studies relating to the linkage of molecular markers with specific genes or QTLs, MAS breeding of oats lags far behind other cereals. Currently, a large hope is being put in genotyping by sequencing techniques, which generate a large number of SNPs and sequence information that may be helpful in identifying the genes important for breeding (Huang et al., 2014).

4.10 Conclusions

The genus *Avena* is characterized by a high degree of genetic diversity. The world's collection of oat germplasm seems to be sufficient for future breeding and utilization. Greater emphasis should be placed on *in situ* conservation of some wild species. Also, some *ex situ* resources of wild species should be expanded to be a sufficiently representative backup of *in situ* diversity. Because oat is a minor cereal, less attention is paid to the development of molecular aspects of breeding than in major ones. Development of oat genomics area is a key point for the future.

References

Aaman, P., Graham, H., 1987. Analysis of total and insoluble mixed-linked (1.fwdarw.3), (1.fwdarw.4)-β-D-glucans in barley and oats. J. Agric. Food Chem. 35 (5), 704–709.

Achleitner, A., Tinker, N., Zechner, E., Buerstmayr, H., 2008. Genetic diversity among oat varieties of worldwide origin and associations of AFLP markers with quantitative traits. Theor. Appl. Genet. 117, 1041–1053.

Ahokas, H., Manninen, M., 2000. Retrospecting genetic variation of Finnish oat (*Avena sativa*) landraces and observations on revived lines grown prior to 1957. Genet. Resour. Crop Evol. 47 (3), 345–352.

Allen, H., Kaufmann, M., 1979. Athabasca oat. Can. J. Plant Sci. 59 (1), 245–246.

Andersson, A., Börjesdotter, D., 2011. Effects of environment and variety on content and molecular weight of β-glucan in oats. J. Cereal Sci. 54 (1), 122–128.

Andersson, A., Carmen de Vincente, M., 2010. Gene flow between crops and their wild relatives. Evol. Appl. 3 (4), 402–403.

Andrews, T., Morrison, I., Penner, G., 1998. Monitoring the spread of ACCase inhibitor resistance among wild oat (*Avena fatua*) patches using AFLP analysis. Weed Sci. 46, 196–199.

Anikster, Y., Feldman, M., Horovitz, A., 1997. The Amiad experiment. In: Maxted, N., Ford-Lloyd, B., Hawkes, J. (Eds.), Plant Genetic Conservation – The *In Situ* Approach. Chapman, Hall, London, pp. 239–253.

Antonova, N., 2005. The Avena collection in Bulgaria. In: Lipman, E., Maggioni, L., Knüpffer, H., Ellis, R., Leggett, J., Kleijer, G., Faberová, I., Le Blanc, A. (Eds.), Cereal Genetic Resources in Europe. International Plant Genetic Resources Institute, Rome, Italy, p. 318.

Aung, T., Thomas, H., Jones, I., 1977. The transfer of the gene for mildew resistance from *Avena barbata* (4*x*) into the cultivated oat *A. sativa* by an induced translocation. Euphytica 26 (3), 623–632.

Aung, T., Chong, J., Leggett, J., 1996. The transfer of crown rust resistance gene Pc 94 from a wild diploid to cultivated hexaploid oat. Paper presented at the Ninth European and Mediterranean Cereal Rusts and Powdery Mildews Conference, September 2–6, Lunteren, The Netherlands.

Bacon, R., 1991. Registration of 'Ozark' Oat. Crop Sci. 31, 1383–1384.

Barr, A., Tasker, S., 1992. Breeding for herbicide resistance in oats: opportunities and risks. Paper presented at the Proceedings of the Fourth International Oat Conference, Adelaide, Australia, vol. 2. Wild Oats in Agriculture, Adelaide, Australia.

Baum, B., 1977. Oats: Wild and Cultivated, a Monograph of the Genus *Avena* L. (Poaceae). Biosystematics Research Institute, Ottawa, Canada.

Baum, B., Fedak, G., 1985. *Avena atlantica*, a new diploid species of the oat genus from Morocco. Can. J. Bot. 63 (6), 1057–1060.

Baum, B., Rajhathy, T., 1976. A study of *Avena macrostachya*. Can. J. Bot. 54 (21), 2434–2439.

Baum, B., Fleischmann, G., Martens, J., Rajhathy, T., Thomas, H., 1972. Notes on the habitat and distribution of *Avena* species in the Mediterranean and Middle East. Can. J. Bot. 50 (6), 1385–1397.

Baum, B., Rajhathy, T., Sampson, D., 1973. An important new diploid *Avena* species discovered on the Canary Islands. Can. J. Bot. 51 (4), 759–762.

Beavis, W., Frey, K., 1987. Expression of nuclear-cytoplasmic interactions and heterosis in quantitative traits of oats (*Avena* spp.). Euphytica 36 (3), 877–886.

Becher, R., 2007. EST-derived microsatellites as a rich source of molecular markers for oats. Plant Breed. 126, 274–278.

Bilz, M., Kell, S., Maxted, N., Lansdown, R., 2011. European Red List of Vascular Plants. Publications Office of the European Union, Luxembourg.

Boczkowska, M., Tarczyk, E., 2013. Genetic diversity among Polish landraces of common oat (*Avena sativa* L.). Genet. Resour. Crop Evol. 60, 2157–2169.

Boczkowska, M., Nowosielski, J., Nowosielska, D., Podyma, W., 2014. Assessing genetic diversity in 23 early Polish oat cultivars based on molecular and morphological studies. Genet. Resour. Crop Evol. 61, 927–941.

Boczkowska, M., Harasimiuk, M., Onyśk, A., 2015. Studies on genetic variation within old Polish cultivars of common oat. Cereal Res. Commun. 43, 12–21, doi: 10.1556/CRC.2014.0025.

Branson, C., Frey, K., 1989a. Correlated response to recurrent selection for groat-oil content in oats. Euphytica 43 (1-2), 21–28.

Branson, C., Frey, K., 1989b. Recurrent selection for groat oil content in oat. Crop Sci. 29 (6), 1382–1387.

Briggle, L., Smith, R., Pomeranz, Y., Robbins, G., 1975. Protein concentration and amino acid composition of *Avena sterilis* L. groats. Crop Sci. 15 (4), 547–549.

Brown, P.D., 1985. The transfer of oat stem rust resistance gene Pg-16 from tetraploid *Avena barbata* Pott. to hexaploid *Avena sativa* L. University of Wisconsin, Madison. Int. B Sci. Eng. 45 (7), 2036B.

Brown, C., Craddock, J., 1972. Oil content and groat weight of entries in the world oat collection. Crop Sci. 12 (4), 514–515.

Burrows, V., 1986. Breeding oats for food and feed: conventional and new techniques and materials. In: Webster, F. (Ed.), Oats: Chemistry and Technology. American Society of Cereal Chemists, St. Paul, MN, USA, pp. 13–46.

Butler-Stoney, T., Valentine, J., 1991. Exploitation of the Genetic Potential of Oats for Use in Feed and Human Nutrition. Project-Report 32 HGCA, London, UK.

Cabral, A., Gnanesh, B., Mitchell Fetch, J., McCartney, C., Fetch, T., Park, R., et al., 2014. Oat fungal diseases and the application of molecular marker technology for their control. In: Goyal, A., Manoharachary, C. (Eds.), Future Challenges in Crop Protection Against Fungal Pathogens. Springer, New York, pp. 343–358.

Castell, C., Stuthman, D., 2002. Phenotypic recurrent selection for improving partial resistance to crown rust in oat populations. Phytopathology 9, S12.

Cavan, G., Biss, P., Moss, S., 1998. Herbicide resistance and gene flow in wild-oats (*Avena fatua* and *Avena sterilis* ssp. *ludoviciana*). Ann. Appl. Biol. 133 (2), 207–217.

CBD, 2011. The Nagoya Protocol on Access to Genetic Resources and the Fair and Equitable Sharing of Benefits Arising From Their Utilization (ABS) to the Convention on Biological. Secretariat of the Convention on Biological Diversity, Canada.

CBD, 2013. Secretariat of the Convention on Biological Diversity. In: Convention on Biological Diversity.

Cervantes-Martinez, C., Frey, K., White, P., Wesenberg, D., Holland, J., 2001. Selection for greater β-glucan content in oat grain. Crop Sci. 41 (4), 1085–1091.

Chen, G., Chong, J., Gray, M., Prashar, S., P.J, D., 2006. Identification of single-nucleotide polymorphisms linked to resistance gene *Pc68* to crown rust in cultivated oat. Can. J. Plant Pathol. 28 (2), 214–222.

Chen, G., Chong, J., Prashar, S., Procunier, J., 2007. Discovery and genotyping of high-throughput SNP markers for crown rust resistance gene *Pc94* in cultivated oat. Plant Breed. 126 (4), 379–384.

Cheng, D., Armstrong, K., Tinker, N., Wight, C., He, S., Lybaert, A., et al., 2002. Genetic and physical mapping of Lrk10-like receptor kinase sequences in hexaploid oat (*Avena sativa* L.). Genome 45 (1), 100–109.

Cho, K., White, P., 1993. Enzymatic analysis of β-glucan content in different oat genotypes. Cereal Chem. 70, 539–542.

Chong, J., Zegeye, T., 2004. Physiologic specialization of *Puccinia coronata* f. sp. *avenae*, the cause of oat crown rust, in Canada from 1999 to 2001. Can. J. Plant Pathol. 26 (1), 97–108.

Clayton, W.D., Vorontsova, M.S., Harman, K.T., Williamson, H., 2006 onwards. GrassBase – The Online World Grass Flora. http://www.kew.org/data/grasses-db.html. (accessed 05.06.2014).

Coffman, F., 1961. Oat and Oat Improvement. American Society of Agronomy, Madison, WI, USA.

Comeau, A., 1982. Geographic distribution of resistance to barley yellow dwarf virus in *Avena sterilis*. Can. J. Plant Pathol. 4 (2), 147–151.

COMECON, 1984. The International COMECON List of Descriptors for the Genus *Avena* L. VIR, Leningrad.

Cox, D., Frey, K., 1984. Improving cultivated oats (*Avena sativa* L.) with alleles for vegetative growth index from *A. sterilis* L. Theor. Appl. Genet. 68 (3), 239–245.

Cox, T., Murphy, J., 1990. The effect of parental divergence on F_2 heterosis in winter wheat crosses. Theor. Appl. Genet. 79 (2), 241–250.

de la Roche, I., Burrows, V., McKenzie, R., 1977. Variation in lipid composition among strains of oats. Crop Sci. 17 (1), 145–148.

Diederichsen, A., 2007. Assessments of genetic diversity within a world collection of cultivated hexaploid oat (*Avena sativa* L.) based on qualitative morphological characters. Genet. Resour. Crop Evol. 55 (3), 419–440.

Diederichsen, A., Timmermans, E., Williams, D., Richards, K., 2001. Holdings of *Avena* germplasm at Plant Gene Resources of Canada and status of the collection. Oat Newsl. 47, 35–42.

Dumlupinar, Z., Dokuyucu, T., Maral, H., Kara, R., Akkaya, A., 2012. Evaluation of Turkish oat landraces based on morphological and phenological traits. Žemdirbystė Agric. 99, 149–158.

EADB. European Avena Database. From ulius Kühn-Institut – Research Centre for Cultivated Plants (JKI). Available from: http://eadb.jki.bund.de/eadb/ (accessed 05.12.2014.).

EC, 1992. Council Directive 92/43/EEC of 21 May 1992 on the Conservation of Natural Habitats and of Wild Fauna and Flora, L 206 22.7.1992 C.F.R.

EC, 1999. Council Directive 98/95/EC of 14 December 1998 amending, in respect of the consolidation of the internal market, genetically modified plant varieties and plant genetic resources, Directives 66/400/EEC, 66/401/EEC, 66/402/EEC, 66/403/EEC, 69/208/EEC, 70/457/EEC and 70/458/EEC on the marketing of beet seed, fodder plant seed, cereal seed, seed potatoes, seed of oil and fibre plants and vegetable seed and on the common catalogue of varieties of agricultural plant species.

EC, 2008. Commission Directive 2008/62/EC of 20 June 2008 providing for certain derogations for acceptance of agricultural landraces and varieties which are naturally adapted to the local and regional conditions and threatened by genetic erosion and for marketing of seed and seed potatoes of those landraces and varieties 21.6.2008.

EC, 2013. Regulation No 1305/2013 of The European Parliament and of The Council of 17 December 2013 on support for rural development by the European Agricultural Fund for Rural Development (EAFRD) and repealing Council Regulation (EC) No 1698/2005 20.12.2013.

EC, 2014a. Common Catalogue of Varieties of Agricultural Plant Species – 33rd complete edition.

EC, 2014b. No. 150/59 20.5.2014.

El Oualidi, J., Khamar, H., Fennane, M., Ibn Tattou, M., Chauvet, S., Taleb, M., 2012. Checklist Des Endémiques Et Spécimens Types de la Flore Vasculaire De l'Afrique Du Nord25Universite Mohammed V-Agdal, Rabat.

Endo, R., Brown, C., 1964. Barley yellow dwarf virus resistance in oats. Crop Sci. 4 (3), 279–283.

EURISCO. EURISCO Catalogue. Available from: http://eurisco.ecpgr.org (accessed 21.12.2014.).

FAO, 2009. International Treaty on Plant Genetic Resources for Food and Agriculture. Food and Agriculture Organization of the United Nations, Rome, Italy.

FAO/IPGRI, 2001. Multi-Crop Passport Descriptors [MCPD].

FAOLEX, 2000. 728/2000, Finland 2000 Act on Seed Trade, Available from: http://www.finlex.fi/fi/laki/ajantasa/2000/20000728#L6P30

FAOSTAT, F.A.O., 2013. Agriculture data. Food and Agriculture Organization, Rome (Italy).

FAOwiews. Available from: http://apps3.fao.org/wiews/germplasm_query.htm?i_l=EN (accessed 16.09.2014.).

Fetch, J., Fetch, T., 2011. Inheritance of resistance to oat stem rust in the cultivars Ronald and AC Gwen. Can. J. Plant Sci. 91 (2), 419–423.

Findlay, W., 1956. Oats. Their Cultivation and Use From Ancient Times to the Present Day. Oliver and Boyd, Edinburgh.

Forsberg, R., Shands, H., 1989. Oat breeding. Janick, J. (Ed.), Plant Breeding Reviews, 6, Timber, Oregon, OR, USA.

Forsberg, R., Brinkman, M., Karow, R., Duerst, R., 1991. Registration of 'Centennial' oat. Crop Sci. 31, 1086–1087.

Francisco-Ortega, J., Santos-Guerra, A., Kim, S., Crawford, D., 2000. Plant genetic diversity in the Canary Islands: a conservation perspective. Am. J. Bot. 87, 909–919.

Frese, L., Henning, A., Neumann, B., Unger, S., 2013. CWR *In Situ* Strategy Helpdesk created and managed by S. Kell (University of Birmingham). Available from: http://www.agrobiodiversidad.org/aegro (accessed 22.01.2013.).

Frey, K., 1982. Multiline breeding. In: Vasil, I., Scowcroft, W., Frey, K. (Eds.), Plant Improvement and Somatic Cell Genetics. Academic Press, New York, USA, pp. 43–71.

Frey, K., 1986. Genetic Resources and Their Use in Oat Breeding. Paper presented at the Second International Oat Conference, Dordrecht.

Frey, K., 1989. The distribution of *Avena strigosa* Schreb. in Poland. Fragmenta Floristica Geobotanica 34, 43–51.

Frey, K., 1991a. Distribution of *Avena strigosa* (Poaceae) in Europe. Fragmenta Floristica Geobotanica 36, 281–288.

Frey, K., 1991b. Genetic Resources of Oats: Use of Plant Introductions in Cultivar Development1CSSA Special Publication, Madison, WI, USA, pp. 15–24.

Frey, K., 1992. Oat Improvement With Genes From *Avena* Species. Paper presented at the Fourth International Oat Conference, Adelaide, Australia.

Frey, K., 1994. Remaking a Crop Gene Pool: the Case History of *Avena.* Paper presented at the SABRAO Seventh International Congress and WSAA Symposium, Academia Sinica.

Frey, K., Holland, J., 1999. Nine cycles of recurrent selection for increased groat-oil content in oat. Crop Sci. 39 (6), 1636–1641.

Frey, K., Hammond, E., Lawrence, P., 1975. Inheritance of oil percentage in interspecific crosses of hexaploid oats. Crop Sci. 15 (1), 94–95.

Fu, Y., Peterson, G., Scoles, G., Rossnagel, B., Schoen, D., Richards, K., 2003. Allelic diversity changes in 96 Canadian oat cultivars released from 1886 to 2001. Crop Sci. 43, 1989–1995.

Fu, Y., Peterson, G., Williams, D., Richards, K., Fetch, J., 2005. Patterns of AFLP variation in a core subset of cultivated hexaploid oat germplasm. Theor. Appl. Genet. 111, 530–539.

Fu, Y., Chong, J., Fetch, T., Wang, M., 2007. Microsatellite variation in *Avena sterilis* oat germplasm. Theor. Appl. Genet. 114 (6), 1029–1038.

Gagkaeva, T., Gavrilova, O., Yli-Mattila, T., Loskutov, I., 2012. Evaluation of oat germplasm for resistance to *Fusarium* Head Blight. Plant Breed. Seed Sci. 64, 15–22.

Garcia, P., Vences, F., Perez de la Vega, M., Allard, R., 1989. Allelic and genotypic composition of ancestral Spanish and colonial Californian gene pools of *Avena barbata*: evolutionary implications. Genetics 122 (3), 687–694.

Garcia, P., Saenz de Miera, L., Vences, F., Benchacho, M., Perez de la Vega, M., 2007. Conservation of Spanish wild oats: *Avena canariensis*, *A. prostrata* and *A. murphyi*. In: Maxted, N., Ford-Lloyd, B., Kell, S., Iriondo, J., Dulloo, M., Turok, J. (Eds.), Crop Wild Relative Conservation and Use. CAB International, North America, pp. 413–428.

García, P., Morris, M., Sáenz-de-Miera, L., Allard, R., Pérez de la Vega, M., Ladizinsky, G., 1991. Genetic diversity and adaptedness in tetraploid *Avena barbata* and its diploid ancestors *Avena hirtula* and *Avena wiestii*. Proc. Natl. Acad. Sci. 88, 1207–1211.

Germeier, C., 2008. Global Strategy for the *Ex Situ* Conservation of Oats (*Avena* spp.). Federal Centre for Breeding Research on Cultivated Plants (BAZ), Quedlinburg, Germany.

Germeier, C., Maggioni, L., Katsiotis, A., Lipman, E., 2011. Report of a Working Group on Avena. Bioversity International, Rome, Italy.

Gnanesh, B., Fetch, J., Menzies, J., Beattie, A., Eckstein, P., McCartney, C., 2013. Chromosome location and allele-specific PCR markers for marker-assisted selection of the oat crown rust resistance gene *Pc91*. Mol. Breed. 32 (3), 679–686.

Gnanesh, B., Fetch, J., Zegeye, T., McCartney, C., Fetch, T., 2014. Oat. Pratap, A., Kumar, J. (Eds.), Alien Gene Transfer in Crop Plants, 2, Springer, New York, pp. 51–73.

Goffreda, J., Burnquist, W., Beer, S., Tanksley, S., Sorrells, M., 1992. Application of molecular markers to assess genetic relationships among accessions of wild oat, *Avena sterilis*. Theor. Appl. Genet. 85 (2–3), 146–151.

GrainGenes. GrainGenes: A Database for Triticeae and *Avena*. from Agricultural Research Service of the US Department of Agriculture. Available from: http://wheat.pw.usda.gov/GG2/Avena/ (accessed 03.10.2014.).

Grau Nersting, L., Bode Andersen, S., von Bothmer, R., Gullord, M., Bagger Jørgensen, R., 2006. Morphological and molecular diversity of Nordic oat through one hundred years of breeding. Euphytica 150 (3), 327–337.

Green, N., 2008. The Scottish Landrace Protection Scheme (SLPS). Paper presented at the AAB Conference on Plant Genetic Resources for Food and Agriculture, Wellesbourne, Warwick, UK.

Griffiths, N., 1984. Studies on Chromosome Manipulation in *Avena*. PhD Thesis, University of Wales, Aberystwyth, UK.

Guarino, L., Chadja, H., Mokkadem, A., 1991. Collection of *Avena macrostachya* Bal Ex Coss. et Dur. (Poaceae) germplasm in Algeria. Econ. Bot. 45 (4), 460–466.

Guerra, S., Betancort, R., 2013. *Avena canariensis*. The IUCN Red List of Threatened Species. Available from: www.iucnredlist.org (accessed 21.12.2014.).

Guma, I., Vega, M., García, P., 2006. Isozyme variation and genetic structure of populations of *Avena barbata* from Argentina. Genet. Resour. Crop Evol. 53 (3), 587–601.

Gupta, S., Frey, K., Skrdla, R., 1987. Selection for grain yield of oats via vegetative growth rates measured at anthesis and maturity. Euphytica 36 (1), 91–97.

Hagberg, P., 1988. *Avena marocanna* – A Genetic Resource Used in Oat Improvement. Paper presented at the Third International Oat Conference. Svalof AB, Lund, Sweden.

Hagberg, P., Mattson, B., 1986. Increased variability in oats from crosses between different species. In: O.G. (Ed.), Svalof 1886–1986: Research and Results of Plant Breeding. LTs forlag, Stockholm, Sweden, pp. 121–127.

Hansen, J., Renfrew, J., 1978. Palaeolithic–Neolithic seed remains at Franchthi Cave, Greece. Nat. Biotechnol. 271, 349–352.

Harder, D., Chong, J., Brown, P., Sebesta, J., Fox, S., 1992. Wild oat as a source of disease resistance: history, utilization, and prospects. Proceedings of the Fourth International Oat Conference, Wild Oats in World Agriculture, vol II, Adelaide, Australia, pp. 71–81.

Harlan, J., 1977. The Origins of Cereals Agriculture in the Old World. In: Reed, C. (Ed.), Origins of Agriculture. Moulton Publishers, Hague, pp. 357–383.

Hayes, J., Jones, I., 1966. Variation in the pathogenicity of *Erysiphe graminis* D.C. F. Sp. *avenae*, and its relation to the development of mildew-resistant oat cultivars. Euphytica 15 (1), 80–86.

Hede, A., 2001. A New Approach to Triticale Improvement. Paper presented at the Research Highlights of the CIMMYT Wheat Program, 1999–2000, Oaxaca, Mexico.

Hillman, G., Colledge, S., Harris, D., 1989. Plant-food economy during the Epipalaeolithic period at Tell Abu Hureyra, Syria: dietary diversity, seasonality and modes of exploitation. In: Harris, D., Hillman, G. (Eds.), Foraging and Farming. Unwin Hyman, Winchester, pp. 2240–2268.

Holland, J., 1997. Oat Improvement. Research Signpost, Trivandrum, India.
Hopf, M., 1969. Plan remains and early farming in Jericho. In: Ucko, P., Dimbleby, G. (Eds.), The Domestication and Exploitation of Plants and Animals. Aldine Publishing Company, Chicago, pp. 355–359.
Hoppe, G., Hoppe, H., 1991. Cluster analyses as breeding aid shown by the example of interspecific hybrids of *Avena*. A Zuchtungsforsch 21, 183–190.
Hoppe, H., Kummer, M., 1991. New productive hexaploid derivatives after introgression of *Avena pilosa* features. Vortr. Pflanzenzucht 20, 56–61.
Howarth, C., Cowan, A., Leggett, J., Valentine, J., 2000. Using molecular mapping to access and understanding valuable traits in wild relatives of oats. Paper presented at the Proceedings of the Sixth International Oat Conference, November 13–16, 2000, Canterbury, New Zealand.
Huang, Y., Poland, J., Wight, C., Jackson, E., Tinker, N., 2014. Using genotyping-by-sequencing (gbs) for genomic discovery in cultivated oat. PLoS ONE 9 (7), e102448.
Hunter, H., 1924. Oats. Their Varieties and Characteristics. A Practical Handbook for Farmers, Seedsmen, and Students. Benn, London.
Iannucci, A, Codianni, P., Cattivelli, L., 2011. Evaluation of genotype diversity in oat germplasm and definition of ideotypes adapted to the mediterranean environment. Int. J. Agron. 2011, 1–8.
IBPGR, 1985. Oat Descriptors. International Board for Plant Genetic Resources, Rome, Italy.
Irigoyen, M., Loarce, Y., Fominaya, A., Ferrer, E., 2004. Isolation and mapping of resistance gene analogs from the *Avena strigosa* genome. Theor. Appl. Genet. 109 (4), 713–724.
IUCN, 1994. IUCN Red List Categories, IUCN Species Survival Commission. International Union for Conservation of Nature and Natural Resources, Gland.
IUCN, 2014. The IUCN Red List of Threatened Species. Available from: www.iucnredlist.org (accessed 31.12.2014.).
Iwig, M., Ohm, H., 1978. Genetic control for percentage groat protein in 424 advanced generation lines from an oat cross. Crop Sci. 18 (6), 1045–1049.
Jia, H., Livingston, D., Murphy, J., Porter, D., 2006. Evaluation of freezing tolerance in advanced progeny from a cross of *Avena sativa* × *A. macrostachya*. Cereal Res. Commun. 34, 1037–1104.
Johnson, H., Broadhurst, D., Kell, D., Theodorou, M., Merry, R., Griffith, G., 2004. High-throughput metabolic fingerprinting of legume silage fermentations via Fourier transform infrared spectroscopy and chemometrics. Appl. Environ. Microbiol. 70, 1583–1592.
Kahle, A., Allard, R., Krzakowa, M., 1980. Associations between isozyme phenotypes and environment in the slender wild oat (*Avena barbata*) in Israel. Theor. Appl. Genet. 56, 31–47.
Kanan, G., Jaradat, A., 1996. Wild Oats in Jordan. Paper presented at the Fifth International Oat Conference, vol 2, 30 July–6 Aug 1996, Saskatoon, Canada.
Katsiotis, A., Ladizinsky, G., 2011. Surveying and conserving European *Avena* species diversity. In: Maxted, N., Ehsan Dulloo, M., Ford-Lloyd, B., Frese, L., Iriondo, J., Pinheiro de, M., Carvalho (Eds.), Agrobiodiversity Conservation: Securing the Diversity of Crop Wild Relatives and Landraces. CABI, pp. 65–71.
Korniak, T., 1997. *Avena strigosa* (Poaceae) in north-eastern Poland. Fragmenta Floristica Geobotanica 42, 201–206.
Kremer, C., Lee, M., Holland, J., 2001. A restriction fragment length polymorphism based linkage map of a diploid *Avena* recombinant inbred line population. Genome 44 (2), 192–204.
Kropač, Z., 1981. *Avena strigosa* – a disappearing synanthropic species in Czechoslovakia. Preslia (Praha) 53, 305–321.

Kubiak, K., 2009. Genetic diversity of Avena strigosa Schreb. ecotypes on the basis of isoenzyme markers. Biodivers. Res. Conserv. 15, 23–28.

Kuenzel, K., Frey, K., 1985. Protein yield of oats as determined by protein percentage and grain yield. Euphytica 34 (1), 21–31.

Kuszewska, K., Korniak, T., 2009. Bristle Oat (*Avena strigosa* Schreb.) – a weed or an useful plant? Herba Polonica 55, 341–347.

Kynast, R., Okagaki, R., Galatowitsch, M., Granath, S., Jacobs, M., Stec, A., et al., 2004. Dissecting the maize genome by using chromosome addition and radiation hybrid lines. Proc. Natl. Acad. Sci. USA 101 (26), 9921–9926.

Ladizinsky, G., 1971. Avena prostrata: a new diploid species of oat. Israel J. Bot. 20, 24–27.

Ladizinsky, G., 1988. Biological Species and Wild Genetic Resources in *Avena*. Paper presented at the Third International Oat Conference, July 4–8, 1988, Lund, Sweden.

Ladizinsky, G., 1995. Domestication via hybridization of the wild tetraploid oats *Avena magna* and *A. murphyi*. Theor. Appl. Genet. 91, 639–646.

Ladizinsky, G., 1998. A new species of *Avena* from Sicily, possibly the tetraploid progenitor of hexaploid oats. Genet. Resour. Crop Evol. 45 (3), 263–269.

Ladizinsky, G., 1999. Cytogenetic relationships between *Avena insularis* ($2n = 28$) and both *A. strigosa* ($2n = 14$) and *A. murphyi* ($2n = 28$). Genet. Resour. Crop Evol. 46 (5), 501–504.

Ladizinsky, G., 2012. Studies in Oat Evolution. A Man's Life with Avena. Springer, Heidelberg, New York, Dordrecht, London.

Ladizinsky, G., Fainstein, R., 1977. Domestication of the protein-rich tetraploid wild oats *Avena magna* and *A. murphyi*. Euphytica 26 (1), 221–223.

Ladizinsky, G., Zohary, D., 1971. Notes on species delimination, species relationships and polyploidy in *Avena* L. Euphytica 20 (3), 380–395.

Landry, B., Comeau, A., Minvielle, F., St-Pierre, C., 1984. Genetic analysis of resistance to barley yellow dwarf virus in hybrids between *Avena sativa* 'Lamar' and virus-resistant lines of *Avena sterilis*. Crop Sci. 24 (2), 337–340.

Łapiński, B., Kała, M., Nakielna, Z., Jellen, R., Livingston, D., 2013. The perennial wild species *Avena macrostachya* as a genetic source for improvement of winterhardiness in winter oat for cultivation in Poland. In: Behl, R., Arseniuk, E. (Eds.), Biotechnology and Plant Breeding – Perspectives. Agrobios (International) Publishers, Jodhpur, India, pp. 51–62.

Łapiński, B., Nita, Z., Szołkowska, A., Wieczorek, P., 2014. A hybrid of cultivated oat with the wild species *Avena macrostachya* as a source of new variation for yield quality improvement in naked oats. Biuletyn Instytutu Hodowli i Aklimatyzacji Roślin 218, 43–55.

Legget, J., 1996. Using and Conserving *Avena* Genetic Resources. Paper presented at the Fifth International Oat Conference, July 30– August 6, 1996, Saskatoon, Sask, Canada.

Leggett, J., 1992. The conservation and exploration of wild oat species. Proceedings of the Fourth International Oat Conference, Wild Oats in World Agriculture, vol II, Adelaide, Australia, pp. 57–60.

Leggett, J., Thomas, H., 1995. Oat evolution and cytogenetics. In: Welch, R. (Ed.), The Oat Crop. Springer, Netherlands, pp. 120–149.

Leggett, J., Ladizinsky, G., Hagberg, P., Obanni, M., 1992. The distribution of nine *Avena* species in Spain and Morocco. Can. J. Bot. 70 (2), 240–244.

Leisova, L., Kucera, L., Dotlacil, L., 2007. Genetic resources of barley and oat characterised by microsatellites. Czech J. Genet. Plant Breed. 43, 97–104.

Leonard, K., Anikster, Y., Manisterski, J., 2004. Oat crown rust virulence in collections from *Avena sativa* and *A. sterilis* in Israel. Cereal Rusts Powdery Mildews Bull., Available from: [www.crpmb.org/] 2004/0607leonard.

Leonova, S., Shelenga, T., Hamberg, M., Konarev, A., Loskutov, I., Carlsson, A., 2008. Analysis of oil composition in cultivars and wild species of oat (*Avena* sp.). J. Agric. Food Chem. 56 (17), 7983–7991.

Levitt, J., 1972. Responses of Plants to Environmental Stresses. Academic Press, New York.

Li, R., Wang, S., Duan, L., Li, Z., Christoffers, M., Mengistu, L., 2007. Genetic diversity of wild oat (*Avena fatua*) populations from China and the United States. Weed Sci. 55, 95–101.

Li, W., Peng, Y., Wei, Y., Baum, B., Zheng, Y., 2009. Relationships among *Avena* species as revealed by consensus chloroplast simple sequence repeat (ccSSR) markers. Genet. Resour. Crop Evol. 56 (4), 465–480.

Linnaeus, C., 1753. Species Plantarum, vol. 1, London.

Linnaeus, C., 1762. Species Plantarum, London.

Lista de especies silvestres de Canarias, 2004. Hongos, plantas y animales terrestres. Gobierno de Canarias: Consejería de Medio Ambiente y Ordenación Territorial.

Lista Roja, 2008. de la Flora Vascular Española. Dirección General de Medio Natural y Política Forestal. Ministerio de Medio Ambiente, y Medio Rural y Marino, y Sociedad Española de Biología de la Conservación de Plantas, Madrid.

Loskutov, I., 1998a. The collection of wild oat species of C. I. S. as a source of diversity in agricultural traits. Genet. Resour. Crop Evol. 45 (4), 291–295.

Loskutov, I., 1998b. Establishment of the core collection of *Avena* wild species. Report of a working group on *Avena*. Fifth meeting, May 7–9, 1998, Lithuania, Vilnius, pp. 34–36.

Loskutov, I., 2001. Interspecific crosses in the genus *Avena* L. Russ. J. Genet. 37 (5), 467–475.

Loskutov, I., 2007. Oat (*Avena* L.). Distribution, Taxonomy, Evolution and Breeding Value. VIR, Sankt-Petersburg.

Loskutov, I., 2008. On evolutionary pathways of *Avena* species. Genet. Resour. Crop Evol. 55 (2), 211–220.

Loskutov, I., Rines, H., 2011. Avena. In: Kole, C. (Ed.), Wild Crop Relatives: Genomic and Breeding Resources. Springer, Berlin, Heidelberg, pp. 109–183.

Mal, B., 1987. Wild Genetic Resource Potential for Forage Oat Improvement. Paper presented at the First Symposium on Crop Improvement, February 23–27, 1987, India.

Malzev, A., 1930. Wild and Cultivated Oats Sectio Euavena Griseb, vol. 38, Leningrad.

Marschall, H., Shaner, G., 1992. Genetics and inheritance in oat. In: Marschall, H., Sorrels, M. (Eds.), Oat Science and Technology. American Society of Agronomy and Crop Science Society of America, Madison, WI, USA, pp. 509–571.

Marshall, A., Cowan, S., Edwards, S., Griffiths, I., Howarth, C., Langdon, T., et al., 2013. Crops that feed the world 9. Oats – a cereal crop for human and livestock feed with industrial applications. Food Secur. 5 (1), 13–33.

Martens, J., Mckenzie, R., Harder, D., 1980. Resistance to *Puccinia graminis avenae* and *P. coronata avenae* in the wild and cultivated *Avena* populations of Iran, Iraq and Turkey. Can. J. Genet. Cytol. 22 (4), 641–649.

Mattsson, B., 1988. The development of oat germplasm at Svalov. Paper presented at the Third International Oat Conference, Svalof AB, Lund, Sweden.

Matzk, F., 1996. Hybrids of crosses between oat and Andropogoneae or Paniceae species. Crop Sci. 36, 17–21.

Maxted, N., Ford-Lloyd, B., Hawkes, J., 1997. Plant Genetic Conservation – The *In Situ* Approach. Springer Science & Business Media.

McDaniel, M., 1974. Registration of TAM 0-31 oats. Crop Sci. 14, 127–128.

McFerso, J., Frey, K., 1991. Recurrent selection for protein yield of oat. Crop Sci. 31 (1), 1–8.

McKenzie, R., Brown, P., Martens, J., Harder, D., Nielsen, J., Gill, C., et al., 1984. Registration of Dumont oats. Crop Sci. 24, 207.

McKenzie, R., Brown, P., Harder, D., Chong, J., Nielsen, J., Gill, C., et al., 1986. Registration of 'Riel' oat. Crop Sci. 26, 1256.

McMullen, M., Patterson, F., 1992. Oat cultivar development in the U.S.A. and Canada. In: Marshall, H., Sorrels, M.E., (Eds.), American Society of Agronomy and Crop Science Society of America, Madison, WI, USA, pp. 573–612.

McMullen, M., Phillips, R., Stuthman, D., 1982. Meiotic irregularities in *Avena sativa* L./*A. sterilis* L. hybrids and breeding implications. Crop Sci. 22 (4), 890–897.

Mengistu, L., Messsersmith, C., Christoffers, M., 2005. Genetic diversity of herbicide-resistant and -susceptible *Avena fatua* populations in North Dakota and Minnesota. Weed Res. 45, 413–423.

Meuwissen, T.H., Hayes, B.J., Goddard, M.E., 2001. Prediction of total genetic value using genome-wide dense marker maps. Genetics 157 (4), 1819–1829.

Miller, S., Wood, P., Pietrzak, L., Fulcher, R., 1993. Mixed linkage β-glucan, protein content and kernel weigh in *Avena* species. Cereal Chem. 70, 231–233.

Moore-Colyer, R., 1995. Oats and oat production in history and pre-history. In: Welch, R. (Ed.), The Oat Crop: Production and Utilization. Chapman & Hall, London, pp. 1–33.

Moreno, S., Martín, J., Ortiz, J., 1998. Inter-simple sequence repeats PCR for characterization of closely related grapevine germplasm. Euphytica 101, 117–125.

Morikawa, T., 1988. New Genes for Dwarfness Transferred From Wild Oats *Avena fatua* Into Cultivated Oat. Paper presented at the Third International Oat Conference, Svalof AB, Lund, Sweden.

Morikawa, T., 1995. Transfer of mildew resistance from the wild oat *Avena prostrata* into the cultivated oat. Bull. Univ. Osaka Prefec. Ser. B 47, 1–10.

Morikawa, T., Sumiya, M., Kuriyama, S., 2007. Transfer of new dwarfing genes from the weed species *Avena fatua* into cultivated oat *A. byzantina*. Plant Breed. 126 (1), 30–35.

Moser, H., Frey, K., 1994. Direct and correlated responses to three S1-recurrent selection strategies for increasing protein yield in oat. Euphytica 78 (1-2), 123–132.

Murphy, J., Hoffman, L., 1992. The origin, history, and production of oat. In: Marshall, H., Sorrells, M. (Eds.), Oat Science and Technology. American Society of Agronomy and Crop Science Society of America, Madison, WI, USA, pp. 1–28.

Murphy, J., Phillips, T., 1993. Isozyme variation in cultivated oat and its progenitor species *Avena sterilis* L. Crop Sci. 33 (6), 1366–1372.

Nielsen, J., 1977. A collection of cultivars of oats immune or highly resistant to smut. Can. J. Plant Sci. 57 (1), 199–212.

Nielsen, J., 1978. Frequency and geographical distribution of resistance to *Ustilago* parasitic on wheat, barley and oats. Can. J. Plant Sci. 58, 1099–1101.

Nielsen, J., 1993. Host specificity of *Ustilago avenae* and *U. hordei* on eight species of *Avena*. Can. J. Plant Pathol. 15 (1), 14–16.

Nikoloudakis, N., Katsiotis, A., 2008. The origin of the C-genome and cytoplasm of *Avena* polyploids. Theor. Appl. Genet. 117 (2), 273–281.

O'Donoughue, L., Wang, Z., Röder, M., Kneen, B., Leggett, J., Sorrells, M., et al., 1992. An RFLP-based linkage map of oats based on a cross between two diploid taxa (*Avena atlantica* × *A. hirtula*). Genome 35 (5), 765–771.

O'Donoughue, L., Sorrells, M., Tanksley, S., Autrique, E., Deynze, A.V., Kianian, S., et al., 1995. A molecular linkage map of cultivated oat. Genome 38 (2), 368–380.

Ohm, H., Patterson, F., 1973. A six-parent diallel cross analysis for protein in *Avena sterilis* L. Crop Sci. 13 (1), 27–30.

Ohm, H., Shaner, G., Buechley, G., Aldridge, W., Bostwick, D., Ratcliffe, R., et al., 1995. Registration of 'IN09201' spring oat. Crop Sci. 35, 940.

Okoń, S., Chrząstek, M., Kowalczyk, K., Koroluk, A., 2014. Identification of new sources of resistance to powdery mildew in oat. Eur. J. Plant Pathol. 139 (1), 9–12.

Oliver, R., Tinker, N., Lazo, G., Chao, S., Jellen, E., Carson, M., et al., 2013. SNP discovery and chromosome anchoring provide the first physically-anchored hexaploid oat map and reveal synteny with model species. PLoS ONE 8 (3), e58068.

Orr, W., Molnar, S., 2007. Development and mapping of PCR-based SCAR and CAPS markers linked to oil QTLs in oat. Crop Sci. 47, 848–850.

Orr, W., Molnar, S., 2008. Development of PCR-based SCAR and CAPS markers linked to β-glucan and protein content QTL regions in oat. Genome 51 (6), 421–425.

Orr, W., De Koeyer, D., Chenier, C., Tinker, N., Molnar, S., 1999. SCAR Markers for Oat Rust Resistance Genes (*Pc68*, *Pg3*, *Pg9*) Designed for Marker Assisted Selection. Paper presented at the Plant Animal Genome VII Conference, San Diego, CA.

Paczos-Grzęda, E., Chrząstek, M., Okoń, S., Grądzielewska, A., Miazga, D., 2009. Zastosowanie markerów ISSR do analizy wewnątrzgatunkowego podobieństwa genetycznego *Avena sterilis* L. Biuletyn Instytutu Hodowli i Aklimatyzacji Roślin 252, 215–223.

Pal, N., Sandhu, J., Domier, L., Kolb, F., 2002. Development and characterization of microsatellite and RFLP-derived PCR markers in oat. Crop Sci. 42, 912–918.

Paton, D., 1986. Oat starch: physical, chemical and structural properties. In: Weber, W.E. (Ed.), Oats: Chemistry and Technology. American Association of Cereal Chemistry, St. Paul, pp. 93–120.

Peng, Y., Baum, B., Ren, C., Jiang, Q., Chen, G., Zheng, Y., et al., 2009. The evolution pattern of rDNA ITS in *Avena* and phylogenetic relationship of the *Avena* species (Poaceae: Aveneae). Hereditas 147 (5), 183–204.

Peterson, D., 2001. Oat antioxidants. J. Cereal Sci. 33 (2), 115–129.

PGRC. Genetic Resource Information Network - Canadian Version (GRIN-CA), from Agriculture and Agri-Food Canada. Available from: http://pgrc3.agr.gc.ca/index_e.html (accessed 12.09.2014.).

Phillips, T., Murphy, J., Goodman, M., 1993. Isozyme variation in germplasm accessions of the wild oat *Avena sterilis* L. Theor. Appl. Genet. 86, 54–64.

Pinheiro de Carvalho, M., Bebeli, P., Bettencourt, E., Costa, G., Dias, S., Santos, T., et al., 2013. Cereal landraces genetic resources in worldwide GeneBanks. A review. Agron. Sust. Dev. 33 (1), 177–203.

Pissarev, V., 1963. Different Approaches in Triticale Breeding. Paper presented at the Second International Wheat Genetics Symposium.

Podyma, W., 1993. Genetic resources and variability of Avena strigosa Schreb. In: Frison, E., Koenig, J., Schittenhelm. S. (Eds.), Report of the Fourth Meeting of the ECP/GR Avena working group. IBPGR, Rome, Italy, pp. 13–29.

Podyma, W., Scholten, M., Bettencourt, E., 2005. *Avena strigosa* (Schreb.) in north-western Europe: a historical landrace without crop wild relatives? Available from: http://www.pgrforum.org/Documents/Conference_posters/Avena_casestudy.pdf

Rajhathy, T., Baum, B., 1972. *Avena damascena*: a new diploid oat species. Can. J. Genet. Cytol. 14 (3), 645–654.

Rajhathy, T., Thomas, H., 1974. Cytogenetics of Oats (*Avena* L.)2The Genetics Society of Canada, Ottawa.

Rayapati, P.J., Gregory, J., Lee, M., Wise, R.P., 1994. A linkage map of diploid *Avena* based on RFLP loci and a locus conferring resistance to nine isolates of *Puccinia coronata* var. *avenae*. Theor. Appl. Genet. 89, 831–837.

Reinbergs, E., 1983. OAC Woodstock oats. Can. J. Plant Sci. 63 (2), 543–545.

Renfrew, J., 1969. The archaeological evidence for the domestication of plants: methods and problems. In: Ucko, P., Dimbleby, G. (Eds.), The Domestication and Exploitation of Plants and Animals. Aldine Publishing Company, Chicago, pp. 149–172.

Rezai, A., Frey, K., 1988. Variation in relation to geographical distribution of wild oats-seed traits. Euphytica 39 (2), 113–118.

Rezai, A., Frey, K., 1989. Variation for physiological and morphological traits in relation to geographical distribution of wild oats. SABRAO J. 21, 1–9.

Rezai, A., Frey, K., 1990. Multivariate analysis of variation among wild oat accessions – seed traits. Euphytica 49 (2), 111–119.

Rines, H., Stuthman, D., Briggle, L., Youngs, V., Jedlinski, H., Smith, D., et al., 1980. Collection and evaluation of *Avena fatua* for use in oat improvement. Crop Sci. 20 (1), 63–68.

Rines, H., Molnar, S.J., Tinker, N., Phillips, R., 2006. Oat genome mapping and molecular breeding in plants v 1. In: Kole, C. (Ed.), Cereals and millets, pp. 349. Available from: http://ezproxy.colostate-pueblo.edu/login?url=http://dx.doi.org/10.1007/978-3-540-34389-9

Rines, H., Porter, H., Carson, M., Ochocki, G., 2007. Introgression of crown rust resistance from diploid oat *Avena strigosa* into hexaploid cultivated oat *A. sativa* by two methods: direct crosses and through an initial 2*x*·4*x* synthetic hexaploid. Euphytica 158 (1–2), 67–79.

Rodionova, N., Soldatov, V., Merezhko, V., Jarosh, N., Kobyljanskij, V., 1994. Flora of Cultivated Plants 2 Kolos, Moscow.

Rooney, W., Rines, H., Phillips, R., 1994. Identification of RFLP markers linked to crown rust resistance genes *Pc 91* and *Pc* in oat. Crop Sci. 34 (4), 940–944.

Rothman, P., 1984. Registration of four stem rust and crown rust resistant oat germplasm lines. Crop Sci. 24, 1217–1218.

Rothman, P., 1986. Adequate Rust Resistance in Oats. Paper presented at the Second International Oat Conference. Martinus, Dordrecht, The Netherlands.

Ryan, D., Kendall, M., Robards, K., 2007. Bioactivity of oats as it relates to cardiovascular disease. Nutr. Res. Rev. 20 (2), 147–162.

Sadanaga, K., Simons, M., 1960. Transfer of crown rust resistance of diploid and tetraploid species to hexaploid oats. Agron. J. 52 (5), 285–288.

Santos, A., Livingston, D., Murphy, J., 2002. Agronomic, Cytological and RAPD Evaluations of *Avena sativa* × *A. macrostachya* Populations. Raleigh, North Carolina, USA.

Schipper, H., Frey, K., 1991. Observed gains from three recurrent selection regimes for increased groat-oil content of oat. Crop Sci. 31 (6), 1505–1510.

Scholten, M., Spoor, B., Green, N., 2009. Machair corn: management and conservation of a historical machair component. Glasgow Natural. 25, 63–71.

Šebesta, J., Kühn, F., 1990. *Avena fatua* L. subsp. *fatua* v. *glabrata* Peterm. subv. *pseudo-basifixa* Thell. as a source of crown rust resistance genes. Euphytica 50 (1), 51–55.

Sebesta, J., Harder, D., Jones, L., Kummer, M., Clifford, B., Zwatz, B., et al., 1986. Pathogenicity of Crown Rust, Stem Rust and Powdery Mildew on Oats in Europe and Sources of Resistance. Paper presented at the Second International Oat Conference, Martinus, Dordrecht, The Netherlands.

Sharma, D., Forsberg, R., 1977. Spontaneous and induced interspecific gene transfer for crown rust resistance in *Avena*. Crop Sci. 17 (6), 855–860.

Shelukhina, O., Badaeva, E., Loskutov, I., Pukhal'sky, V., 2007. A comparative cytogenetic study of the tetraploid oat species with the A and C genomes: *Avena insularis*, *A. magna*, and *A. murphyi*. Russ. J. Genet. 43 (6), 613–626.

Shelukhina, O., Badaeva, E., Brezhneva, T., Loskutov, I., Pukhalsky, V., 2008. Comparative analysis of diploid species of *Avena* L. using cytogenetic and biochemical markers: *Avena canariensis* Baum et Fedak and *A. longiglumis* Dur. Russ. J. Genet. 44 (6), 694–701.

Shmida, A., Pollak, G., Fragman-Sapir, O., 2011. Red Data Book. Endangered Plants of Israel. Israel Nature and Parks Authority.
Simmonds, N., 1993. Introgression and incorporation. Strategies for the use of crop genetic resources. Biol. Rev. 68 (4), 539–562.
Simons, M., Robertson, L., Frey, K., 1985. Association of host cytoplasm with reaction to *Puccinia coronata* in progeny of crosses between wild and cultivated oats. Plant Dis. 69, 969–971.
Simons, M., Michel, L., Frey, K., 1987. Registration of 3 oat germplasm lines resistant to the crown rust fungus. Crop Sci. 27, 369.
Smekalova, T., Maslovky, O., 2013. *Avena volgensis*. The IUCN Red List of Threatened Species. Available from: www.iucnredlist.org (accessed 21.12.2014.).
Souza, E., Sorrells, M.E., 1989. Pedigree analysis of North American oat cultivars released from 1951 to 1985. Crop Sci. 29 (3), 595–601.
Souza, E., Sorrells, M.E., 1991. Relationships among 70 North American oat germplasms: II. Cluster analysis using qualitative characters. Crop Sci. 31 (3), 605–612.
Souza, E., Sorrells, M.E., 1991. Relationships among 70 North American oat germplasms: I. Cluster analysis using quantitative characters. Crop Sci. 31 (3), 599–605.
Stevens, J., Brinkman, M., 1986. Performance of *Avena sativa* L./*Avena fatua* L. Backcross lines. Euphytica 35 (3), 785–792.
Stuthman, D., 1995. Oat breeding and genetics. In: Welch, R.W. (Ed.), The Oat Crop: Production and Utilization. Chapman and Hall, London, U.K, pp. 150–176.
Suneson, C., 1967. Registration of Rapida oats. Crop Sci. 7, 168.
Suttie, M., Reynolds, S., 2004. Fodder Oats: A World Overview. Food and Agriculture Organization of the United Nations, Rome, Italy.
Takeda, K., Frey, K., 1976. Contributions of vegetative growth rate and harvest index to grain yield of progenies from *Avena sativa* × *A. sterilis* crosses. Crop Sci. 16 (6), 817–821.
Thomas, H., 1968. The addition of single chromosomes of *Avena hirtula* to the cultivated hexaploid oat *A. sativa*. Can. J. Genet. Cytol. 10 (3), 551–563.
Thomas, H., 1970. Chromosome relationships between the cultivated oat *Avena sativa* (6*x*) and *A. ventricosa* (2*x*). Can. J. Genet. Cytol. 12 (1), 36–43.
Thomas, H., Griffiths, N., 1985. Oat cytogenetics. Alien chromosome substitution line. Annual Report of Welsh Plant Breeding Station 1984, University College of Wales, Aberystwyth, 102–103.
Thomas, H., Powell, W., Aung, T., 1980a. Interfering with regular meiotic behaviour in *Avena sativa* as a method of incorporating the gene for mildew resistance from *A. barbata*. Euphytica 29 (3), 635–640.
Thomas, H., Haki, J., Arangzeb, S., 1980b. The introgression of characters of the wild oat *Avena magna* ($2n=4x=28$) into the cultivated oat *A. sativa* ($2n=6x=42$). Euphytica 29 (2), 391–399.
Thompson, R., 1966. Mesa, new oat for southern Arizona. Agric. Ariz. 18, 8.
Thompson, R., 1967. Registration of Mesa oats. Crop Sci. 7, 167.
Thro, A., 1982. Feasibility of oats (*Avena sativa* L.) as an oilseed crop. Available from: http://lib.dr.iastate.edu/rtd/7550 (Retrospective Theses and Dissertations, Paper 7550).
Thro, A., Frey, K., 1985. Inheritance of groat-oil content and high-oil selection in oats (*Avena sativa* L.). Euphytica 34 (2), 251–263.
Tinker, N., Kilian, A., Wight, C., Heller-Uszynska, K., Wenzl, P., Rines, H., et al., 2009. New DArT markers for oat provide enhanced map coverage and global germplasm characterization. BMC Genomics 10, 39.
Trofimovskaya, A., Pasynkov, V., Rodionova, N., Soldatov, V., 1976. Genetic potential of section true oat of genus *Avena* and it importance for breeding. Works Appl. Bot. Genet. Plant Breed. 58, 83–109.
UPOV, 1994a. Guidelines for the Conduct of Tests for Distinctness, Uniformity and Stability. Oats. (*Avena sativa* L. & *Avena nuda* L.). UPOV, Geneva.

UPOV, 1994b. International Convention for the Protection of New Varieties of Plants. UPOV, Geneva.

USDA, 2014. USDA, ARS, National Genetic Resources Program. Germplasm Resources Information Network - (GRIN). Available from: National Germplasm Resources Laboratory. http://www.ars-grin.gov/ (accessed 21.12.2014.).

Valdés, B., Scholz, H., Raab-Straube, E., Parolly, G., 2009. Valdés, B., Scholz, H.; with contributions from Raab-Straube, E. von, Parolly, G. Available from: http://ww2.bgbm.org/EuroPlusMed/PTaxonDetail.asp?NameCache=Poaceae&PTRefFk=7100000 (accessed 31.12.2014.).

Vavilov, N., 1926. Sentry proiskhozhdeniya kulturnykh rastenii (Centres of origin of cultivated plants). Works Appl. Bot. Genet. Plant Breed. 17, 9–107.

Vavilov, N., 1965. Centers of Origin of Cultivated Plants5Nauka, Moscow–Leningrad, USSR.

Vavilov, N., 1992. Origin and Geography of Cultivated Plants (D. Love, Trans.). Cambridge University Press, London, UK.

Veteläinen, M., Negri, V., Maxted, N., 2009. European Landraces on Farm Conservation, Management and Use15Bioversity International, Rome, Italy.

Vivero, J., Ensermu, K., Sebsebe, D., 2006. Progress on the red list of plants of Ethiopia and Eritrea: conservation and biogeography of endemic flowering taxa. In: Ghazanfar, S., HJ, B. (Eds.), Taxonomy and Ecology of African Plants, Their Conservation and Sustainable Use. Royal Botanic Gardens, Kew, pp. 761–778.

Weibull, J., 1986. Screening for resistance against *Rhopalosiphum padi* (L.). I. *Avena* species and breeding lines. Euphytica 35 (3), 993–999.

Weibull, J., Johansen Bojensen, L., Rasomavicius, V., 2001. *Avena strigosa* Schreb. in Denmark and Lithuania. Plant Genet. Resour. Newsl. 131, 1–4.

Welch, R., 1975. Fatty acid composition of grain from winter and spring sown oats, barley and wheat. J. Sci. Food Agric. 26, 429–435.

Welch, R., Leggett, J., Lloyd, J., 1991. Variation in the kernel (13) (14)-β-D-glucan content of oat cultivars and wild *Avena* species and its relationship to other characteristics. J. Cereal Sci. 13, 73–117.

Welch, R., Brown, J., Leggett, J., 2000. Interspecific and intraspecific variation in grain and groat characteristics of wild oat (*Avena*) species: very high groat (1→3), (1→4)-β-D-glucan in an *Avena atlantica* genotype. J. Cereal Sci. 31 (3), 7–17.

Wight, C.P., Tinker, N.A., Kianian, S.F., Sorrells, M.E., O'Donoghue, L.S., Hoffmann, D.L., 2003. A molecular marker map in 'Kanota' × 'Ogle' hexaploid oat (*Avena* spp.) enhanced by additional markers and a robust framework. Genome 46, 28–47.

Wilson, W., McMullen, M., 1997. Dosage dependent genetic suppression of oat crown rust resistance gene *Pc-62*. Crop Sci. 37 (6), 1699–1705.

WTO, 1994. Agreement On Trade-Related Aspects of Intellectual Property Rights. World Trade Organization. Available from: http://www.wto.org/english/docs_e/legal_e/27-trips_01_e.htm

Yan, H., Baum, B., Zhou, P., Zhao, J., Wei, Y., Ren, C., et al., 2014. Phylogenetic analysis of the genus *Avena* based on chloroplast intergenic spacer psbA–trnH and single-copy nuclear gene *Acc1*. Genome 57 (5), 267–277.

Youngs, V., 1986. Oat lipids and lipid-related enzymes. In: Webster, F. (Ed.), Oats: Chemistry and Technology. American Association of Cereal Chemistry, St. Paul.

Yu, J., Herrmann, M., 2006. Inheritance and mapping of a powdery mildew resistance gene introgressed from *Avena macrostachya* in cultivated oat. Theor. Appl. Genet. 113 (3), 429–437.

Yu, G., Wise, R., 2000. An anchored AFLP- and retrotransposon-based map of diploid *Avena*. Genome 43 (5), 736–749.

Zhou, X., Jellen, E., Murphy, J., 1999. Progenitor germplasm of domisticated hexaploid oat. Crop Sci. 39 (4), 1208–1214.
Zillinsky, F., Borlaug, N., 1971. Progress in developing triticale as an economic crop. CIMMYT Res. Bull. 17, 1–27.
Zillinsky, F., Murphy, H., 1967. Wild oat species as source of disease resistance for improvement of cultivated oats. Plant Dis. Rep. 51, 391–395.
Zillinsky, F., Sadanaga, K., Simons, M., Murphy, H., 1959. Rust-resistant tetraploid derivatives from crosses between *Avena abyssinica* and *A. strigosa*. Agron. J. 51 (6), 343–345.
Zohary, D., 1971. Origin of southwest Asiatic cereals: wheat, barley, oats, and rye. In: Davis, P., Harper, P., Hedge, I. (Eds.), Plant Life of South-West Asia. Botanical Society of Edinburgh, Edinburgh, pp. 235–260.
Zohary, D., Hopf, M., 1988. Domestication of Plants in the Old World. Clarendon Press, Oxford.

Sorghum

5

Yi-Hong Wang, Hari D. Upadhyaya[†], Ismail Dweikat[§]*
*Department of Biology, University of Louisiana at Lafayette, Lafayette, Louisiana, USA; [†]International Crops Research Institute for the Semi-Arid Tropics (ICRISAT), Genebank, Patancheru, Telangana, India; [§]Department of Horticulture, University of Nebraska, Lincoln, Nebraska, USA

5.1 Introduction

Sorghum (*Sorghum bicolor* L. Moench) is widely used as food, feed, fiber, and bioenergy crop. The grain is used as food or feed; the stem can be used as a source of fiber, fuel, and lately as feedstock for cellulosic ethanol. According to the Food and Agriculture Organization (FAO), the top sorghum producing countries have been the United States, Nigeria, India, Mexico, and Argentina (Table 5.1). In fact, the United States has been at the top for sorghum production from 1961 to 2011, when Nigeria and Mexico surpassed it. Among the top five producing countries, Argentina is also a top per capita producing country (Table 5.1). We will assume here that sorghum is more important as food or feed in countries with high per capita production. We noticed that the 20 top per capita producing countries in Table 5.1 roughly fall in two groups. Those in Africa grow the crop mostly for food while those in the Americas and Australia use it mostly for animal feed, beef, and pork industries (FAO, 1995).

Interestingly, there is also a huge yield gap between these two groups (Fig. 5.1). This gap is largely due to doubling of yield in the Americas from 1961 to 2013 (or an increase of 88%) while yield increases during the same period in Africa was noted only at 25%. For 2013, the latest data available, an average yield in the Americas and Australia is four times that of the African group; in 1961, it was 2.7 times. In the Americas, the highest yield averaged from 1961 to 2013 was achieved in the United States (3665 kg/ha) and Argentina (3253 kg/ha). In Africa, the lowest yield is found in Niger (357 kg/ha) and Somalia (385 kg/ha), about 10% of the average yield in the United States or Argentina. Obviously, sorghum is and will continue to play a critical role in food security in tropical and northeastern Africa.

5.2 Origin, distribution, and diversity

Based on De Wet (1978), the genus *Sorghum* Moench consists of sections *Chaeotosorghum, Heterosorghum, Parasorghum, Stiposorghum,* and *Sorghum*. Section *Sorghum* contains rhizomatous *Sorghum halepense* (L.) Pers. ($2n = 40$) and *Sorghum propinquum* (Kunth) Hitchcock ($2n = 20$), and the annual *S. bicolor* (L.) Moench ($2n = 20$).

Genetic and Genomic Resources for Grain Cereals Improvement. http://dx.doi.org/10.1016/B978-0-12-802000-5.00005-8

Table 5.1 World sorghum production in 2012

Top 20 per capita producers			Top 20 total producers	
Continent	**Country**	**kg per capita**	**Country**	**Production (1000 kg)**
South America	Argentina	126.4622	Mexico	6,969,502
Africa	Burkina Faso	110.0561	Nigeria	6,900,000
Africa	Chad	101.4285	USA	6,272,360
Oceania	Australia	97.68803	India	6,010,000
Africa	Mali	74.29622	Argentina	5,200,000
Africa	Niger	60.08171	Ethiopia	3,951,294
North America	Mexico	60.00587	Australia	2,238,912
Africa	Cameroon	53.83751	Brazil	2,038,767
South America	Bolivia	46.64325	China, mainland	2,000,000
Africa	Ethiopia	45.65911	Burkina Faso	1,924,000
Africa	Nigeria	41.40936	Sudan (former)	1,883,000
Africa	Sudan (former)	41.18368	Mali	1,212,440
Africa	Togo	37.40251	Chad	1,200,000
Central America	Belize	37.03704	Cameroon	1,102,000
South America	Uruguay	30.96432	Niger	1,000,000
Central America	El Salvador	21.80316	Egypt	900,000
South America	Paraguay	21.62203	Tanzania	838,717
North America	USA	19.86238	Venezuela	500,000
Africa	Somalia	18.78126	Bolivia	478,000
Africa	Botswana	18.5095	Yemen	459,241

FAO 2012 data. From http://faostat.fao.org/site/339/default.aspx.

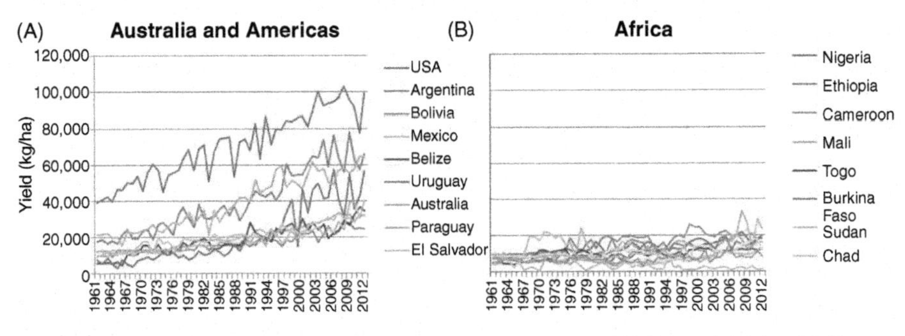

Figure 5.1 Sorghum grain yield in the Americas and Australia (A) and Africa (B) from 1961 to 2013. Shown only the top 20 per capita producing countries in Table 5.1. Data from http://faostat3.fao.org/faostat-gateway/go/to/download/Q/QC/E.

S. bicolor ssp. *bicolor* includes all domesticated grain sorghums; ssp. *drummondii* (Steud.) de Wet comb. nov. includes hybrids among grain sorghums and their closest wild relatives; ssp. *arundinaceum* (Desv.) De Wet et Harlan includes the wild progenitors of grain sorghums. Four races of ssp. *arundinaceum* are recognized: (1) *aethiopicum* of the arid African Sahel, (2) *virgatum* of northeastern Africa, (3) *arundinaceum* of the African tropical forest, and (4) *verticilliflorum* of the African Savanna. Grain sorghums fall into five basic races, bicolor, caudatum, durra, guinea, and kafir, and 10 intermediate races (Harlan and De Wet, 1972). Phylogenetic study based on rDNA internal transcribed spacer region suggest that *S. propinquum*, *S. halepense,* and *S. bicolor* subsp. *arundinaceum* race *aethiopicum* are the closest wild relatives of cultivated sorghum (Sun et al., 1994). The distribution of *S. bicolor* is shown in Fig. 5.2.

Domesticated sorghums originated from wild members of *S. bicolor* subsp. *arundinaceum* from Sudan and Ethiopia (De Wet, 1978) where wild sorghum grains may have been consumed as food 8000 years ago (Wendorf et al., 1992; Dahlberg and Wasylikowa, 1996). In support of central and northeast Africa as the origin, Aldrich et al. (1992) and Aldrich and Doebley (1992) demonstrated that wild sorghum from northeast and central Africa exhibits greater genetic similarities to cultivated sorghums compared to wild sorghum of northwest or southern Africa. Using molecular markers, Aldrich and Doebley (1992) also showed that cultivated sorghum is derived from the wild ssp. *arundinaceum*. That Sudan is the center of origin for sorghum is also supported by molecular marker analysis. Assar et al. (2005) showed that Sudanese sorghum varieties, both landraces and improved varieties, contained higher genetic diversity than sorghums from the International Crops Research Institute for the Semi-Arid Tropics (ICRISAT) and the USA.

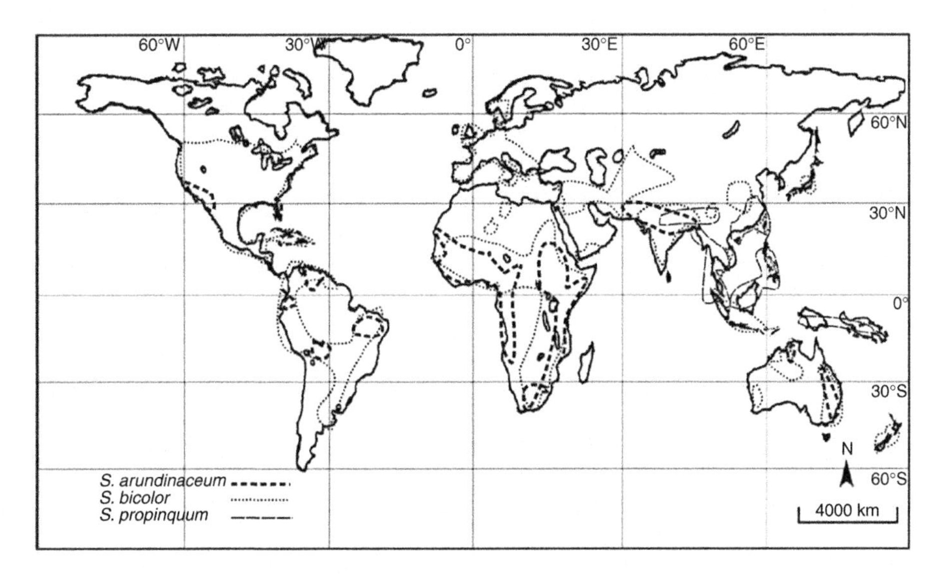

Figure 5.2 Geographic distribution of *S. arundinaceum* (*S. bicolor* subsp. *arundinaceum*), *S. bicolor*, and *S. propinquum*.
From Liu et al. (2014).

From its primitive form, cultivated sorghum has evolved into 5 major (bicolor, guinea, caudatum, kafir, and durra) and 10 intermediate races, which include all combination hybrids of the major races based on panicle architecture and spikelets as described by Harlan and De Wet (1972). This is supported by phylogenetic analysis using molecular markers (Tao et al., 1993; Deu et al., 1994; Cui et al., 1995; Agrama and Tuinstra, 2004; Deu et al., 2006; Perumal et al., 2007; Brown et al., 2011; Wang et al., 2013b) although this is not always the case (Ritter et al., 2007). The following description of these races is based on Harlan and De Wet (1972), De Wet (1978), and Dahlberg (2000).

Race bicolor sorghums are the most closely related to the wild sorghums of all the cultivated races. They are the most primitive of the five races because of their long, clasping glumes, elongate seed, and open panicles although their ability to naturally disperse seed is lost. They are low-yielding but may be grown for their sweet stems (other races also contain sweet sorghums). Bicolor sorghums can be recurrently produced by crossing grain sorghum with Sudangrass (*S. bicolor* [L.] Moench ssp. *drummondii* [Nees ex Steud.] De Wet & Harlan). Therefore, their primitive characteristics may be because they are either an ancient domestication or recent and current introgression with wild or weedy sorghums. The race is heterogeneous, and consists of several distinct subraces such as Sudangrass, sorgo, broomcorn, and bicolor. Bicolor race has contributed to the breeding of forage sorghum (Kamesawara Rao et al., 2004).

In contrast to bicolor, race kafir has more compact and often cylindrical panicles. Kafir sorghums are important staples across the eastern and southern savannah from Tanzania to South Africa. Agronomically, this is an important race as it includes some of the hybrid races such as guinea-kafir in India, kafir-caudatum in the United States, and kafir-durra have been important sources of breeding materials. In addition, the cytoplasmic male sterile system used in hybrid grain sorghum breeding uses durra cytoplasm and kafir derivatives as maintainers. Race kafir is adapted to a more temperate environment; therefore, it is more likely to be photoperiod insensitive than other races (Grenier et al., 2001).

Like kafir sorghums, race caudatum is also agronomically important. It has turtle-backed grains – flat on one side and curved on the opposite side. The grains are usually exposed at maturity. The panicles range from compact to open. The compact panicles are usually found in the drier areas and the open ones are most common in areas with high rainfall. It is widely grown in Chad, Sudan, northeast Nigeria, and Uganda. For hybrid races, most of the modern American hybrid grain sorghums are kafir-caudatum. The durra-caudatum race is a source of yellow endosperm and large seed size.

Race durra also has compact panicles with flattened ovate sessile spikelets. Durra sorghums are widely grown along the fringes of the southern Sahara, across West Africa, the Near East, and parts of India. The race includes some of the most drought-tolerant sorghums that can be grown in the driest regions.

Race guinea has long and gaping glumes exposing the grain at maturity. They have large and open panicles, with the branches often loosely hanging at maturity. It is most commonly grown in West Africa where the larger-seeded guineas (5.0–9.0 mm long) are grown in the drier zone, and the very small-seeded ones (3.0–5.0 mm long) in the wettest zones at the fringe of the forest. The open and hanging panicles and gaping glumes reduce mold damage under wet conditions. Race guinea in temperate regions is more likely to be photoperiod sensitive (Grenier et al., 2001). The guinea race from West Africa has also provided resistance to grain mold (Kamesawara Rao et al., 2004). The fact that guinea sorghums were spread from West Africa to South Asia has been supported by SSR marker analysis (Folkertsma et al., 2005).

The earliest domesticated sorghums were probably bicolor-like from which modern domesticated races were derived. Sorghum was first cultivated along a broad band of the savannah between the Sudan and Nigeria where the wild race *verticilliflorum* is abundant as wild grass (Harlan, 1971). From here the cultivation spread to tropical West Africa, the arid northeast, and southeast Africa. Selection for adaptation to a wet tropical habitat produced race guinea (De Wet et al., 1972). Caudatum sorghums are limited to the original regions of bicolor domestication. Durra sorghums probably originated outside Africa from bicolor sorghums in the Sind-Punjab region of India some 3000 years ago (De Wet and Huckabay, 1967).

5.3 Erosion of genetic diversity from the traditional areas

Genetic diversity of crop plants can be negatively impacted by the practice of modern agriculture. Modern agriculture adopts improved commercial crop varieties to maximize food production. The high-yielding potential of these varieties promotes their widespread cultivation. For example, in India an estimated 400,000 rice varieties were grown before colonialism; this number dropped to 30,000 in the mid-nineteenth century (Heal et al., 2004) and today 75% of the rice land is occupied by just 10 varieties (McNeely, 2005). Similar loss of landrace to modern varieties in rice was also reported in Sri Lanka (Hargrove et al., 1988; Rhoades, 1991), Bangladesh, and Indonesia (Hargrove et al., 1988). These represent perfect examples of how modern varieties replace landraces in crop plants. This loss of crop landraces is especially serious in developed countries. In North America and northwestern Europe, landraces have become almost absent because of widespread adoption of modern varieties (van de Wouw et al., 2010).

The loss of landraces is minimum in places where there is no modern variety invasion. For example, in southern Africa, only 14% of the total sorghum growing area was planted with modern varieties based on 1995 data (Maredia et al., 2000). Similarly in Niger, Deu et al. (2010) surveyed sorghum varietal changes in 71 villages with 28 SSR markers and found no major loss of sorghum landraces from 1976 to 2003. Another study on sorghum landrace loss in Niger also failed to find major loss at the national level during the same period (Bezançon et al., 2009).

Studies on Ethiopia, sorghum's center of origin, seem to support the conclusions. To assess genetic erosion in sorghum, Mekbib (2008) used interviews with farmers and compared sorghums collected in 1960 and 2000 in Ethiopia. It was found that at individual farmers' level, the five most important factors for varietal loss were reduced benefit from the varieties, drought, Khat (*Catha edulis*) expansion, reduced land size, and introduction of other food crops. There was a complementation, not rivalry, between farmer varieties (FVs) and improved varieties (IVs). The prediction in the late 1970s that complete erosion of FVs by IVs would happen by the end of the

1980s, the principle of genetic erosion that competition between IVs and FVs favors the former and results in the replacement of the latter did not occur (Mekbib, 2008). Teshome et al. (2007) also surveyed sorghum varietal changes in Ethiopia by interviewing farmers in an 8-year span (1992/1993 to 2000/2001). They found that over the 8-year period, the total area planted to sorghum decreased drastically in all five farming communities studied and 51–72% of the farmers in the communities decreased the field size planted to sorghum because of population growth, land redistribution policy, seasonal changes, and stagger cropping followed by interspecies crop displacement. Landrace richness increased significantly in two but decreased significantly in three of the communities. Farmers' selection criteria significantly increased landrace richness. Neither studies reported significant sorghum landrace loss in Ethiopia.

Our example of sorghum landrace loss in developed countries is Australia where sorghum hybrids with resistance to sorghum midge have been grown in more than 80% of the sorghum growing area before 1995 (Jordan et al., 1998). Since selection for resistance to sorghum midge is one of the primary objectives of Australian sorghum breeding programs, the relationship between resistance and genetic diversity was assessed using RFLP analysis among 26 grain sorghum hybrids grown commercially in Australia. The genetic distances between each sorghum hybrid and a standard highly resistant hybrid were found to be strongly negatively correlated to hybrid midge resistance ratings ($r = -0.77$, $p < 0.001$). Furthermore, the results showed that adopting midge-resistant hybrids has been associated with a narrowing of the genetic diversity. This reduction in genetic diversity will have implications for the genetic vulnerability of sorghum in Australia and breeding for sorghum yield (Jordan et al., 1998).

5.4 Status of germplasm resource conservation

Genetic resources are fundamental to genetic improvement and study of crop plants. Conserving the germplasm in its natural habitat is known as *in situ* (as on farm) and conserving the germplasm away from its natural habitat is *ex situ* conservation, such as a genebank (Upadhyaya et al., 2014b). *Ex situ* conservation represents the most significant and widespread means of conserving sorghum germplasm. According to FAO (2009), the world sorghum collections stand at 235,711 accessions. The six biggest *ex situ* sorghum germplasm collections are listed in Table 5.2. ICRISAT serves as the world's depository for sorghum germplasm materials.

The largest sources of the 45,192 accessions in the United States were from Ethiopia, Sudan, Yemen, Mali, India, and the United States (http://www.ars-grin.gov/cgi-bin/npgs/html/tax_stat.pl). About 16% of the world collection of sorghum (235,711 accessions) is conserved in ICRISAT's genebank at Patancheru, India (FAO, 2009). This collection of 37,949 accessions is from 92 countries and comprises of 32,578 landraces, 4814 advanced breeding lines, 99 cultivars, and 458 wild and weedy relatives (Upadhyaya et al., 2014b). Among the five basic races, durra accounted for 21.21%, caudatum 20.12%, guinea 12.89%, bicolor 4.59%, and kafir 3.49% of the ICRISAT collection (Upadhyaya et al., 2014b).

The ICRISAT collection is divided into active and base collections (Upadhyaya et al., 2014b). The active collection is stored under medium-term storage (10–20 years)

Table 5.2 Holders of the six largest *ex situ* sorghum germplasm collections in the world

Country	Institution*	No. of accessions	Total holdings (%)
USA	PGRCU (S9) and NSSL**	45,192	19
Global	ICRISAT	37,949	16
China	ICGR-CAAS	18,856	8
India	NBPGR	16,499	7
Ethiopia	IBC	9,428	4
Brazil	CNPMS	7,071	3
World		235,711	100

* ICRISAT, International Crops Research Institute for the Semi-Arid Tropics; PGRCU, Plant Genetic Resources Conservation Unit (S9); NSSL, National Seed Storage Laboratory; ICGR-CAAS, Institute of Crop Genetic Resources of the Chinese Academy of Agricultural Science; NBPGR, National Bureau of Plant Genetic Resources; IBC, Institute of Biodiversity Conservation; CNPMS, Embrapa Milho e Sorgo.
** PGRCU, NSSL, and Plant Variety Protection Office (PVPO) hold 37,799; 7,342; and 51 accessions, respectively.
Data from FAO (2009) and the USA data from http://www.ars-grin.gov/cgi-bin/npgs/html/tax_acc.pl?taxno=454806&unavail=on&rownum=0 (accessed September 2014).

condition (4°C and 20–30% relative humidity). The collection is also used for distribution, utilization, and multiplication. For each accession, 400 g of sorghum seed is harvested and dried to 8% moisture and stored in a screw-capped aluminum container. The base collection is kept for long-term storage at −20°C. For accessions in this collection, 75 g seed is cleaned and dried to 5–7% moisture content at 15°C and 15% RH for approximately 3–4 weeks. The dried seed is vacuum-sealed in an aluminum foil pouch and stored after confirming that germination is >90%. Seed viability is monitored at 5 and 10 in medium-term and 10–20 years interval in long-term storage. Any sample having seed stocks <50 g or seed viability <85% is taken out for regeneration (Upadhyaya et al., 2014b). Over 30,000 of the accessions have also been conserved in the Svalbard Global Seed Vault, Norway (Upadhyaya et al., 2014b).

5.5 Germplasm evaluation and maintenance

ICRISAT has evaluated 94–99% of its 37,949 accessions for most of the morphologic traits; 42–44% evaluated for shoot fly, downy mildew, and stem borer resistance; 18–22% to grain mold, leaf blight, rust, and *Striga* resistance; and 10% to anthracnose resistance (Upadhyaya et al., 2014b). For grain quality, 26–29% accessions have been evaluated for protein and lysine contents. The whole collection has been evaluated for plant color: 4.48% had no pigmentation while the rest are pigmented. The midrib color in the collection varied from white, dull green, yellow and brown, and brown midrib germplasm is about 0.03% of the total collection. About 12% of the accessions had their three quarters of the grains fully covered by the glume while the grains in 3.66% of the accessions are totally uncovered (Upadhyaya et al., 2014b).

The US's germplasm collections are also rather extensively evaluated (Table 5.3). For example, evaluating 36,017 accessions for 100 seed weight found 5 with seed

Table 5.3 Evaluation of sorghum germplasm maintained in the United States

No. of accessions*	Trait evaluated	Result	Sources
32,796	Flowering time	197 flowers in <50 days (PI 537456 also has high-yield potential)	http://www.ars-grin.gov/cgi-bin/npgs/html/desc.pl?69073
22,107	Short-day anthesis		http://www.ars-grin.gov/cgi-bin/npgs/html/desc.pl?69070
1,533	Photoperiod sensitivity		http://www.ars-grin.gov/cgi-bin/npgs/html/desc.pl?69031
13,298	Vigor	Most vigor: PI 501545	http://www.ars-grin.gov/cgi-bin/npgs/html/desc.pl?69091
24,540	Overall plant desirability		http://www.ars-grin.gov/cgi-bin/npgs/html/desc.pl?69039
1,390	Primary plant usage		http://www.ars-grin.gov/cgi-bin/npgs/html/desc.pl?69021
24,801	Height uniformity		http://www.ars-grin.gov/cgi-bin/npgs/html/desc.pl?69060
31,008	Height	20 < 50 cm tall (PI 643035 and PI 642995 also have high-yield potential)	http://www.ars-grin.gov/cgi-bin/npgs/html/desc.pl?69001
24,519	Basal tillers		http://www.ars-grin.gov/cgi-bin/npgs/html/desc.pl?69036
24,509	Nodal tillers		http://www.ars-grin.gov/cgi-bin/npgs/html/desc.pl?69049
26,842	Midrib juiciness		http://www.ars-grin.gov/cgi-bin/npgs/html/desc.pl?69003
26,527	Leaf midrib color		http://www.ars-grin.gov/cgi-bin/npgs/html/desc.pl?69009
24,448	Panicle length		http://www.ars-grin.gov/cgi-bin/npgs/html/desc.pl?69050
23,531	Panicle erectness		http://www.ars-grin.gov/cgi-bin/npgs/html/desc.pl?69062

Table 5.3 Evaluation of sorghum germplasm maintained in the United States *(cont.)*

No. of accessions*	Trait evaluated	Result	Sources
15,740	Panicle compactness		http://www.ars-grin.gov/cgi-bin/npgs/html/desc.pl?69061
14,753	Panicle shape		http://www.ars-grin.gov/cgi-bin/npgs/html/desc.pl?69066
14,730	Panicle branch angle		http://www.ars-grin.gov/cgi-bin/npgs/html/desc.pl?69037
15,732	Awns		http://www.ars-grin.gov/cgi-bin/npgs/html/desc.pl?69027
24,107	Exsertion		http://www.ars-grin.gov/cgi-bin/npgs/html/desc.pl?69059
23,817	Plant color		http://www.ars-grin.gov/cgi-bin/npgs/html/desc.pl?69063
23,481	Stalk waxiness		http://www.ars-grin.gov/cgi-bin/npgs/html/desc.pl?69055
36,017	Seed weight	Seed weight >7.0 g: PI 651202, PI 495919, PI 474852, PI 465603, PI 465443	http://www.ars-grin.gov/cgi-bin/npgs/html/eval.pl?495020
22,840	Seed sprouting tendency		http://www.ars-grin.gov/cgi-bin/npgs/html/desc.pl?69054
14,643	Seed type		http://www.ars-grin.gov/cgi-bin/npgs/html/desc.pl?69033
14,690	Seed shattering		http://www.ars-grin.gov/cgi-bin/npgs/html/desc.pl?69052
14,785	Testa		http://www.ars-grin.gov/cgi-bin/npgs/html/desc.pl?69056
14,240	Transverse wrinkle		http://www.ars-grin.gov/cgi-bin/npgs/html/desc.pl?69057

(Continued)

Table 5.3 Evaluation of sorghum germplasm maintained in the United States *(cont.)*

No. of accessions*	Trait evaluated	Result	Sources
14,807	Kernel cover		http://www.ars-grin.gov/cgi-bin/npgs/html/desc.pl?69045
14,564	Kernel plumpness		http://www.ars-grin.gov/cgi-bin/npgs/html/desc.pl?69046
14,692	Kernel shape		http://www.ars-grin.gov/cgi-bin/npgs/html/desc.pl?69047
18,928	Kernel color		http://www.ars-grin.gov/cgi-bin/npgs/html/desc.pl?69005
14,767	Mesocarp		http://www.ars-grin.gov/cgi-bin/npgs/html/desc.pl?69048
14,425	Pericarp color		http://www.ars-grin.gov/cgi-bin/npgs/html/desc.pl?69051
14,542	Endosperm color		http://www.ars-grin.gov/cgi-bin/npgs/html/desc.pl?69038
14,723	Endosperm texture		http://www.ars-grin.gov/cgi-bin/npgs/html/desc.pl?69040
14,551	Endosperm type		http://www.ars-grin.gov/cgi-bin/npgs/html/desc.pl?69041
14,925	Glume color		http://www.ars-grin.gov/cgi-bin/npgs/html/desc.pl?69042
14,807	Glume pubescence		http://www.ars-grin.gov/cgi-bin/npgs/html/desc.pl?69043
14,861	Grain weathering		http://www.ars-grin.gov/cgi-bin/npgs/html/desc.pl?69044
24,466	Yield potential	One early (PI 537456) and two dwarf (PI 643035 and PI 642995) accessions also have high-yield potential	http://www.ars-grin.gov/cgi-bin/npgs/html/desc.pl?69058

Table 5.3 **Evaluation of sorghum germplasm maintained in the United States *(cont.)***

No. of accessions*	Trait evaluated	Result	Sources
2,882	Acid detergent fiber, protein, fat, dry matter digestibility, phosphorus, starch, metabolizable energy, net energy for gain (cattle), net energy for lactation (cattle), net energy for maintenance (cattle), total digestable nutrients percentage		http://www.ars-grin.gov/cgi-bin/npgs/html/desc.pl?69084
1,245	Brix		http://www.ars-grin.gov/cgi-bin/npgs/html/eval.pl?493202
1,211	Sucrose		http://www.ars-grin.gov/cgi-bin/npgs/html/eval.pl?493203
1,058	Lodging		http://www.ars-grin.gov/cgi-bin/npgs/html/desc.pl?69023
47,033	Aluminum tolerance	Most tolerant: PI 494884, PI 510763, PI 513626, PI 513653; PI 513671, PI 513739, PI 513760, PI 513834	http://www.ars-grin.gov/cgi-bin/npgs/html/desc.pl?69067
7,338	Manganese tolerance		http://www.ars-grin.gov/cgi-bin/npgs/html/desc.pl?69068
14,580	Greenbug resistance	Resistant to greenbug biotype-E: PI 264453, PI 220248, PI 229828, PI 266965, PI 494893; PI 524770, PI 302136	http://www.ars-grin.gov/cgi-bin/npgs/html/desc.pl?69102

(Continued)

Table 5.3 Evaluation of sorghum germplasm maintained in the United States *(cont.)*

No. of accessions*	Trait evaluated	Result	Sources
5,564	Yellow sugarcane aphid	Most tolerant: PI 457709	http://www.ars-grin.gov/cgi-bin/npgs/html/desc.pl?69014
8,940	Fall army worm		http://www.ars-grin.gov/cgi-bin/npgs/html/desc.pl?69015
3,937	Downy mildew	175 resistant	http://www.ars-grin.gov/cgi-bin/npgs/html/desc.pl?69093
25,756	Anthracnose		http://www.ars-grin.gov/cgi-bin/npgs/html/desc.pl?69011
26,750	Rust		http://www.ars-grin.gov/cgi-bin/npgs/html/desc.pl?69064
306	Gray leafspot		http://www.ars-grin.gov/cgi-bin/npgs/html/desc.pl?69018
40	Downy mildew, anthracnose		Prom et al. (2007)
40	Ergot resistance		Prom et al. (2008)
98	Downy mildew resistance		Prom et al. (2011)
154	Anthracnose resistance		Erpelding (2011)
1,452	Ladder spot		http://www.ars-grin.gov/cgi-bin/npgs/html/desc.pl?69075
1,467	Zonate leaf spot		http://www.ars-grin.gov/cgi-bin/npgs/html/desc.pl?69076
427	Sugarcane mosaic virus		http://www.ars-grin.gov/cgi-bin/npgs/html/desc.pl?69077

* The number of accessions evaluated for each trait was downloaded in September 2014. Phenotypic listed in the table can be downloaded from the web address given.

weight over 7.0 g; out of 13,298 accessions, one was found to have the most vigor; two dwarf (PI 643035 and PI 642995) and one early (PI 537456) accessions also had high-yield potential (Table 5.3). What is also valuable is biotic or abiotic resistant accessions: seven of 14,580 accessions were resistant to greenbug biotype-E; eight accessions were found to be extremely tolerant to aluminum toxicity (Table 5.3). In those efforts, a large number of accessions were evaluated to identify a few accessions that may be valuable in breeding (Table 5.3). Therefore, germplasm evaluation increases the value of collected materials to plant breeders and other researchers.

Sorghum is self-pollinated with varying degrees of outcrossing (Pedersen et al., 1998; Barnaud et al., 2008). Germplasm accessions are maintained and multiplied by selfing. Individual panicles are covered with well-labeled selfing paper bags ($L \times W \times H$; 35×10×5 cm) as soon as heads emerge from flag leaf before anthesis. Heads are kept covered for at least 21 days (i.e., up to dough stage) and then removed, but the bags are tied around the peduncles to identify the selfed plants once the seeds become mature and dry (Upadhyaya et al., 2014b). At harvest, seeds from at least 50 selfed plants are bulked to maintain an accession. Wild relatives are also maintained by selfing; however, some wild species do not set seeds, which are maintained as live plant in a field genebank (Upadhyaya et al., 2014b).

5.6 Use of germplasm in crop improvement

One of the major functions of a germplasm collection is its utilization. However, maximum use of germplasm is more likely if a core collection representing the diversity of the whole collection is created (Upadhyaya et al., 2014b). Upadhyaya and Ortiz (2001) postulated the concept of minicore collection, which consists of 1% of the entire collection to enhance utilization of germplasm when size of the core collection is too large and precise phenotyping is difficult and not cost-effective. Phenotypic data of the entire germplasm collection of a given species is used to identify 10% of the accessions to form the core collection. This core is further evaluated for morphoagronomic and seed quality traits to select 10% of the core collection accessions for forming the minicore collection. At both stages, standard clustering procedures are used to create groups of similar accessions (Upadhyaya et al., 2009). Core collection in sorghum consists of 2247 accessions (Grenier et al., 2001) while minicore collection contains 242 accessions (Upadhyaya et al., 2009). The selected accessions in minicore represent all 5 basic and 10 intermediate races, as well 10 geographic regions. However, the entire collection as well as both subsets is dominated by caudatum, durra, and guinea, the basic races, and caudatum-bicolor and guinea-caudatum, the intermediate races (Upadhyaya et al., 2014b). The minicore collection has been extensively phenotyped and genotyped. Accessions with outstanding biomass, grain, and sugar yield (Upadhyaya et al., 2014a) and resistance to pests and diseases (Sharma et al., 2010; Sharma et al., 2012) have been identified, which may be used in sorghum breeding programs.

The most efficient use of collected germplasm is to use them directly as varieties. To date, 34 ICRISAT accessions have been directly released as varieties in 17 countries, with a few of these released in more than one country (Upadhyaya et al., 2014b). One such variety is IS 18758, a popular guinea-caudatum landrace from Ethiopia. It has been released as E 35-1 in Burkina Faso and as Gambella 1107 in Burundi thanks to its desirable plant type, high grain yield, good grain quality, straw glume color, resistance to leaf diseases, and tolerance to grain weathering. IS 18758 has been extensively used in breeding programs at ICRISAT and in national breeding programs elsewhere (Upadhyaya et al., 2014b). Another accession, IS 33844, is an excellent maldandi-type with large and lustrous grains and high-yield potential. Selection from IS 33844 has been released as "Parbhani Moti" for postrainy season in Maharashtra, India (Reddy et al., 2006). The overall use of the ICRISAT's germplasm in genetic improvement is at least partially reflected in their germplasm distribution, which totals 261,521 samples to 109 countries since 1973 (Upadhyaya et al., 2014b). That is about 6500 samples per year on average or about 171 samples per 1000 accessions based on ICRISAT's total collection today. According to Marshall (1989), 20 samples distributed per 1000 accessions collected is considered adequate for a collection. Based on this criterion, ICRISAT's germplasm collection has been put into very efficient use.

5.7 Limitations in germplasm use

Though a wide range of germplasm are available nationally and internationally, breeders are more likely to use only adapted and improved accessions, avoiding wild and weedy relatives, and landraces in their breeding program (Marshall, 1989; Nass and Paterniani, 2000). Plant breeders are aware of the limitation of working with exotic germplasm. Use of wild and landrace accessions in breeding programs is low because of the lack of knowledge about the genetic value of the materials or the linkage drag associated with the transfer of beneficial traits from such germplasm (Upadhyaya et al., 2014b).

5.8 Germplasm enhancement through wide crosses

Wild species of cultivated sorghum could be used to improve sorghum germplasm. This is facilitated by the fact that domesticated sorghum is sexually compatible with all of its wild relatives in section *Sorghum* (Andersson and Carmen, 2010) including *S. halepense* (Johnsongrass), *S. propinquum*, *S. bicolor* ssp. *drummondii*, *S. bicolor* ssp. *arundinaceum*, and its four races (*aethiopicum*, *virgatum*, *arundinaceum*, and *verticilliflorum*).

Artificial crossing between sorghum and Johnsongrass has been carried out to create useful germplasm. There are two types of hybrid plants: (1) sterile plants with 30 chromosomes and vigorous rhizomes and (2) fertile plants with 40 chromosomes and weak rhizomes (cited in Dweikat, 2005). Sangduen and Hanna (1984) reported that hybrids between sorghum and Johnsongrass were vigorous, leafy, and more closely resembled Johnsongrass than *S. bicolor* in perennial growth habit, open panicle, seed color/shape, and seed shattering. Stem thickness, rhizome expression, and seed size were intermediate to the parents. Piper and Kulakow (1994) conducted a similar experiment (with two backcross generations added) in an effort to produce perennial sorghum varieties. In their study, as seed mass increased with backcross generation, rhizome expression decreased, that is, backcross progeny became more like the *S. bicolor* parent although there were exceptions. In the cross between sorghum and Johnsongrass reported by Dweikat (2005), progenies were analyzed by genetic markers to confirm that the two genomes behaved normally during meiosis as demonstrated by 1:2:1 segregation ratio of 34 markers. Valuable traits, such as resistance to greenbug and chinch bug and adaptability to cold temperature, were found in the progenies (Dweikat, 2005).

Another source of useful traits is Sudangrass. Sudangrass is a cultivated form of *S. bicolor* subsp. *drummondii* (Sahoo et al., 2010), which is an obnoxious weed (shattercane) that closely resembles cultivated sorghum (Defelice, 2006). *S. bicolor* ssp. *drummondii* is a natural hybrid between ssp. *bicolor* and *verticilliflorum* (De Wet, 1978). *S. bicolor* ssp. *drummondii* grouped mainly with bicolor race sorghums based on genetic markers (Casa et al., 2005). Sudangrass has been shown to suppress the nematodes, *Meloidogyne chitwoodi* (Mojtahedi et al., 1993) and *Meloidogyne hapla* (Viaene and Abawi, 1998). It turns out that Sudangrass cultivars contain the cyanogenic glucoside dhurrin that is degraded through an intermediate step to *p*-hydroxybenzal-dehyde (*p*-HBA) and HCN. Incubating *M. hapla* eggs in Sudangrass extract resulted in a 55% reduction in the number of juveniles (J2) penetrating lettuce roots, but juveniles exposed to the extracts were not affected (Widmer and Abawi, 2000). Sorghum-Sudangrass hybrids are crosses between forage-type sorghums and Sundangrass. They are tall, fast-growing summer annual grasses and can also suppress weeds and nematodes (Clark, 2007). The hybrid accounts for 3/4 of the total hybrid sorghum seed market (Smith and Frederiksen, 2000).

Wild relatives have the potential to enhance a number of economically important traits in cultivated sorghum. They are a source of resistance to *Striga* ssp. (Rich et al., 2004), a parasitic weed on sorghum. As mentioned earlier, Johnsongrass has been found to be a source of resistance to greenbug and chinch bug and adaptability to cold temperature (Dweikat, 2005). Johnsongrass (accession IS 14212) is also a source of resistance to sorghum shoot fly (Kamala et al., 2009), as is *Sorghum dimidiatum* (accession IS 18945) (Nwanze et al., 1995). *S. bicolor* ssp. *arundinaceum* has been shown to improve grain yield in hybrid grain sorghum (Jordan et al., 2004). Similarly, *S bicolor* ssp. *drummondii* and *Sorghum purpureosericeum* also produced progenies that performed better than the cultivated sorghum parent from a single backcross (Jordan et al., 2011a). Texas A&M University has employed *S. propinquum* to produce cultivars with early maturity and good yields but these have yet to be released (cited in Hajjar and Hodgkin, 2007). Cultivated sorghum has been successfully crossed with *Sorghum macrospermum* ($2n = 40$) (Kuhlman et al., 2008, 2010), a source of resistance to sorghum midge (Franzmann and Hardy, 1996; Sharma and Franzmann, 2001), sorghum downy mildew (Kamala et al., 2002), and shoot fly (Sharma et al., 2005). Finally, both Johnsongrass (Sangduen and Hanna, 1984) and *S. propinquum* (Washburn et al., 2013) have been used to introduce perenniality (rhizomatousness) to cultivated sorghum.

5.9 Integration of genomic and genetic resources in crop improvement

5.9.1 *Molecular markers and genotyping, genetic maps and trait mapping, molecular breeding*

In recent years, marker development, genome mapping, and tagging of economically important traits have taken off thanks to high throughput marker systems such as the single nucleotide polymorphism (SNP) marker. A large number of SNPs can be identified through whole genome resequencing in sorghum (Nelson et al., 2011; Zheng et al., 2011; Morris et al., 2013; Mace et al., 2013). For example, resequencing of 44 diverse sorghum genotypes yielded 4,946,038 SNPs with a missing data rate of <50% and 1,982,971 small-to-medium length indels (Mace et al., 2013). Although it is still expensive to genotype a mapping or breeding population, such high density of markers (>1 marker per 150 bp on average) will unleash unprecedented power in genetic mapping of quantitative traits. An excellent review of mapping of the sorghum genome and major genes can be found in Madhusudhana (2014). An update on mapping of major genes and QTL is provided in Table 5.4.

Since most economically important traits are controlled by multiple genes or QTL (quantitative trait loci; Table 5.4) and the environment, it is essential that these traits be mapped so that underlying genes can be transferred to more desirable genetic background through molecular breeding. One such example is resistance to *Striga hermonthica* (Del.) Benth., a major biotic constraint to sorghum production in Africa and difficult to control. Haussmann et al. (2004) mapped five resistance QTL in a recombinant inbred line (RIL) population from N13 (resistant) × E36-1 (susceptible) cross. These QTL were stable across environments and explained 12–30% of the observed genetic variation for *Striga* resistance. Linked markers were used to transfer the QTL to popular local varieties. When 32 of the transfer lines were evaluated for *Striga* resistance, lines at least as resistant as N13 were identified and resulted in commercial release of four varieties in the genetic background of popular but susceptible varieties. In *Striga* infested field, these varieties outperformed recurrent parents by 80–198% in grain yield (Deshpande et al., 2013; Mohamed et al., 2014). The project succeeded in using genetic markers that are physically not very close to the respective QTL although more tightly linked markers will be more efficient in monitoring the chromosome transfer.

The *Striga*-susceptible E36-1 is very tolerant to drought and has been used as a donor of drought tolerance QTL in molecular breeding. Five simple sequence repeat (SSR) markers were used to select the three stay-green QTL of E36-1 in SBI-01, SBI-07, and SBI-10 linkage groups. In the F_1 generation, two of these QTL were transferred into three genotypes. In the BC_1F_1 generation, 32 genotypes had at least one QTL incorporated. From a population of 157 BC_2F_1 progenies, 45 genotypes had incorporated either one or two of the stay-green QTL. The results showed that stay-green QTL and consequently drought tolerance can be transferred successfully into farmer preferred sorghum varieties through molecular breeding (Ngugi et al., 2013). Hash et al. (2003) selected six stay-green QTL, including *Stg1*, *Stg2*, *Stg3*, and *Stg4*

Table 5.4 Mapping of major genes in sorghum since 2010

Gene	Trait	Linkage group	References
Major genes			
SbBADH2	Fragrance	7	Yundaeng et al. (2013)
lgs	Low *Striga* germination stimulant activity	5	Satish et al. (2012a)
RMES1	Sorghum aphid resistance	6	Wang et al. (2013a)
Rf2	Pollen fertility restoration	2	Jordan et al. (2010)
Rf5	Pollen fertility restoration	5	Jordan et al. (2011b)
Major QTL			
QPh-dsr09-2	Plant height	9	Reddy et al. (2013)
QTdu.dsr-10.1	Trichome density on upper leaf surface, component of shoot fly resistance	10	Aruna et al. (2011); Satish et al. (2012b)
qRT7	Brace root	7	Li et al. (2014)
qHD6a	Heading date	6	Zou et al. (2012)
PHT	Plant height between SbAGF06–Xcup19	7	Guan et al. (2011)
3 QTL	Transpiration ratio	9, 10	Kapanigowda et al. (2014)
GDR21	Greenbug resistance	9	Punnuri et al. (2013)
hDPW4.1	Grain yield heterosis	4	Ben-Israel et al. (2012)
Rhizome QTL	Rhizome expression (fine mapping)	1	Washburn et al. (2013)

(Subudhi et al., 2000; Sanchez et al., 2002; Harris et al., 2007) as well as *StgA* and *StgB* for transfer into a number of genetically diverse, tropically adapted elite sorghum varieties, which already have a range of drought tolerance. They have generated single-QTL introgression lines that can now be used individually as donor parents of specific stay-green QTL; these QTL have been shown to have the largest favorable phenotypic effects across several genetic backgrounds (Vadez et al., 2013). Marker-assisted selection in both *Striga* resistance and drought tolerance demonstrated the power of the technique if robust markers are identified close to the underlying QTL.

5.9.2 Association mapping

Association is an efficient method to identify the genetic loci underlying traits at a relatively high resolution. With the completion of reference genome sequence, the advent of high throughput sequencing technology now enables rapid and accurate sequencing of a large number of crop genomes to detect the genetic basis of phenotypic variations in crops (Huang and Han, 2014). More recent studies using GWAS revealed significant marker–trait associations in sorghum, that is, days to flowering, culm length, number of tillers, number of panicles, and panicle length (Shehzad et al., 2009; Bhosale et al., 2012); kernel weight and tiller number (Upadhyaya et al., 2012a); plant height (Murray et al., 2009; Wang et al., 2012; Upadhyaya et al., 2012b; Upadhyaya et al., 2013a); stem sugar (brix) (Murray et al., 2009); anthracnose resistance (Upadhyaya et al., 2013b), rust and grain mold resistance (Upadhyaya et al., 2013c); and maturity (Upadhyaya et al., 2012b; Upadhyaya et al., 2013a), with many of these markers comapped on the same linkage groups previously reported as harboring QTL or candidate gene for anthracnose, rust, and grain mold resistance, tiller, plant height, and maturity (Upadhyaya et al., 2014b). Morris et al. (2013) quantified variation in nucleotide diversity, linkage disequilibrium, and recombination rates across the genome using genome-wide SNP map (971 sorghum accessions characterized with 265000 SNPs by using genotyping-by-sequencing). Further GWAS reveals several SNPs associated with total plant height (or height components, that is, preflag height, which quantifies elongation in the lower portion of the stem, and flag-to-apex length, which quantifies elongation in the upper portion of the stem) and candidate genes for inflorescence architecture, and independent spread of multiple haplotypes carrying alleles for short stature or long inflorescence branches. Such genome-wide map of SNP variation provides a basis for crop improvement through marker-assisted breeding and genomic selection in sorghum (Upadhyaya et al., 2014b).

5.10 Conclusions

Sorghum bicolor is one of the most variable species. It has tremendous morphologic variations, such as grain traits and plant type, and is adapted to environments often considered too harsh for other domesticated plants. The variation is also reflected in seed size. Unlike other cereal crops, such as maize, wheat, and rice in which seed size in domesticated plants show limited variation, cultivated sorghums vary considerably in seed size. These variations may be partly attributed to the widespread coexistence with its wild relatives in the center of origin. Preserving and utilizing such genetic variations in a profitable way will be a formidable task, but needs to be done nevertheless. The use of representative subsets, such as minicore collection, is helping researchers find new genetic variations associated with agronomically beneficial traits for use in breeding and genomics research of sorghum (Upadhyaya et al., 2014a). Sorghum is a genomic resource-rich crop and its increasing use will guide breeders to develop targeted populations/cultivars with specific adaptation.

References

Agrama, H.A., Tuinstra, M.R., 2004. Phylogenetic diversity and relationships among sorghum accessions using SSRs and RAPDs. Afr. J. Biotechnol. 2, 334–340.

Aldrich, P.R., Doebley, J., 1992. Restriction fragment variation in the nuclear and chloroplast genomes of cultivated and wild *Sorghum bicolor*. Theor. Appl. Genet. 85, 293–302.

Aldrich, P.R., Doebley, J., Schertz, K.F., Stec, A., 1992. Patterns of allozyme variation in cultivated and wild *Sorghum bicolor*. Theor. Appl. Genet. 85, 451–460.

Andersson, M.S., Carmen, M., 2010. Gene Flow Between Crops and Their Wild Relatives. Johns Hopkins University Press, Baltimore, MD, pp. 443–464.

Aruna, C., Bhagwat, V.R., Madhusudhana, R., Sharma, V., Hussain, T., Ghorade, R.B., et al., 2011. Identification and validation of genomic regions that affect shoot fly resistance in sorghum [*Sorghum bicolor* (L.) Moench]. Theor. Appl. Genet. 122, 1617–1630.

Assar, A.H., Uptmoor, R., Abdelmula, A.A., Salih, M., Ordon, F., Friedt, W., 2005. Genetic variation in sorghum germplasm from Sudan, ICRISAT, and USA assessed by simple sequence repeats (SSRs). Crop Sci. 45, 1636–1644.

Barnaud, A., Trigueros, G., Mckey, D., Joly, H.I., 2008. High outcrossing rates in fields mixed with sorghum landraces: how are landraces maintained? Heredity 101, 445–452.

Ben-Israel, I., Kilian, B., Nida, H., Fridman, E., 2012. Heterotic trait locus (HTL) mapping identifies intra-locus interactions that underlie reproductive hybrid vigor in *Sorghum bicolor*. PLoS ONE 7, e38993.

Bhosale, S.U., Stich, B., Rattunde, H.F.W., Weltzien, E., Haussmann, B.I.G., Hash, C.T., et al., 2012. Association analysis of photoperiodic flowering time genes in west and central African sorghum [*Sorghum bicolor* (L.) Moench]. BMC Plant Biol. 12 (1), 32.

Bezançon, G., Pham, J.L., Deu, M., Vigouroux, Y., Sagnard, F., Mariac, C., et al., 2009. Changes in the diversity and geographic distribution of cultivated millet (*Pennisetum glaucum* (L.) R. Br.) and sorghum (*Sorghum bicolor* (L.) Moench) varieties in Niger between 1976 and 2003. Genet. Resour. Crop Evol. 56, 223–236.

Brown, P.J., Myles, S., Kresovich, S., 2011. Genetic support for phenotype based racial classification in Sorghum. Crop Sci. 51, 224–230.

Casa, A.M., Mitchell, S.E., Hamblin, M.T., Sun, H., Bowers, J.E., Paterson, A.H., et al., 2005. Diversity and selection in sorghum: simultaneous analyses using simple sequence repeats. Theor. Appl. Genet. 111, 23–30.

Clark, A., 2007. Managing Cover Crops Profitably, third ed Sustainable Agriculture Network, Beltsville, MD, p. 244.

Cui, Y.X., Xu, G.W., Magill, C.W., Schertz, K.F., Hart, G.E., 1995. RFLP-based assay of *Sorghum bicolor* (L.) Moench genetic diversity. Theor. Appl. Genet. 90, 787–796.

Dahlberg, J.A., 2000. Classification and characterization of sorghum. In: Smith, C.W., Frederiksen, R.A. (Eds.), Sorghum: Origin, History, Technology, and Production. John Wiley & Sons Inc, New York, pp. 99–130.

Dahlberg, J.A., Wasylikowa, K., 1996. Image and statistical analyses of early sorghum remains (8000 B. P.) from the Nabta Playa archaeological site in the Western Desert, southern Egypt. Veget. History Archaeobot. 5, 293–299.

Defelice, M.S., 2006. Shattercane, *Sorghum bicolor* (L.) Moench ssp. *drummondii* (Nees ex Steud.) de Wet ex Davidse – black sheep of the family. Weed Technol. 20, 1076–1083.

Deshpande, S.P., Mohamed, A., Hash, Jr., C.T., 2013. Molecular breeding for *Striga* resistance in sorghum. In: Varshney, R.K, Tuberosa, R. (Eds.), Translational Genomics for Crop Breeding, vol. I. Biotic Stress. Wiley-Blackwell, Oxford, UK, pp. 77–93.

Deu, M., Gonzalez-de-Leon, D., Glazsmann, J.C., Degremont, I., Chantereau, J., Lanaud, C., et al., 1994. RFLP diversity in cultivated sorghum in relation to racial differentiation. Theor. Appl. Genet. 88, 838–844.

Deu, M., Rattunde, F., Chantereau, J., 2006. A global view of genetic diversity in cultivated sorghums using a core collection. Genome 49, 168–180.

Deu, M., Sagnard, F., Chantereau, J., Calatayud, C., Vigouroux, Y., Pham, J.L., et al., 2010. Spatio-temporal dynamics of genetic diversity in *Sorghum bicolor* in Niger. Theor. Appl. Genet. 120, 1301–1313.

De Wet, J.M.J., 1978. Systematics and evolution of sorghum sect sorghum (Gramineae). Am. J. Bot. 65, 477–484.

De Wet, J.M.J., Harlan, J.R., Kurmarohita, B., 1972. Origin and evolution of guinea sorghums. East Afr. Agric. Forest. J. 38, 114–119.

De Wet, J.M.J., Huckabay, J.P., 1967. The origin of *Sorghum bicolor*. II. Distribution and domestication. Evolution 21, 787–802.

Dweikat, I., 2005. A diploid, interspecific, fertile hybrid from cultivated sorghum, *Sorghum bicolor*, and the common Johnsongrass weed *Sorghum halepense*. Mol. Breed. 16, 93–101.

Erpelding, J.E., 2011. Anthracnose field evaluation of sorghum germplasm from Botswana. Plant Protect. Sci. 47, 149–156.

FAO, 1995. Sorghum and Millets in Human Nutrition (FAO Food and Nutrition Series, No. 27). Rome. Available from: http://www.fao.org/docrep/t0818e/t0818e04.htm

FAO, 2009. Commission on Genetic Resources for Food and Agriculture. Draft Second Report on the World's Plant Genetic Resources for Food and Agriculture (Final Version). Rome, 330 p. Available from: http://www.fao.org/3/a-k6276e.pdf (accessed September 2014.).

Folkertsma, R.T., Rattunde, H.F.W., Chandra, S., Raju, G.S., Hash, C.T., 2005. The pattern of genetic diversity of Guinea-race *Sorghum bicolor* (L.) Moench landraces as revealed with SSR markers. Theor. Appl. Genet. 111, 399–409.

Franzmann, B.A., Hardy, A.T., 1996. Testing the host status of Australian indigenous sorghums for the sorghum midge. Third Australian Sorghum Conference, Tamworth, Australia, pp. 365–367.

Grenier, C., Bramel-Cox, P.J., Harmon, P., 2001. Core collection of sorghum: I. Stratification based on eco-geographical data. Crop Sci. 41, 234–240.

Guan, Y.A., Wang, H.L., Qin, L., Zhang, H.W., Yang, Y.B., Gao, F.J., et al., 2011. QTL mapping of bio-energy related traits in *Sorghum*. Euphytica 182, 431–440.

Hajjar, R., Hodgkin, T., 2007. The use of wild relatives in crop improvement: a survey of developments over the last 20 years. Euphytica 156, 1–13.

Hargrove, T.R., Cabanilla, V.L., Coffman, W.R., 1988. Twenty years of rice breeding. BioScience 38, 675–681.

Harlan, J.R., 1971. Agricultural origins: centers and noncenters. Science 174, 468–474.

Harlan, J.R., De Wet, J.M.J., 1972. A simplified classification of cultivated sorghum. Crop Sci. 12, 172–176.

Harris, K., Subudhi, P.K., Borrell, A., Jordan, D., Rosenow, D., Nguyen, H., et al., 2007. Sorghum stay-green QTL individually reduce post-flowering drought-induced leaf senescence. J. Exp. Bot. 58, 327–338.

Hash, C.T., Bhasker Raj, A.G., Lindup, S., Sharma, A., Beniwal, C.R., Folkertsma, R.T., et al., 2003. Opportunities for marker-assisted selection (MAS) to improve the feed quality of crop residues in pearl millet and sorghum. Field Crops Res. 84, 79–88.

Haussmann, B.I., Hess, D.E., Omanya, G.O., Folkertsma, R.T., Reddy, B.V., Kayentao, M., et al., 2004. Genomic regions influencing resistance to the parasitic weed *Striga hermonthica* in two recombinant inbred populations of sorghum. Theor. Appl. Genet. 109, 1005–1016.

Heal, G., Walker, B., Levin, S., Arrow, K., Dasgupta, P., Daily, G., et al., 2004. Genetic diversity and interdependent crop choices in agriculture. Resour. Energy Econ. 26, 175–184.

Huang, X., Han, B., 2014. Natural variations and genome-wide association studies in crop plants. Annu. Rev. Plant Biol. 65, 531–551.

Jordan, D.R., Tao, Y.Z., Godwin, I.D., Henzell, R.G., Cooper, M., McIntyre, C.L., 1998. Loss of genetic diversity associated with selection for resistance to sorghum midge in Australian sorghum. Euphytica 102, 1–7.

Jordan, D., Butler, D., Henzell, B., Drenth, J., McIntyre, L., 2004. Diversification of Australian sorghum using wild relatives. In: New Directions for a Diverse Planet: Proceedings of the Fourth International Crop Science Congress, September 26–October 1, 2004, Brisbane, Australia.

Jordan, D.R., Mace, E.S., Henzell, R.G., Klein, P.E., Klein, R.R., 2010. Molecular mapping and candidate gene identification of the *Rf2* gene for pollen fertility restoration in sorghum [*Sorghum bicolor* (L.) Moench]. Theor. Appl. Genet. 120, 1279–1287.

Jordan, D.R., Mace, E.S., Cruickshank, A.W., Hunt, C.H., Henzell, R.G., 2011a. Exploring and exploiting genetic variation from unadapted sorghum germplasm in a breeding program. Crop Sci. 51, 1444–1457.

Jordan, D.R., Klein, R.R., Sakrewski, K.G., Henzell, R.G., Klein, P.E., Mace, E.S., 2011b. Mapping and characterization of *Rf5*: a new gene conditioning pollen fertility restoration in A1 and A2 cytoplasm in sorghum (*Sorghum bicolor* (L.) Moench). Theor. Appl. Genet. 123, 383–396.

Kamala, V., Singh, S.D., Bramel, P.J., Manohar Rao, D., 2002. Sources of resistance to downy mildew in wild and weedy sorghums. Crop Sci. 42, 1357–1360.

Kamala, V., Sharma, H.C., Manohar Rao, D., Varaprasad, K.S., Bramel, P.J., 2009. Wild relatives of sorghum as sources of resistance to sorghum shoot fly, *Atherigona soccata*. Plant Breed. 128, 137–142.

Kamesawara Rao, N., Bramel, P.J., Reddy, V.G., Deb, U.K., 2004. Conservation, utilization and distribution of sorghum germplasm. In: Bantilan, M.C.S., Deb, U.K., Gowda, C.L.L., Reddy, B.V.S., Obilana, A.B., Evenson, R.E. (Eds.), Sorghum Genetic Enhancement: Research Process, Dissemination and Impacts. International Crops Research Institute for the Semi-Arid Tropics, Patancheru, AP, India, pp. 43–62.

Kapanigowda, M., Payne, W., Rooney, W., Mullet, J., Balota, M., 2014. Quantitative trait locus (QTL) mapping of transpiration ratio related to pre-flower drought tolerance in sorghum [*Sorghum bicolor* (L.) Moench]. Funct. Plant Biol. 41, 1049–1065.

Kuhlman, L.C., Burson, B.L., Klein, P.E., Klein, R.R., Stelly, D.M., Price, H.J., et al., 2008. Genetic recombination in *Sorghum bicolor*× *S. macrospermum* interspecific hybrids. Genome 51, 749–756.

Kuhlman, L.C., Burson, B.L., Stelly, D.M., Klein, P.E., Klein, R.R., Price, H.J., et al., 2010. Early-generation germplasm introgression from *Sorghum macrospermum* into sorghum (*S. bicolor*). Genome 53, 419–429.

Li, R., Han, Y., Lv, P., Du, R., Liu, G., 2014. Molecular mapping of the brace root traits in sorghum (*Sorghum bicolor* L. Moench). Breed. Sci. 64, 193–198.

Liu, H., Zeng, F.Y., Liu, Q., 2014. Geographical distribution of *Sorghum* Moench (Poaceae). J. Trop. Subtrop. Bot. 22, 1–11.

Mace, E.S., Tai, S., Gilding, E.K., Li, Y., Prentis, P.J., Bian, L., et al., 2013. Whole-genome sequencing reveals untapped genetic potential in Africa's indigenous cereal crop sorghum. Nat. Commun. 4, 2320.

Madhusudhana, R., 2014. Genetic mapping in sorghum. In: Wang, Y.H., Upadhyaya, H.D., Kole, C. (Eds.), Genetics, Genomics and Breeding of Sorghum. CRC Press, New York, pp. 141–168.

Maredia, M.K., Byerlee, D., Pee, P., 2000. Impacts of food crop improvement research: evidence from sub-Saharan Africa. Food Policy 25, 531–559.

Marshall, D.R., 1989. Limitations to the use of germplasm collections. In: Brown, A.H.D., Marshall, D.R., Frankel, O.H., Williams, J.T. (Eds.), The Use of Plant Genetic Resources. Cambridge University Press, New York, pp. 105–120.

McNeely, J.A., 2005. Mainstreaming agrobiodiversity. In: Petersen, C., Huntley, B. (Eds.), Mainstreaming Biodiversity in Production Landscapes. Global Environmental Facility, Washington, DC, pp. 36–50.

Mekbib, F., 2008. Genetic erosion of sorghum (*Sorghum bicolor* (L.) Moench) in the centre of diversity, Ethiopia. Genet. Resour. Crop Evol. 55, 351–364.

Mohamed, A., Ali, R., Elhassan, O., Suliman, E., Mugoya, C., Masiga, C.W., et al., 2014. First products of DNA marker-assisted selection in sorghum released for cultivation by farmers in sub-Saharan Africa. Afr. J. Plant Sci. Mol. Breed. 3, 3.

Mojtahedi, H., Santo, G.S., Ingham, R.E., 1993. Suppression of *Meloidogyne chitwoodi* with Sudangrass cultivars as green manure. J. Nematol. 25, 303–311.

Morris, G.P., Ramu, P., Deshpande, S.P., Hash, C.T., Shah, T., Upadhyaya, H.D., et al., 2013. Population genomic and genome-wide association studies of agroclimatic traits in sorghum. Proc. Natl Acad. Sci. USA 110, 453–458.

Murray, S.C., Rooney, W.L., Hamblin, M.T., Mitchell, S.E., Kresovich, S., 2009. Sweet sorghum genetic diversity and association mapping for Brix and height. Plant Genome 2 (1), 48–62.

Nass, L.L., Paterniani, E., 2000. Pre-breeding: a link between genetic resources and maize breeding. Agric. Sci. 57, 581–587.

Ngugi, K., Kimani, W., Kiambi, D., Mutitu, E.W., 2013. Improving drought tolerance in *Sorghum bicolor* L. Moench: marker-assisted transfer of the stay-green quantitative trait loci (QTL) rom a characterized donor source into a local farmer variety. Int. J. Sci. Res. Knowledge 1, 154–162.

Nelson, J.C., Wang, S., Wu, Y., Li, X., Antony, G., White, F.F., et al., 2011. Single-nucleotide polymorphism discovery by high-throughput sequencing in sorghum. BMC Genomics 12, 352.

Nwanze, K.F., Seetharama, N., Sharma, H.C., Stenhouse, J.W., 1995. Biotechnology in pest management: improving resistance in sorghum to insect pests. Afr. Crop Sci. J. 3, 209–215.

Perumal, R., Krishnaramanujam, R., Menz, M.A., Katilé, S., Dahlberg, J., Magill, C.W., et al., 2007. Genetic diversity among sorghum races and working groups based on AFLPs and SSRs. Crop Sci. 47, 1375–1383.

Pedersen, J.F., Toy, J.J., Johnson, B., 1998. Natural outcrossing of sorghum and Sudangrass in the Central Great Plains. Crop Sci. 38, 937–939.

Piper, J.K., Kulakow, P.A., 1994. Seed yield and biomass allocation in *Sorghum bicolor* and F_1 and backcross generations of *S. bicolor* × *S. halepense* hybrids. Can. J. Bot. 72, 468–474.

Prom, L.K., Erpelding, J.E., Montes-Garcia, N., 2007. Chinese sorghum germplasm evaluated for resistance to downy mildew and anthracnose. Commun. Biometry Crop Sci. 2, 26–31.

Prom, L.K., Erpelding, J.E., Montes-Garcia, N., 2008. Evaluation of sorghum germplasm from China against *Claviceps africana*, causal agent of sorghum [*Sorghum bicolor* (L.) Moench] ergot. Plant Health Prog., doi:10.1094/PHP-2008-0519-01-RS.

Prom, L.K., Montes-Garcia, N., Erpelding, J.E., Perumal, R., Medina-Ocegueda, M., 2011. Response of sorghum accessions from Chad and Uganda to natural infection by the downy mildew pathogen, *Peronosclerospora sorghi* in Mexico and the USA. J. Plant Dis. Protect. 117, 2–8.

Punnuri, S., Huang, Y., Steets, J., Wu, Y., 2013. Developing new markers and QTL mapping for greenbug resistance in sorghum [*Sorghum bicolor* (L.) Moench]. Euphytica 191, 191–203.

Reddy, V.G., Upadhyaya, H.D., Gowda, C.L.L., 2006. Current status of sorghum genetic resources at ICRISAT: their sharing and impacts. Int. Sorghum Millets Newsl. 47, 9–13.

Reddy, R.N., Madhusudhana, R., Mohan, S.M., Chakravarthi, D.V.N., Mehtre, S.P., Seetharama, N., et al., 2013. Mapping QTL for grain yield and other agronomic traits in post-rainy sorghum [*Sorghum bicolor* (L.) Moench]. Theor. Appl. Genet. 126, 1921–1939.

Rhoades, R.E., 1991. The world's food supply at risk. Natl. Geogr. 179 (4), 74–103.

Rich, P.J., Grenier, C., Ejeta, G., 2004. *Striga* resistance in the wild relatives of sorghum. Crop Sci. 44, 2221–2229.

Ritter, K.B., McIntyre, C.L., Godwin, I.D., Jordan, D.R., Chapman, S.C., 2007. An assessment of the genetic relationship between sweet and grain sorghums, within *Sorghum bicolor* ssp. *bicolor* (L.) Moench, using AFLP markers. Euphytica 157, 161–176.

Sahoo, L., Schmidt, J.J., Pedersen, J.F., Lee, D.J., Lindquist, J.L., 2010. Growth and fitness components of wild× cultivated *Sorghum bicolor* (Poaceae) hybrids in Nebraska. Am. J. Bot. 97, 1610–1617.

Sanchez, A.C., Subudhi, P.K., Rosenow, D.T., Nguyen, H.T., 2002. Mapping QTLs associated with drought resistance in sorghum (*Sorghum bicolor* L. Moench). Plant Mol. Biol. 48, 713–726.

Sangduen, N., Hanna, W.W., 1984. Chromosome and fertility studies on reciprocal crosses between two species of autotetraploid sorghum *Sorghum bicolor* (L.) Moench and *S. halepense* (L.) Pers. J. Hered. 75, 293–296.

Satish, K., Gutema, Z., Grenier, C., Rich, P.J., Ejeta, G., 2012a. Molecular tagging and validation of microsatellite markers linked to the low germination stimulant gene (*lgs*) for *Striga* resistance in sorghum [*Sorghum bicolor* (L.) Moench]. Theor. Appl. Genet. 124, 989–1003.

Satish, K., Madhusudhana, R., Padmaja, P.G., Seetharama, N., Patil, J.V., 2012b. Development, genetic mapping of candidate gene-based markers and their significant association with the shoot fly resistance quantitative trait loci in sorghum [*Sorghum bicolor* (L.) Moench]. Mol. Breed. 30, 1573–1591.

Sharma, H.C., Franzmann, B.A., 2001. Host-plant preference and oviposition responses of the sorghum midge, *Stenodiplosis sorghicola* (Coquillett) (Dipt., Cecidomyiidae) towards wild relatives of sorghum. J. Appl. Entomol. 125, 109–114.

Sharma, H.C., Reddy, B.V.S., Dhillon, M.K., Venkateswaran, K., Singh, B.U., Pampapathy, G., et al., 2005. Host plant resistance to insects in sorghum: present status and need for future research [online]. J. Semi-Arid Trop. Agric. Res. 1, 1–8.

Sharma, R., Upadhyaya, H.D., Manjunatha, S.V., Rao, V.P., Thakur, R.P., 2012. Resistance to foliar diseases in a mini-core collection of sorghum germplasm. Plant Dis. 96, 1629–1633.

Sharma, R., Rao, V.P., Upadhyaya, H.D., Reddy, V.G., Thakur, R.P., 2010. Resistance to grain mold and downy mildew in a mini-core collection of sorghum germplasm. Plant Dis. 94, 439–444.

Shehzad, T., Iwata, H., Okuno, K., 2009. Genome-wide association mapping of quantitative traits in sorghum (*Sorghum bicolor* (L.) Moench) by using multiple models. Breed. Sci. 59, 217–227.

Smith, C.W., Frederiksen, R.A., 2000. History of cultivar development in the United States: from "Memoirs of A. B. Maunder-Sorghum Breeder". In: Smith, C.W., Frederiksen, R.A. (Eds.), Sorghum-Origin, History, Technology, and Production. John Wily & Sons, New York, pp. 191–224.

Subudhi, P.K., Rosenow, D.T., Nguyen, H.T., 2000. Quantitative trait loci for the stay green trait in sorghum (*Sorghum bicolor* L. Moench): consistency across genetic backgrounds and environments. Theor. Appl. Genet. 101, 733–741.

Sun, Y., Skinner, D.Z., Liang, G.H., Hulbert, S.H., 1994. Phylogenetic analysis of *Sorghum* and related taxa using internal transcribed spacers of nuclear ribosomal DNA. Theor. Appl. Genet. 89, 26–32.

Tao, Y., Manners, J.M., Ludlow, M.M., Henzell, R.G., 1993. DNA polymorphisms in grain sorghum (*Sorghum bicolor* (L.) Moench). Theor. Appl. Genet. 86, 679–688.

Teshome, A., Patterson, D., Asfew, Z., Torrance, J.K., Arnason, J.T., 2007. Changes of *Sorghum bicolor* landrace diversity and farmers' selection criteria over space and time, Ethiopia. Genet. Resour. Crop Evol. 54, 1219–1233.

Upadhyaya, H.D., Ortiz, R., 2001. Mini core subset for capturing diversity and promoting utilization of chickpea genetic resources in crop improvement. Theor. Appl. Genet. 102, 1292–1298.

Upadhyaya, H.D., Pundir, R.P.S., Dwivedi, S.L., Gowda, C.L.L., Reddy, V.G., Singh, S., 2009. Developing a mini core collection of sorghum for diversified utilization of germplasm. Crop Sci. 49, 1769–1780.

Upadhyaya, H.D., Wang, Y.H., Sharma, S., Singh, S., Hasenstein, K.H., 2012a. SSR markers linked to kernel weight and tiller number in sorghum identified by association mapping. Euphytica 187, 401–410.

Upadhyaya, H.D., Wang, Y.H., Sharma, S., Singh, S., 2012b. Association mapping of height and maturity across five environments using the sorghum mini core collection. Genome 55, 471–479.

Upadhyaya, H.D., Wang, Y.H., Gowda, C.L.L., Sharma, S., 2013a. Association mapping of maturity and plant height using SNP markers with the sorghum mini core collection. Theor. Appl. Genet. 126, 2003–2015.

Upadhyaya, H.D., Wang, Y.H., Sharma, R., Sharma, S., 2013b. Identification of genetic markers linked to anthracnose resistance in sorghum using association analysis. Theor. Appl. Genet. 126, 1649–1657.

Upadhyaya, H.D., Wang, Y.H., Sharma, R., Sharma, S., 2013c. SNP markers linked to leaf rust and grain mold resistance in sorghum. Mol. Breed. 32, 451–462.

Upadhyaya, H.D., Dwivedi, S.L., Ramu, P., Singh, S.K., Singh, S., 2014a. Genetic variability and effect of postflowering drought on stalk sugar content in sorghum mini core collection. Crop Sci. 54, 2120–2130.

Upadhyaya, H.D., Sharma, S., Dwivedi, S.L., Singh, S.K., 2014b. Sorghum genetic resources: conservation and diversity assessment for enhanced utilization in sorghum improvement. In: Wang, Y.H., Upadhyaya, H.D., Kole, C. (Eds.), Genetics, Genomics and Breeding of Sorghum. CRC Press, New York, pp. 28–55.

Vadez, V., Deshpande, S., Kholova, J., Ramu, P., Hash, C.T., 2013. Molecular breeding for stay-green: progress and challenges in sorghum. In: Varshney, R.K., Tuberosa, R. (Eds.), Translational Genomics for Crop Breeding, vol. 2. Improvement for Abiotic Stress, Quality and Yield Improvement. Wiley-Blackwell, Oxford, UK, pp. 125–141.

van de Wouw, M., Kik, C., van Hintum, T., van Treuren, R., Visser, B., 2010. Genetic erosion in crops: concept, research results and challenges. Plant Genet. Resour. 8, 1–15.

Viaene, N.M., Abawi, G.S., 1998. Management of *Meloidogyne hapla* on lettuce in organic soil with Sudangrass as a cover crop. Plant Dis. 82, 945–952.

Wang, Y.H., Bible, P., Loganantharaj, R., Upadhyaya, H.D., 2012. Identification of SSR markers associated with height using pool-based genome-wide association mapping in sorghum. Mol. Breed. 30, 281–292.

Wang, F., Zhao, S., Han, Y., Shao, Y., Dong, Z., Gao, Y., et al., 2013a. Efficient and fine mapping of *RMES1* conferring resistance to sorghum aphid *Melanaphis sacchari*. Mol. Breed. 31, 777–784.

Wang, Y.H., Upadhyaya, H.D., Burrell, A.M., Sahraeian, S.M., Klein, R.R., Klein, P.E., 2013b. Genetic structure and linkage disequilibrium in a diverse, representative collection of the C4 model plant, *Sorghum bicolor*. G3 Genes Genomes Genet. 3, 783–793.

Washburn, J.D., Murray, S.C., Burson, B.L., Klein, R.R., Jessup, R.W., 2013. Targeted mapping of quantitative trait locus regions for rhizomatousness in chromosome SBI-01 and analysis of overwintering in a *Sorghum bicolor* × *S. propinquum* population. Mol. Breed. 31, 153–162.

Wendorf, F., Close, A.E., Schild, R., Wasylikowa, K., Housley, R.A., Harlan, J.R., et al., 1992. Saharan exploitation of plants 8,000 years bp. Nature 359, 721–724.
Widmer, T.L., Abawi, G.S., 2000. Mechanism of suppression of *Meloidogyne hapla* and its damage by a green manure of Sudan grass. Plant Dis. 84, 562–568.
Yundaeng, C., Somta, P., Tangphatsornruang, S., Wongpornchai, S., Srinives, P., 2013. Gene discovery and functional marker development for fragrance in sorghum (*Sorghum bicolor* (L.) Moench). Theor. Appl. Genet. 126, 2897–2906.
Zheng, L.Y., Guo, X.S., He, B., Sun, L.J., Peng, Y., Dong, S.S., et al., 2011. Genome-wide patterns of genetic variation in sweet and grain sorghum (*Sorghum bicolor*). Genome Biol. 12, R114.
Zou, G., Zhai, G., Feng, Q., Yan, S., Wang, A., Zhao, Q., et al., 2012. Identification of QTLs for eight agronomically important traits using an ultra-high-density map based on SNPs generated from high-throughput sequencing in sorghum under contrasting photoperiods. J. Exp. Bot. 63, 5451–5462.

Pearl millet

6

Santosh K. Pattanashetti, Hari D. Upadhyaya, Sangam Lal Dwivedi, Mani Vetriventhan, Kothapally Narsimha Reddy
International Crops Research Institute for the Semi-Arid Tropics (ICRISAT), Genebank, Patancheru, Telangana, India

6.1 Introduction

Pearl millet [*Pennisetum glaucum* (L.) R. Br., Syn. *Cenchrus americanus* (L.) Morrone] is the sixth most important cereal worldwide and the main food source in the semiarid regions of Asia and Africa. Globally, pearl millet is cultivated on 30 million ha with majority of the crop in Africa (~18 million ha) and Asia (>10 million ha) (Yadav and Rai, 2013). It is mainly grown for food and forage in India and Africa, while as a forage crop in the Americas. India is the single largest producer of pearl millet (7.95 million ha, 8.90 Mt) with seven major growing states (AICPMIP, 2014). Pearl millet is a C_4 plant with very high photosynthetic efficiency, dry matter production capacity, short duration, and high degree of tolerance to heat and drought. It is also adapted on saline, acidic, and aluminum toxic soils (Yadav and Rai, 2013).

Pearl millet grain is a staple food for around 90 million people in the Sahelian region of Africa and northwestern India, with majority of the produce being used as food (www.icrisat.org). It is consumed primarily as a thick porridge (toh), but it is also milled into flour to prepare unfermented breads and cakes (roti), steam-cooked dishes (couscous), fermented foods (kisra and gallettes), nonalcoholic beverages, and snacks. Roasted young earheads are a popular food for children. In the Sahelian countries like Senegal, Mali, Niger, and Burkina Faso, pearl millet is consumed in preference to sorghum. In northern Nigeria, pearl millet is used in making a popular fried cake known as *masa* (Girgi and O'Kennedy, 2007).

Pearl millet is a "high-energy" cereal that contains carbohydrates, protein, and fat, rich in vitamins B and A, high in calcium, iron, and zinc, and also contains potassium, phosphorus, magnesium, zinc, copper, and manganese. Feeding trials conducted in India have shown that pearl millet is nutritionally superior to maize and rice (NRC, 1996; DeVries and Toenniessen, 2001). Pearl millet grains are gluten-free and the biological value of its protein is superior over wheat. Grain has no tannin, contains oil (5–7%), and has higher protein and energy levels than maize or sorghum and more balanced amino acid profile compared to sorghum and maize (Rai et al., 2008). In India, although grains are mainly used for human consumption, alternative uses have increased (55%) mainly as animal feed for dairy (in rural parts of western Rajasthan) and to some extent in poultry, alcohol industry, starch industry, processed food industry, and export demand (Basavaraj et al., 2010). Both green fodder and dry stover are used for cattle.

Genetic and Genomic Resources for Grain Cereals Improvement. http://dx.doi.org/10.1016/B978-0-12-802000-5.00006-X

Stover has excellent forage quality with lower hydrocyanin content; green fodder is rich in protein, calcium, phosphorus, and other minerals with safer limits of oxalic acid.

6.2 Origin, distribution, and diversity

P. glaucum subsp. *monodii*, found in the Sahelian region of Africa, is the progenitor of cultivated pearl millet (Harlan, 1975; Brunken, 1977). Domestication of pearl millet has been suggested to have occurred through single or multiple events (Harlan, 1975; Portères, 1976) and the earliest archeologic evidence for pearl millet domestication is from northern Ghana, ~3000 years BP (D'Andrea and Casey, 2002). Domestication followed by migration of aboriginal populations in sub-Saharan regions and secondary diversification in the eastern Sahel has been suggested (Tostain et al., 1987; Tostain, 1992). Early-flowering types may have domesticated around Lake Chad, which is considered as a secondary center of diversification. Furthermore, the late flowering, new early-flowering varieties, pearl millet varieties from East and South Africa and India could have originated from this region (Robert et al., 2011). Monophyletic origin of pearl millet has been suggested based on similarity among domestic and wild forms for isozymes and simple sequence repeats (SSRs) (Tostain, 1993; Mariac et al., 2006a, 2006b; Oumar et al., 2008; Kapila et al., 2008).

Molecular profiling revealed that cultivated pearl millet contains 81% of the alleles and 83% of the genetic diversity of the wild pearl millet (Oumar et al., 2008), suggesting that domestication has decreased genetic diversity. The gene flow between domestic and wild forms could later have contributed to increase in genetic diversity of the domestic genepool (Robert et al., 2011). A large number of spontaneously occurring weedy and intermediate form plants, which mimic the cultivated plants in their vegetative and floral morphologies called "shibras" (=*Pennisetum stenostachyum*), are observed throughout West African Sahel. The *stenostachyum* is considered as a product of postdomestication hybridization between pearl millet and its wild ancestor (Brunken et al., 1977). Its widespread occurrence outside the natural area of its wild progenitor suggests its persistence over long times.

Quantitative trait loci (QTL) analyses using F_2 populations of crosses between pearl millet and its wild progenitor (*P. glaucum* subsp. *monodii*) identified two genomic regions on linkage groups (LG) 6 and 7, which control most of the key morphologic differences between wild and cultivated types (Poncet et al., 1998, 2000, 2002). These genomic regions have a prime role in the developmental control of spikelet structure and domestication process of pearl millet, which may also correspond with QTLs involved in domestication of other cereals like maize and rice (Poncet et al., 2000, 2002). Furthermore, Miura and Terauchi (2005) proposed the occurrence of a putative supergene or gene complex during domestication that differentiated weedy and cultivated pearl millet.

6.2.1 Taxonomy and diversity

The genus *Pennisetum* (bristle grass) is the largest and important genera belonging to the subtribe Panicinae, tribe Paniceae, subfamily Panicoideae of the family Poaceae. The genus is mainly characterized by its inflorescence, a false spike with spikelet on contracted axes, or spikelets fascicled in false spikes, always surrounded by involucres, which are crowded with slender, basally free, glabrous to plumose bristles; the spikelets are sessile or pedicillate, falling with the involucres, only persistent in the cultivated species (Watson and Dallwitz, 1992). The *Pennisetum* genus includes over 80–140 species (Brunken, 1977; Clayton and Renvoize, 1986), which differ in their duration (annual or perennial), reproduction (sexual or asexual/apomictic), somatic chromosome number ($2n$ = 10–78), basic chromosome number (x = 5, 7, 8, or 9; Jauhar, 1981a, 1981b), chromosome size (92–395 Mbp), genome size (1Cx = 0.75–2.49 pg) (Martel et al., 1997), and ploidy levels (diploid to octoploid) (Table 6.1). Aneuploids are fairly common (Schmelzer, 1997) and frequent occurrence of B chromosomes was reported in many species (Vari et al., 1999).

Diversity in basic chromosome number, different chromosome sizes between species with larger size for those with low basic chromosome number (Jauhar, 1981a), reflect "chromosome repatterning" within the genus during evolution. Different hypotheses were proposed for ancestral basic chromosome number in *Pennisetum*. Jauhar (1981a) and Rao et al. (1989) based on chromosome pairing observations in *Pennisetum,* suggested an increase in basic chromosome number from $x = 5$. But, frequent occurrence of $x = 9$ in *Pennisetum* and closely related genera like *Cenchrus*, *Panicum*, and *Setaria* suggests a decrease in "x" from ancestral number ($x = 9$) (Robert et al., 2011). In the genus *Pennisetum*, nearly 3/4th of the species are polyploids (Jauhar, 1981a). Cultivated species (*Pennisetum glaucum*) is essentially a diploid, though both spontaneous and induced polyploids have been reported. Based on meiotic chromosome pairing studies in section *Brevivulvula*, species *Pennisetum pedicellatum* and *Pennisetum polystachyon* were reported to be autoploid or autoalloploid or segmental alloploid. Similarly, many species such as *Pennisetum subangustatum*, *Pennisetum atrichum*, *Pennisetum setosum*, and *Pennisetum squamulatum*, were suggested to be autoalloploids (for details see Robert et al., 2011). However, *P. squamulatum* is suggested to be an octoploid (Kaushal et al., 2008).

6.2.1.1 Sections

Based on morphological differences, genus *Pennisetum* is divided into five sections (Stapf and Hubbard, 1934), that is, *Penicillaria* (tropical Africa and India), *Brevivulvula* (pan-tropical), *Gymnothrix* (pan-tropical), *Heterostachya* (northeast Africa), and *Eupennisetum* (tropical and subtropical Africa and India), each having several species (Table 6.1). The differences between sections were often not very strong as more than one morphological section consisted of taxa with one common basic chromosome number and also one section included taxa from different basic chromosome number groups. Section *Penicillaria* represents species with $x = 7$ and include cultivated pearl millet (*P. glaucum*) and napier/elephant grass (*P. purpureum*), while Brunken (1977) included them under a different section *Pennisetum*. Other sections comprise 22 species in *Gymnothrix*, 5 in *Pennisetum,* and 3 each in *Heterostachya* and *Brevivulvula*.

6.2.1.2 Genepool

The classification of genepool for pearl millet and its wild relatives was proposed based on the genetic relationship between species (Harlan and de Wet, 1971). Later, it was

Table 6.1 Characteristics of *Pennisetum* species

Genepool/section/species/subspecies/common name	Mode of reproduction, plant habit*	Geographic range	Chromosome number**			2C value (pg)
			n	$2n$	x	
I. Primary genepool (GP1)						
Section *Penicillaria/Pennisetum*[†]			7	14	7	
P. glaucum (L.) R. Br.						
1. *P. glaucum* ssp. *glaucum* (Pearl millet)	Sex, A	Wide	7	14(2x)	7	4.71
2. *P. glaucum* ssp. *monodii*						
Two ecotypes:						
(a) *P. violaceum*	Sex, A	Africa	7	14(2x)	7	4.52
(b) *P. mollissimum*	A	West Africa	–	14(2x)	–	4.51
II. Secondary genepool (GP2)						
Section *Penicillaria/Pennisetum*[†]			7–28	21–28	7	
P. purpureum (Elephant grass)	Sex/Apo, Inb, P, SP	Wide	7, 14, 28	28(4x), 27, 21	7	4.59
Section *Heterostachya*				54–56	7, 9	
P. squamulatum	Apo, Inb, P, C, RSE	Africa	–	54(6x), 56(8x)	7, 9	9.56
III. Tertiary genepool (GP3)						
Section *Brevivulvula*			9–36	18–78	9	
Pennisetum hordeoides	Apo, A	Africa, Asia-tropical	9	36, 54	9	–
P. pedicellatum (Hairy fountaingrass)	Apo, Inb, A	Africa, Asia-tropical, Australasia, South America	14, 18, 27	54(6x), 36(4x), 24, 30, 32, 35, 42, 45, 48, 53	9	5.61
1. ssp. *pedicellatum*						
2. ssp. *unispiculum*						
P. polystachion (Missiongrass)	A/P, C	Wide	27	54(6x)	9	5.66
1. ssp. *polystachion*	Apo	–	18, 36	18, 24, 32, 36, 45, 48, 52, 53, 54, 56, 63, 78	9	–
2. ssp. *atrichum*	P, A	South, West and East Africa	–	36(4x)	9	4.25

Section *Eupennisetum/Pennisetum*[†]			9–27	18–68	9, 17	
Pennisetum clandestinum (Kikiyugrass)	Apo/Sex, P, MF, RE, SP	Africa, Asia-temperate, Asiatropical, Australasia, Pacific, America, Antarctic	18, 27	36	9	2.30
P. flaccidum (Himalayan fountaingrass)	Apo/Sex, P, C, RS	Asia-temperate, Asia-tropical	9, 18, 27	18, 36	9	–
Pennisetum foermeranum	P, RE	Southern Africa	–	–	–	–
P. setaceum (Tender fountaingrass)	Apo, Inb, P, C	Europe, Africa, Asia-temperate, Australasia, Pacific	18	27(3*x*), 54(6*x*), 68	9, 17	2.78, 5.28
Pennisetum sieberianum	A	Africa, Asia-temperate	–	–	–	–
Pennisetum villosum (Feathertop)	Apo, Inb, P, MF, RE	Cosmopolitan	27	36(4*x*), 18, 27, 45, 54	9	3.47
Section *Gymnothri*			5–18	10–72	5, 8, 9	
Pennisetum alopecuroides (Chinese fountaingrass)	Sex, P, C	Asia-temperate, Asia-tropical, Australasia	18	18(2*x*), 22	9	1.90
Pennisetum basedowii	A, C, CL	Australasia	–	54	9	–
Pennisetum chilense	P, C, RE	South America	–	–	–	–
Pennisetum frutescens	Apo, P, CE, RE	South America	–	63	9	–
Pennisetum hohenackeri	P, C, CD	Africa, Asia-tropical	9	18 (2*x*)	–	1.69
Pennisetum latifolium (Uruguay fountaingrass)	Apo, P, RS	Australasia, South America	–	36	9	–

(Continued)

Table 6.1 Characteristics of *Pennisetum* species *(cont.)*

Genepool/section/species/subspecies/common name	Mode of reproduction, plant habit*	Geographic range	Chromosome number**			2C value (pg)
			n	2*n*	*x*	
Pennisetum macrourum (Waterside Reed)	Apo, P, RE	Africa, Asia-temperate, Australasia	–	36, 54	9	–
Pennisetum massaicum	Apo/Sex, P, RS, K	Africa	–	16, 32	8	–
Pennisetum mezianum	Apo, Inb, P, RS, K	Africa, Asia-tropical	–	16, 32	8	3.01 (4*x*)
Pennisetum montanum	P	South America	16	32	8	–
Pennisetum nervosum (Bentspike fountaingrass)	P, RSE	America	–	36, 72	9	–
P. ramosum (unresolved sp.)	Sex/Apo, Inb, A, C	Africa	5	10(2*x*), 20	5	4.04
Pennisetum sphacelatum	P, C	Africa	–	18	9	–
Pennisetum thunbergii	P, C, RE	Africa, Asia-temperate, Asia-tropical, Australasia	–	18	9	–
Pennisetum trachyphyllum	P	Africa	–	–	–	–
Pennisetum tristachyum	P, C	America	–	–	–	–
Pennisetum unisetum	P, C	Africa, Asia-temperate	–	18	9	–
Section *Heterostachya*			9–18	14–56	7, 9	
P. orientale (White fountaingrass)	Sex/Apo, Inb, P, C, RS	Africa, Asia-temperate, Asia-tropical, Pacific, South America	9, 18	36(4*x*), 45, 27, 56	9	3.77
P. schweinfurthii (=*Pennisetum tetrastachyum*)	Sex, A	Africa	–	14(2*x*)	7	4.97

Section unknown						
Pennisetum advena (Foreign fountaingrass)	P, C	North America	–	–	–	–
Pennisetum annuum	A	South America	–	–	–	–
Pennisetum articulare	P, C	Pacific	–	–	–	–
Pennisetum bambusiforme	P, C	America	–	36	–	–
Pennisetum beckeroides	P, RS	Africa	–	–	–	–
Pennisetum caffrum	P, C, CD	Africa Indian ocean	–	–	–	–
Pennisetum centrasiaticum	P	Asia-temperate	–	36	–	–
P. ciliare (syn. *Cenchrus ciliaris*) (Buffelgrass)	Apo/Sex, P, C, RE	Africa, West Asia, India	9, 18	32, 34, 36, 40, 45, 52, 54	9	–
Pennisetum complanatum	P, RE	Pacific, North America	–	–	–	–
Pennisetum crinitum	P, C	North America, Mexico	–	–	–	–
Pennisetum distachyum	P, C	North America, South America	–	36	–	–
Pennisetum divisum	P, RS, K	Africa, Asia-temperate, Asia-tropical	18, 27	36	–	–
Pennisetum domingense	P	South America	–	–	–	–
Pennisetum durum	P	North America, Mexico	–	–	–	–
Pennisetum glaucifolium	P, CE, RS	Africa, Asia-temperate	–	–	–	–
Pennisetum gracilescens	P, C, RS	Africa	–	–	–	–
Pennisetum henryanum	P, C	Pacific	–	–	–	–
Pennisetum humile	P, C, RE	Africa	–	–	–	–
Pennisetum intectum	P	South America	–	–	–	–

(Continued)

Table 6.1 Characteristics of *Pennisetum* species *(cont.)*

Genepool/section/species/subspecies/common name	Mode of reproduction, plant habit*	Geographic range	Chromosome number**			2C value (pg)
			n	2*n*	*x*	
Pennisetum lanatum	P, C, RE	Asia-temperate, Asia-tropical	18	–	9	–
Pennisetum laxius	A	Africa	–	–	–	–
Pennisetum ledermannii	P	Africa	–	–	–	–
Pennisetum longissimum	P	Asia-temperate	–	54	–	–
Pennisetum longistylum	P, MF, CE, RE	Africa	–	45	–	–
Pennisetum macrostachyum (Pacific fountaingrass)	P, C	Asia-tropical, Pacific	–	54	–	–
Pennisetum mildbraedii	P, C, RE	Africa	–	–	–	–
Pennisetum monostigma	P, C	Africa	–	18	–	–
Pennisetum nodiflorum	P	Africa	–	–	–	–
Pennisetum nubicum	A	Africa, Asia-temperate	–	–	–	–
Pennisetum occidentale	P	South America	–	–	–	–
Pennisetum pauperum	P, C	South America	–	–	–	–
Pennisetum peruvianum	P	South America	–	–	–	–
Pennisetum petiolare	A	Africa	–	–	–	–
Pennisetum pirottae	P	Africa	–	–	–	–
Pennisetum procerum	P	Africa	–	–	–	–
Pennisetum prolificum	P, C	North America	–	–	–	–
Pennisetum pseudotriticoides	P, C	Africa	–	–	–	–
Pennisetum pumilum	P, C	Africa	–	–	–	–
Pennisetum qianningense	P, C	Asia-temperate	–	36	–	–
Pennisetum rigidum	P, RE	South America	–	–	–	–
Pennisetum riparium	P, MF, RE	Africa	–	–	–	–
Pennisetum rupestre	P, RE	South America	–	–	–	–
Pennisetum sagittatum	P	South America	–	–	–	–

Pennisetum setigerum (syn. *Cenchrus setiger*) (Birdwoodgrass)	P	North-east Africa, India	–	34 (2*x*)	–	–
Pennisetum shaanxiense	P	Asia-temperate	–	–	–	–
Pennisetum sichuanense	P, RS	Asia-temperate	–	–	–	–
Pennisetum stramineum	P, C	Africa, Asia-temperate	–	–	–	–
Pennisetum tempisquense	P, C	South America	–	–	–	–
Pennisetum thulinii	P, C, CD	Africa	–	–	–	–
Pennisetum trisetum	P	Africa	36	–	–	–
Pennisetum uliginosum	P, C, CL, RE	Africa	–	–	–	–
Pennisetum weberbaueri	P	South America	–	–	–	–
Pennisetum yemense	P, C, CD, RSE, RS/RE	Africa, Asia-temperate	–	–	–	–

* Mode of reproduction: Sex, sexual; Apo, apomictic; Inb, inbreeder. Plant habit: A, annual; P, perennial; C, caespitose; CD, clumped densely; CE, cataphylls evident; CL, clumped loosely; K, knotty; MF, mat forming; RE, rhizomes elongated; RS, rhizomes short; RSE, root stock evident; SP, stolons present.
** Range of chromosome numbers given for sections.
† Brunken (1977) grouped them under section *Pennisetum*.
Sources: Robert et al. (2011); Chemisquy et al. (2010); Clayton et al. (2006); www.theplantlist.org.

extended to the concept of "complex of species" (Pernès, 1984) based on crossability of wild species with domesticated form and also the amount of gene flow occurring between all the members of the genepool. This classification includes primary, secondary, and tertiary genepools. Primary genepool (GP1) includes two subspecies: (1) cultivated pearl millet (*Pennisetum glaucum* ssp. *glaucum*, $2n = 2x = 14$, AA) with distinctly stalked and persistent involucres at maturity, and (2) wild progenitor (*P. glaucum* subsp. *monodii*: two ecotypes, *Pennisetum violaceum* and *Pennisetum mollissimum*) with subsessile, deciduous involucres at maturity (Robert et al., 2011). The intermediate weedy forms, "shibras" (=*P. stenostachyum*) are found throughout West African Sahel. Members of GP1 are not reproductively isolated and can cross-hybridize in sympatric condition forming fertile hybrids with normal chromosome pairing. Despite this gene flow, mechanisms like linkage, gametophytic competition, and phenology contribute to the maintenance of the genetic structure of the species (Robert et al., 2011).

The secondary genepool (GP2) includes napier/elephant grass (*Pennisetum purpureum* Schum., $2n = 4x = 28$, A′A′BB), and recently included apomictic, octoploid species *P. squamulatum* ($2n = 8x = 56$), whose genome size is similar to *P. purpureum*. Cytogenetic analysis like crossability, DNA content, genetic relatedness, and cytology of advanced hybrids supported the placement of *P. squamulatum* in GP2 (Kaushal et al., 2008). Napier grass is a rhizomatous perennial with desirable traits like resistance to most pests, vigorous growth, and outstanding forage potential; most of these characteristics appear to be on B genome (Hanna, 1987). *P. purpureum* and *P. squamulatum* can be easily crossed with cultivated pearl millet but their hybrids are highly sterile. The tertiary genepool (GP3) includes the remaining *Pennisetum* spp., which are true biological species compared to GP1 and GP2 species. Several *Pennisetum* species in the GP3 are of economic importance, namely *Pennisetum flaccidum*, *P. mezianum*, and *P. setaceum* (ornamental), *P. squamulatum*, *P. polystachyon*, *P. pedicellatum*, and *Pennisetum orientale* (forage). A strong reproductive barrier occurs between members of GP3 with GP1 and GP2 affecting gene flow and occurrence of hybrids. Gene transfer is possible through radical manipulations involving *in vitro* techniques or by using complex hybrid bridges. GP3 species with $x = 9$ cross more readily with pearl millet than those with $x = 5$ (*P. ramosum*) and $x = 8$ (*P. mezianum*) (Dujardin and Hanna, 1989a).

6.2.1.3 Races

Brunken et al. (1977) classified the world collection of cultivated pearl millet based on seed shapes into four races:

1. *typhoides*: It occurs from Senegal to Ethiopia, and northern to southern Africa. It is the only basic race found outside Africa, and predominantly grown in India. It is widely distributed and morphologically most variable among the four races. Caryopses are obovate in frontal and profile views, and obtuse and terete in cross-section. Grains are occasionally shorter and enclosed by floral bracts. Inflorescences are variable in length and more or less cylindrical, but rarely short inflorescences with elliptic shape are also found (Appa Rao and de Wet, 1999).
2. *nigritarum*: It is found from western Sudan to northern Nigeria, probably native to eastern Sahel and its isolated occurrences in western Sahel due to recent migration events. The caryopsis is angular in cross-section with three and six facets. Grains apex is usually

truncate with purple tinge and mature grain is elongated and protrudes beyond the floral bracts. Inflorescence is candle-shaped.

3. *globosum*: It occurs from central Burkina Faso to western Sudan and is most commonly found in central Nigeria, Niger, Ghana, Togo, and Benin. This large-seeded race has experienced a great degree of migration in recent times. The caryopses are characteristically spherical with each of its dimensions equal, otherwise terete and obtuse, and grain depth always exceeds 2.4 mm. It has candle-shaped inflorescence.
4. *leonis*: It is mainly found in Sierra Leone, infrequently observed in Senegal, southern Mauritania, and the hilly areas of southern India. Caryopsis is oblanceolate and terete with acute apex due to remnants of stylar base. Grains are elongated and nearly oblanceolate from all lateral perspectives. At maturity, 1/3 of the grain protrudes beyond the floral bracts. It has candle-shaped inflorescence.

There are arguments that patterns of racial variation observed in crops are the result of several independent domestications (Portères, 1976). Although independent domestications cannot be ruled out, it appears that migration events, very early in the history of pearl millet, followed by a combination of geographic and ethnographic isolation are responsible for the present-day pattern of variation in seed morphology of the crop (Appa Rao and de Wet, 1999).

6.2.1.4 Genomic relationships

The study of genomic relationships provides information on phylogenetic relations and evolution, and it is also useful in breeding programs seeking gene introgression. GP1 species (*P. glaucum*, *P. violaceum*, and *P. mollissimum*) are reported to have similar amount of DNA per basic chromosome set (Martel et al., 1997). Genomic relationships between *Pennisetum* species were investigated by studying variations in chloroplast DNA (Clegg et al., 1984; Renno et al., 2001), mitochondrial DNA (Chowdhury and Smith, 1988), and repetitive DNA sequences (Ingham et al., 1993). Significant relationships were noticed between *P. glaucum*, *P. purpureum*, and *P. squamulatum* suggesting a common origin. Homeology between A (pearl millet) and A′ (*P. purpureum*) genomes has been established earlier by conventional cytogenetic techniques (Jauhar, 1981a), and B genome was found dominant over A′ genome that masks genetic variability (Hanna, 1987). Recent investigations using genomic *in situ* hybridization revealed homeology between the genomes of pearl millet (A) and *P. purpureum* (A′, B), and also differences in the distribution and proportion of homologous regions (dos Reis et al., 2014). Despite differences in ploidy levels, *P. purpureum* ($4x$) and *P. glaucum* ($2x$) have almost the same 2C values (4.59, 4.71 pg, respectively), but different monoploid genome size ($1Cx = 1.15, 2.35$ pg, respectively). The *P. purpureum* chromosomes (~80 Mbp) are half the size of *P. glaucum* (~165 Mbp) showing important chromosomal changes linked to divergence of these species (Martel et al., 1997).

Phylogenetic relationships in *Pennisetum* suggest an ancestral small chromosome ($x = 9$), but reduced chromosome number (decreasing dysploidy) ($x = 5, 7$, or 8) and increased chromosome size (Martel et al., 2004; Robert et al., 2011) rather than increase in basic chromosome number (ascendant dysploidy) (Jauhar, 1981b;

Jauhar et al., 2006). Molecular phylogenetic studies using ribosomal and chloroplast DNA sequences suggest the genus *Pennisetum* as paraphyletic (a taxonomic group that includes some but not all of the descendants of a common ancestor) (Martel et al., 2004; Donadio et al., 2009) with the species *Cenchrus ciliaris* (=*Pennisetum cenchroides*/*Pennisetum ciliare*) nested within it (Doust et al., 2007). Close genomic relationships were observed between *P. squamulatum* and aposporous apomictic species *Cenchrus ciliaris*, which also showed complete macrosynteny at the apospory-specific genomic region governing apomixes in these species (Goel et al., 2006). Chromosomal and genomic characteristics combined with phylogenetic relationships favor the inclusion of *Cenchrus* species within the genus *Pennisetum*. Hence, it was proposed to reconsider the taxonomic position of *Cenchrus* species and to rename them as *Pennisetum* species as previously known (Robert et al., 2011).

6.3 Erosion of genetic diversity and gene flow

The success of any plant-breeding program depends on the germplasm diversity available in the crop. Conserving the rich diversity of crop varieties and related wild species is essential for providing farmers and plant breeders with diverse sources to improve and adapt crops to meet future challenges. The genetic variability accumulated over centuries is fast eroding, mainly due to the replacement of landraces by improved cultivars, natural catastrophes (droughts, floods, fire hazards, etc.), industrialization, human settlement, overgrazing, and destruction of plant habitats for irrigation projects and dams (Upadhyaya and Gowda, 2009). Changes in the diversity of landraces in centers of diversity of cultivated plants need to be assessed in order to monitor and conserve agrobiodiversity. This notably applies in tropical areas where factors, such as increased population, climate change, and shifts in cropping systems, are hypothesized to cause varietal erosion.

Pearl millet landraces are fast disappearing in most states of India due to large-scale adoption of improved cultivars. For example, a popular landrace from Punjab, "Gullisita" is no longer found in its original form. Recurring droughts in a region enforce the farmers to introduce new cultivars from elsewhere, resulting in loss of local landraces. With the availability of irrigation, pearl millet is being replaced by more remunerative crops like rice, pigeonpea, and cotton in Punjab, India; coconut in Srikakulam district of Andhra Pradesh, India; cassava and cotton in the Central African Republic; wheat, maize, and other commercial crops in Yemen; and early maturing maize in northern Togo, southern Africa, and Nigeria (Appa Rao et al., 1993, 1994; Appa Rao, 1999).

A rapid survey was undertaken (2000–2005) by AICPMIP (pearl millet) in Rajasthan (India) to measure and monitor the genetic erosion of pearl millet landraces through participatory approach under an FAO project (www.globalplanofaction.org), which revealed narrowing of genetic diversity. Diversity in major cereals/millet crops including pearl millet was found decreasing at a local level in India (NBPGR, 2007). Genetic erosion of crop landraces was found increasing in southern Tunisia (Mohamed, 1998). Improved cultivars were found replacing traditional varieties in different geographical regions of Tanzania (Appa Rao et al., 1989a) and Yemen (Muallem, 1987). Based on comparison of diversity in pearl millet varieties collected in 79 villages spanning the entire cereal-growing zone of Niger over a 26 year period (1976–2003), high diversity of pearl millet varieties was confirmed (Bezançon et al., 2009). Genetic erosion was not observed suggesting that farmers' management could preserve the diversity of millet varieties in Niger despite recurrent and severe drought periods and major social changes.

Gene flow within the GP1 species of *Pennisetum* has been assessed by some workers. Weedy plants with intermediate (domesticated × wild) phenotypes occur in most pearl millet fields in West Africa, even in the absence of wild populations. Under experimental conditions, hybrid (domesticated × wild) frequency decreased steadily during germination (42%), at emergence (37%), and thinning (17%) by the farmer, and finally only 11% of mature plants were hybrids (Couturon et al., 2003). This shows that under the combined pressures of natural and human selection, the frequency of hybrids in the field declined drastically up to maturity without completely preventing the introgression of wild pearl millet genes into the cultivated genome. Morphologic and AFLP marker data suggest some introgression from the wild to the weedy population but very low gene flow between the parapatric wild and domesticated populations (Mariac et al., 2006b). Analyses of introgressions between cultivated and wild accessions using microsatellite markers showed modest introgressions. Wild accessions in the central region of Niger showed introgressions of cultivated alleles, while accessions of cultivated pearl millet showed introgressions of wild alleles in the western, central, and eastern parts of Niger (Mariac et al., 2006a).

6.4 Germplasm resources conservation

Germplasm of *Pennisetum* spp. (cultivated and wild) is conserved as 66,682 accessions in 97 genebanks across 65 countries, while germplasm of related genera *Cenchrus* spp. is conserved (3758 accessions) in 50 genebanks across 32 countries (www.genesys-pgr.org). Eleven major genebanks hold almost 90% of the pearl millet germplasm (Table 6.2). The ICRISAT genebank at Patancheru (India) has the largest collection (22,888 accessions) from 52 countries, including 794 wild germplasm belonging to 27 *Pennisetum* species. At Patancheru, majority of these accessions are conserved as base/active collections under cold storage. However, several accessions of wild species, which either do not produce seed or produce very less seed, are maintained alive in a field genebank. In addition, three regional genebanks of ICRISAT (Niamey, Niger; Nairobi, Kenya and Bulawayo, Zimbabwe) together conserve 14,272 accessions of pearl millet. ICRISAT has contributed 20,527 accessions of pearl millet as safety duplication to the Svalbard Global Seed Vault (SGSV), Norway, while nine other genebanks with small collections have deposited 1459 accessions of *Pennisetum* and *Cenchrus* spp. at SGSV.

Table 6.2 Major genebanks conserving cultivated and wild pearl millet germplasm

Institute/genebank	*Pennisetum* spp. Cultivated	Wild	Total
Embrapa Milho e Sorgo, Sete Lagoas, Brazil	7,225	–	7,225
Plant Gene Resources of Canada, Saskatoon Research Centre, Agriculture and Agri-Food Canada, Saskatoon, Canada	3,543	263	3,806
Laboratoire des Ressources Génétiques et Amélioration des Plantes Tropicales, ORSTOM, Montpellier Cedex, France	3,620	798	4,418
International Crop Research Institute for the Semi-Arid Tropics, Patancheru, India	22,094	794	22,888
National Bureau of Plant Genetic Resources, New Delhi & Jodhpur, India	9,144	327	9,471
National Plant Genetic Resources Centre, National Botanical Research Institute, Windhoek, Namibia	1,419	2	1,421
Institut national de la recherche agronomique du Niger, Niamey, Niger	2,052	–	2,052
International Crops Research Institute for Semi-Arid Tropics, Niamey, Niger	2,817	–	2,817
Plant Genetic Resources Program, Islamabad, Pakistan	1,377	–	1,377
Serere Agriculture and Animal Production Research Institute, Serere, Uganda	2,142	–	2142
Plant Genetic Resources Conservation Unit, Southern Regional Plant Introduction Station, University of Georgia, USDA-ARS, Griffin, USA	1,090	973	2,063

Source: https://www.genesys-pgr.org/; genebanks with >1000 accessions considered for listing.

At ICRISAT, the short-term storage is maintained at 18–20°C, 30–40% RH, and used for temporary holding of seeds for drying and preparation for subsequent transfer to medium- and long-term storage. Active collections are conserved under medium term (4°C, 20% RH) for 15–20 years by drying seeds up to 7–9% moisture content and stored in aluminum cans with screw caps having rubber gaskets. Base collections are conserved under long term (–20°C) for >50 years by drying seeds up to 4–6% moisture content and storing in hermetically sealed laminated aluminum foil packets to extend viability (Upadhyaya and Gowda, 2009).

6.5 Germplasm characterization and evaluation

Germplasm collection is of little value unless it is characterized, evaluated, and documented properly toward their enhanced utilization in crop improvement.

6.5.1 *Agronomic traits*

A total of 21,461 accessions have been characterized at ICRISAT for 23 morphoagronomic traits using the descriptors for pearl millet (IBPGR and ICRISAT, 1993). These accessions showed large phenotypic diversity for almost all quantitative traits (Table 6.3). Likewise, diversity for qualitative traits, such as panicle shape (nine classes), seed shape (five classes), and seed color (10 classes), was also observed. Predominant types found were those with candle-shaped panicles, short-bristled panicles, globular seed shape, grey seed color, and seeds with partly corneous endosperm texture. Several accessions were identified as a promising source for green fodder yield (147), while some others for seed yield potential (6). Enormous diversity reported for morphoagronomic traits among landraces and wild relatives from India, west and central Africa, Cameroon, Yemen, and Ghana has been summarized by Dwivedi et al. (2012).

Table 6.3 Range variation for morphoagronomic traits observed in core collection and entire collection evaluated during rainy season at Patancheru, India

	Range		Mean	
Characters	Entire collection*	Core collection (Upadhyaya et al., 2009a)**	Entire collection	Core collection (Upadhyaya et al., 2009a)
Days to flowering	33–159	33–157	72.7	72.7
Plant height (cm)	30–490	35–490	248.5	243.3
Total tillers (no.)	1–35	1–35	2.7	2.7
Productive tillers (no.)	1–19	1–19	2.1	2.1
Panicle exsertion (cm)	−45 to 29	−32 to 22	3.5	3.5
Panicle length (cm)	5–135	5–120	28.9	28.2
Panicle width (cm)	8–58	10–55	24.0	23.9
1000 grain weight (g)	1.5–21.3	2.9–19.3	8.5	8.5

* Entire collection: 21,461 accessions.
** Core collection: 2,094 accessions.

6.5.2 Abiotic stress tolerance

Drought is the primary constraint for pearl millet production in the drier semi-arid and arid regions of south Asia and Africa. Traditional landraces from drier regions are good sources of drought adaptation (Table 6.4) and they could produce significantly greater biomass, grain, and stover yields than elite populations (Yadav, 2008). Under severe moisture stress, high tillering and small-panicled landraces produce higher grain yield than low tillering and large-panicled landraces (Van Oosterom et al., 2006). Drought-tolerant genotypes extract less water prior to anthesis and more water after anthesis resulting in better yield (Vadez et al., 2013). High-temperature stress at seedling and reproductive stages has an impact on crop establishment and yield of pearl millet. Genetic variation has been observed for heat tolerance at seedling and reproductive stage among germplasm (Table 6.4). A recent study for reproductive-stage heat tolerance over 3–4 years could identify tolerant breeding lines and germplasm line (IP 19877) having equivalent seed set as that of tolerant check 9444 (Gupta et al., 2015). Low-temperature stress at vegetative stage causes increased basal tillering and grain yield; at elongation stage, it leads to reduced spikelet fertility, inflorescence length, and decreased grain yield; at grain development stage, it leads to increase in grain yield (Fussell et al., 1980). Pearl millet germplasm tolerant to salinity have been reported (Table 6.4). Pearl millet is often cultivated in less-fertile soils with low amount of organic matter and low to medium levels of available phosphorus (Yadav and Rai, 2013). Pearl millet production on acid sandy soils of the Sahel is limited by low-P status. For low-P tolerance, both seedling and mature plant traits were found useful for selection as secondary traits (Gemenet et al., 2015). Diverse germplasm including breeding lines, varieties, and landraces with resistance to abiotic stresses has been summarized by Dwivedi et al. (2012).

6.5.3 Biotic stress tolerance

Resistance to major diseases like downy mildew, blast, smut, ergot, and rust have been reported in pearl millet germplasm (Table 6.4). At ICRISAT, evaluation of a large number of germplasm accessions has led to the identification of resistant/tolerant sources for downy mildew (54), smut (397), ergot (283), and rust (332) (Upadhyaya et al., 2007). Several germplasms with multiple disease resistance to major diseases have also been identified (Table 6.4). Diverse germplasm including breeding lines, varieties, and landraces showing resistance to biotic stresses have been summarized by Dwivedi et al. (2012). GP1 wild species, *P. glaucum* ssp. *monodii* is a good source for smut resistance, while ssp. *stenostachyum* is a good source for rust and leaf blast resistance. Some of the accessions of GP3 species were found immune to rust, while a group of accessions showed resistance to several fungal diseases including blast (Wilson and Hanna, 1992). A large number of accessions (223) belonging to 12 wild *Pennisetum* species were free from downy mildew, while most of the *Pennisetum schweinfurthii* accessions were free from downy mildew and also found resistant to rust (Singh and Navi, 2000). *Striga* [*Striga hermonthica* (Del.) Benth.] is a serious constraint for pearl millet production in West Africa. Few landraces with

Table 6.4 **Sources of resistance to major abiotic and biotic stress among pearl millet landraces and derivatives of landraces or wild relatives**

Resistant landraces or derivatives*	References
Abiotic stress	
Drought: CZMS 44A (landrace 3072); IP 8210; landraces 220, 184, 235, 238	Manga and Yadav (1997); Kusaka et al. (2005); Yadav (2010)
Heat: IP 3201; IP 19877	Howarth et al. (1997); Gupta et al. (2015)
Salinity: 93613, KAT/PM-2, Kitui, Kitui local, 93612; 10876, 10878, 18406, 18570; IP 3757, 3732; Birjand pearl millet; IP 6112; IP 3616, 6104, 6112; ZZ ecotype	Ashraf and McNeilly (1992); Ali et al. (2004); Krishnamurthy et al. (2007); Kafi et al. (2009); Esechie and Al-Farsi (2009); Nadaf et al. (2010); Radhouane (2013)
Biotic stress	
Downy mildew: 18 resistant progenies (IP 2696); P 310, 472, 1564, 700516, D322/1/-2-2, P1449-3, P 8695-1, 8896-3, 3281; DMRP 292 (IP 18292); Gwagwa; 62 accessions resistant to 2 or more pathotypes	Singh et al. (1988); Singh (1990); Singh and Talukdar (1998); Wilson et al. (2008); Sharma et al. (2015)
Blast: Tift $85D_2B_1$ (*monodii*); landrace 122, 162, 192; Acc. 36, 41, 46, 71; 32 IP acc. (resistant to one or more pathotypes), IP 7846, 11036, 21187 (resistant to four pathotypes)	Hanna et al. (1987); Wilson et al. (1989a); Wilson et al. (1989b); Sharma et al. (2013)
Smut: Selections from germplasm (6); landrace 133, 224, 192; SSC 46-2-2-1, SC 77-7-2-3-1, SC 18-7-3-1	Thakur et al. (1986); Wilson et al. (1989a); Yadav and Duhan (1996)
Ergot: 27 F_6 lines (intermating of 20 acc.); 16 acc.	Thakur et al. (1982); Kumar et al. (1997)
Rust: *monodii* (3 acc.); 2696-1-4; landrace 192; ICML 17, 18, 19, 20, 21 (selected from bulk germplasm); Tift 3, Tift 4 (landrace from Burkina Faso); Tift $89D_2$ (Se Fa); Tift 65 (*monodii*); 7042-1-4-4, IP 8695-4, 700481-27-5	Hanna et al. (1985); Andrews et al. (1985); Wilson et al. (1989a); Singh et al. (1990); Wilson and Burton (1991); Hanna and Wells (1993); Burton and Wilson (1995); Pannu et al. (1996)
Striga: Serere 2A9, 80S224, P2671, P2950; Buduma-Chad; PS 202, 637, 639, 727	Roger and Ramaiah (1983); Gworgwor (2001); Wilson et al. (2004)
Multiple disease resistant*	
DM, ER, RU, BL: Tift #5S1 (Bulk of 114 acc. of *monodii*)	Hanna et al. (1993)
DM, ER, RU: 7 inbreds and 6 bulk populations (Intermating between landraces and pedigree selection)	Thakur et al. (1985)
DM, SM: ICML 5 to 10	Thakur and King (1988)
DM, RU: IP 1481-L-2, P2895-3, IP 8877-3, 700481-5-3; ICML 11 (IP 2696)	Singh (1990); Singh et al. (1987b)
BL, RU: Acc. 122, 162; Tift 86DB/A; Tift 65	Wilson et al. (1989a); Hanna and Wells (1989); Burton and Wilson (1995)
BL, SM: Acc. 192	Wilson et al. (1989a)
SM, RU: Acc. 133, 224	Wilson et al. (1989a)

* Content in bracket indicates derived from the landrace or wild relative.

** DM, downy mildew; ER, ergot; RU, rust; BL, blast; SM, smut.

less susceptibility and some *monodii* accessions with resistance to *Striga* have been identified (Table 6.4). Shibras have been found to show resistance to *Striga* in the field (Parker and Wilson, 1983).

6.5.4 Seed nutritional quality

Enormous variability has been found in pearl millet germplasm collection for protein (up to 24.3%) among 260 accessions (Singh et al., 1987a) and micronutrient concentrations, namely Fe (51–121 mg/kg) and Zn (46–87 mg/kg) among 191 accessions (Rai et al., 2015).

6.5.5 Source of male sterility

Cytoplasmic male sterility (CMS) sources in pearl millet were identified in populations of crosses involving genetically diverse parents, for example, A_1 (Burton, 1958), A_3 CMS (Athwal, 1966); or derived from diverse germplasm accessions, for example, A_2 (Athwal, 1961) and several unclassified CMS sources (Appa Rao et al., 1989b); or identified from broad-based genepools, for example, A_{egp} CMS in early genepool (Sujata et al., 1994), A_5 CMS in large-seeded genepool (Rai, 1995). CMS sources have also been identified in populations derived from crosses between GP1 wild relative [*P. glaucum* subsp. *monodii* (=*P. violaceum*)] as female parent and cultivated pearl millet as male parent, for example, germplasm accessions from Senegal were identified as source of A_v (Marchais and Pernès, 1985) and A_4 CMS (Hanna, 1989).

6.6 Germplasm regeneration and documentation

Germplasm regeneration is essential to produce uncontaminated and representative seed samples of the original accessions in sufficient quantities for subsequent evaluation, distribution, and conservation. Methods suggested for regeneration of pearl millet germplasm (Burton, 1979) are described next.

6.6.1 Intercrossing

Intermating of 100 or more plants in isolation would help to maintain an accession close to its original form. Limited availability of isolations does not permit this. Hence, manual intercrossing among 100 or more plants of an accession without allowing for outcrossing from other accessions is practiced. Pollination of receptive panicle on each plant with mixture of pollen from an accession is an effective method (Appa Rao, 1999).

6.6.2 Cluster bagging

To regenerate thousands of landraces at ICRISAT, cluster-bagging method is being used to prevent unwanted genetic contamination and avoid hand pollination

(Upadhyaya and Gowda, 2009). About 160 plants per accession are grown in four rows. Before stigma emergence, the main stem panicle from two to four adjacent plants in a row is covered in a single parchment paper bag. All the plants of an accession are covered with bags in this fashion. Cross-pollination takes place among the diverse plants covered in a single bag, thereby reducing the inbreeding depression. During harvesting, one such bagged panicle per plant from at least 120 plants per accession is taken and an equal quantity of seeds from each plant is used to reconstitute the accession.

6.6.3 Selfing

Selfing of plants that are used to describe an accession is the most convenient way to obtain large amounts of seed of a new accession (Burton, 1985). At ICRISAT, inbreds and genetic stocks are maintained by selfing (Upadhyaya and Gowda, 2009).

6.6.4 Genepools

Various accessions of similar maturity coming from the same region/country are grown as a mixture of equal quantities of seeds from all accessions in isolation to intermate and seed is harvested from a part or all of the plants. Germplasm pools that are increased each year and contain many accessions offer an easy way to handle germplasm (Burton, 1979). It allows breaking undesirable linkages, resulting in the emergence of new recombinants and improves adaptation. To ensure proper representation of accessions entering the genepools, it is desirable to group accessions based on maturity, specific morphologic characters, and region of origin. Accessions are evaluated and chosen to create trait-specific genepools (TSG). For genetic resource purposes, TSGs are best maintained without any selection pressure to allow for local adaptation. At ICRISAT, TSGs have been developed by random mating (four to six generations) of a large number of germplasm accessions, for example, early genepool (1143 acc.), high tillering genepool (1093), large-grain genepool (887), large-panicle genepool (804); INMG 1 (123 acc.), INMG 2 (208), INMG 3 (73), INMG 4 (51), INMG 5 (69) (Rai et al., 1997; Singh and Jika, 1988).

At ICRISAT, good-quality seed is regenerated for conservation during the post-rainy season with regular monitoring by pathologists to have healthy seeds. Physiologically mature seeds are harvested, seed moisture brought down to 5–7% by drying in cool and dry atmospheres, or in the drying cabinets at 18°C and 16% RH. For pearl millet, the recommended sample size per accession in a base collection is 12,000 seeds, with a minimum of 3,000 seeds (IBPGR, 1985). At ICRISAT, about 12,000 seeds of pearl millet are used for long-term conservation and stored in hermetically sealed containers to maintain low seed moisture content with a view to control humidity. Prior to conservation, seed viability is recorded initially for all accessions and also randomly monitored during storage by drawing samples at regular intervals of 6 months for their germinability. Accessions that show <85% viability or below critical quantity of 50 g are routinely regenerated. Proper documentation allows rapid accessioning of new samples, answer queries on the conserved germplasm, and monitor quality and quantity of stored material to carry out regeneration and distribution. Hence, a computerized data handling system is ideal for a genebank.

6.7 Gap analyses of germplasm

Identifying the gaps in germplasm collections of genebanks helps to plan for collecting landraces that are not represented in genebanks. Toward this at ICRISAT, gap analyses of cultivated pearl millet germplasm from different geographical regions like west and central Africa (WCA), Asia, and east and southern Africa (ESA) and also for wild relatives like *P. glaucum* ssp. *monodii* and *P. pedicellatum* (Deenanath grass) were made to identify the missing diversity in germplasm assembled in its genebank (Upadhyaya et al., 2009b, 2010, 2012, 2014a, 2014b). The WCA region, being the center of diversity for pearl millet, has germplasm sources for resistance to biotic and abiotic stresses. Gap analysis using passport and characterization data and geographical information system tools could identify gaps in cultivated germplasm in the provinces of Chad, Ghana, Nigeria, Burkina Faso, Mali, and Mauritania in WCA; provinces of India and Pakistan in Asia; provinces of Sudan, Uganda, and seven countries in southern Africa. Similarly, gaps were identified in 86 provinces of 8 countries in the primary center of origin for *P. glaucum* subsp. *monodii*, while 194 provinces in 21 countries of Asia and Africa for *P. pedicellatum*. This information would be helpful in assembling poorly represented or totally missing germplasm from these regions of diversity to avoid genetic erosion and ensure their conservation and utilization in crop improvement.

6.8 Limitations in germplasm use

The use of available genetic resources in crop improvement is the most neglected part of germplasm conservation (de Wet, 1989). A very large gap exists between actual utilization of the germplasm and availability of collection in the genebanks (Wright, 1997; Upadhyaya et al., 2006). Germplasm resources would not be used if the information needed by crop improvement scientists is not readily available. Extensive use of fewer and closely related parents and their derivatives in crop improvement is contrary to the very purpose of establishing large germplasm collections. Representative subsets, in the form of core collection (Frankel, 1984) or minicore collection (Upadhyaya and Ortiz, 2001), has been suggested as an entry point for germplasm utilization in crop breeding. Core and minicore collections (Upadhyaya et al., 2009a, 2011) and reference sets (Upadhyaya, 2009) are now available in pearl millet. Evaluation of core collection has led to the identification of new sources of variation for grain and fodder traits (Khairwal et al., 2007), while evaluation of minicore collection led to identification of sources of resistance to blast and downy mildew (Sharma et al., 2013, 2015). More targeted evaluation of these subsets for morphoagronomic traits, resistance to various abiotic and biotic stresses, and their molecular profiling will lead to mining of allelic variation associated with agronomically beneficial traits.

6.9 Germplasm uses in pearl millet improvement

6.9.1 Cultivated germplasm

The greatest achievement is conserving the vanishing pearl millet germplasm and making it available for crop improvement. At ICRISAT, to enhance the utilization of pearl millet germplasm, core and minicore collections and reference sets have been developed (Bhattacharjee et al., 2007; Upadhyaya et al., 2009a, 2011; Upadhyaya, 2009), which would allow their extensive evaluation for identification of trait-specific germplasm and widely adaptable accessions across locations. Trait-specific genepools have been developed earlier for early maturity, high tillering, large panicle, and large grain to provide breeders with useful variability for utilization. Recently, a *Striga*-resistant genepool was developed by recurrent selection (Kountche et al., 2013).

In general, Indian pearl millet landraces have earliness, high tillering, high harvest index, and local adaptation, whereas African landraces are good sources of high head volume, large seed size, and disease resistance. Significant progress has been made with regard to utilization of germplasm in pearl millet improvement. Early maturing accessions like IP 4021 and IP 3122 were supplied most frequently. Pearl millet germplasm has been widely used in developing composites, which include a wide range of germplasm and improved breeding lines. The *Iniadi* germplasm from the Togo–Ghana–Burkina Faso–Benin regions of western Africa is most commonly used in pearl millet breeding programs worldwide (Andrews and Anand Kumar, 1996). The most successful example of the use of pearl millet landraces is ICTP 8203, a large-seeded and high-yielding open-pollinated variety bred at ICRISAT, Patancheru as a selection within the large-seeded *Iniadi* landrace (IP 17862) from northern Togo (Rai et al., 1990). This variety was released in India as MP 124 in Maharashtra and Andhra Pradesh and as PCB 138 in Punjab, while as Okashana 1 in Namibia and as Nyankhombo [ICMV 88908] in Malawi.

Direct selection within the landrace has led to the development of a large-seeded and downy mildew resistant male sterile line ICMA 88004 in India (Rai et al., 1995); varieties IKMP 3 (IP 11381) and IKMP 5 (IP 11317) in Burkina Faso; open-pollinated variety ICMV-IS 88102 (IP 6426) in Burkina Faso, while as Benkadi Nio in Mali; IP 6104 and IP 19586 released in Mexico as high forage yielding varieties. Some varieties were developed by crosses involving at least one of the parents as landrace, for example, Okashana 2 and Kangara (SDMV 92040) varieties in Namibia and PMV 3 in Zimbabwe (Obilana et al., 1997). ICRISAT has contributed genetic material to public and private institutions, which has helped in breeding high-yielding varieties/hybrids with resistance to biotic and abiotic factors. Using pearl millet germplasm from ICRISAT, a large number of varieties/hybrids have been released by NARS across the world (163) including India (80) as of December 2010 (Anonymous, 2011).

Donor parents like 863B (IP 22303), P 1449-2 (IP 21168), ICMB 90111 (IP 22319), ICMP 451 (IP 22442), and IP 18293 were identified as sources of gene for resistance against different pathotypes of downy mildew in India (Upadhyaya et al., 2007). Several promising germplasm accessions have been identified for resistance to downy mildew (IP #9645, 14537, 18292), blast (IP 7846), smut and heat tolerance (IP #19799, 19843, 19877), multiple disease resistance (IP #21268, 21296), salinity tolerance (IP #3732, 3757, 22269), high Fe and/or Zn content (IP #9198, 11535, 12364, 17672), and yellow endosperm with high beta-carotene content (IP #15533, 15536) in seed. Germplasm with earliness, resistance to abiotic and biotic stresses, new dwarfing genes, high iron and zinc, sweet stalk, yellow endosperm, and so on, are widely used in crop improvement programs in different countries. Several new and useful traits, such as narrow leaf, glossy leaf, brown midrib leaf, and leaf color variants, have been used extensively in academic studies.

6.9.2 Wild relatives

There has been a general lack of interest in using wild species because of the large amount of genetic variability already available in pearl millet landraces. However, *P. glaucum* ssp. *monodii* for new source of cytoplasmic-nuclear male sterility (CMS), *P. purpureum* for forage, stiff stalk and restorer genes of the A_1 CMS system, *P. mezianum* for drought tolerance, *Pennisetum orientale* for drought tolerance and forage, *P. schweinfurthii* for large seeds, *P. pedicellatum* and *P. polystachion* for downy mildew resistance, and *P. squamulatum* for apomictic gene are useful (Rai et al., 1997).

Intersubspecific, interspecific, and intergeneric hybridization have been attempted to expand the cultigen genepool in pearl millet (Table 6.5). Intersubspecific crosses involving ssp. *glaucum* (cultivated), *monodii* (wild relative), and *stenostachyum* (weedy relative) have been successful in transferring desirable traits like rust resistance, male sterility, and alleles for enhancing yield components from GP1 to pearl millet. Interspecific hybridization between *P. glaucum* and *P. purpureum* has led to development of forage hybrids with high biomass and better quality (Table 6.5). The A′ genome from *P. purpureum* contributes excellent genetic variability for inflorescence and plant types, maturity, and fertility restoration of the sterile cytoplasm, which has been transferred to pearl millet (Hanna, 1983, 1990). *P. squamulatum* has desirable traits like perenniality, apomixis, disease resistance, tolerance to drought and frost, and so on, which could be transferred to pearl millet. Crosses between members of GP1 with GP3 showed incompatibility at pollen germination or at stylar or ovarian level, but in some cases, hybrids could be recovered through embryo rescue but often showing male sterility (Table 6.5). Efforts of intergeneric hybridization and somatic hybridization have shown little success. Concerted efforts need to be made for utilization of GP2 and GP3 species through the development of wide crosses by utilizing special techniques and generation of prebreeding populations for use by breeders.

Table 6.5 **Attempts of wide crosses involving *Pennisetum* spp.**

Crosses	Details	References
Intersubspecific/interspecific		
Within GP1		
*P. glaucum** × *P. glaucum* ssp. *monodii*	Normal pollen germination and growth in stigmatic tract	Kaushal and Sidhu (2000)
	Transferred rust resistance and thermosensitive genetic male sterility	Hanna et al. (1985), Kaushal et al. (2004)
	Frequency of natural hybrids decreased up to 11% at maturity	Couturon et al. (2003)
P. glaucum ssp. *monodii* × *P. glaucum*	Transferred cytoplasmic male sterility	Marchais and Pernès (1985), Hanna (1989), Rai et al. (1996)
P. glaucum × *P. mollissimum*	Inheritance of domestication traits studied; Transgressive segregants obtained for panicle length and width	Poncet et al., 1998, 2000
P. glaucum × *P. stenostachyum*	Some introgression from wild to weedy population, but very low gene flow between wild and domesticated populations	Mariac et al. (2006b)
Between GP1 and GP2		
P. glaucum × *P. purpureum*	Homeology between A and A′ genome confirmed based on meiotic studies and genomic *in situ* hybridization	Jauhar (1981b), Jauhar and Hanna (1998), dos Reis et al. (2014)
	High frequency of meiotic abnormalities in F_1's resulting in sterile pollen	Techio et al. (2006)
	Hybrids with improved forage potential and quality produced	Hanna et al. (1984), Obok et al. (2012), Obok (2013), Kannan et al. (2013)
	Genes controlling earliness, long inflorescence, leaf size, and male fertility restoration transferred from *P. purpureum*	Hanna (1983)
P. glaucum (2*x*/4*x*) × *P. squamulatum*	Recovered partially fertile hybrids	Dujardin and Hanna (1983, 1989a), Marchais and Tostain (1997)
	Transfer of gene/s controlling apomixis through crosses and backcrosses	Dujardin and Hanna (1985)
(*P. glaucum* × *P. squamulatum*) × (*P. glaucum* × *P. purpureum*)	Double cross hybrid used to transfer apomixis from *P. squamulatum* into a backcross derivative	Dujardin and Hanna (1989b)

(Continued)

Table 6.5 Attempts of wide crosses involving *Pennisetum* spp. *(cont.)*

Crosses	Details	References
Between GP1 and GP3		
*P. glaucum/P. glaucum ssp. monodii** × *P. orientale/P. setaceum/P. schweinfurthii/Pennisetum ramosum/P. pedicellatum/P. polystachyon*	Incompatibility observed at pollen germination, pollen tube growth or fertilization level	Kaushal and Sidhu (2000), Chaix and Marchais (1996)
P. glaucum × *P. schweinfurthii/P. ramosum*	Hybrids recovered by embryo rescue but normally male sterile and partially female fertile	Marchais and Tostain (1997), Nagesh and Subrahmanyam (1996)
P. glaucum × *P. mezianum*	Hybrids recovered by embryo rescue/embryo cloning and showed low pollen fertility	Nagesh and Subrahmanyam (1996), Marchais and Tostain (1997)
P. glaucum × *P. macrourum*	Failed to get any hybrid although germinating pollen abundant on stigma	Dujardin and Hanna (1989a)
P. glaucum (2*x*/4*x*) × *P. ramosum*	Embryo rescue could recover hybrids only from tetraploid crosses	Nagesh and Subrahmanyam (1996), Marchais and Tostain (1997)
P. glaucum × *P. orientale*	Partial homology between the genomes	Jauhar (1981b)
	Male fertility of F_1 restored by chromosome doubling	Dujardin and Hanna (1987)
	Embryo rescue used to obtain hybrids	Nagesh and Subrahmanyam (1996)
	Backcrossing resulted in hybrids with varying chromosome numbers	Hanna and Dujardin (1982), Dujardin and Hanna (1989b)
P. glaucum × *P. setaceum*	Obligate apomictic interspecific hybrids recovered but did not allow further progress	Dujardin and Hanna (1989a)
P. glaucum × *P. pedicellatum/P. polystachyon*	Miniature hybrid seeds formed that failed to germinate; embryo culture might be useful for recovering hybrids	Dujardin and Hanna (1989a)
P. glaucum × *P. dubium*	Single hybrid obtained but no homology between parental genomes	Gildenhuys and Brix (1961)
P. glaucum × *P. cenchroides*	Incompatibility for pollen tube growth in stylar region	Mohindra and Minocha (1991)
	High frequency of proembryos with large undifferentiated endosperm; hybrid obtained was completely sterile	Read and Bashaw (1974), Chaix and Marchais (1996); Marchais and Tostain (1997)

Table 6.5 **Attempts of wide crosses involving *Pennisetum* spp. *(cont.)***

Crosses	Details	References
Within GP3		
Cenchrus ciliaris × *P. orientale*	Hybrid with excellent forage potential	Hussey et al. (1993)
P. flaccidum × *P. mezianum*	Hybrids of (2*n* + *n*) type were more winter hardy and produced more forage than (*n* + *n*) type	Burson and Hussey (1996)
Between GP2 and GP3		
P. purpureum × *P. squamulatum*/*P. schweinfurthii*/*P. ramosum*/*P. pedicellatum*/*P. polystachyon*	Incompatibility at pollen germination, pollen tube growth or fertilization level	Chaix and Marchais (1996)
P. schweinfurthii × *P. purpureum*	Hybrid obtained due to fertilization between unreduced (2*n*) and normal (*n*) gamete	Vidhya and Fazlullah Khan (2003)
Intergeneric		
P. glaucum × *Panicum maximum*/ *Zea mays*	Low compatibility and pollen tubes arrested in the style	Chaix and Marchais (1996)
Oryza sativa × *P. orientale*	Hybrid recovered were regenerated vegetatively and completely sterile	Wu and Tsai (1963)
Triticum aestivum × *P. glaucum*	Hybrid embryos produced haploid wheat due to elimination of pearl millet chromosomes	Laurie (1989), Gernand et al. (2005)
Hordeum vulgare × *P. glaucum*	Globular embryos obtained	Zenkteler and Nitzsche (1984)
Avena sativa × *P. glaucum*	Hybrids obtained at low frequency but pearl millet chromosomes lost in embryo or plant	Matzk (1996)
P. glaucum × *Zea mays*/*Triticale*/ *Sorghum bicolor*	Haploid induction rate (HIR) varied in crosses	Gugsa et al. (2013)

* *Pennisetum glaucum* syn. *Pennisetum typhoides*/*Pennisetum americanum*; *P. glaucum* ssp. *monodii* syn. *P. violaceum*; *Cenchrus ciliaris* syn. *P. cenchroides*; 2*x*, diploid; 4*x*, tetraploid.

6.9.3 Chromosome segment substitution lines

Development of chromosome segment substitution lines (CSSLs), originating from crosses between cultivated and wild types, will be a most useful resource as each of these lines will have different genomic regions from wild germplasm introgressed in cultivated genotypic background. Characterization of these populations for different traits and genotyping helps in exploitation of useful alleles found in wild germplasm. CSSLs based on crosses between cultivated pearl millet and its wild relatives are not yet available. However, at ICRISAT, a set of CSSLs (1492) based on a cross between two agronomically elite parental lines differing for several important agronomic traits have been developed (Ramana Kumari et al., 2014). CSSLs will be an ideal set of genetic stocks for mapping and fine-mapping the multitude of traits for which their parents differ.

6.10 Genomic resources in management and utilization of germplasm

Pearl millet has a genome of ~2350 Mb (Bennett et al., 2000) and haploid genome content of 1Cx = 2.36 pg (Martel et al., 1997). Genomic resources like molecular markers, genetic linkage maps, and genes associated with specific traits developed in pearl millet using diverse genetic material and their utilization for assessing diversity or population structure and genetic mapping have been summarized by Dwivedi et al. (2012). Toward utilization of molecular markers in management and utilization of germplasm, they have been employed for assessing diversity among and within pearl millet landraces. Genetic diversity analysis of representative accessions of Indian origin (10) from core collection (504 accessions) using highly polymorphic RFLP probes (16) revealed high variability within-accession (30.9%) and also between accessions (69.1%) (Bhattacharjee et al., 2002). Landraces from western and eastern Rajasthan revealed higher variation within landrace population than between regional samples as assessed by AFLP markers (Vom Brocke et al., 2003). Genetic diversity assessed among wild (46) and cultivated (421) pearl millet accessions from Niger using microsatellites (25) revealed lower number of alleles and lower gene diversity in cultivated than wild accessions and a strong differentiation between cultivated and wild groups (Mariac et al., 2006a). Similarly, wild (84) and cultivated (355) accessions originating from the whole pearl millet distribution area in Africa and Asia analyzed by microsatellite loci (27) revealed higher diversity in the wild pearl millet group (Oumar et al., 2008). The cultivated pearl millet sample possessed 81% of the alleles and 83% of the genetic diversity of the wild pearl millet sample.

Next-generation sequencing (NGS) technologies offer efficient and cost-effective sequencing of a large number of samples for organisms, which provide opportunity to relate sequence differences with phenotypes of diverse accessions in comparison with decoded reference genome. The role of NGS technologies to enhance efficiency of different genebank activities need to be established. For collection management, NGS technologies could be useful to basically support all management areas (Van Treuren and Van Hintum, 2014). For example, DNA sequence data of genebank accessions may be used to determine the genetic structure of collections and to improve the composition thereof by eliminating redundancies (Van Treuren et al., 2009). Ample sequence data of the existing collection allows genebank curators to take more informed decisions about acquisition by evaluating potentially interesting materials for their added value to the genetic diversity already present in the collection (Van Treuren et al., 2008). One can consider removal of accessions that contribute least to the genetic diversity of the collection (Van Treuren et al., 2009). NGS data could also be used to monitor the regeneration of accessions in order to ensure the maintenance of genetic integrity thereof, for example, by comparing sequence data of samples before and after regeneration (Van Hintum et al., 2007).

6.11 Conclusions

Pearl millet is gaining importance as a climate-resilient and health-promoting nutritious crop. Recent evidences using microsatellites suggest the monophyletic origin of pearl millet and its further migration and secondary diversification leading to enormous diversity. Genetic erosion of landraces has been evident in different pearl millet growing regions due to replacement with modern cultivars. Large variability found in pearl millet germplasm has been conserved in several genebanks. Toward enhancing the utilization of pearl millet germplasm, available subsets like core and minicore collections and reference sets should be extensively evaluated to identify trait-specific germplasm and also develop genomic resources to associate sequence differences with trait variations. Although transfer of desirable traits from primary wild relatives has been successful, concerted efforts are needed to broaden cultivated genepool by utilizing secondary and tertiary genepool toward developing climate-resilient cultivars. Development of genomic resources is expected to rise as the genome sequence of pearl millet is due for release and also due to faster developments in NGS technologies that could be efficiently utilized for management and utilization of pearl millet germplasm and in turn for crop improvement.

References

AICPMIP, 2014. Project Coordinator's Review. Available from: www.aicpmip.res.in/pcr2014.pdf

Ali, G.M., Khan, N.M., Hazara, R., McNeilly, T., 2004. Variability in the response of pearl millet (*Pennisetum americanum* (L.) Leeke) accessions to salinity. Acta Agron. Hung. 52, 277–286.

Andrews, D.J., Anand Kumar, K., 1996. Use of the West African pearl millet landrace *Iniadi* in cultivar development. Plant Genet. Resour. Newsl. 105, 15–22.

Andrews, D.J., Rai, K.N., Singh, S.D., 1985. A single dominant gene for rust resistance in pearl millet. Crop Sci. 25 (3), 565–566.

Anonymous, 2011. Available from: http://www.thehindubusinessline.com/industry-and-economy/agri-biz/icrisat-germplasm-crops-contributing-to-global-food-security-nars-report/article2047506.ece

Appa Rao, S., 1999. Genetic Resources. In: Khairwal, I.S., Rai, K.N., Andrews, D.J., Harinarayana, G. (Eds.), Pearl Millet Breeding. Oxford and IBH Publishing Co. Pvt. Ltd, New Delhi, pp. 49–81.

Appa Rao, S., de Wet, J.M.J., 1999. Taxonomy and Evolution. In: Khairwal, I.S., Rai, K.N., Andrews, D.J., Harinarayana, G. (Eds.), Pearl Millet Breeding. Oxford and IBH Publishing Co. Pvt. Ltd, New Delhi, pp. 29–47.

Appa Rao, S., House, L.R., Gupta, S.C., 1989a. A Review of Sorghum, Pearl Millet and Finger Millet Improvement in SADCC Countries. SADCC/ICRISAT, Bulawayo, Zimbabwe.

Appa Rao, S., Mengesha, M.H., Rajagopal Reddy, C., 1989b. Development of cytoplasmic male-sterile lines of pearl millet from Ghana and Botswana germplasm. In: Manna, G.K., Sinha, U. (Eds.), Perspectives in Cytology and Genetics. Kalyani Publishers, India, pp. 817–823.

Appa Rao, S., Murked, A.W., Mengesha, M.H., Amer, H.M., Reddy, K.N., Alsurai, A., et al., 1993. Collecting crop germplasm in Yemen. Plant Genet. Resour. Newsl. 94/95, 28–31.

Appa Rao, S., Mengesha, M.H., Nwasike, C., Ajayi, O., Olabanji, O.G., Aba, D., 1994. Collecting crop germplasm in Nigeria. Plant Genet. Resour. Newsl. 97, 63–66.

Ashraf, M., McNeilly, T.M., 1992. The potential for exploiting variation in salinity tolerance in pearl millet (*Pennisetum americanum* (L.) Leeke). J. Plant Breed. 108 (3), 234–240.

Athwal, D.S., 1961. Recent developments in the breeding and improvement of bajra (pearl millet) in the Punjab. Madras Agric. J. 48, 18–19.

Athwal, D.S., 1966. Current plant breeding research with special reference to *Pennisetum*. Indian J. Genet. Plant Breed. 26A, 73–85.

Basavaraj, G., Parthasarathy Rao, P., Bhagavatula, S., Ahmed, W., 2010. Availability and utilization of pearl millet in India. J. SAT Agric. Res. 8, 1–6.

Bennett, M.D., Bhandol, P., Leitch, I.J., 2000. Nuclear DNA amounts in angiosperms and their modern uses-807 new estimates. Ann. Bot. 86, 859–909.

Bezançon, G., Pham, J.-L., Deu, M., Vigouroux, Y., Sagnard, F., Mariac, C., et al., 2009. Changes in the diversity and geographic distribution of cultivated millet (*Pennisetum glaucum* (L.) R. Br.) and sorghum (*Sorghum bicolor* (L.) Moench) varieties in Niger between 1976 and 2003. Genet. Resour. Crop Evol. 56, 223–236.

Bhattacharjee, R., Bramel, J., Hash, T., Kolesnikova-Allen, A., Khairwal, S., 2002. Assessment of genetic diversity within and between pearl millet landraces. Theor. Appl. Genet. 105 (5), 666–673.

Bhattacharjee, R., Khairwal, I.S., Bramel, P.J., Reddy, K.N., 2007. Establishment of a pearl millet [*Pennisetum glaucum* (L.) R. Br.] core collection based on geographical distribution and quantitative traits. Euphytica 155 (1–2), 35–45.

Brunken, J.N., 1977. A systematic study of *Pennisetum* sect. *Pennisetum* (Gramineae). Am. J. Bot. 64 (2), 161–176.

Brunken, J.N., de Wet, J.M.J., Harlan, J.R., 1977. The morphology and domestication of pearl millet. Econ. Bot. 31, 163–174.

Burson, B.L., Hussey, M.A., 1996. Breeding apomictic forage grasses. In: Williams, M.J. (Ed.), Proceedings of the American Forage and Grassland Council, Vancouver BC, Canada, June 13–15, AFGC, Georgetown, TX, pp. 226–230.

Burton, G.W., 1958. Cytoplasmic male sterility in pearl millet (*Pennisetum glaucum*) (L.) R. Br. Agron. J. 50, 230–231.

Burton, G.W., 1979. Handling cross-pollinated germplasm efficiently. Crop Sci. 19, 685–690.

Burton, G.W., 1985. Collection, evaluation and storage of pearl millet germplasm. Field Crops Res. 11, 123–129.

Burton, G.W., Wilson, J.P., 1995. Registration of Tift 65 parental inbred line of pearl millet. Crop Sci. 35 (4), 1244.

Chaix, G., Marchais, L., 1996. Diversity of Pennicillarian millets (*Pennisetum glaucum* and *P. purpureum*) as for the compatibility between their gynoecia and pollens from some other Poaceae. Euphytica 88, 97–106.

Chemisquy, M.A., Giussani, L.M., Scataglini, M.A., Kellogg, E.A., Morrone, O., 2010. Phylogenetic studies favour the unification of *Pennisetum*, *Cenchrus* and *Odontelytrum* (Poaceae): a combined nuclear, plastid and morphological analysis, and nomenclatural combinations in *Cenchrus*. Ann. Bot. 106 (1), 107–130.

Chowdhury, M.K.V., Smith, R.L., 1988. Mitochondrial DNA variation in pearl millet and related species. Theor. Appl. Genet. 76, 25–32.

Clayton, W.D., Renvoize, S.A., 1986. Genera Graminum, Grasses of the World, Kew Bulletin. Her Majesty's Stationery Office, London, p. 389, Additional Series XIII.

Clayton, W.D., Harman, K.T., Williamson, H., 2006. GrassBase – The Online World Grass Flora. Available from: http://www.kew.org/data/grasses-db.html

Clegg, M.T., Rawson, J.R.Y., Thomas, K., 1984. Chloroplast DNA variation in pearl millet and related species. Genetics 106, 449–461.

Couturon, E., Mariac, C., Bezançon, G., Lauga, J., Renno, J.F., 2003. Impact of natural and human selection on the frequency of the F_1 hybrid between cultivated and wild pearl millet (*Pennisetum glaucum* (L.) R. Br.). Euphytica 133, 329–337.

D'Andrea, A.C., Casey, J., 2002. Pearl millet and Kintampo subsistence. Afr. Arachaeol. Rev. 19, 147–173.

de Wet, J.M.J., 1989. Cereals for the semi-arid tropics. In: Plant Domestication by Induced Mutation. Proceedings of an Advisory group meeting on the possible use of mutation breeding for rapid domestication of new crop plants. Joint FAO/IAEA Division of Nuclear techniques in Food and Agriculture, 17–21 Nov. 1986, Vienna, p. 79–88.

DeVries, J., Toenniessen, G., 2001. Securing the Harvest: Biotechnology, Breeding and Seed Systems for African Crops. CAB International, Wallingford.

Donadio, S., Giussani, L.M., Kellogg, E.A., Zuloaga, F.O., Morrone, O., 2009. A preliminary molecular phylogeny of *Pennisetum* and *Cenchrus* (Poaceae-Paniceae) based on the *trnL-F*, *rpl16* chloroplast markers. Taxonomy 58, 392–404.

dos Reis, G.B., Mesquita, A.T., Torres, G.A., Andrade-Vieira, L.F., Pereira, A.V., Davide, L.C., 2014. Genomic homeology between *Pennisetum purpureum* and *Pennisetum glaucum* (Poaceae). Comp. Cytogenet. 8 (3), 199–209.

Doust, A.N., Penly, A.M., Jacobs, S.W.L., Kellogg, E.A., 2007. Congruence, conflict, and polyploidization shown by nuclear and chloroplast markers in the monophyletic "bristle clade" (Paniceae, Panicoideae, Poaceae). Syst. Bot. 32, 531–544.

Dujardin, M., Hanna, W.W., 1983. Apomictic and sexual pearl millet × *Pennisetum squamulatum* hybrids. J. Hered. 74, 277–279.

Dujardin, M., Hanna, W.W., 1985. Cytology and reproductive behavior of pearl millet – Napier grass hexaploids × *Pennisetum squamulatum* trispecific hybrids. J. Hered. 76, 382–384.

Dujardin, M., Hanna, W.W., 1987. Inducing male fertility in crosses between pearl millet and *Pennisetum orientale* Rich. Crop Sci. 27, 65–68.

Dujardin, M., Hanna, W.W., 1989a. Crossability of pearl millet with wild *Pennisetum* species. Crop Sci. 29 (1), 77–80.

Dujardin, M., Hanna, W.W., 1989b. Developing apomictic pearl millet – characterization of a BC_3 plant. J. Genet. Breed. 43, 145–149.

Dwivedi, S., Upadhyaya, H., Senthilvel, S., Hash, C., Fukunaga, K., Diao, X., et al., 2012. Millets: genetic and genomic resources. In: Janick, J. (Ed.), Plant Breeding Reviews, vol. 35, John Wiley & Sons, USA, pp. 247–375.

Esechie, H.A., Al-Farsi, S.M., 2009. Performance of elite pearl millet genotypes under salinity stress in Oman. Crop Res. (Hisar) 37 (1/3), 28–33.

Frankel, O.H., 1984. Genetic perspectives of germplasm conservation. In: Arber, W., Llimensee, K., Peacock, W.J., Starlinger, P. (Eds.), Genetic Manipulation: Impact on Man and Society. Cambridge University Press, Cambridge, Part III, Paper No. 15.

Fussell, L.K., Pearson, C.J., Norman, M.J.T., 1980. Effect of temperature during various growth stages on grain development and yield of *Pennisetum americanum*. J. Exp. Bot. 31 (121), 621–633.

Gemenet, D.C., Hash, C.T., Sanogo, M.D., Sy, O., Zangre, R.G., Leiser, W.L., et al., 2015. Phosphorus uptake and utilization efficiency in West African pearl millet inbred lines. Field Crops Res. 171, 54–66.

Gernand, D., Rutten, T., Varshney, A., Rubtsova, M., Prodanovic, S., Bru, C., et al., 2005. Uniparental chromosome elimination at mitosis and interphase in wheat and pearl millet crosses involves micronucleus formation, progressive heterochromatinization, and DNA fragmentation. Plant Cell 17, 2431–2438.

Gildenhuys, P., Brix, K., 1961. Cytogenetic evidence of relationship between the $x = 7$ and $x = 9$ groups of *Pennisetum* species. Zuchter 31, 125–127.

Girgi, M., O'Kennedy, M., 2007. Pearl millet. Transgenic crops. Biotechnol. Agric. Forest. 59, 119–127.

Goel, S., Chen, Z., Akiyama, Y., Conner, J.A., Basu, M., Gualtieri, G., et al., 2006. Comparative physical mapping of the apospory-specific genomic region in two apomictic grasses: *Pennisetum squamulatum* and *Cenchrus ciliaris*. Genetics 173 (1), 389–400.

Gugsa, L., Haussmann, B.I.G., Kumlehn, J., Melchinger, A., 2013. Towards a protocol for double haploid production in pearl millet using wide hybridisation. In: Agricultural development within the rural-urban continuum, Tropentag, Stuttgart-Hohenheim, September 17–19, p. 1.

Gupta, S.K., Rai, K.N., Singh, P., Ameta, V.L., Gupta, S.K., Jayalekha, A.K., et al., 2015. Seed set variability under high temperatures during flowering period in pearl millet (*Pennisetum glaucum* (L.) R. Br.). Field Crops Res. 171, 41–53.

Gworgwor, N.A., 2001. Evaluation of pearl millet varieties for resistance to *Striga hermonthica*. Int. Sorghum Millets Newsl. 42, 85–87.

Hanna, W.W., 1983. Germplasm transfer from *P. purpureum* to *P. americanum*. Agronomy Abstracts. ASA, Madison, WI, p. 66.

Hanna, W.W., 1987. Utilization of wild relatives of pearl millet. In: Proceedings of the International Pearl Millet Workshop, April 7–11, 1986, ICRISAT, pp. 33–42.

Hanna, W.W., 1989. Characteristics and stability of a new cytoplasmic-nuclear male sterile source in pearl millet. Crop Sci. 29, 1457–1459.

Hanna, W.W., 1990. Transfer of germplasm from the secondary to the primary gene pool in *Pennisetum*. Theor. Appl. Genet. 80, 200–204.

Hanna, W.W., Dujardin, M., 1982. Apomictic interspecific hybrids between pearl millet and *Pennisetum orientale* L.C. Rich. Crop Sci. 22, 857–859.

Hanna, W.W., Wells, H.D., 1989. Inheritance of *Pyricularia* leaf spot resistance in pearl millet. J. Hered. 80 (2), 145–147.

Hanna, W.W., Wells, H.D., 1993. Registration of parental line Tift $89D_2$, rust resistant pearl millet. Crop Sci. 33 (2), 361–362.

Hanna, W.W., Gaines, T.P., Gonzalez, B., Monson, W.G., 1984. Effect of ploidy on yield and quality of pearl millet × napiergrass hybrids. Agron. J. 76 (6), 969–971.

Hanna, W.W., Wells, H.D., Burton, G.W., 1985. Dominant gene for rust resistance in pearl millet. J. Hered. 76, 134.

Hanna, W.W., Wells, H.D., Burton, G.W., 1987. Registration of pearl millet inbred parental lines, Tift $85D_2A_1$ and Tift $85D_2B_1$. Crop Sci. 27 (6), 1324–1325.

Hanna, W.W., Wilson, J.P., Wells, H.D., Gupta, S.C., 1993. Registration of Tift #5 S-1 pearl millet germplasm. Crop Sci. 33 (6), 1417–1418.

Harlan, J.R., 1975. Crops and Man. American Society of Agronomy – Crop Science Society of America, Madison, WI, USA.

Harlan, J.R., de Wet, J.M.J., 1971. Toward a rational classification of cultivated plants. Taxonomy 20, 509–517.

Howarth, C.J., Pollock, C.J., Peacock, J.M., 1997. Development of laboratory-based methods for assessing seedling thermotolerance in pearl millet. New Phytol. 137 (1), 129–139.

Hussey, M.A., Bashaw, E.C., Hignight, K.W., Wipff, J., Hatch, S.L., 1993. Fertilisation of unreduced female gametes: a technique for genetic enhancement within the *Cenchrus*–*Pennisetum* agamic complex. In: Baker, M.J., et al. (Eds.), Proceedings of the Seventeenth International Grassland Congress, Palmerston North, New Zealand, February 8–21, pp. 404–405.

IBPGR, 1985. Documentation of genetic resources: information handling systems for genebank management. In: Konopka, J., Hanson, J. (Eds.), Proceedings of a Workshop at the Nordic Gene Bank, Alnarp, Sweden, November 21–23, 1984, p. 87.

IBPGR and ICRISAT, 1993. Descriptors for Pearl Millet [*Pennisetum glaucum* (L.) R. Br.]. IBPGR/ICRISAT, Rome.

Ingham, L.D., Hanna, W.W., Baier, J.W., Hannah, L.C., 1993. Origin of the main class of repetitive DNA within selected *Pennisetum* species. Mol. Genet. 238, 350–358.

Jauhar, P.P., 1981a. Cytogenetics and Breeding of Pearl Millet and Related Species. Alan R. Liss, New York, p. 91.

Jauhar, P.P., 1981b. Cytogenetics of pearl millet. Adv. Agron. 34, 407–470.

Jauhar, P.P., Hanna, W.W., 1998. Cytogenetics and genetics of pearl millet. Adv. Agron. 64, 1–26.

Jauhar, P.P., Rai, K.D., Ozias-Akins, P., Chen, Z., Hanna, W.W., 2006. Genetic improvement of pearl millet for grain and forage production: cytogenetic manipulation and heterosis breeding. In: Singh, R.J., Jauhar, P.P. (Eds.), Genetic Resources, Chromosome Engineering, and Crop Improvement, vol. 2, Cereals. CRC, Taylor & Francis, Boca Raton, FL, pp. 281–307.

Kafi, M., Zamani, G., Ghoraishi, S.G., 2009. Relative salt tolerance of south Khorasan millets. Desert 14 (1), 63–70.

Kannan, B., Valencia, E., Altpeter F., 2013. Interspecific hybridization between elephantgrass and pearl millet and selection of hybrids with high-biomass production and enhanced biosafety. International Annual Meetings: Water, Food, Energy & Innovation for a Sustainable World. American Society of America (ASA), Crop Science Society of America (CSSA) and Soil Science Society of America (SSSA), Tampa, Florida, pp. 165.

Kapila, R.K., Yadav, R.S., Plaha, P., Rai, K.N., Yadav, O.P., Hash, C.T., et al., 2008. Genetic diversity among pearl millet maintainers using microsatellite markers. Plant Breed. 127, 33–37.

Kaushal, P., Sidhu, J.S., 2000. Pre-fertilization incompatibility barriers to interspecific hybridizations in *Pennisetum* species. J. Agric. Sci. 134, 199–206.

Kaushal, P., Roy, A.K., Zadoo, S.N., Choubey, R.N., 2004. Cytogenetic analysis of thermosensitive genic male sterility (TGMS) recovered from *Pennisetum glaucum* (L.) R. Br. × *P. violaceum* (Lam.) Rich cross. Cytologia 69, 409–418.

Kaushal, P., Khare, A., Zadoo, S.N., Roy, A.K., Malaviya, D.R., Agrawal, A., et al., 2008. Sequential reduction of *Pennisetum squamulatum* genome complement in *P. glaucum* ($2n = 28$) × *P. squamulatum* ($2n = 56$) hybrids and their progenies revealed its octoploid status. Cytologia 73 (2), 151–158.

Khairwal, I.S., Yadav, S.K., Rai, K.N., Upadhyaya, H.D., Kachhawa, D., Nirwan, B., et al., 2007. Evaluation and identification of promising pearl millet germplasm for grain and fodder traits. J. SAT Agric. Res. 5 (1), 1–6.

Kountche, B.A., Hash, C.T., Dodo, H., Laoualy, O., Sanogo, M.D., Timbeli, A., et al., 2013. Development of a pearl millet *Striga*-resistant genepool: response to five cycles of recurrent selection under *Striga*-infested field conditions in West Africa. Field Crops Res. 154, 82–90.

Krishnamurthy, L., Serraj, R., Rai, K.N., Hash, C.T., Dakheel, A.J., 2007. Identification of pearl millet [*Pennisetum glaucum* (L.) R. Br.] lines tolerant to soil salinity. Euphytica 158 (1–2), 179–188.

Kumar, S., Thakur, D.P., Singh, R., Arya, S., 1997. Reaction of pearl millet germplasm to ergot disease under natural field conditions. Ann. Biol. (Ludhiana) 13 (2), 297–298.

Kusaka, M., Ohta, M., Fujimura, T., 2005. Contribution of inorganic components to osmotic adjustment and leaf folding for drought tolerance in pearl millet. Physiol. Plant. 125 (4), 474–489.

Laurie, D.A., 1989. The frequency of fertilization in wheat × pearl millet crosses. Genome 32, 1063–1067.

Manga, V.K., Yadav, O.P., 1997. Development of a landrace based male-sterile line (CZMS 44A) of pearl millet. Crop Improv. 24 (1), 125–126.

Marchais, L., Pernès, J., 1985. Genetic divergence between wild and cultivated pearl millets (*Pennisetum typhoides*) I. Male sterility. Z. Pflanzenzüchtg 95, 103–112.

Marchais, L., Tostain, S., 1997. Analysis of reproductive isolation between pearl millet (*Pennisetum glaucum* (L.) R. Br.) and *P. ramosum*, *P. schweinfurthii*, *P. squamulatum*, and *Cenchrus ciliaris*. Euphytica 93, 97–105.

Mariac, C., Luong, V., Kapran, I., Mamadou, A., Sagnard, F., Deu, M., et al., 2006a. Diversity of wild and cultivated pearl millet accessions (*Pennisetum glaucum* [L.] R. Br.) in Niger assessed by microsatellite markers. Theor. Appl. Genet. 114, 49–58.

Mariac, C., Robert, T., Allinne, C., Remigereau, M.S., Luxereau, A., Tidjani, M., et al., 2006b. Genetic diversity and gene flow among pearl millet crop/weed complex: a case study. Theor. Appl. Genet. 113 (6), 1003–1014.

Martel, E., De Nay, D., Siljak-Yakovlev, S., Brown, S., Sarr, A., 1997. Genome size variation and basic chromosome number in pearl millet and fourteen related *Pennisetum* species. J. Hered. 88, 139–143.

Martel, E., Poncet, V., Lamy, F., Siljak-Yakovlev, S., Lejeune, B., Sarr, A., 2004. Chromosome evolution of *Pennisetum* species (Poaceae): implication of ITS phylogeny. Plant Syst. Evol. 249, 139–149.

Matzk, F., 1996. Hybrids of crosses between oat and Andropogoneae or Paniceae species. Crop Sci. 36, 17–21.

Miura, R., Terauchi, R., 2005. Genetic control of weediness traits and the maintenance of sympatric crop-weed polymorphism in pearl millet (*Pennisetum glaucum*). Mol. Ecol. 14, 1251–1261.

Mohamed, L., 1998. Inventory of some cultivated landraces threatened by genetic erosion in southern Tunisia. Plant Genet. Resour. Newsl. 113, 8–12.

Mohindra, V., Minocha, J.L., 1991. Pollen-pistil interactions and interspecific incompatibility in *Pennisetum*. Euphytica 56, 1–5.

Muallem, A.S., 1987. Genetic resources of cereal crops in PDR Yemen. 2. Barley, millet and maize. Plant Genet. Resour. Newsl. 72, 32–33.

Nadaf, S.K., Al-Hinai, S.A., Al-Farsi, S.M., Al-Lawati, A.H., Al-Bakri, A.N., 2010. Differential response of salt tolerant pearl millet genotypes to irrigation water salinity. In:

Mushtaque, A., Al-Rawahy, S.A., Hussain, N. (Eds.), A Monograph on Management of Salt-Affected Soils and Water for Sustainable Agriculture. Sultan Qaboos University, Oman, pp. 47–60.

Nagesh, C.H., Subrahmanyam, N.C., 1996. Interspecific hybridization of *Pennisetum glaucum* (L.) R. Br. with wild relatives. J. Plant Biochem. Biotechnol. 5, 1–5.

NBPGR, 2007. State of Plant Genetic Resources for Food and Agriculture in India (1996–2006): A Country Report. National Bureau of Plant Genetic Resources, New Delhi.

NRC, 1996. Lost Crops of Africa, vol. 1: Grains. Board on Science and Technology for International Development. National Research Council/National Academy Press, Washington, DC, pp. 372.

Obilana, A.B., Monyo, E.S., Gupta, S.C., 1997. Impact of genetic improvement in sorghum and pearl millet: developing country experiences. In: Proceedings of the International Conference on Genetic improvement of Sorghum and Pearl Millet, September 22–27, 1996, Lubbock, Texas, USA. Collaborative Research Support Program on Sorghum and Pearl Millet, Lincoln, Nebraska, USA, pp. 119–141.

Obok, E.E., 2013. Mineral contents of selected pearl millet (*Pennisetum glaucum* (L.) R. Br.) × elephant grass (*Pennisetum purpureum* (Schum.)) interspecific hybrids of Nigerian origin. J. Plant Stud. 2 (2), 22–27.

Obok, E.E., Aken'Ova, M.E., Iwo, G.A., 2012. Forage potentials of interspecific hybrids between elephant grass selections and cultivated pearl millet genotypes of Nigerian origin. J. Plant Breed. Crop Sci. 4 (9), 136–143.

Oumar, I., Mariac, C., Pham, J.L., Vigouroux, Y., 2008. Phylogeny and origin of pearl millet (*Pennisetum glaucum* [L.] R. Br) as revealed by microsatellite loci. Theor. Appl. Genet. 117, 489–497.

Pannu, P.P.S., Sokhi, S.S., Aulakh, K.S., 1996. Resistance in pearl millet against rust. Indian Phytopathol. 49 (3), 243–246.

Parker, C., Wilson, A.K., 1983. *Striga*-resistance identified in semi-wild 'Shibra' millet (*Pennisetum* sp.). Mededelingen van de Faculteit Landbouwwetenschappen, Rijksuniversiteit Genetics 48 (4), 1111–1117.

Pernès, J., 1984. Plant Genetic Resources Management vol. 1. ACCT, Paris, France.

Poncet, V., Lamy, F., Enjalbert, J., Joly, H., Sarr, A., Robert, T., 1998. Genetic analysis of the domestication syndrome in pearl millet (*Pennisetum glaucum* L., Poaceae): inheritance of the major characters. Heredity 81, 648–658.

Poncet, V., Lamy, F., Devos, K.M., Gale, M.D., Sarr, A., Robert, T., 2000. Genetic control of domestication traits in pearl millet (*Pennisetum glaucum* L., Poaceae). Theor. Appl. Genet. 100, 147–159.

Poncet, V., Martel, E., Allouis, S., Devos, M., Lamy, F., Sarr, A., et al., 2002. Comparative analysis of QTLs affecting domestication traits between two domesticated × wild pearl millet (*Pennisetum glaucum* L., Poaceae) crosses. Theor. Appl. Genet. 104 (6–7), 965–975.

Portères, R., 1976. African cereals: *Eleusine*, fonio, black fonio, teff, *Bracharia*, *Paspalum*, *Pennisetum* and African rice. In: Harlan, J.R., de Wet, J.M.J., Stemler, A.B.L. (Eds.), Origins of African Plant Domestication. Mouton, The Hague, The Netherlands.

Radhouane, L., 2013. Agronomic and physiological responses of pearl millet ecotype (*Pennisetum glaucum* (L.) R. Br.) to saline irrigation. Emirates J. Food Agric. 25 (2), 109–116.

Rai, K.N., 1995. A new cytoplasmic-nuclear male sterility system in pearl millet. Plant Breed. 114, 445–447.

Rai, K.N., Anand Kumar, K., Andrews, D.J., Rao, A., Raj, A.G.B., Witcombe, J.R., 1990. Registration of ICTP 8203 pearl millet. Crop Sci. 30, 959.

Rai, K.N., Rao, A.S., Hash, C.T., 1995. Registration of pearl millet parental lines ICMA 88004 and ICMB 88004. Crop Sci. 35, 1242.

Rai, K.N., Virk, D.S., Harinarayana, G., Appa Rao, S., 1996. Stability of male-sterile sources and fertility restoration of their hybrids in pearl millet. Plant Breed. 115, 494–500.

Rai, K.N., Appa Rao, S., Reddy, K.N., 1997. Pearl millet. In: Fuciillo, D., Sears, L., Stapleton, P. (Eds.), Biodiversity in Trust: Conservation and Use of Plant Genetic Resources in CGIAR Centers. Cambridge University Press, Cambridge, pp. 243–258.

Rai, K.N., Gowda, C.L.L., Reddy, B.V.S., Sehgal, S., 2008. The potential of sorghum and pearl millet in alternative and health food uses. Compr. Rev. Food Sci. Food Saf. 7, 340–352.

Rai, K.N., Velu, G., Govindaraj, M., Upadhyaya, H.D., Rao, A.S., Shivade, H., et al., 2015. Iniadi pearl millet germplasm as a valuable genetic resource for high grain iron and zinc densities. Plant Genet. Resour. 13 (1), 75–82.

Ramana Kumari, B., Kolesnikova-Allen, M.A., Hash, C.T., Senthilvel, S., Nepolean, T., Kavi Kishor, P.B., et al., 2014. Development of a set of chromosome segment substitution lines in pearl millet [*Pennisetum glaucum* (L.) R. Br.]. Crop Sci. 54 (5), 2175–2182.

Rao, Y.S., Rao, S.A., Mengesha, M.H., 1989. New evidence on the phylogeny of basic chromosome number in *Pennisetum*. Curr. Sci. 58 (15), 869–871.

Read, J.C., Bashaw, E.C., 1974. Intergeneric hybrid between pearl millet and buffelgrass. Crop Sci. 14 (3), 401–403.

Renno, J.F., Mariac, C., Poteaux, C., Bezancon, G., Lumaret, R., 2001. Haplotype variation of cpDNA in the agamic grass complex *Pennisetum* section *Brevivalvula* (Poaceae). Heredity 86, 537–544.

Robert, T., Khalfallah, N., Martel, E., Poncet, V., Remigereau, M., Rekima, S., et al., 2011. *Pennisetum*. In: Kole, C. (Ed.), Wild Crop Relatives: Genomic and Breeding Resources, vol. 9, Springer, Heidelberg, Berlin, pp. 217–255.

Roger, Z.G., Ramaiah, K.V., 1983. Screening of pearl millet cultivars for resistance to *Striga hermonthica*. Proceedings of the Second International Workshop on *Striga*, Ouagadougou, Upper Volta, October 5–8, 1981, pp. 77–81, 83–86.

Schmelzer, G.H., 1997. Review of *Pennisetum* section *Brevivalvula* (Poaceae). Euphytica 97, 1–20.

Sharma, R., Upadhyaya, H.D., Manjunatha, S.V., Rai, K.N., Gupta, S.K., Thakur, R.P., 2013. Pathogenic variation in the pearl millet blast pathogen *Magnaporthe grisea* and identification of resistance to diverse pathotypes. Plant Dis. 97 (2), 189–195.

Sharma, R., Upadhyaya, H.D., Sharma, S., Gate, V.L., Raj, C., 2015. Identification of new sources of resistance to multiple pathotypes of *Sclerospora graminicola* in the pearl millet mini core germplasm collection. Crop Sci. 55 (4), 1619–1628.

Singh, S.D., 1990. Sources of resistance to downy mildew and rust in pearl millet. Plant Dis. 74 (11), 871–874.

Singh, B.B., Jika, N., 1988. Five pearl millet genepools in Niger. Plant Genet. Resour. Newsl. 73/74, 29–30.

Singh, S.D., Navi, S.S., 2000. Genetic resistance to pearl millet downy mildew. II. Resistance in wild relatives. J. Mycol. Plant Pathol. 30 (2), 167–171.

Singh, S.D., Talukdar, B.S., 1998. Inheritance of complete resistance to pearl millet downy mildew. Plant Dis. 82 (7), 791–793.

Singh, P., Singh, U., Eggum, B.O., Anand Kumar, K., Andrews, D.J., 1987a. Nutritional evaluation of high protein genotypes of pearl millet (*Pennisetum americanum* (L.) Leeke). J. Sci. Food Agric. 38, 41–48.

Singh, S.D., Andrews, D.J., Rai, K.N., 1987b. Registration of ICLM 11 rust resistant pearl millet germplasm. Crop Sci. 27, 367–368.

Singh, S.D., Williams, R.J., Reddy, P.M., 1988. Isolation of downy mildew resistant lines from a highly susceptible cultivar of pearl millet. Indian Phytopathol. 41 (3), 450–456.

Singh, S.D., King, S.B., Reddy, P.M., 1990. Registration of five pearl millet germplasm sources with stable resistance to rust. Crop Sci. 30 (5), 1165.

Stapf, O., Hubbard, C.E., 1934. *Pennisetum*. In: Prain, D. (Ed.), Flora of Tropical Africa, vol. 9, Reeve & Co. Ltd, Ashford, Kent, pp. 954–1070, Part 6.

Sujata, V., Sivaramakrishnan, S., Rai, K.N., Seetha, K., 1994. A new source of cytoplasmic male sterility in pearl millet: RFLP analysis of mitochondrial DNA. Genome 37, 482–486.

Techio, V.H., Davide, L.C., Pereira, A.V., 2006. Meiosis in elephant grass (*Pennisetum purpureum*), pearl millet (*Pennisetum glaucum*) (Poaceae, Poales) and their interspecific hybrids. Genet. Mol. Biol. 29 (2), 353–362.

Thakur, R.P., King, S.B., 1988. Registration of six smut resistant germplasms of pearl millet. Crop Sci. 28 (2), 382–383.

Thakur, R.P., Williams, R.J., Rao, V.P., 1982. Development of resistance to ergot in pearl millet. Phytopathology 72 (4), 406–408.

Thakur, R.P., Rao, V.P., Williams, R.J., Chahal, S.S., Mathur, S.B., Pawar, N.B., et al., 1985. Identification of stable resistance to ergot in pearl millet. Plant Dis. 69 (11), 982–985.

Thakur, R.P., Rao, K.V.S., Williams, R.J., Gupta, S.C., Thakur, D.P., Nafade, S.D., et al., 1986. Identification of stable resistance to smut in pearl millet. Plant Dis. 70 (1), 38–41.

Tostain, S., 1992. Enzyme diversity in pearl millet (*Pennisetum glaucum*), wild millet. Theor. Appl. Genet. 83, 733–742.

Tostain, S., 1993. Evaluation de la diversité des mils pénicillaires diploïdes (*Pennisetum glaucum* (L.) R. Br.) au moyen de marqueurs enzymatiques. Etudes des relations entre formes sauvages et cultivées. PhD Thesis, Université Paris XI, Orsay, France.

Tostain, S., Riandey, M.F., Marchais, L., 1987. Enzyme diversity in pearl millet (*Pennisetum glaucum*), West Africa. Theor. Appl. Genet. 74, 188–193.

Upadhyaya, H.D., 2009. Reference set of pearl millet. Grain Legumes, 1, http://oar.icrisat.org/4197/1/Web_Art_2009_Reference_Set_of_Pearl_Millet.pdf.

Upadhyaya, H.D., Gowda, C.L.L., 2009. Managing and Enhancing the Use of Germplasm – Strategies and Methodologies. Technical Manual No. 10. ICRISAT, Patancheru.

Upadhyaya, H.D., Ortiz, R., 2001. A mini-core collection for capturing diversity and promoting utilization of chickpea genetic resources in crop improvement. Theor. Appl. Genet. 102, 1292–1298.

Upadhyaya, H.D., Gowda, C.L.L., Buhariwalla, H.K., Crouch, J.H., 2006. Efficient use of crop germplasm resources: identifying useful germplasm for crop improvement through core and mini-core collections and molecular marker approaches. Plant Genet. Resour. 4 (1), 25–35.

Upadhyaya, H.D., Reddy, K.N., Gowda, C.L.L., 2007. Pearl millet germplasm at ICRISAT genebank – status and impact. J. SAT Agric. Res. 3 (1), 1–5.

Upadhyaya, H.D., Gowda, C.L.L., Reddy, K.N., Singh, S., 2009a. Augmenting the pearl millet core collection for enhancing germplasm utilization in crop improvement. Crop Sci. 49 (2), 573.

Upadhyaya, H.D., Reddy, K.N., Irshad Ahmed, M., Gowda, C.L.L., Haussmann, B.I.G., 2009b. Identification of geographical gaps in the pearl millet germplasm conserved at ICRISAT genebank from West and Central Africa. Plant Genet. Resour. 8 (1), 45–51.

Upadhyaya, H.D., Reddy, K.N., Irshad Ahmed, M., Gowda, C.L.L., 2010. Identification of gaps in pearl millet germplasm from Asia conserved at the ICRISAT genebank. Plant Genet. Resour. 8 (3), 267–276.

Upadhyaya, H.D., Yadav, D., Reddy, K.N., Gowda, C.L.L., Singh, S., 2011. Development of pearl millet minicore collection for enhanced utilization of germplasm. Crop Sci. 51 (1), 217.

Upadhyaya, H.D., Reddy, K.N., Irshad Ahmed, M., Gowda, C.L.L., 2012. Identification of gaps in pearl millet germplasm from East and Southern Africa conserved at the ICRISAT genebank. Plant Genet. Resour. 10 (3), 202–213.

Upadhyaya, H.D., Reddy, K.N., Singh, S., Gowda, C.L.L., Irshad Ahmed, M., Kumar, V., 2014a. Diversity and gaps in *Pennisetum glaucum* subsp. *monodii* (Maire) Br. germplasm conserved at the ICRISAT genebank. Plant Genet. Resour. 12 (2), 226–235.

Upadhyaya, H.D., Reddy, K.N., Singh, S., Irshad Ahmed, M., Vinod, K., Ramachandran, S., 2014b. Geographical gaps and diversity in Deenanath grass (*Pennisetum pedicellatum* Trin.) germplasm conserved at the ICRISAT genebank. Indian J. Plant Genet. Resour. 27 (2), 93–101.

Vadez, V., Kholová, J., Yadav, R.S., Hash, C.T., 2013. Small temporal differences in water uptake among varieties of pearl millet (*Pennisetum glaucum* (L.) R. Br.) are critical for grain yield under terminal drought. Plant Soil 371 (1–2), 447–462.

Van Hintum, T.J.L., van de Wiel, C.C.M., Visser, D.L., van Treuren, R., Vosman, B., 2007. The distribution of genetic variation in a *Brassica oleracea* genebank collection related to the effects on diversity of regeneration, as measured with AFLPs. Theor. Appl. Genet. 114, 777–786.

Van Oosterom, E.J., Weltzien, E., Yadav, O.P., Bidinger, F.R., 2006. Grain yield components of pearl millet under optimum conditions can be used to identify germplasm with adaptation to arid zones. Field Crops Res. 96 (2/3), 407–421.

Van Treuren, R., Van Hintum, T.J.L., 2014. Next-generation genebanking: plant genetic resources management and utilization in the sequencing era. Plant Genet. Resour. 12 (3), 298–307.

Van Treuren, R., Van Hintum, T.J.L., Van de Wiel, C.C.M., 2008. Marker-assisted optimization of an expert-based strategy for the acquisition of modern lettuce varieties to improve a genebank collection. Genet. Resour. Crop Evol. 55, 319–330.

Van Treuren, R., Engels, J.M.M., Hoekstra, R., Van Hintum, T.J.L., 2009. Optimization of the composition of crop collections for *ex situ* conservation. Plant Genet. Resour. 7, 185–193.

Vari, A.K., Sidhu, J.S., Minocha, J.L., 1999. Cytogenetics. In: Khairwal, I.S., Rai, K.N., Andrews, D.J., Harinarayana, G. (Eds.), Pearl Millet Breeding. Oxford and IBH Publishing Co. Pvt. Ltd, New Delhi, pp. 83–117.

Vidhya, K., Fazlullah Khan, A.K., 2003. Hybrid between *P. schweinfurthii* and Napier grass. Cytologia 68 (2), 183–190.

Vom Brocke, K., Christinck, A., Eva-Weltzien, R., Presterl, T., Geiger, H.H., 2003. Farmers seed system and management practices determine pearl millet genetic diversity in semiarid regions of India. Crop Sci. 43, 1680–1689.

Watson, L., Dallwitz, M.J., 1992. The Grass Genera of the World. CAB International, Wallingford, pp. 674–676.

Wilson, J.P., Burton, G.W., 1991. Registration of Tift 3 and Tift 4 rust resistant pearl millet germplasms. Crop Sci. 31 (6), 1713.

Wilson, J.P., Hanna, W.W., 1992. Disease resistance in wild *Pennisetum* species. Plant Dis. 76 (11), 1171–1175.

Wilson, J.P., Burton, G.W., Wells, H.D., Zongo, J.D., Dicko, I.O., 1989a. Leaf spot, rust and smut resistance in pearl millet landraces from central Burkina Faso. Plant Dis. 73 (4), 345–349.

Wilson, J.P., Wells, H.D., Burton, G.W., 1989b. Inheritance of resistance to *Pyricularia grisea* in pearl millet accessions from Burkina Faso and inbred Tift 85DB. J. Hered. 80 (6), 499–501.

Wilson, J.P., Hess, D.E., Hanna, W.W., Kumar, K.A., Gupta, S.C., 2004. *Pennisetum glaucum* subsp. *monodii* accessions with *Striga* resistance in West Africa. Crop Prot. 23, 865–870.

Wilson, J.P., Sanogo, M.D., Nutsugah, S.K., Angarawai, I., Fofana, A., Traore, H., et al., 2008. Evaluation of pearl millet for yield and downy mildew resistance across seven countries in sub-Saharan Africa. Afr. J. Agric. Res. 3 (5), 371–378.

Wright, B., 1997. Crop genetic resource policy: the role of *ex situ* genebanks. Aust. J. Agric. Resour. Econ. 41, 81–115.

Wu, S., Tsai, C., 1963. Cytological studies on the intergeneric F_1 hybrid between *Oryza sativa* L. × *Pennisetum* sp. Acta Bot. Sin. 11 (4), 293–307.

Yadav, O.P., 2008. Performance of landraces, exotic elite populations and their crosses in pearl millet (*Pennisetum glaucum*) in drought and non-drought conditions. Plant Breed. 127 (2), 208–210.

Yadav, O.P., 2010. Evaluation of landraces and elite populations of pearl millet for their potential in genetic improvement for adaptation to drought-prone environments. Indian J. Genet. Plant Breed. 70 (2), 120–124.

Yadav, M.S., Duhan, J.C., 1996. Screening of pearl millet genotypes for resistance to smut disease. Plant Dis. Res. 11 (1), 95–96.

Yadav, O.P., Rai, K.N., 2013. Genetic improvement of pearl millet in India. Agric. Res. 2 (4), 275–292.

Zenkteler, M., Nitzsche, W., 1984. Wide hybridization experiments in cereals. Theor. Appl. Genet. 68, 311–315.

Finger and foxtail millets

7

Mani Vetriventhan, Hari D. Upadhyaya, Sangam Lal Dwivedi, Santosh K. Pattanashetti, Shailesh Kumar Singh
International Crops Research Institute for the Semi-Arid Tropics (ICRISAT), Genebank, Patancheru, Telangana, India

7.1 Introduction

Foxtail and finger millets are the second and third most important crops among millets after pearl millet. Foxtail millet is widely cultivated in Asia, Europe, North America, Australia, and North Africa for grains or forage, and an essential food for human consumption in China, India, Korea, and Japan (Austin, 2006). China ranks top in foxtail millet production with the annual cultivating area of about 2 million ha and an annual total grain production of about 6 Mt (Diao, 2011). Finger millet accounts for 12% of the global millets area and is grown in more than 25 countries in eastern and southern Africa, and across Asia from the Near East to the Far East. The major finger millet producing countries are Uganda, India, Nepal, and China (www.cgiar.org/our-research/crop-factsheets/millets/).

Foxtail and finger millets are good sources for micro and macronutrients with high nutraceutical and antioxidant properties. These crops are rich in protein, fat, crude fiber, iron, and other minerals and vitamins. Foxtail millet contains almost twice the amount of protein (11.2%) and fat (4%) as compared to rice, while finger millet contains over >10-fold higher calcium as compared to other cereals including rice and wheat (Saleh et al., 2013). Upadhyaya et al. (2011a) identified grain nutrients rich accessions in finger millet core collection (Upadhyaya et al., 2006a) having 37.66–65.23 mg/kg of Fe, 22.46–25.33 mg/kg of Zn, 3.86–4.89 g/kg of Ca, and 8.66–11.09% of protein. Similarly, Upadhyaya et al. (2011b) identified grain nutrients rich accessions in foxtail millet core collection (Upadhyaya et al., 2008) having 171.2–288.7 mg/kg of Ca, 58.2–68.0 mg/kg of Fe, 54.5–74.2 mg/kg of Zn, and 15.6–18.5% of protein. The husked grains of foxtail millet are used as food in Asia, southeastern Europe, and Africa. The flour is used for making cakes, porridges, and puddings. Foxtail millet is used in the preparation of beer and alcohol, especially in Russia and Myanmar, and for vinegar and wine in China, and primarily grown as bird feed, hay, and silage in Europe and the United States, while in China, the straw is an important fodder. Similarly, finger millet is used as food in Asia and Africa, and flour is used to prepare porridge and usually served with a side dish of vegetables, meat, or fish. In Africa, finger millet provides malt for making local beer and other alcoholic or nonalcoholic beverages. Finger millet straw is used as forage for cattle, sheep, and

Genetic and Genomic Resources for Grain Cereals Improvement. http://dx.doi.org/10.1016/B978-0-12-802000-5.00007-1

goats. In Uganda, the by-products of finger millet beer production are fed to chickens, pigs, and other animals (www.protabase.org).

Finger and foxtail millets are important ancient crops of dryland agriculture and the potential climate-resilient crops for food and nutritional security in the climate change scenario. However, mostly farmers cultivate unimproved varieties or traditional landraces that yields poorly. It is mainly because of unavailability of improved varieties, limited research efforts, and funding for these crops. Assessing genetic variability of germplasm collections, development, and use of genetic and genomic resources for breeding high-yielding cultivars, developing crop production and processing technologies, value addition for improving consumption, public private partnerships, and policy recommendations are needed to upscale these crops to make them more remunerative to farmers.

7.2 Origin, distribution, diversity, and taxonomy

7.2.1 Finger millet

Finger millet (*Eleusine coracana* (L.) Gaertn.) is an allotetraploid evolved from its wild progenitor, *E. coracana* subsp. *africana*. The genus *Eleusine* contains about 10 species, both annuals and perennials, with three basic chromosome numbers 8, 9, and 10. Four are tetraploids, namely, *E. coracana* ($2n = 4x = 36$, AABB), *Eleusine africana* ($2n = 4x = 36$, AABB) and *Eleusine kigeziensis* ($2n = 4x = 36$, AADD), and *Eleusine reniformis* ($2n = 4x = 36$); Seven are diploids with a basic chromosome number of 8 in *Eleusine multiflora* ($2n = 2x = 16$, CC), 9 in *Eleusine indica* ($2n = 2x = 18$, AA), *Eleusine tristachya* ($2n = 18$, AA), *Eleusine floccifolia* ($2n = 18$, BB), *Eleusine intermedia* ($2n = 18$, AB), and *Eleusine verticillata* ($2n = 2x = 18$), and 10 in *Eleusine jaegeri* ($2n = 2x = 20$, DD) (Hiremath and Chennaveeraiah, 1982; Neves et al., 2005; Dwivedi et al., 2012). *E. coracana* subsp. *africana* is considered as a putative progenitor to cultivated finger millet, *E. coracana* subsp. *coracana*, and are completely cross-compatible and produce fertile hybrids (Mehra, 1962; Hiremath and Salimath, 1992).

Domestication of cultivated finger millet, *E. coracana* started around 5000 years ago in Western Uganda and the Ethiopian highlands and the crop extended to the Western Ghats of India around 3000 BC (Hilu et al., 1979; Hilu and de Wet, 1976). Cytologic analyses of hybrids, chloroplast DNA restriction analysis, and *in situ* hybridization of diploid and polyploidy species shows that *E. indica* is the "A" genome donor, while *E. floccifolia* is the "B" genome donor of cultivated *E. coracana* (Bisht and Mukai, 2001; Hiremath and Salimath, 1992; Hilu, 1988). Contrary to this, Liu et al. (2014) suggest *E. indica* as the primary A genome parent, while *E. tristachya* or its extinct sister or ancestor as the secondary A genome parent for derivation of *E. coracana*, while B genome donor is extinct. This is also supported by the close phylogenetic relationships of diploids, *E. indica* and *E. tristachya* with *E. africana*, *E. coracana*, and *E. kigeziensis* for cpDNA and nrDNA *Pepc4* (Neves et al., 2005; Liu et al., 2011b).

Isozyme and DNA marker analyses have revealed that cultivated finger millet has a narrow genetic base, but variation in the wild subspecies is considerably higher (Werth et al., 1994; Muza et al., 1995; Salimath et al., 1995a; Dagnachew et al., 2014). Considerable diversity is found in finger millet, wherein based on inflorescence morphology they can be grouped into races and subraces (Prasada Rao et al., 1993). The species *E. coracana* consists of two subspecies, *africana* (wild) and *coracana* (cultivated). The subsp. *africana* has two wild races, *africana* and *spontanea*, while subsp. *coracana* has four cultivated races; *elongata*, *plana*, *compacta*, and *vulgaris*. These cultivated races are further divided into subraces; *laxa*, *reclusa*, and *sparsa* in race *elongata*; *seriata*, *confundere*, and *grandigluma* in race *plana*; and *liliacea*, *stellata*, *incurvata*, and *digitata* in race *vulgaris*. The race *compacta* has no subraces (de Wet et al., 1984; Prasada Rao and de Wet, 1997).

7.2.2 Foxtail millet

Foxtail millet (*Setaria italica* (L.) P. Beauv.) is a member of the subfamily Panicoideae and the tribe Paniceae with chromosome number of $2n = 2x = 18$ (AA). It is an important ancient crop of dry land agriculture, grown since >10,500 years ago in China (Yang et al., 2012). The green foxtail, *Setaria viridis* ($2n = 2x = 18$, AA), is a wild ancestor of cultivated foxtail millet. The genus *Setaria* is organized into three gene pools based on observations drawn from interspecific hybridization and hybrid pollen fertility. The primary gene pool is composed of cultivated foxtail (*S. italica*) and its putative wild ancestor *S. viridis* (Harlan and de Wet, 1971). The secondary gene pool contains *Setaria adhaerans* ($2n = 2x = 18$) and two allotetraploids *Setaria verticillata* and *Setaria faberii* ($2n = 4x = 36$) (Li et al., 1942; Benabdelmouna et al., 2001). The tertiary gene pool contains *Setaria glauca* (or *Setaria pumila*, $4x$ to $8x$) in addition to many other wild species. Morphological and molecular studies on cultivated and green foxtail revealed large genetic diversity (Reddy et al., 2006; Upadhyaya et al., 2008; Vetriventhan et al., 2012; Wang et al., 2010a, 2012; Jia et al., 2013b).

Several hypotheses regarding the origin and domestication have been proposed and a multiple domestication hypothesis has been widely accepted (de Wet et al., 1979; Li et al., 1995). Li et al. (1995) suggest a multiple domestication hypothesis with three centers, that is, China, Europe, and Afghanistan–Lebanon. A study by Hirano et al. (2011) on the geographical genetic structure of 425 landraces of foxtail millet and 12 accessions of green foxtail by transposon display (TD) as a genome-wide marker shows two clear genetic borders: (1) between accession from East Asia and those from other regions including Central, South, or Southeast Asia, and the Middle East, and (2) between West Europe and East Europe suggesting strong regional differentiations and a long history of the cultivation in each region, supporting multiple domestications events of foxtail millet.

Foxtail millet has abundant within-species diversity. Prasada Rao et al. (1987) suggested three races of foxtail millet based on the comparative morphology of the foxtail millet accessions: (1) race *moharia* is common in Europe, southeast Russia, Afghanistan, and Pakistan; (2) race *maxima* is common in eastern China, Georgia (Eurasia), Japan, Korea, Nepal, and northern India (it has also been introduced in the United States); and (3) race *indica* is found in the remaining parts of India and Sri Lanka. These races can be further divided into 10 subraces (*aristata*, *fusiformis*, and *glabra* in *moharia*; *compacta*, *spongiosa*, and *assamense* in *maxima*; and *erecta*, *glabra*, *nana*, and *profusa* in *indica*). Later, Li et al. (1995) added the race *nana* along with *maxima*, *moharia*, and *indica* and described the plants that resemble the wild green millet, and are very short and slender, with many tillers, very short panicles with poor yield performance, and early maturity as a separate race *nana*.

7.3 Erosion of genetic diversity from the traditional areas

Loss of genetic diversity (genetic erosion), including the loss of individual genes or particular combinations of genes, and loss of varieties and crops occur rapidly in crops mainly because of replacement of traditional landraces by modern, high-yielding cultivars, natural catastrophes, and large-scale destruction and modification of natural habitats harboring wild species. Genetic erosion of foxtail and finger millets occurs mostly due to their neglect and often replacement with commercial or nonfood crops. Decline in finger millet cultivation in Socotra (an island in Yemen) and Kabale Highlands, Uganda has been reported (Bawazir and Bamousa, 2014; Mbabwine et al., 2005). Assessment of the status of plant genetic resources in Kabale Highlands, Uganda revealed that finger millet is one of the threatened crops where only few farmers cultivate finger millet and many have stopped its cultivation (Mbabwine et al., 2005). In India, the area under cultivation of foxtail and finger millets and other small millets declined mainly due to poor yield, unavailability of improved cultivars, and policy shift that focuses on rice and wheat.

7.4 Status of germplasm resource conservation

Large numbers of foxtail and finger millets germplasm accessions are available to the scientific community. Globally >46,000 foxtail millet and >37,000 finger millet germplasm accessions have been conserved *ex situ* in genebanks. The major collections of foxtail millet germplasm accessions are housed at China, India, France, and Japan, while India and African countries such as Kenya, Ethiopia, Uganda, and Zambia conserve major finger millet collections (Table 7.1).

7.5 Germplasm evaluation and maintenance

Foxtail and finger millets are highly self-pollinating crops, so there is no special regeneration and maintenance practice as in the case of cross-pollinated crops such as pearl millet. The field used for regeneration should not have grown the same crops in the previous year in order to avoid volunteer plants. Individual accessions can be planted in rows (4 m length) and harvested panicles by hand will be bulked to make up the accession. Considerable efforts have been made in foxtail and finger millets

Table 7.1 **Major genebanks across the globe conserving foxtail and finger millet germplasm**

Country	Institute	Germplasm accessions		
		Cultivated	Wild	Total
Finger millet				
India	National Bureau of Plant Genetic Resources (NBPGR)	9511	11	9522
	AICRP on Small Millets	6257	–	6257
	ICRISAT	5880	204	6084
	The Ramaiah Gene Bank, Tamil Nadu Agricultural University	2219	–	2219
	University of Agricultural Science, Banglore	1019	–	1019
Kenya	National Genebank of Kenya, Crop Plant Genetic Resources Centre, Muguga (KARI-NGBK)	2854	77	2931
Ethiopia	Ethiopian Institute of Biodiversity (EIB)	2173	–	2173
Uganda	Serere Agriculture and Animal Production Research Institute (SAARI)	1231	–	1231
Zambia	SADC Plant Genetic Resources Centre (SRGB)	1037	3	1040
Foxtail millet				
China	Institute of Crop Science, Chinese Academy of Agricultural Sciences (ICS-CAAS)	26233	–	26233
India	NBPGR	4384	8	4392
	AICRP on Small Millets	2512	–	2512
	ICRISAT	1488	54	1542
France	Laboratoire des Ressources Génétiques et Amélioration des Plantes Tropicales, (ORSTOM-MONTP)	3500	–	3500
Japan	Department of Genetic Resources I, National Institute of Agrobiological Sciences (NIAS)	2505	26	2531

Source: http://www.fao.org/wiews-archive/germplasm_query.htm; Institutes/genebanks with >1000 accessions considered for listing.

germplasm evaluation for various traits of economic interest, including biotic and abiotic stresses tolerance and grain nutritional content, and are discussed hereunder.

7.5.1 Agronomic traits

Large genetic variation for morphoagronomic traits has been found in foxtail (Li et al. 1995; Upadhyaya et al., 2008; Nirmalakumari and Vetriventhan, 2010) and finger millets (Upadhyaya et al., 2006a, 2007; Suryanarayana et al., 2014). For example, at the ICRISAT genebank, finger millet germplasm accessions conserved have large variation for days to flowering (50–120 days), plant height (30–240 cm), basal tillers (1–70),

Table 7.2 Diversity for agronomic traits in finger and foxtail millet active collection conserved at ICRISAT, Patancheru

Crop/trait	Mean	Range
Finger millet		
Days to flowering-rainy	80.4	50–120
Plant height (cm)-rainy	100.7	30–240
Basal tillers number	5.2	1–70
Flag leaf blade length (mm)	358.1	100–750
Flag leaf blade width (mm)	12.6	5–20
Flag leaf sheath length (mm)	102.5	8–280
Peduncle length (mm)	215.5	18–450
Panicle exsertion (mm)	113.5	0–360
Inflorescence length (mm)	93.1	10–320
Inflorescence width (mm)	78.4	7–460
Longest finger length (mm)	72.6	10–250
Longest finger width (mm)	11.6	2–50
Panicle branches number	7.7	2–27
Foxtail millet		
Days to flowering-rainy	53.5	32–135
Plant height (cm)-rainy	110.0	20–215
Basal tillers number	7.5	1–52
Flag leaf blade length (mm)	284.7	30–520
Flag leaf blade width (mm)	20.2	5–40
Flag leaf sheath length (mm)	138.5	50–260
Peduncle length (mm)	299.6	80–500
Panicle exsertion (mm)	162.5	10–360
Inflorescence length (mm)	163.1	10–390
Inflorescence width (mm	19.2	5–120
Weight of 5 panicles (g)	30.1	0.6–117

inflorescence length (10–320 mm), and so on; similarly in foxtail millet for days to flowering (32–135 days), plant height (20–215 cm), basal tillers number (1–52), inflorescence length (10–390 mm), and so on (Table 7.2). Upadhyaya et al. (2011b) identified 21 accessions of foxtail millet with higher grain yield compared to the best control cultivar. The ICRISAT global finger millet composite collection was evaluated for morphoagronomic traits and identified best-performing accessions for grain yield, early flowering, more number of fingers, high basal tiller number, and ear head length (Table 7.3).

7.5.2 Grain nutrients

At ICRISAT, finger millet core collection accessions assessed for genetic variability for grain nutrient contents and identified 15 promising accessions each for grain Fe (37.66–65.23 mg/kg), Zn (22.46–25.33 mg/kg), Ca (3.86–4.89 g/kg), and protein (8.66–11.09%) contents, and 24 accessions were selected based on their superiority over

Table 7.3 Germplasm/cultivars reported as sources for agronomic and nutritional traits and resistant/tolerant to biotic and abiotic stresses in finger millet and foxtail millet

Trait	Germplasm/cultivar sources	References
Finger millet		
Early flowering	IE# 49, 120, 189, 196, 234, 501, 509, 581, 588, 600, 641, 694, 847, 2030, 2093, 2158, 2275, 2293, 2322, 2323, 2957, 3104, 3537, 3543, 4425, 4431, 4432, 4442, 4711, 4734, 4755, 4759, 6013, 6550	Bharathi (2011)
Basal tillers	IE# 2296, 2034, 4711, 2293, 2299, 2608, 2619, 5145, 6553, 847, 2408, 2534, 3987, 1013, 120, 2042, 2091, 2106, 2139, 2146, 2233, 2288, 2367, 2410, 2504, 2645, 2657, 2674	Bharathi (2011)
Finger number	IE# 6033, 3790, 4586, 6059, 3111, 4476, 3106, 2914, 4677, 5733, 5875, 5877, 4257, 5105, 5563, 6510, 4297, 2957, 5689, 5956, 4563, 3120, 2816, 6013, 2303, 2591, 6252, 6241, 4866	Bharathi (2011)
Head length	IE# 2223, 2621, 2789, 6553, 3581, 3431, 3722, 6512, 2108, 2781, 3046, 2486, 5321, 3704, 798, 3489, 5022, 2591, 2608, 4476, 2611, 3531, 2336, 4125, 4658, 6546	Bharathi (2011)
Forage yield	IE# 2117, 24, 2568, 2651, 2753, 2796, 2811, 2880, 2942, 2979, 3789, 50, 672, 715, 860, 908, 916, 96, 99	Bharathi (2011)
Grain yield	IEs 94, 2340, 2498, 2578, 2587, 2683, 2773, 2903, 2983, 2992, 3194, 3790, 3802, 4600, 4974, 5198, 5472, 3663, 3693, 3744, 4121, 4310, 4679, 5862, 6142, 6236, 667, 1010, 2299, 2590, 2678, 2684, 2698, 2712, 2756, 2827, 2872, 3135, 3136, 3270	Bharathi (2011)
Iron	IE# 4708, 2921, 4709, 588, 5736, 4476, 942, 4734, 5794, 4107, 7338, 2093, 5870, 4443, 817	Upadhyaya et al. (2011a)
Zinc	IE #3120, 7508, 6546, 3025, 7386, 7407, 615, 712, 5788, 633, 2008, 1023, 886, 4817, 510	Upadhyaya et al. (2011a)
Calcium	IE# 4476, 2030, 6546, 4708, 2568, 2957, 6537, 2608, 2572, 2921, 4443, 2780, 4866, 7386, 4709, CO# 9, 11, GE 2491, Malawi # 1305, 1314, 1861, 1866, 1895, 1907, 1915, 1940, 1952, IE# 3156, 3184, 3799, 3802	Upadhyaya et al. (2011a); Vadivoo et al. (1998)
Protein	IE #6537, 9, 4709, 4708, 6541, 2921, 6546, 4476, 4443, 588, 6013, 2093, 4817, 3120, 3101, Malawi# 1305, 1314, 1861, 1907, 1958, 2049, MS# 174, 887, 1168, 2777, 2784, 2869, GE# 37, 60, 1106, 2491, 2500, CO# 7, 9, IE# 3156, 3184	Upadhyaya et al. (2011a); Vadivoo et al. (1998)

(Continued)

Table 7.3 Germplasm/cultivars reported as sources for agronomic and nutritional traits and resistant/tolerant to biotic and abiotic stresses in finger millet and foxtail millet *(cont.)*

Trait	Germplasm/cultivar sources	References
Blast	ED 201-5A, ICM 401, PRM 9802, SANJI 1, TNAU 1009, VL# 234, 324, 328, 330, 332, 333, Genotype no. 2400, 4313, 4914, 4915, 4929, 4966, 5102, 5126, 5148, IE #1055, 2821, 2872, 4121, 4491, 4570, 5066, 5091, 5537	Kumar and Kumar (2009); Mantur et al. (2001); Babu et al. (2013b);
Drought	PR202, VL315, PES 400, PRM# 8107, 8112, VL 315	Bhatt et al. (2011); Gupta et al. (2014a)
Heat stress	GP # 3, 111, 153	Babu et al. (2013a)
Salinity	GPU 48, Indaf 5, Co 12, Trichy 1, IE #518, 2034, 2217, 2790, 2872, 3045, 3077, 3391, 3470, 3973, 4073, 4329, 4671, 4673, 4757, 4789, 4795, 4797, 5066, 6154, 6165, 6326	Shailaja and Thirumeni (2007); Vijayalakshmi et al. (2014); Krishnamurthy et al. (2014b)
Foxtail millet		
Grain yield	ISe# 710, 969, 1820, 388, 842, 49, 1888, 90, 364, 1767, 362, 1808, 846, 869, 1511, 909, 1846, 1610, 795, 1458, 1704	Upadhyaya et al. (2011b)
Early flowering	ISe# 1312, 1151, 1227, 1201, 1234, 1335, 1286, 1161, 1320, 1647, 1638, 1037, 1181, 1563, 1254, 1204, 1547, 1187, 403, 1118, 1163	Upadhyaya et al. (2011b)
Calcium	ISe# 1227, 1181, 1059, 1419, 827, 751, 1474, 1685, 900, 840, 1629, 1851, 769, 1581, 1286, 1136, 1161, 1773, 931, 869, 663	Upadhyaya et al. (2011b)
Iron	ISe# 1151, 1286, 1400, 1305, 1332, 1059, 1581, 1320, 1312, 144, 1163, 1460, 160, 1037, 1597, 1009, 1161, 1704, 1187, 1745, 838, GPUS# 14, 18, S 130, SiA# 2619, 326, 2599, ATPS 83, ISC 247, TNAU 43	Upadhyaya et al. (2011b); Philip and Maloo (1996)
Zinc	ISe# 1286, 748, 1387, 195, 1134, 1408, 1419, 1161, 900, 1820, 1320, 1654, 1704, 1605, 403, 1808, 751, 1674, 144, 1234, 985	Upadhyaya et al. (2011b)
Protein	ISe# 1312, 1227, 1789, 1254, 1541, 827, 748, 1305, 1647, 1335, 751, 1118, 1134, 1151, 195, 1234, 1067, 1419, 144, 735, 1161	Upadhyaya et al. (2011b)
Blast	GPUS-6, AZJ-11, SIA-2592, SIA-2593, SIA-2596, SIA-2606, SiA-2608, RSE-62, Niangu 1, Jinan 8337, ISe #375, 376, 748, 751, 769, 771, 785, 846, 1059, 1067, 1137, 1181, 1187, 1201, 1204, 1258, 1286, 1320, 1335, 1387, 1419, 1541, 1547, 1563, 1575, 1593, 1685, 1704	Jain et al. (1991); Tian and Quan (1995); Zhang and Guan (1995); Sharma et al. (2014)

Table 7.3 Germplasm/cultivars reported as sources for agronomic and nutritional traits and resistant/tolerant to biotic and abiotic stresses in finger millet and foxtail millet *(cont.)*

Trait	Germplasm/cultivar sources	References
Brown spot	GPUS# 26, 27, SiA #3039, 3059, 3064, 3066, 3088, TNAU #213, 235, 225, DHGR# 2061, 2062	Kumar and Kumar (2009)
Banded leaf and sheath blight	RFM# 82, 83, 84, 85, 87, 88, 90, 93, 94, 95, 96, 97	Jain et al. (2014)
Downy mildew	Luyu 2, Luyu 6, Wanchi 1, Yugengze and Baisu from Japan, Pingrangsu, Qiushusu and Duolangsu, ISe# 25, 30-1, 172, 274, 465, Meera (SR 16), Tie Gu 7, Longgu 28, Jingu 16, Jingu 11, Lugu No 7, Yugo No 3, Lujin 3, Beihuang, Zhenggu 2	Wang et al. (1997); Maloo et al. (2001); Wang et al. (1998a); Jiyaju (1989); Jiyaju and Yuzhi (1993); Dwivedi et al. (2012)
Smut	Jingu 16, Lugu No 7, K8763, Sarativskoye 2, Sarativskoye 3, Sarativskoye 6, Veselepodolyanskoye 632, Barnaulskoye 80, Gorilinka, Tie Gu 7	Jiyaju (1989); Jiyaju and Yuzhi (1993); Wang et al. (1998a)
Rust	Lugu No 7, Yugu No 2, Yugu No 3, Niangu 1	Jiyaju and Yuzhi (1993); Tian and Quan (1995)
Drought	BSi-1, EM 15/BSi 467, EM 8/BSi 467, Tie Gu 7, Jinan 8337	Begum et al. (2013); Wang et al. (1998a); Zhang and Guan (1995)
Lodging	Tie Gu 7, Longgu 28, Nenxian 13, Jingu 11, Yugu No. 1, Yugu No. 2, Yegu 5, Yanggu, Liuyuexian 2, Cang 155, Gufeng 1, An 4844, Heng 8735, Ji 9409, Pin 324, Zheng 9188, Pin 540, Cang 409, An 7169, An 9217, Bao 182	Wang et al. (1998a); Jiyaju (1989); Chen and Qi (1993); Tian et al. (2010); Dwivedi et al. (2012)
Salinity	ISe #254, 869, 1851, 96, 388, 480, 995, 1629, 969, 1888, Honggu, Xiaohuanggu, and Sanbianchou, ICERI 5, ICERI 6	Ardie et al. (2015); Krishnamurthy et al. (2014a); Tian et al. (2008)

control cultivars for two or more grain nutrients (Upadhyaya et al., 2011a) (Table 7.3). Vadivoo et al. (1998) reported a wide range of variation for protein (6.7–12.3%) and calcium (162–487 mg/100 g of grain) in finger millet and identified 20 and 16 genotypes with significantly higher protein and calcium content, respectively (Table 7.3).

Similarly in foxtail millet, Upadhyaya et al. (2011b) identified 21 diverse accessions with agronomically (earliness and high grain yield) and nutritionally (high seed protein, 15.6–18.5%; Ca, 171.2–288.7 mg/kg; Fe, 58.2–68.0 mg/kg; and Zn, 54.5–74.2 mg/kg) superior traits (Table 7.3). Seed protein, fat, starch, and amino acids content in 259 foxtail millet cultivars from six provinces in China showed a wide range of variation (g/100 g) for protein (11.85–20.58), fat (2.82–4.47), starch (65.59–74.12), and amino acids (Yang et al., 2013). Philip and Maloo (1996) evaluated 40 genotypes of foxtail millet varieties for their Fe content and grouped the varieties with high Fe content. Li et al. (2009) evaluated vitamin E contents of 400 foxtail millet accessions, and identified accessions with high vitamin E content. Shao et al. (2014) evaluated folic acid (vitamin B9) content in 245 foxtail millet traditional varieties originating from different regions in Shanxi, China showing wide genetic variations (0.37–2.37 μg/g) and a total of 24 varieties with higher folic acid content were identified, among them, Jingu 21, a major leading cultivar, recorded folic acid content of 2 μg/g.

7.5.3 Biotic stress

Blast caused by *Pyricularia grisea* is a very prominent disease in finger millet, which affects the productivity, utilization, and trade in Eastern and Southern Africa and South Asia. The average loss due to blast has been reported to be around 28–36% (Nagaraja et al., 2007), and in certain areas yield losses could be as high as 80–90% (Vishwanath et al., 1986; Rao, 1990). At ICRISAT, a comprehensive disease severity assessment (rating) scale has been developed for leaf, neck, and finger blast based on the qualitative and quantitative differences of lesions observed on plants infected with blast pathogen (Babu et al., 2013b). In mini core collection, 66 accessions with combined resistance to leaf, neck, and finger blast have been identified, of which nine genotypes also have desirable agronomic traits such as early flowering, medium plant height, and semicompact to compact inflorescence (Babu et al., 2013b). In addition, many researchers have also evaluated finger millet genotypes and reported sources of resistance to blast disease (Table 7.3).

The most serious diseases of foxtail millet are blast (*Pyricularia setariae*), downy mildew (*Sclerospora graminicola*), rust (*Uromyces setariae-italiae*), and smut (*Ustilago crameri*) (Dwivedi et al., 2012; http://database.prota.org/PROTAhtml/Setaria%20italica_En.htm). At ICRISAT, foxtail millet core collection accessions were evaluated for blast resistance in field and greenhouse under artificial inoculation conditions and identified 21 accessions resistant to neck and head blast under field evaluation and 11 accessions had seedling leaf blast resistance in the greenhouse. Further evaluation against four isolates of blast pathogen led to the identification of 16 accessions with resistance to leaf, sheath, neck, and head blast to at least one isolate, and two accessions (ISe 1181 and ISe 1547) showed free from head blast infection and are resistant to leaf, neck, and sheath blast against four isolates (Sharma et al., 2014). Many studies on screening foxtail millet germplasm accessions or cultivars against blast, brown spot, banded leaf, and sheath blight diseases have been carried out and sources for resistance have been reported (Table 7.3).

7.5.4 Abiotic stress

7.5.4.1 Drought

Foxtail millet is a relatively drought-tolerant crop compared to other cereals. Significant correlations between agronomic traits like panicle weight, grain weight per panicle, plant height, length of rachis, and 1000-grain-weight, and physiologic parameters like relative chlorophyll, soluble protein, malondialdehyde (MDA), and superoxide dismutase (SOD) with drought-resistant index (DRI) under drought condition were reported (Zhang et al., 2012b). Various drought screening studies employing different methods had enabled identification of drought-tolerant genotypes in foxtail millet (Table 7.3). Screening of 17,313 accessions of foxtail millet genotypes from China for drought tolerance using seedling survival under repeated drought stress led to grouping of accessions into five grades with grade 1 being the most drought tolerant, which included the cultivar, Yugu1 (Li, 1997). A quick and simple screening technique for screening foxtail millet drought tolerance using mannitol or polyethylene glycol (PEG-6000) tests has been used and suggested relative water content and germination rate under osmotic stress as indicators of drought tolerance at the seedling stage (Zhang et al., 2005; Zhu et al., 2008). Lata et al. (2011) used lipid peroxidation measure to assess membrane integrity under stress as biochemical marker to screen 107 cultivars and classified the genotypes as highly tolerant, tolerant, sensitive, and highly sensitive.

In finger millet, drought reduced leaf area, dry matter accumulation, seed weight, radiation use efficiency, biomass, and yield (Maqsood and Ali, 2007). Drought stress induced increase in the activity of superoxide dismutase, ascorbate peroxidase, and glutathione reductase in tolerant varieties (PR 202 and VL 315), while lower in susceptible varieties (PES 400 and VR 708) (Bhatt et al., 2011). Ascorbate peroxidase:superoxide dismutase ratio, which is a crucial factor in alleviating drought stress, was higher in drought-tolerant varieties compared to susceptible varieties under stress. The susceptible varieties recorded maximum stress-induced damage, wherein higher accumulation of MDA and hydrogen peroxide was found (Bhatt et al. 2011). Neshamba (2010) studied the variability of drought tolerance in finger millet and identified 16 drought-tolerant accessions than the best check.

7.5.4.2 Heat stress

In finger millet, high-temperature stress caused significant influence on growth, development, and yield. Traits that influence yield, such as panicle number, finger number, finger length, seed number, and seed weight, were significantly reduced by high-temperature stress (Opole et al., 2010; Babu et al., 2013a). Babu et al. (2013a) established a temperature induction response (TIR) technique for screening high-temperature tolerance at seedling stage in finger millet, wherein they standardized the sublethal, that is, challenging temperatures 38–54°C (for 5 h) and lethal temperatures as 57°C (for 2 h) and found some thermotolerant genotypes (Table 7.3).

7.5.4.3 Salinity

Foxtail and finger millets are the potential crops for salt-affected soils (Krishnamurthy et al., 2014a, 2014b). Most recently at ICRISAT, Krishnamurthy et al. (2014b) studied the finger millet crop response to salinity in terms of total shoot or grain biomass at maturity using mini core collection and grouped the accessions into tolerant (22), moderately tolerant (20), sensitive (21), and the sensitive and late ones (5) based on yield under saline condition (Table 7.3). Similarly, Krishnamurthy et al. (2014a) screened the foxtail millet core collection under saline condition in pot culture, which revealed a large variation for salinity tolerance and identified salinity-tolerant accessions (Table 7.3).

7.5.4.4 Lodging

Lodging is a constraint in many crops, including finger and foxtail millets mainly due to soft stalk that are prone to lodging and crop management and environmental factors. In foxtail millet, Tian et al. (2010) suggested lodging coefficient as a suitable indicator for field selection for lodging resistance. Foxtail millet landraces and improved varieties that resist lodging have been reported from China (Table 7.3). Reddy et al. (2009) reported that 25.29% of the finger millet germplasm from East African countries assembled at ICRISAT were of nonlodging types, mostly from Uganda and Kenya, which could be used as sources to transfer the trait into new breeding lines after multienvironment evaluation.

7.5.4.5 Waterlogging

Finger millet is relatively tolerant to waterlogging as well as drought, while foxtail millet is susceptible to waterlogging but tolerant to drought (Kono et al., 1987). Prolonged waterlogging decreased total root number in finger millet, but increased in foxtail millet (Kono et al., 1988). Zegada-Lizarazu and Iijima (2005) reported significant reduction of water use efficiency (WUE) in these crops under waterlogging, while under drought condition significant reduction was for shoot dry weight and leaf area. Few varieties were reported as tolerant to waterlogging in foxtail millet like Jinan 8337 (Zhang and Guan, 1995) and Lugu 7 (Chen and Qi, 1993); however, extensive screening has not been reported in both crops.

7.6 Use of germplasm in crop improvement

Germplasm resources provide a pool of genes for breeding high-yielding, biotic- and abiotic-resistant cultivars. Systematic breeding efforts and utilization of genetic resources are limited in finger and foxtail millets. Use of germplasm accessions can be enhanced if small subsets of a few hundred germplasm accessions, which represent the entire diversity present in the crop species, are available. At ICRISAT, Upadhyaya et al. (2006a, 2008) formed core collections in finger and foxtail millets (Table 7.4). For establishing core collection in foxtail millet, entire germplasm accessions were stratified into three taxonomic races (*indica*, *maxima*, and *moharia*). Principal coordinate analysis was performed on 12 qualitative traits for each of the biological races separately that resulted in the formation of 29 clusters. From each cluster, 10% of the accessions were selected to constitute a core collection of 155 accessions (Upadhyaya et al., 2008). Similarly, the entire germplasm collection of finger millet was stratified into four regions: Africa, Asia, America, and Europe. The information on country of origin was not available for 181 accessions, which were grouped into "unknown"

Table 7.4 Representative germplasm subsets of foxtail and finger millets germplasm collection

Crop	Germplasm subsets	No. of accessions used	No. of traits/SSRs involved	No. of accessions in subset	References
Finger millet	Core	5940	14	622	Upadhyaya et al. (2006a)
	Core	4511		551	Gowda et al. (2007b)
	Core	1000	23	77	Haradari et al. (2012)
	Mini core	622	20	80	Upadhyaya et al. (2010)
	Composite collection	–	–	1000	Upadhyaya et al. (2005)
	Reference set	1000	19	300	Upadhyaya (2008a)
Foxtail millet	Core	1474	23	155	Upadhyaya et al. (2008)
	Core	1478	23	156 and 78	Gowda et al. (2007a)
	Mini core	155	21	35	Upadhyaya et al. (2011b)
	Composite collection	–	–	500	Upadhyaya et al. (2006b)
	Reference set	500	20 SSRs	200	Upadhyaya (2008b)

region. A principal component analysis (PCA) was performed on the accessions from each region. A hierarchical cluster analysis was conducted on the first five PCA scores in each region separately. From each cluster, about 10% accessions were randomly selected to form the core collection (Upadhyaya et al. 2006a). Core collections of finger and foxtail millets developed at ICRISAT were further evaluated under multiple environments for morphoagronomic traits and formed core of core called mini core collections (Upadhyaya et al., 2010, 2011b, Table 7.4). Gowda et al. (2007a, 2007b) developed core collections in foxtail and finger millets. The core and mini core collection approaches provide an effective mechanism for proper exploitation of germplasm resources for trait identification and allele mining.

ICRISAT currently conserves 6,804 accessions (including 204 wild accessions) of finger millet and 1,542 accessions (including 54 wild accessions) of foxtail millet, and a total of 41,956 and 16,435 seed samples of finger and foxtail millets, respectively, were supplied to 50 and 53 countries, respectively. It includes a total of 35 sets of finger millet, and 24 sets of foxtail millet core/mini core collection/reference set. Among the germplasm accessions supplied, five accessions of finger millet were released directly as varieties in Zambia (IE 2929, IE 2947), Uganda (IE 2440, IE 4625), and Kenya (IE 4115).

Wild foxtail genotypes that are highly exposed to herbicides evolved to be resistant to some herbicides were utilized to transfer the herbicide resistance into cultivated foxtail millet, which enables the use of herbicides in foxtail millet cultivation. For example, the herbicide sethoxydim-resistant green foxtail millet collected in a cultivated field in Manitoba, Canada (population UM131) was found to be 3000 times more resistant to sethoxydim than the wild type (Wang and Darmency, 1997) and 54, 29, 11 times more resistant to tralkoxydim, diclofop-methyl, and fenoxaprop-*p*-ethyl than the wild type, respectively (Heap and Morrison, 1996). Wang and Darmency (1997) transferred sethoxydim resistance from green foxtail to breed foxtail millet with improved herbicide resistance.

Various male-sterile lines of foxtail millet have been identified having dominant, recessive genes and photo/thermosensitive nuclear system (Cui et al., 1979; Wang et al., 1993, 2002, 2010b; Zhao et al., 1996; Hao et al., 2009), gene interaction male-sterile lines (Hu et al., 1986), cytoplasmic male sterility (Zhu et al., 1991), and cytoplasmic-nuclear male-sterile type (Zhi et al., 2007). These lines are potential sources for heterosis breeding in foxtail millet and have been used in developing hybrid cultivars in China. For example, Zhangzagu 5, a high yielding hybrid cultivar, was released from Zhangjiakou Academy of Agricultural Sciences, Hebei Province (Liu et al. 2014). Most of the currently released Chinese spring foxtail millet male-sterile lines were derived from Chang10A, whose cytoplasm was contributed by Qinyuanmujizui (Liu et al., 2011a; Wang et al., 1998b), while most summer foxtail millet male-sterile lines were derived from Huangmi1A with the cytoplasm from Dahuanggu (Liu et al., 1996, 2006). In the case of finger millet, Gupta et al. (1997) reported a genetic male-sterile line, INFM 95001, identified through treating the finger millet germplasm accession IE 3318 (=SDFM 63 from Zimbabwe) with ethyl methane sulfonate (EMS), would facilitate crossing for the production of finger millet hybrid progenies to generate new segregants to enhance genetic recombination in recurrent selection programs for finger millet improvement.

7.7 Limitations in germplasm use

Though considerable numbers of germplasm accessions are available nationally and internationally in foxtail and finger millets (Table 7.1), breeders continue to use only a limited number of accessions. Therefore, a large number of valuable accessions remain unexplored mainly due to nonavailability of precise data on traits of economic interest, limited research efforts, and funding. Forming representative core and mini core collections from the entire collection can enhance the use of germplasm in breeding programs because of the reduced size of collections that capture the diversity of entire collections of the particular species. Interestingly, representative germplasm subsets like core, mini core, and composite collections, and genotyping-based reference sets are available in both crops (Table 7.4) that can serve as potential resources for crop improvement and for genomic studies.

7.8 Germplasm enhancement through wide crosses

Foxtail and finger millets are self-fertilizing crops with some amount of cross-pollination occurring in foxtail millet (4%) (Li et al., 1935) and finger millet (1%) mediated by wind (Jansen and Ong, 1996; Purseglove, 1972). Floral morphology and anthesis behavior make them the most difficult species for hybridization; however, emasculation and crossing techniques have been reported (Li et al., 1935; Richardson, 1958; Siles et al., 2001). The interspecific hybridization studies in finger millet were mostly with the view to determine genome relationship; however, these studies provide basic information on crossability and barriers, genome relationship, and so on. Interspecific F_1 hybrids, *E. indica* × *E. floccifolia* and *E. tristachya* × *E. indica* were made to investigate species genome affinity and crossing barriers, suggesting that genomes of *E. floccifolia* and *E. tristachya* are homologous to *E. indica* (Salimath et al., 1995b). Later, Mallikharjun et al. (2005) made an interspecific hybridization involving two tetraploid species (*E. coracana* and *E. africana*) and four diploid species, namely, *E. intermedia, E. indica, E. floccifolia,* and *E. tristachya* ($2n = 2x = 18$), and suggested *E. indica* with AA genome as the pivotal donor species in the evolution of *E. africana* and *E. coracana*. The crosses *E. intermedia* × *E. coracana*, *E. tristachya* × *E. coracana*, *E. africana* × *E. indica*, *E. africana* × *E. floccifolia* and *E. intermedia* × *E. africana* showed crossability of 3.2–8.3%, and the resultant triploid hybrids showed normal growth and substantial flowering, but the anthers were mostly shriveled with little content in them resulting in drastic reduction of pollen stainability (2–8%) in F_1 hybrids and all these plants were completely sterile (Mallikharjun et al., 2005). The hybrid between *E. coracana* × *E. africana* was found to be intermediate between parents for most of the traits such as productive tillers, finger length, finger number and days to 50% flowering, and exhibits reduced pollen fertility compared to their parents (Shet et al., 2010b). Interspecific hybrids were successfully produced, when *E. intermedia* was used as a female parent in two crosses involving *E. indica* and *E. floccifolia* and as male parent with *E. tristachya*; however, all F_1's were completely sterile and reciprocal crosses in all the three combinations (*E. intermedia* × *E. indica*, *E. intermedia* × *E. floccifolia* and *E. tristachya* × *E. intermedia*) did not yield F_1 hybrids. The F_1 hybrids showed characteristics that were intermediate in nature or similar to one of the parents (Devarumath et al., 2005). Evaluation and utilization of wild gene pool for crop improvement has received limited research attention. *E. africana* is a close relative of the cultivated species of *E. coracana*. Earlier reports indicate that gene transfer from *E. africana* to *E. coracana* is feasible and useful in breeding. *E. africana* has more tillering ability (15–20), high drought tolerance capacity, matures early (95–100 days), and has more fingers per ear and long finger length (Shet et al., 2010b). Interspecific hybridization between *E. africana* and cultivated finger millet varieties (HR 911, PR 201, Indaf 8, HR911, and PR 202) were attempted in order to transfer some of the desirable characters from wild species to the popular cultivars. The hybrids were intermediate for productive tillers, finger length, finger number, and days to 50% flowering and exhibited reduced pollen fertility (Shet et al., 2009, 2010a).

The green millet (*S. viridis*) could be an interesting source for improvement of foxtail millet without a complex and time-consuming breeding strategy as only two backcross generations associated with selection are enough to eliminate weedy characters and to return to the cultivated type (Naciri et al., 1992). Zangre and Darmency (1993) reported that it could be possible to recover the cultivated type using a simple selection procedure in F_2 and F_3. Darmency et al. (1987) demonstrated the potential of interspecific hybridization between cultivated foxtail millet and its wild progenitor green foxtail, and polyploidization to improve the traits of foxtail millet. The tetraploidization resulted in a shift in characteristics toward the crop species; especially, a twofold increase in seed weight was noticed. Nonadditive effects were found for most characters, except for the seed shedding, which was found to be encoded by at least four loci. Cultivated type plants were easily recovered in the diploid and the tetraploid F_2 (Darmency et al., 1987). Interspecific hybridizations between foxtail millet cultivars and a green foxtail resistant to the herbicides have also been made to transfer herbicide resistance into cultivated foxtail millet and obtained resistant genotypes (Darmency and Pernès, 1985; Wang and Darmency, 1997; Darmency et al., 1987). Recently, Qie et al. (2014) made an interspecific hybridization between *S. italica* × *S. viridis* and identified QTLs that contribute to germination and early seedling drought tolerance. They found that both *S. viridis* and *S. italica* contributed favorable alleles for drought tolerance indicating that wild *S. viridis* populations may serve as a reservoir for novel stress tolerance alleles, which could be employed in foxtail millet breeding.

7.9 Integration of genomic and genetic resources in crop improvement

7.9.1 Molecular markers and genome sequence

Genomic resources, like availability of molecular markers, linkage maps, and genome sequence, are essential for gene tagging, quantitative trait loci (QTL) mapping, and marker-assistant selection for rapid crop improvement. However, these genomic resources are limited in finger millet. As a result, limited numbers of markers have been developed (Table 7.5). The Bio-Innovate Program has initiated a finger millet genome-sequencing project by partnering with the African Orphan Crop Consortium. This initiative is being coordinated by ICRISAT, Nairobi in partnership with Biosciences eastern and central Africa (BecA) Hub, University of California, University of Georgia (UGA), and the Swedish University of Agricultural Sciences (SLU).

In foxtail millet, large numbers of molecular markers have been developed like SSR (simple sequence repeat), expressed sequence tag–simple sequence repeat (EST–SSR), intron length polymorphic (ILP), transposable element (TE), and microRNA (miRNA)-based markers during pregenome sequence era. The major breakthrough in the area of *Setaria* genomics is the release of two reference genome sequences (Bennetzen et al., 2012; Zhang et al., 2012a). Zhang et al. (2012a) sequenced the

Table 7.5 Genomic resources of finger and foxtail millets

DNA markers	No. of markers	References
Finger millet		
Nucleotide binding site (NBS); leucine-rich repeat (LRR)	9	Panwar et al. (2011)
EST–SSR	17	Arya et al. (2009)
	11	Panwar et al. (2011)
	30	Naga et al. (2012)
	545	Babu et al. (2014d)
	58	Babu et al. (2014c)
	74	Babu et al. (2014b)
	56	Nirgude et al. (2014)
SSR	82	Dida et al. (2007)
	27	Reddy et al. (2011)
	49	Musia (2013)
Foxtail millet		
EST–SSRs	26	Jia et al. (2007)
	12	Obidiegwu et al. (2013)
	447	Kumari et al. (2013)
SSR	269	Jia et al. (2009)
	143	Zhao et al. (2012)
	172	Gupta et al. (2012)
	64	Gupta et al. (2013)
	45	Lin et al. 2011
	35	Sato et al. (2013)
	21,294	Pandey et al. (2013)
	788	Zhang et al. (2014)
Intron length polymorphism (ILP) markers	98	Gupta et al. (2011)
	5,123	Muthamilarasan et al. (2014)
Transposable elements (TE) based markers	20,278	Yadav et al. (2014a)
SV (Structural variants)	152	Bai et al. (2013)
microRNA-based genetic markers	176	Yadav et al. (2014b)

foxtail cv. "Zhang gu" and a predicted genome size of ~485 Mb. Zhang et al. (2012a) also sequenced another genotype of foxtail millet named "A2," a photothermosensitive male-sterile line and identified 542,322 single nucleotide polymorphisms (SNPs), 33,587 small insertion and deletions (indels), and 10,839 structural variants (SV) between A2 and Zhang gu. The sequence analysis showed the presence of 38,801 genes, out of which ~82% were expressed. Bennetzen et al. (2012) sequenced foxtail millet accession "Yugu1" and green foxtail accession "A10." The final genomic sequence assembly contains ~400 Mb of assembly covering ~80% of the genome, showing the presence of 24,000–29,000 expressed genes. The availability of foxtail millet genome sequence in the public domain has enabled large-scale development of genomic resources through sequence scanning for microsatellite repeat-motifs and physical mapping of markers on chromosomes (Pandey et al., 2013; Zhang et al., 2014; Xu et al., 2013). Next-generation sequencing technologies enable large-scale genotyping and sequencing of germplasm accessions, which enables genome-wide variation analysis and scanning sequence variations like SNP, indels polymorphisms, and SV (Bai et al., 2013; Jia et al., 2013a) (Table 7.5). Foxtail millet core and finger millet mini core collections of ICRISAT germplasm accessions were genotyped for rapid SNP characterization through genotyping-by-sequencing (GBS) approach – a new low-cost, high-throughput sequencing technology. It will enable the study of the population genetics and structure for effective use of these materials for genetic and genomic research (H.D. Upadhyaya, personal communication).

7.9.2 Genetic maps

The first genetic map of foxtail millet genome was reported by Wang et al. (1998c) using RFLP markers. They constructed intra- and inter-specific maps. The intraspecific map was based on the cross, Longgu 25 × Pagoda flower green comprised of nine linkage groups aligned with nine foxtail millet chromosomes using trisomic lines, and it spanned 964 cM. Furthermore, the intraspecific map was compared to an interspecific map developed based on *S. italica* (cv.B100) × *S. viridis* (acc.A10), which showed that the order of the markers and the genetic distances between the loci are highly conserved. Later these intra- and interspecific genetic maps were enriched with additional markers (Devos et al., 1998; Jia et al., 2009; Mauro-Herrera et al., 2013). In addition, many other groups also developed linkage maps, for example, Zhang et al. (2012a) constructed a genetic map from a cross between Zhang gu and A2, and mapped 751 markers including 118 SNPs and 641 SVs; Qie et al. (2014) constructed a genetic map based on an interspecific cross between *S. italica* cv. Yugu 1 and *S. viridis* acc. W53; and Sato et al. (2013) developed genetic map using two F_2 populations (JP. No. 73913 × JP No. 222613 and JP. No. 73913 × JP No. 71640).

In finger millet, Dida et al. (2007) generated the first genetic map based on interspecific F_2 population of the cross between *E. coracana* subsp. *coracana* cv. Okhale-1 and its wild progenitor *E. coracana* subsp. *africana* acc. MD20 using RFLP, AFLP, EST, and SSR markers. Assignment of linkage groups to the A and B genome was done by comparing the hybridization patterns of probes in Okhale-1, MD 20, and *E. indica* (A genome donor to *E. coracana*) acc. MD-36. The maps span 721 and 787 cM for the A and B genome, respectively, covering all 18 finger millet chromosomes at least partially.

A number of studies on comparative genomics involving foxtail millet and finger millet with other members of the grass family have been previously reported (Devos et al., 1998, 2000; Doust et al., 2004; Srinivasachary et al., 2007; Kumari et al., 2013; Muthamilarasan et al., 2014; Pandey et al., 2013). These studies revealed close relationships of the *gramineous* crops indicating syntenic relationship among the chromosomes of foxtail millet with other *gramineous* crops like sorghum, maize, rice, and so on.

7.10 Utilization of genetic and genomic resources

In finger and foxtail millet, core and mini core collections have been developed at ICRISAT (Upadhyaya et al., 2006a, 2008, 2010, 2011b). In addition, with the support from the Generation Challenge Programme, ICRISAT scientists have developed composite collections of finger millet consisting of 1000 accessions (Upadhyaya et al., 2005), and foxtail millet consisting of 500 accessions (Upadhyaya et al., 2006b). Composite collections have been genotyped with SSR markers and genotype-based reference sets have been established in these crops (Upadhyaya, 2008a, 2008b) (Table 7.4). These germplasm subsets represent diversity of the entire collections of the particular species, and are ideal genetic resources for genomic studies and are being used.

Molecular markers have been utilized successfully for phylogeny, genetic structure and diversity, and QTL identification in foxtail millet. Genetic loci responsible for tiller number, axillary branch number, *spikelet-tipped bristles, Waxy* gene, and male sterility in foxtail millet have been reported (Wang et al., 1998c, 2013; Devos et al., 1998; Doust et al., 2004, 2005; Doust and Kellogg, 2006; Sato et al. 2013). Qie et al. (2014) reported the genomic regions controlling germination and early seedling drought tolerance in foxtail millet, where the wild green foxtail genotype and the foxtail millet cultivar contributed favorable alleles for the traits, indicating that wild *S. viridis* populations may serve as a reservoir of novel alleles for stress tolerance, which could be employed in foxtail millet breeding. Association of an SNP in a novel DREB2-like gene SiDREB2 with stress tolerance was reported in foxtail millet, and an allele-specific marker for dehydration tolerance has been developed (Lata et al., 2011). Through association mapping approach, genomic regions linked with agronomic traits have been reported in foxtail millet using SSR markers (Vetriventhan, 2011; Gupta et al., 2014b). The advent of next-generation sequencing technologies has enabled large-scale genotyping and resequencing of germplasm accessions. Most recently, Jia et al. (2013a) identified 512 loci associated with 47 agronomic traits through genome-wide association mapping by sequencing 916 diverse foxtail millet varieties (Jia et al., 2013a).

Considerable efforts have also been made in the case of finger millet on utilization of markers and QTL identification. Panwar et al. (2011) designed functional markers based on the nucleotide sequences of different NBS-LRR domain containing blast-resistant genes of cereals. Several primers gave unique bands linked with disease resistance exhibiting clear polymorphism between blast resistant and susceptible genotypes of finger millet. Nirgude et al. (2014) developed EST–SSRs utilizing nucleotide sequences of different candidate genes like *opaque2 modifiers* and calmodulin (CALcium-MODULated proteIN) cereals. The *opaque2 modifier* specific EST–SSRs could differentiate the finger millet genotypes into high, medium, and low protein containing genotypes, while calcium-dependent candidate gene based EST–SSRs could differentiate the genotypes based on the calcium content with few exceptions. These results indicate the possible role of genic-SSRs in governing trait variations and could be utilized in germplasm characterization for particular traits like disease resistance and calcium content. Using association-mapping approach, QTLs for agronomic traits, blast disease, and tryptophan and protein content have been identified in finger millet (Babu et al., 2014c, 2014a, 2014b).

7.11 Conclusions

Finger and foxtail millets are the most important crops mostly grown under marginal areas with limited resources. Foxtail millet is a major crop in India, China, and Japan, while finger millet is an important crop in India and African countries. Foxtail and finger millet germplasm accessions conserved globally, reported to have large genetic variation at both phenotypic and molecular level, enable mining of alleles for important traits for yield improvement. Germplasm sets, such as core, mini core, composite, and reference sets, established in these crops have huge applicability in genetic and genomic research, and crop improvement. These crops along with other small millets are considered as "nutri-cereals" owing to their rich nutritional and health benefits; however, few attempts have been made to understand the genetics and genomics of the nutritional trait. Foxtail millet has received more research attention and is considered as a model species to explore plant architectural traits, evolutionary genomics, and physiologic attributes of the C_4 Panicoid crops. In the case of finger millet, genome sequencing has been initiated, and will be available to researchers. Increased use of genetic and genomic resources will guide breeders in developing climate-resilient cultivars in these "nutri-cereals" to enhance food and nutritional security.

References

Ardie, S.W., Khumaida, N., Nur, A., Fauziah, N., 2015. Early identification of salt tolerant foxtail millet (*Setaria italica* L. Beauv.). Proc. Food Sci. 3, 303–312.

Arya, L., Verma, M., Gupta, V.K., Karihaloo, J.L., 2009. Development of EST–SSRs in finger millet (*Eleusine coracana* ssp. *coracana*) and their transferability to pearl millet (*Pennisetum glaucum*). J. Plant Biochem. Biotechnol. 18 (1), 97–100.

Austin, D.F., 2006. Fox-tail millets (*Setaria*: Poaceae) – abandoned food in two hemispheres. Econ. Bot. 60 (2), 143–158.

Babu, D.V., Sudhakar, P., Reddy, Y.S.K., 2013a. Screening of thermotolerant ragi genotypes at seedling stage using TIR technique. Bioscan 8 (4), 1493–1495.

Babu, T.K., Thakur, R.P., Upadhyaya, H.D., Reddy, P.N., Sharma, R., Girish, A.G., et al., 2013b. Resistance to blast (*Magnaporthe grisea*) in a mini-core collection of finger millet germplasm. Eur. J. Plant Pathol. 135, 299–311.

Babu, B.K., Agrawal, P.K., Pandey, D., Jaiswal, J.P., Kumar, A., 2014a. Association mapping of agro-morphological characters among the global collection of finger millet genotypes using genomic SSR markers. Mol. Biol. Rep. 41 (8), 5287–5297.

Babu, B.K., Agrawal, P.K., Pandey, D., Kumar, A., 2014b. Comparative genomics and association mapping approaches for opaque2 modifier genes in finger millet accessions using genic, genomic and candidate gene-based simple sequence repeat markers. Mol. Breed. 34 (3), 1261–1279.

Babu, B.K., Dinesh, P., Agrawal, P.K., Sood, S., Chandrashekara, C., Bhatt, J.C., et al., 2014c. Comparative genomics and association mapping approaches for blast resistant genes in finger millet using SSRs. PloS ONE 9 (6), e99182.

Babu, B.K., Pandey, D., Agrawal, P.K., Sood, S., Kumar, A., 2014d. *In-silico* mining, type and frequency analysis of genic microsatellites of finger millet (*Eleusine coracana* (L.)

Gaertn.): a comparative genomic analysis of NBS-LRR regions of finger millet with rice. Mol. Biol. Rep. 41 (5), 3081–3090.

Bai, H., Cao, Y., Quan, J., Dong, L., Li, Z., Zhu, Y., et al., 2013. Identifying the genome-wide sequence variations and developing new molecular markers for genetics research by re-sequencing a landrace cultivar of foxtail millet. PLoS ONE 8, e73514.

Bawazir, A.A., Bamousa, A.S., 2014. Biodiversity and conservation of plant genetic resources on Socotra. Discourse J. Agric. Food Sci. 2 (7), 217–224.

Begum, F., Sultana, R., Nessa, A., 2013. Screening of drought tolerant foxtail millet (*Setaria italica* Beauv.) germplasm. Bangladesh J. Sci. Ind. Res. 48 (4), 265–270.

Benabdelmouna, A., Abirached-Darmency, M., Darmency, H., 2001. Phylogenetic and genomic relationships in *Setaria italica* and its close relatives based on the molecular diversity and chromosomal organization of 5S and 18S-5.8S-25S rDNA genes. Theor. Appl. Genet. 103, 668–677.

Bennetzen, J.L., Schmutz, J., Wang, H., Percifield, R., Hawkins, J., Pontaroli, A.C., et al., 2012. Reference genome sequence of the model plant *Setaria*. Nat. Biotechnol. 30, 555–561.

Bharathi, A., 2011. Phenotypic and genotypic diversity of global finger millet (*Eleusine coracana* (L.) gaertn.) composite collection. PhD Thesis. Tamil Nadu Agricultural University, Coimbatore, Tamil Nadu.

Bhatt, D., Negi, M., Sharma, P., Saxena, S.C., Dobriyal, A.K., Arora, S., 2011. Responses to drought induced oxidative stress in five finger millet varieties differing in their geographical distribution. Physiol. Mol. Biol. Plants 17 (4), 347–353.

Bisht, M.S., Mukai, Y., 2001. Genomic *in situ* hybridization identifies genome donor of finger millet (*Eleusine coracana*). Theor. Appl. Genet. 102, 825–832.

Chen, J., Qi, Y., 1993. Recent developments in foxtail millet cultivation and research in China. In: Riley, K.W., Gupta, S.C., Seetharam, A., Mushonga, J.N. (Eds.), Advances in Small Millets. Oxford and IBH Publishing Co, New Delhi, pp. 101–107.

Cui, W.S., Du, G., Zhao, Z.H., 1979. The selection and utilization of male sterile line/Shanxi 280 of foxtail millet. Sci. Agric. Sin. 12, 43–46, (in Chinese).

Dagnachew, L., De Villiers, S., Sewalem, T., Dida, M., Masresha, F., Kimani, W., et al., 2014. Genetic diversity and eco-geographical distribution of *Eleusine* species collected from Ethiopia. Afr. Crop Sci. J. 22 (1), 45–58.

Darmency, H., Pernès, J., 1985. Use of wild *Setaria viridis* (L.) Beauv. to improve triazine resistance in cultivated *S. italica* (L.) by hybridization. Weed Res. 25, 175–179.

Darmency, H., Ouin, C., Pernès, J., 1987. Breeding foxtail millet (*Setaria italica*) for quantitative traits after interspecific hybridization and polyploidization. Genome 29, 453–456.

de Wet, J.M.J., Oestry-Stidd, L.L., Cubero, J.L., 1979. Origins and evolution of foxtail millet. J. d'agriculture traditionnelle et de botanique appliquée 26, 53–63.

de Wet, J.M.J., Prasada Rao, K.E., Brink, D.E., Mengesha, M.H., 1984. Systematic and evolution of *Eleusine coracana* (Gramineae). Am. J. Bot. 71, 550–557.

Devarumath, R.M., Hiremath, S.C., Rao, S.R., Kumar, A., Bewal, S., 2005. Genome analysis of finger millet *E. coracana* by interspecific hybridization among diploid wild species of *Eleusine* (Poaceae). Cytologia 70 (4), 427–434.

Devos, K.M., Wang, Z.M., Beales, C.J., Sasaki, T., Gale, M.D., 1998. Comparative genetic maps of foxtail millet (*Setaria italica*) and rice (*Oryza sativa*). Theor. Appl. Genet. 96, 63–68.

Devos, K.M., Pittaway, T.S., Reynolds, A., Gale, M.D., 2000. Comparative mapping reveals a complex relationship between the pearl millet genome and those of foxtail millet and rice. Theor. Appl. Genet. 100, 190–198.

Diao, X.M., 2011. Current status of foxtail millet production in China and future development directions. The Industrial Production and Development System of Foxtail Millet in China. Chinese Agricultural Science and Technology Press, Beijing, pp. 20–30.

Dida, M.M., Srinivasachary, Ramakrishnan, S., Bennetzen, J.L., Gale, M.D., Devos, K.M., 2007. The genetic map of finger millet, *Eleusine coracana*. Theor. Appl. Genet. 114, 321–332.

Doust, A.N., Kellogg, E.A., 2006. Effect of genotype and environment on branching in weedy green millet (*Setaria viridis*) and domesticated foxtail millet (*Setaria italica*) (Poaceae). Mol. Ecol. 15, 1335–1349.

Doust, A.N., Devos, K.M., Gadberry, M.D., Gale, M.D., Kellogg, E.A., 2004. Genetic control of branching in foxtail millet. Proc. Natl. Acad. Sci. USA 101 (24), 9045–9050.

Doust, A.N., Devos, K.M., Gadberry, M.D., Gale, M.D., Kellogg, E.A., 2005. The genetic basis for inflorescence variation between foxtail and green millet (Poaceae). Genetics 169 (3), 1659–1672.

Dwivedi, S., Upadhyaya, H.D., Senthilvel, S., Hash, C.T., Fukunaga, K., Diao, X., et al., 2012. Millets: genetic and genomic resources. Plant Breed. Rev. 35, 247–375.

Gowda, J., Rekha, D., Krishnappa, M., 2007a. Methods of constructing core set using agro-morphological traits in foxtail millet [*Setaria italica* (L.) Beauv.]. J. Plant Genet. Res. 20 (3), 193–198.

Gowda, J., Somu, G., Mathur, P.N., 2007b. Formation of core set in finger millet (*Eleusine coracana* (L.) Gaertn.) germplasm using geographical origin and morpho-agronomic characters. Indian J. Plant Genet. Res. 20, 38–42.

Gupta, S.C., Muza, F.R., Andrews, D.J., 1997. Registration of INFM 95001 finger millet genetic male sterile line. Crop Sci. 37, 1409.

Gupta, S., Kumari, K., Das, J., Lata, C., Puranik, S., Prasad, M., 2011. Development and utilization of novel intron length polymorphic markers in foxtail millet (*Setaria italica* (L.) P. Beauv.). Genome 54, 586–602.

Gupta, S., Kumari, K., Sahu, P.P., Vidapu, S., Prasad, M., 2012. Sequence-based novel genomic microsatellite markers for robust genotyping purposes in foxtail millet [*Setaria italica* (L.) P. Beauv.]. Plant Cell Rep. 31, 323–337.

Gupta, S., Kumari, K., Muthamilarasan, M., Subramanian, A., Prasad, M., 2013. Development and utilization of novel SSRs in foxtail millet [*Setaria italica* (L.) P. Beauv.]. Plant Breed. 132 (4), 367–374.

Gupta, S., Garg, R., Kumar, S., Sengar, R.S., 2014a. Assessment of finger millet (*Eleusine coracana* L.) genotypes for drought tolerance at early seedling growth stages. Vegetos 27 (3), 227–230.

Gupta, S., Kumari, K., Muthamilarasan, M., Parida, S.K., Prasad, M., 2014b. Population structure and association mapping of yield contributing agronomic traits in foxtail millet. Plant Cell Rep. 33, 881–893.

Hao, X.F., Wang, J.Z., Wang, G.Q., Wang, L.Y., Wang, X.Y., 2009. AFLP analysis of photo sensitive male sterile gene in millet. J. Shanxi Agric. Sci. 37, 10, (in Chinese).

Haradari, C., Gowda, J., Ugalat, J., 2012. Formation of core set in Indian and African finger millet [*Eleusine coracana* (L.) Gaertn.] germplasm accessions. Indian J. Genet. 72 (3), 358–363.

Harlan, J.R., de Wet, J.M.J., 1971. Towards a rational taxonomy of cultivated plants. Taxonomy 20, 509–517.

Heap, I.M., Morrison, I.N., 1996. Resistance to aryloxyphenoxypropionate and cyclohexanedione herbicides in green foxtail (*Setaria viridis*). Weed Sci. 44, 25–30.

Hilu, K.W., 1988. Identification of the 'A' genome of finger millet using chloroplast DNA. Genetics 118, 163–167.

Hilu, K.W., de Wet, J.M.J., 1976. Domestication of *Eleusine coracana*. Econ. Bot. 30, 199–208.

Hilu, K.W., de Wet, J.M.J., Harlan, J.R., 1979. Archeobotany and the origin of finger millet. Am. J. Bot. 66, 330–333.

Hirano, R., Naito, K., Fukunaga, K., Watanabe, K.N., Ohsawa, R., Kawase, M., 2011. Genetic structure of landraces in foxtail millet (*Setaria italica* (L.) P. Beauv.) revealed with transposon display and interpretation to crop evolution of foxtail millet. Genome 54, 498–506.

Hiremath, S.C., Chennaveeraiah, M.S., 1982. Cytogeneticals studies in wild and cultivated species of *Eleusine* (Gramineae). Caryologia 35 (1), 57–69.

Hiremath, S.C., Salimath, S.S., 1992. The 'A' genome donor of *Eleusine coracana* (L.) Gaertn. (Gramineae). Theor. Appl. Genet. 84, 747–754.

Hu, H.K., Ma, S.Y., Shi, Y.H., 1986. The discovery of a dominant male-sterile gene in millet (*Setaria italica*). Acta Agron. Sin. 12, 73–78, (in Chinese).

Jain, A.K., Gupta, J.C., Yadava, H.S., 1991. Stability of resistance to blast in foxtail millet. Mysore J. Agric. Sci. 25, 221–223.

Jain, A.K., Joshi, R.P., Singh, G., 2014. Identification of host resistance against banded leaf and sheath blight of foxtail millet. JNKVV Res. J. 48, 170–174.

Jansen, P.C.M., Ong, H.C., 1996. *Eleusine coracana* (L.) Gaertn. In: Grubben, G.H.J., Partohardjono, S. (Eds.), Plant Resources of South-East Asia, No. 10 Cereals. Backhuys Publishers, Leiden, The Netherlands.

Jia, X., Shi, Y., Song, Y., Wang, G., Tian-yu, W., Li, Y., 2007. Development of EST–SSR in foxtail millet (*Setaria italica*). Genet. Resour. Crop Evol. 54, 233–236.

Jia, X., Zhang, Z., Liu, Y., Zhang, C., Shi, Y., Song, Y., et al., 2009. Development and genetic mapping of SSR markers in foxtail millet [*Setaria italica* (L.) P. Beauv.]. Theor. Appl. Genet. 118, 821–829.

Jia, G., Huang, X., Zhi, H., Zhao, Y., Zhao, Q., Li, W., et al., 2013a. A haplotype map of genomic variations and genome-wide association studies of agronomic traits in foxtail millet (*Setaria italica*). Nat. Genet. 45, 957–961.

Jia, G., Shi, S., Wang, C., Niu, Z., Chai, Y., Zhi, H., et al., 2013b. Molecular diversity and population structure of Chinese green foxtail [*Setaria viridis* (L.) Beauv.] revealed by microsatellite analysis. J. Exp. Bot. 64 (12), 3645–3656.

Jiyaju, C., 1989. Importance and genetic resources of small millets with emphasis on foxtail millet (*Setaria italica*) in China. In: Seetharam, A., Riley, K.W., Harinarayana, G. (Eds.), Small Millets in Global Agriculture. Oxford and IBH publishing, New Delhi, India, pp. 101–104.

Jiyaju, C., Yuzhi, Q., 1993. Recent developments in foxtail millet cultivation and research in China. In: Seetharam, A., Riley, K.W., Harinarayana, G. (Eds.), Small Millets in Global Agriculture. Oxford and IBH publishing, New Delhi, India, pp. 101–107.

Kono, Y., Yamauchi, A., Kawamura, N., Tatsumi, J., Nonoyama, T., Inagaki, N., 1987. Interspecific differences of the capacities of waterlogging and drought tolerances among summer cereals. Japan J. Crop Sci. 56, 115–129.

Kono, Y., Yamauchi, A., Nonoyama, T., 1988. Comparison of growth response to waterlogging of summer cereals with special reference to rooting ability. Japan J. Crop Sci. 57, 321–331.

Krishnamurthy, L., Upadhyaya, H.D., Gowda, C.L.L., Kashiwagi, J., Purushothaman, R., Singh, S., et al., 2014a. Large variation for salinity tolerance in the core collection of foxtail millet (*Setaria italica* (L.) P. Beauv.) germplasm. Crop Pasture Sci. 65, 353–361.

Krishnamurthy, L., Upadhyaya, H.D., Purushothaman, R., Gowda, C.L.L., Kashiwagi, J., Dwivedi, S.L., et al., 2014b. The extent of variation in salinity tolerance of the mini core collection of finger millet (*Eleusine coracana* L. Gaertn.) germplasm. Plant Sci. 227, 51–59.

Kumar, B., Kumar, J., 2009. Evaluation of small millet genotypes against endemic diseases in mid-western Himalayas. Indian Phytopathol. 62 (4), 518–521.

Kumari, K., Muthamilarasan, M., Misra, G., Gupta, S., Subramanian, A., Parida, S.K., et al., 2013. Development of eSSR-Markers in *Setaria italica* and their applicability in studying genetic diversity, cross-transferability and comparative mapping in millet and non-millet species. PloS ONE 8 (6), e67742.

Lata, C., Bhutty, S., Bahadur, R.P., Majee, M., Prasad, M., 2011. Association of an SNP in a novel DREB2-like gene SiDREB2 with stress tolerance in foxtail millet [*Setaria italica* (L.)]. J. Exp. Bot. 62 (10), 3387–3401.

Li, Y.M., 1997. Breeding for foxtail millet drought tolerant cultivars (in Chinese). In: Li, Y. (Ed.), Foxtail Millet Breeding. Chinese Agriculture Press, Beijing, China, pp. 421–446.

Li, H., Meng, W.J., Liu, T.M., 1935. Problems in the breeding of millet [*Setaria italica* (L.) Beauv.]. J. Am. Soc. Agron. 27, 426–438.

Li, C.H., Pao, W.K., Li, H.W., 1942. Interspecific crosses in *Setaria*. II. Cytological studies of interspecific hybrids involving 1, *S. faberii* and *S. italica*, and 2. A three way cross, F_2 of *S. italica* × *S. viridis* and *S. faberii*. J. Hered. 33, 351–355.

Li, Y., Wu, S., Cao, Y., 1995. Cluster analysis of an international collection of foxtail millet (*Setaria italica* (L.) P. Beauv.). Euphytica 83, 79–85.

Li, G.Y., Fan, Z.Y., Liu, F., Zhang, P., Li, W.X., Liu, Q.S., et al., 2009. Studies on vitamin E in foxtail millet determined by HPLC method. J. Agric. Sci. Technol. 11 (1), 129–133.

Lin, H., Chiang, C., Chang, S., Kuoh, C., 2011. Development of simple sequence repeats (SSR) markers in *Setaria italica* (Poaceae) and cross-amplification in related species. Int. J. Mol. Sci. 12, 7835–7845.

Liu, Z.L., Cheng, R.H., Li, X.Y., 1996. The pedigree analysis and evaluation of north China summer millets. Crops 5, 24, (in Chinese).

Liu, Z.L., Cheng, R.H., Zhang, F.L., Xia, X.Y., Shi, Z.G., Hou, S.L., 2006. Millet variety in boreali-sinica summer millets region and its pedigree evolution and analysis on genetic foundation. Acta Agric. Boreali-Sin. Suppl. 21, 103–109, (in Chinese).

Liu, Z., Bai, G., Zhang, D., Zhu, C., Xia, X., Cheng, R., et al., 2011a. Genetic diversity and population structure of elite foxtail millet [*Setaria italica* (L.) P. Beauv.] germplasm in China. Crop Sci. 51, 1655–1663.

Liu, Q., Triplett, J.K., Wen, J., Peterson, P.M., 2011b. Allotetraploid origin and divergence in *Eleusine* (Chloridoideae, Poaceae): evidence from low-copy nuclear gene phylogenies and a plastid gene chronogram. Ann. Bot. 108 (7), 1287–1298.

Liu, Q., Iang, B., Wen, J., Peterson, P.M., 2014. Low-copy nuclear gene and McGISH resolves polyploid history of *Eleusine coracana* and morphological character evolution in *Eleusine*. Turkish J. Bot. 38, 1–12.

Mallikharjun, D.R., Hiremath, S.C., Rao, S.R., Kumar, A., Sheelavanthmath, S.S., 2005. Genome interrelationship in the genus *Eleusine* (Poaceae) as revealed through heteroploid crosses. Caryologia 58 (4), 300–307.

Maloo, S.R., Saini, D.P., Paliwal, R.V., 2001. Meera (SR 16) – a dual purpose variety of foxtail millet. Int. Sorghum Millets Newsl. 42, 75–76.

Mantur, S.G., Viswanath, S., Kumar, T.B.A., 2001. Evaluation of finger millet genotypes for resistance to blast. Indian Phytopathol. 54 (3), 387.

Maqsood, M., Ali, S.N.A., 2007. Effects of environmental stress on growth, radiation use efficiency and yield of finger millet (*Eleusine coracana*). Pak. J. Bot. 39 (2), 463–474.

Mauro-Herrera, M., Wang, X., Barbier, H., Brutnell, T.P., Devos, K.M., Doust, A.N., 2013. Genetic control and comparative genomic analysis of flowering time in *Setaria* (Poaceae). Genes Genome Genet. 3, 283–295.

Mbabwine, Y., Sabiiti, E.N., Kiambi, D., 2005. Assessment of the status of plant genetic resources in Kabale highlands, Uganda: a case of cultivated crop species. Commissioned by GMP to IPGRI, p. 71.

Mehra, K.K., 1962. Natural hybridization between *Eleusine coracana* and *E. africana* in Uganda. J. Indian Bot. Soc. 41, 531–539.

Musia, G.D., 2013. Identification of microsatellite markers for finger millet (*Eleusine coracana*) by analysis of roche 454 gs-flx titanium sequence data. MSc Thesis. School of Pure and Applied Sciences of Kenyatta University, Nairobi.

Muthamilarasan, M.E., Suresh, B.V., Pandey, G.A., Kumari, K.A., Parida, S.W.K.U., Prasad, M.A., 2014. Development of 5123 intron-length polymorphic markers for large-scale genotyping applications in foxtail millet. DNA Res. 21, 41–52.

Muza, F.R., Lee, D.J., Andrews, D.J., Gupta, S.C., 1995. Mitochondrial DNA variation in finger millet (*Eleusine coracana*). Euphytica 81, 199–205.

Naciri, Y., Darmency, H., Belliard, J., Dessaint, F., Pernès, J., 1992. Breeding strategy in foxtail millet, *Setaria italica* (L. P. Beauv.) following interspecific hybridization. Euphytica 60 (2), 97–103.

Naga, B.L.R.I., Mangamoori, L.N., Subramanyam, S., 2012. Identification and characterization of EST–SSRs in finger millet (*Eleusine coracana* (L.) Gaertn.). J. Crop Sci. Biotechnol. 15 (1), 9–16.

Nagaraja, A., Jagadish, P.S., Ashok, E.G., Gowda, K.T.K., 2007. Avoidance of finger millet blast by ideal sowing time and assessment of varietal performance under rain fed production situation in Karnataka. J. Mycopathol. Res. 45 (2), 237–240.

Neshamba, S.M., 2010. Variability for drought tolerance in finger millet (*Eleusine coracana* (L.) Gaertn.] accessions from Zambia. MSc Thesis. School of Agricultural Science of the University of Zambia, Lusaka.

Neves, S.S., Swire-Clark, G., Hilu, K.W., Baird, W.V., 2005. Phylogeny of *Eleusine* (Poaceae: Chloridoideae) based on nuclear *ITS* and plastid *trnT-trnF* sequences. Mol. Phylogenet. Evol. 35, 395–419.

Nirgude, M., Babu, B.K., Shambhavi, Y., Singh, U.M., Upadhyaya, H.D., Kumar, A., 2014. Development and molecular characterization of genic molecular markers for grain protein and calcium content in finger millet (*Eleusine coracana* (L.) Gaertn.). Mol. Biol. Rep. 41 (3), 1189–1200.

Nirmalakumari, A., Vetriventhan, M., 2010. Characterization of foxtail millet germplasm collections for yield contributing traits. Electron. J. Plant Breed. 1, 140–147.

Obidiegwu, O.N., Obidiegwu, J.E., Parzies, H., 2013. Development of SSR for foxtail millet (*Setaria italica* (L.) P. Beauv.) and its utility in genetic discrimination of a core set. Genes Genomics 35, 609–615.

Opole, R., Prasad, P.V.V., Kirkham, M.B., 2010. Effect of high temperature stress on growth, development and yield of finger millet. In: ASA, CSSA, and SSSA 2010 International Annual Meetings. Oct 31 – Nov 4. Long Beach, CA.

Pandey, G., Misra, G., Kumari, K., Gupta, S., Parida, S.K., Chattopadhyay, D., et al., 2013. Genome-wide development and use of microsatellite markers for large-scale genotyping applications in foxtail millet [*Setaria italica* (L.)]. DNA Res. 20 (2), 197–207.

Panwar, P., Jha, A.K., Pandey, P.K., Gupta, A.K., Kumar, A., 2011. Functional markers based molecular characterization and cloning of resistance gene analogs encoding NBS-LRR disease resistance proteins in finger millet (*Eleusine coracana*). Mol. Biol. Rep. 38 (5), 3427–3436.

Philip, J., Maloo, S.R., 1996. An evaluation of *Setaria italica* for seed iron content. Int. Sorghum Millets Newsl. 82, 82–83.

Prasada Rao, K.E., de Wet, J.M.J., 1997. Small millets. In: Fuccilo, D., Sears, L., Stapleton, P. (Eds.), Biodiversity in Trust. Cambridge University Press, UK, pp. 259–272.

Prasada Rao, K.E., de Wet, J.M.J., Brink, D.K., Mengesha, M.H., 1987. Intraspecific variation and systematics of cultivated *Setaria italica*, foxtail millet (Poaceae). Econ. Bot. 41, 108–116.

Prasada Rao, K.E., de Wet, J.M.J., Reddy, V.G., Mengesha, M.H., 1993. Diversity in the small millets collection at ICRISAT. In: Riley, K.W., Gupta, S.C., Seetharam, A., Mushonga, J.N. (Eds.), Advances in Small Millets. Oxford & IBH Publishing Co, New Delhi, India, pp. 331–346.

Purseglove, J.W., 1972. *Eleusine coracana* (L.) Gaertn. Tropical Crops. Monocotyledons. Longman Group Limited, London, UK, pp. 147–156.

Qie, L., Jia, G., Zhang, W., Schnable, J., Shang, Z., Li, W., et al., 2014. Mapping of quantitative trait locus (QTLs) that contribute to germination and early seedling drought tolerance in the interspecific cross *Setaria italica* × *Setaria viridis*. PloS ONE 9 (7), e101868.

Rao, A.N.S., 1990. Estimates of losses in finger millet (*Eleusine coracana*) due to blast disease (*Pyricularia grisea*). J. Agric. Sci. 24, 57–60.

Reddy, V.G., Upadhyaya, H.D., Gowda, C.L.L., 2006. Characterization of world's foxtail millet germplasm collections. SAT eJ. 2, 1–3.

Reddy, V.G., Upadhyaya, H.D., Gowda, C.L.L., Singh, S., 2009. Characterization of eastern African finger millet germplasm for qualitative and quantitative characters at ICRISAT. SAT eJ. 7, 1–9.

Reddy, I.N.B.L., Reddy, D.S., Reddy, V.P., Narasu, M.L., Sivaramakrishnan, S., 2011. Efficient microsatellite enrichment in finger millet (*Eleusine coracana* (L.) Gaertn.) – an improved procedure to develop microsatellite markers. Asian Austral. J. Plant Sci. Biotechnol. 5 (1), 47–51.

Richardson, W.L., 1958. A technique of emasculating small grass florets. J. Genet. 18, 69–73.

Saleh, A.S.M., Zhang, Q., Chen, J., Shen, Q., 2013. Millet grains: nutritional quality, processing, and potential health benefits. Compr. Rev. Food Sci. Food Saf. 12 (3), 281–295.

Salimath, S.S., de Oliveira, A.C., Godwin, I.D., Bennetzen, J.L., 1995a. Assessment of genome origins and genetic diversity in the genus *Eleusine* with DNA markers. Genome 38, 757–763.

Salimath, S.S., Hiremath, S.C., Murthy, H.N., 1995b. Genome differentiation patterns in diploid species of *Eleusine* (Poaceae). Hereditas 122, 189–195.

Sato, K., Mukainari, Y., Naito, K., Fukunaga, K., 2013. Construction of a foxtail millet linkage map and mapping of spikelet-tipped bristles 1 (*stb1*) by using transposon display markers and simple sequence repeat markers with genome sequence information. Mol. Breed. 31 (3), 675–684.

Shailaja, H.B., Thirumeni, S., 2007. Evaluation of salt-tolerance in finger millet (*Eleusine coracana*) genotypes at seedling stage. Indian J. Agric. Sci. 77, 672–674.

Shao, L.H., Wang, L., Bai, W.W., Liu, Y.J., 2014. Evaluation and analysis of folic acid content in millet from different ecological regions in Shanxi province. Sci. Agric. Sin. 47 (7), 1265–1272.

Sharma, R., Girish, A.G., Upadhyaya, H.D., Humayun, P., Babu, T.K., Rao, V.P., et al., 2014. Identification of blast resistance in a core collection of foxtail millet germplasm. Plant Dis. 98 (4), 519–524.

Shet, R.M., Gireesh, C., Jagadeesha, N., Lokesh, G.Y., Gowda, J., 2009. Genetic variability in segregating generation of interspecific hybrids of finger millet (*Eleusine coracana* (L.) Gaertn.). Environ. Evol. 27, 1013–1016.

Shet, R.M., Jagadeesha, N., Lokesh, G.Y., Gireesh, C., Gowda, J., 2010a. Genetic variability, association and path coefficient studies in two interspecific crosses of finger millet [*Eleusine coracana* (L.) Gaertn.]. Int. J. Plant Sci. 5 (1), 24–29.
Shet, R.M., Mallikarjuna, N.M., Kumar, K.S.N., Gowda, J., Kumar, M.V.M., 2010b. Cytomorphological studies in interspecific hybrids of finger millet. Gregor Mendel Found. J. 1, 68–72.
Siles, M.M., Nelson, L.A., Baltensperger, D.D., 2001. Technique for artificial hybridization of foxtail millet [*Setaria italica* (L.) Beauv.]. Crop Sci. 41 (5), 1408–1412.
Srinivasachary, S., Dida, M.M., Gale, M.D., Devos, K.M., 2007. Comparative analyses reveal high levels of conserved colinearity between the finger millet and rice genomes. Theor. Appl. Genet. 115, 489–499.
Suryanarayana, L., Sekhar, D., Rao, N.V., 2014. Genetic variability and divergence studies in finger millet. Int. J. Curr. Microbiol. Appl. Sci. 3 (4), 931–936.
Tian, R.Z., Quan, J.Z., 1995. Niangu 1, a new summer millet cultivar. Crop Genet. Res. 1, 53–54.
Tian, B.H., Wang, S.Y., Li, Y.J., Wang, J.G., Zhang, L.X., Liang, F., et al., 2008. Response to sodium chloride stress at germination and seedling and identification of salinity tolerant genotypes in foxtail millet landraces originated from China. Acta Agron. Sin. 34 (12), 2218–2222.
Tian, B., Wang, J., Zhang, L., Li, Y., Wang, S., Li, H., 2010. Assessment of resistance to lodging of landrace and improved cultivars in foxtail millet. Euphytica 172, 295–302.
Upadhyaya, H.D., 2008a. Genotyping of composite collection of finger millet [*Eleusine coracana* (L.) Gaertn.]. Generation Challenge Program: Cultivating Plant Diversity for the Resource Poor (Abstract). CIMMYT, Mexico, pp. 64–65.
Upadhyaya, H.D., 2008b. Genotyping of composite collection of foxtail millet [*Setaria italica* (L). P. Beauv.]. Generation Challenge Program: Cultivating Plant Diversity for the Resource Poor (Abstract). CIMMYT, Mexico, pp. 66–67.
Upadhyaya, H.D., Pundir, R.P.S., Hash, C.T., Hoisington, D., Chandra, S., Gowda, C.L.L., et al., 2005. Genotyping finger millet germplasm – developing composite collection. Generation Challenge Program Review Meeting, September 2005, Rome, Italy.
Upadhyaya, H.D., Gowda, C.L.L., Pundir, R.P.S., Reddy, V.G., Singh, S., 2006a. Development of core subset of finger millet germplasm using geographical origin and data on 14 quantitative traits. Genet. Resour. Crop Evol. 53, 679–685.
Upadhyaya, H.D., Varshney, R.K., Hash, C.T., Hoisington, D.A., Gowda, C.L.L., Reddy V.G., et al., 2006b. Development of composite collection and genotyping of foxtail millet (*Setaria italica* (L.) Beauv.) composite collection, Generation Challenge Program Annual Research Meeting, September 2006, Sao Paulo, Brazil.
Upadhyaya, H.D., Gowda, C.L.L., Reddy, V.G., 2007. Morphological diversity in finger millet germplasm introduced from Southern and Eastern Africa. SAT eJ. 3 (1), 1–3.
Upadhyaya, H.D., Pundir, R.P.S., Gowda, C.L.L., Reddy, V.G., Singh, S., 2008. Establishing a core collection of foxtail millet to enhance the utilization of germplasm of an underutilized crop. Plant Genet. Resour. 7 (2), 177–184.
Upadhyaya, H.D., Sarma, N.D.R.K., Ravishankar, C.R., Albrecht, T., Narasimhudu, Y., Singh, S.K., et al., 2010. Developing a mini-core collection in finger millet using multilocation data. Crop Sci. 50, 1924–1931.
Upadhyaya, H.D., Ramesh, S., Sharma, S., Singh, S.K., Varshney, S.K., Sarma, N.D.R.K., et al., 2011a. Genetic diversity for grain nutrients contents in a core collection of finger millet (*Eleusine coracana* (L.) Gaertn.) germplasm. Field Crops Res. 121, 42–52.

Upadhyaya, H.D., Ravishankar, C.R., Narasimhudu, Y., Sarma, N.D.R.K., Singh, S.K., Varshney, S.K., et al., 2011b. Identification of trait-specific germplasm and developing a mini core collection for efficient use of foxtail millet genetic resources in crop improvement. Field Crops Res. 124, 459–467.

Vadivoo, A.S., Joseph, R., Ganesan, N.M., 1998. Genetic variability and diversity for protein and calcium contents in finger millet (*Eleusine coracana* (L.) Gaertn.) in relation to grain color. Plant Foods Hum. Nutr. 52 (4), 353–364.

Vetriventhan, M., 2011. Phenotypic and genetic diversity in the foxtail millet (*Setaria italica* (L.) P. Beauv.) core collection. PhD Thesis. Agricultural College and Research Institute, Tamil Nadu Agricultural University, Madurai, Tamil Nadu, India.

Vetriventhan, M., Upadhyaya, H.D., Anandakumar, C.R., Senthilvel, S., Parzies, H.K., Bharathi, A., et al., 2012. Assessing genetic diversity, allelic richness and genetic relationship among races in ICRISAT foxtail millet core collection. Plant Genet. Resour. 10 (03), 214–223.

Vijayalakshmi, D., Ashok, S.K., Raveendran, M., 2014. Screening for salinity stress tolerance in rice and finger millet genotypes using shoot Na+/K+ ratio and leaf carbohydrate contents as key physiological traits. Indian J. Plant Physiol. 19, 156–160.

Vishwanath, S., Gowda, S.S., Seetharam, A., Gowda, B.T.S., 1986. Reaction to blast disease of released and pre-released varieties of finger millet from different states. Millet Newsl. 5, 31.

Wang, T.Y., Darmency, H., 1997. Inheritance of sethoxydim resistance in foxtail millet, *Setaria italica* (L.) Beauv. Euphytica 94, 69–73.

Wang, T.Y., Du, R.H., Hao, F., 1993. Studies and utilization of highly male sterility of summer millet (*Setaria italica*). Sci. Agric. Sin. 26, 88, (in Chinese).

Wang, Y.R., Chu, J.H., Cao, X.J., Liu, H.X., Duan, C.L., Song, Y.C., 1997. Studies on resistance of foxtail millets (*Setaria italica*) to downy mildew (*Sclerospora graminicola*). Int. Sorghum Millets Newsl. 38, 133–134.

Wang, X., Yan, T., Zhou, H., 1998a. New millet cultivar Tie Gu 7. Crop Genet. Resour. 1, 55–56.

Wang, Y.W., Li, H.X., Wang, G.H., Tian, G., 1998b. Breeding of foxtail millet highly sterile line "Chang10A". Gansu Agric. Sci. Technol. 12, 12–13, (in Chinese).

Wang, Z.M., Devos, K.M., Liu, C.J., Wang, R.Q., Gale, M.D., 1998c. Construction of RFLP-based maps of foxtail millet. Theor. Appl. Genet. 96, 31–36.

Wang, R.Q., Gao, J.H., Mao, L.P., Du, R.H., Diao, X.M., Sun, J.S., 2002. Chromosome location of the male-sterility and yellow seedling gene in line 1066A of foxtail millet. Acta Bot. Sin. 44, 1209–1212, (in Chinese).

Wang, C., Chen, J., Zhi, H., Yang, L., Li, W., Wang, Y., et al., 2010a. Population genetics of foxtail millet and its wild ancestor. BMC Genet. 11 (1), 90.

Wang, Y.W., Li, H.X., Tian, G., Shi, Q.X., 2010b. Study on innovation and application of highly-male-sterile line with high outcrossing rate in millet. Sci. Agric. Sin. 43, 680–689, (in Chinese).

Wang, C., Jia, G., Zhi, H., Niu, Z., Chai, Y., Li, W., et al., 2012. Genetic diversity and population structure of Chinese foxtail millet [*Setaria italica* (L.) Beauv.] landraces. Genes Genomes Genet. 2 (7), 769–777.

Wang, J., Wang, Z., Yang, H., Yuan, F., Guo, E., Tian, G., et al., 2013. Genetic analysis and preliminary mapping of a highly male-sterile gene in foxtail millet (*Setaria italica* L. Beauv.) using SSR markers. J. Integr. Agric. 12, 2143–2148.

Werth, C.R., Hilu, K.W., Langner, C.A., 1994. Isozymes of *Eleusine* (Gramineae) and the origin of finger millet. Am. J. Bot. 81, 1186–1197.

Xu, J., Li, Y., Ma, X., Ding, J., Wang, K., Wang, S., et al., 2013. Whole transcriptome analysis using next-generation sequencing of model species *Setaria viridis* to support C4 photosynthesis research. Plant Mol. Biol. 83, 77–87.

Yadav, C.B., Bonthala, V.S., Muthamilarasan, M., Pandey, G., Khan, Y., Prasad, M., 2014a. Genome-wide development of transposable elements-based markers in foxtail millet and construction of an integrated database. DNA Res. 22, 79–90.

Yadav, C.B., Muthamilarasan, M., Pandey, G., Khan, Y., Prasad, M., 2014b. Development of novel microRNA-based genetic markers in foxtail millet for genotyping applications in related grass species. Mol. Breed. 34 (4), 2219–2224.

Yang, X., Wan, Z., Perry, L., Lu, H., Wang, Q., Zhao, C., et al., 2012. Early millet use in northern China. Proc. Natl. Acad. Sci. USA 109 (10), 3726–3730.

Yang, X.S., Wang, L.L., Zhou, X.R., Shuang, S.M., Zhu, Z.H., Li, N., et al., 2013. Determination of protein, fat, starch, and amino acids in foxtail millet [*Setaria italica* (L.) Beauv.] by Fourier transform near-infrared reflectance spectroscopy. Food Sci. Biotechnol. 22 (6), 1495–1500.

Zangre, G.R., Darmency, H., 1993. Potential for selection in the progeny of an interspecific hybrid in foxtail millet. Plant Breed. 110, 172–175.

Zegada-Lizarazu, W., Iijima, M., 2005. Deep root water uptake ability and water use efficiency of pearl millet in comparison to other millet species. Plant Prod. Sci. 8, 454–460.

Zhang, W.L., Guan, Y.A., 1995. Summer millet Jinan 8337 with high yield and quality. Crop Genet. Resour. 3, 28.

Zhang, J.P., Wang, M.Y., Bai, Y.F., Jia, J.P., Wang, G.Y., 2005. Rapid evaluation on drought tolerance of foxtail millet at seedling stage. J. Plant Genet. Resour. 6, 59–62, (in Chinese, English abstract).

Zhang, Y., Wang, J., Wang, W., Zeng, P., Han, C., Quan, Z., et al., 2012a. Genome sequence of foxtail millet (*Setaria italica*) provides insights into grass evolution and biofuel potential. Nat. Biotechnol. 30, 549–554.

Zhang, W., Zhi, H., Liu, B.-H., Xie, J., Li, J., Li, W., et al., 2012b. Screening of indexes for drought tolerance test at booting stage in foxtail millet. J. Plant Genet. Resour. 13 (5), 765–772.

Zhang, S., Tang, C., Zhao, Q., Li, J., Yang, L., Qie, L., et al., 2014. Development of highly polymorphic simple sequence repeat markers using genome-wide microsatellite variant analysis in foxtail millet [*Setaria italica* (L.) P. Beauv.]. BMC Genomics 15, 78.

Zhao, Z.H., Cui, W.S., Du, G., Yang, S.Q., 1996. The selection of millet photo- (thermo-) sensitive line 821 and study on the relation of sterility to illumination and temperature. Sci. Agric. Sin. 29, 23–31, (in Chinese).

Zhao, W., Lee, G.A., Kwon, S.W., Ma, K.H., Lee, M.C., Park, Y.J., 2012. Development and use of novel SSR markers for molecular genetic diversity in Italian millet (*Setaria italica* L.). Genes Genomics 34, 51–57.

Zhi, H., Wang, Y.Q., Li, W., Wang, Y.F., Li, H.Q., Lu, P., et al., 2007. Development of CMS material from intra-species hybridization between green foxtail and foxtail millet. J. Plant Genet. Resour. 8, 261–264.

Zhu, G.Q., Wu, Q.M., Ma, Y.T., 1991. Breeding of Ve type CMS in foxtail millet. Shaanxi Agric. Sci. 1, 7, (in Chinese).

Zhu, X.-H., Song, Y.-C., Zhao, Z.-H., Shi, Y.-S., Liu, Y.-H., Li, Y., et al., 2008. Methods for identification of drought tolerance at germination period of foxtail millet by osmotic stress. J. Plant Genet. Resour. 9, 62–67, (in Chinese, English abstract).

Proso, barnyard, little, and kodo millets

Hari D. Upadhyaya, Mani Vetriventhan, Sangam Lal Dwivedi, Santosh K. Pattanashetti, Shailesh Kumar Singh
International Crops Research Institute for the Semi-Arid Tropics (ICRISAT), Genebank, Patancheru, Telangana, India

8.1 Introduction

Proso, barnyard, little, and kodo millets belong to the group called small millets, sometimes also referred to as minor millets. Proso millet (*Panicum miliaceum* L.) is commonly known as broomcorn millet, common millet, hog millet, Russian millet, and so on, in different parts of the world. Barnyard millet is generally well-known as Japanese barnyard millet (*Echinochloa crus-galli* (L.) P. Beauv.), Indian barnyard millet (*Echinochloa colona* (L.) Link), cockspur grass, Korean native millet, prickly millet, sawa millet, watergrass, and so on. Kodo millet (*Paspalum scrobiculatum* L.) is also known by different names in different languages in India (kodo in Hindi, khoddi in Urdu, arugu in Telugu, varagu in Tamil), African bastard millet grass, arika, haraka, ditch millet in New Zealand, and mandal in Pakistan. Similarly, little millet (*Panicum sumatrense* Roth. ex. Roem. & Schult.) is also commonly known as samai, gindi, mutaki, kutki, and so on, in different Indian languages.

These crops are cultivated in the marginal areas, and are adapted to a wide range of growing environments. Proso millet is currently grown in Asia, Australia, North America, Europe, and Africa (Rajput et al., 2014), and used for feeding birds and as livestock feed in the developed countries and for food in some parts of Asia. Barnyard millet is mainly grown in India, China, Japan, and Korea for human consumption as well as fodder (Upadhyaya et al., 2014). Kodo and little millets are largely cultivated throughout India by tribal people in small areas. All these crops have superior nutritional properties including high micronutrients, dietary fiber content, and low glycemic index (GI) with potential health prospective (Chandel et al., 2014; Dwivedi et al., 2012; Saleh et al., 2013). Research evidences support that the low-GI carbohydrate diets help in the prevention of obesity, diabetes, and cardiovascular disease (Brand-Miller et al., 2009). Proso, barnyard, kodo, and little millets together with finger and foxtail millets, are used as an ingredient in multigrain and gluten-free cereal products and serve as a major food component for various traditional foods and beverages, such as bread, porridges, and snack foods, while grains are feed to animals, including pigs, fowls, and cage birds.

These crops are under-researched and underutilized compared to foxtail and finger millets and other cereals, and are being neglected in terms of support for

Genetic and Genomic Resources for Grain Cereals Improvement. http://dx.doi.org/10.1016/B978-0-12-802000-5.00008-3

production, promotion, research, and development. More research efforts on proso, barnyard, little, and kodo millets are required for developing high-yielding varieties and to diversify food habits for healthy lives and to face the global threats of malnutrition and climate change. In this chapter, we mainly focus on four small millets, that is, proso, little, barnyard, and kodo millets and provide an overview of their origin, history, domestication, and diversity; the status germplasm collections conserved in genebanks worldwide and at ICRISAT; germplasm evaluation for agronomic and nutritional traits, and for biotic and abiotic stresses; way to enhance the use of germplasm through core collection approach; and genomic resources and their use for germplasm characterization, and genomic research in these crops.

8.2 Origin, distribution, taxonomy, and diversity

Proso millet is an annual herbaceous plant in the genera *Panicum*, and it has a chromosome number of $2n = 36$ with basic chromosome number of $x = 9$. Vavilov (1926) suggested China as the center of diversity for proso millet, while Harlan (1975) opined that proso millet probably was domesticated in China and Europe. The earliest records come from the Yellow River valley site of Cishan, China dated between 10,300 cal years BP and 8,700 cal years BP (Lu et al., 2009). Evidence of proso millet also occurs at a number of pre-7000 cal years BP sites in Eastern Europe, in the form of charred grains and grain impressions in pottery (Hunt et al., 2008; Zohary and Hopf, 2000). These two centers of earlier records suggest independent domestication of proso millet in eastern Europe or Central Asia, or may have also originated from domestication within China and then spread westward across the Eurasian steppe (Hunt et al., 2011; Jones, 2004). Most recently, Hunt et al. (2014) used nuclear and chloroplast DNA sequences from proso millet and a range of diploid and tetraploid relatives to unveil the phylogenies of the diploid and tetraploid species, and suggested the allotetraploid origin of *P. miliaceum*, with the maternal ancestor being *Panicum capillare* (or a close relative) and the other genome being shared with *Panicum repens*; however, further studies of the *Panicum* species, particularly from the Old World are required. Cultivated proso millet can be divided into five races: *miliaceum*, *patentissimum*, *contractum*, *compactum*, and *ovatum* (de Wet, 1986). Race *miliaceum* resembles wild *P. miliaceum* in inflorescence morphology, characterized by large, open inflorescences with suberect branches that are sparingly subdivided. Race *patentissimum* is characterized by slender and diffused panicle branches, which is often difficult to distinguish from race *miliaceum*. Cultivars with more or less compact inflorescences are classified into races *contractum*, *compactum*, and *ovatum*. Cultivars in race *contractum* have compact, drooping inflorescences while the race *compactum* have cylindrical inflorescences that are essentially erect, whereas the cultivars with compact and slightly curved inflorescences that are ovate in shape belong to race *ovatum* (de Wet, 1986).

Little millet belongs to the genus *Panicum* having a chromosome number of $2n = 36$, with basic chromosome number of $x = 9$. It was domesticated in India (de Wet et al., 1983a), particularly in the Eastern Ghats of India, where it forms an important part of tribal agriculture. Little millet is grown across India, Sri Lanka, Nepal, and western Burma. The species *Panicum sumatrense* is divided into subsp. *sumatrense* (cultivated little millet), and subsp. *psilopodium* (wild progenitor). These two subspecies cross where they are sympatric to produce fertile hybrids, derivatives of which are often weed in little millet field (de Wet et al., 1983a). *P. sumatrense* subsp. *sumatrense* has two races, *nana* and *robusta*, and two subraces each, *laxa* and *erecta* in *nana*, and *laxa* and *compacta* in *robusta*. The race *nana* includes plants with decumbent to almost prostrate culms that become erect at the time of flowering. Inflorescences are large, open with the upper branches sometimes clumped and curved at the time of maturity. The *robusta* includes erect plants with large, strongly branched, open, or compact inflorescences (de Wet et al., 1983a).

The genus *Echinochloa* comprises of approximately 25 species and two species, namely, *E. crus-galli* and *E. colona* are cultivated as cereals. *E. crus-galli* is native to temperate Eurasia and was domesticated in Japan around 4000 years ago, while *E. colona* is widely distributed in the tropics and subtropics of the Old World, and was domesticated in India. Both the cultivated species are hexaploids ($2n = 54$) and are morphologically related, but hybrid between these two species is sterile. In general, cultivated plants of *E. colona* are erect or geniculate ascending, often tufted, annual and can grow up to 242 cm tall and awnless spikelets with membranaceous glumes, while plants of *E. crus-galli* are erect, tufted, annual, grow up to 100 cm tall, and awned spikelets with chartaceous glumes (de Wet et al., 1983b). The species *E. crus-galli* is classified into two subspecies (*crus-galli* and *utilis*) and four races (*crus-galli* and *macrocarpa* in subsp. *crus-galli*, and *utilis* and *intermedia* in subsp. *utilis*). Similarity, *E. colona* has two subspecies, *colona* and *frumentacea*. The subsp. *colona* has no races and subsp. *frumentacea* is divided into four races: *stolonifera, intermedia, robusta,* and *laxa* (de Wet et al., 1983b). More recently, Wallace et al. (2015) investigated the patterns of population structure and phylogeny among the accessions belonging to two species, *E. crus-galli* and *E. colona*, through genotyping-by-sequencing (GBS) approach suggested distinct phylogenetic structure within and between two species, four subpopulations within *E. colona* accessions, and three such clusters within *E. crus-galli.*

Kodo millet belongs to the genus *Paspalum,* a diverse genus comprising about 400 species, most of which are native to the tropical and subtropical regions of the Americas, and the main center of origin and diversity of the genus is considered to be South American tropics and subtropics (Chase, 1929). The chromosome number of the kodo millet is reported to be $2n = 4x = 40$ (Hiremath and Dandin, 1975). Kodo millet was domesticated in India around 3000 years ago and cultivated by tribal people in small areas throughout India, from Kerala and Tamil Nadu in the south, to Rajasthan, Uttar Pradesh, and West Bengal in the North. It occurs in moist or shady places across the tropics and subtropics of the Old World (de Wet et al., 1983c). Kodo millet has three races, namely, *regularis, irregularis*, and *variabilis*. The most common race is *regularis*, characterized by racemes with the spikelets arranged in two rows on one side of a flattened rachis. In the case of the race *irregularis*, the spikelets are arranged along the rachis in two to four irregular rows; whereas in the race *variabilis*, the lower part of each raceme is characterized by irregularly arranged spikelets, while spikelets arrangement becomes more regularly two-rowed in the upper part of the raceme (de Wet et al., 1983c).

8.3 Erosion of genetic diversity from the traditional areas

Genetic variation found in traditional landraces and wild species is important for continued progress of crop improvement. Genetic erosion refers to the loss of genetic diversity, sometimes used in a narrow sense, that is, the loss of genes or alleles, as well as more broadly, referring to the loss of varieties, and crop species, mainly because of the replacement of traditional landraces by modern, high-yielding cultivars, natural devastations, and large-scale destruction and modification of natural habitats sheltering wild species. Proso, barnyard, little, and kodo millets continue to be grown largely by the traditional practices using traditional landraces under subsistence farming. Area under cultivation of these crops is decreasing around the world mainly due to increasing importance of a few selected crop species causing genetic erosion of small millets and narrowing the food security basket. In China, proso millet continued to be a very important crop until the beginning of the twentieth century, but recently there has been significant reduction in its cultivation due to the adoption of modern high-yielding varieties of major crops like rice, wheat, and maize. However, it is still produced because of their adaptation to areas that are too dry or too cold for other crops (Bonjean, 2010). Drastic decline in cultivated area (5.34 million ha during 1955–1956 to 0.80 million ha during 2011–2012), and production (2.07 Mt during 1955 to 0.46 Mt during 2011–2012) under six small millets (finger, foxtail, proso, little, barnyard, and kodo millets) was noticed in India (NAAS, 2013). About 72% reduction in barnyard millet area was reported from 11 villages in Garhwal Himalayas (Maikhuri et al., 2001). Decline in cultivation of small millets is mainly due to low productivity, nonavailability of high-yielding varieties, lack of production and processing technologies, and introduction of high-yielding commercial crops.

8.4 Status of germplasm resource conservation

Ex situ conservation is the widely used method to conserve millet genetic resources. Globally >29,000 accessions of proso millet, >8,000 accessions each of barnyard and kodo millets, and >3,000 accessions of little millet have been assembled and conserved (Fig. 8.1). The major genebanks conserving proso, barnyard, kodo, and little millets are presented in Table 8.1. The major collections of proso millet germplasm accessions are assembled in the Russian Federation, China, Ukraine, and India; barnyard millet in Japan and India; kodo millet in India and USA; and little millet in India.

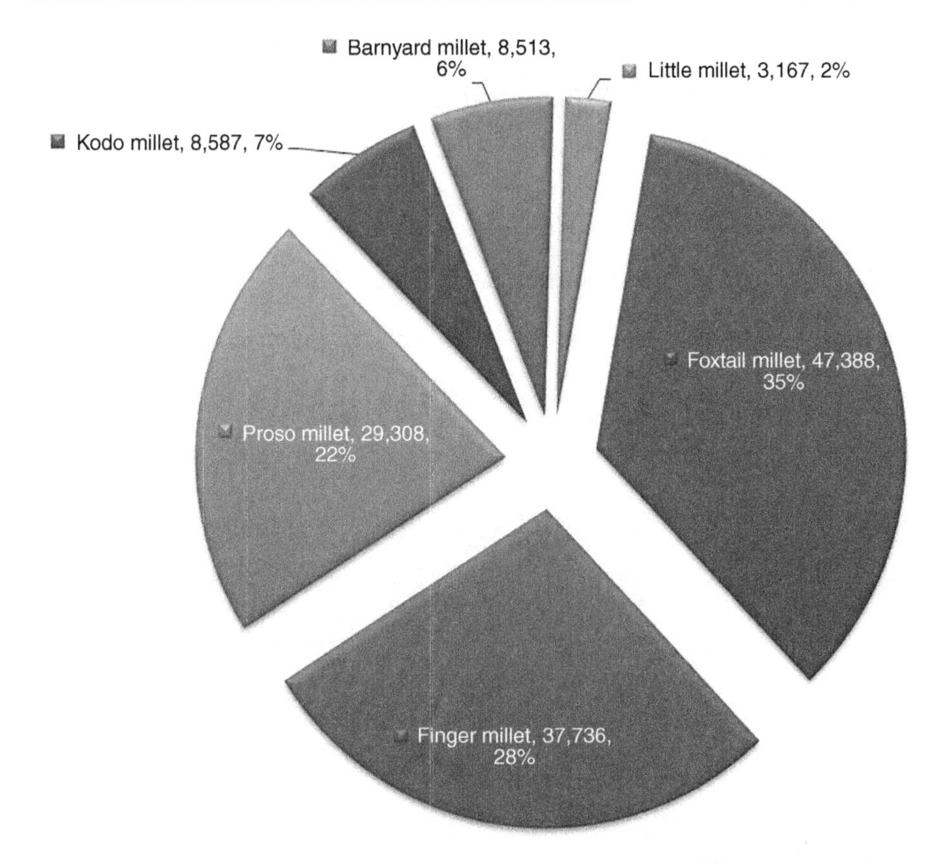

Figure 8.1 Global status of small millet germplasm maintained in different genebanks.

8.5 Germplasm evaluation and maintenance

Proso, barnyard, kodo, and little millets are highly self-pollinating crops, so there is no special regeneration and maintenance practice as in the case of cross-pollinated crops like pearl millet. The field used for regeneration should not have grown the same crops in the previous year in order to avoid volunteer plants. Individual accessions can be planted in rows (4 m length) and harvested panicles by hand will be bulked to make up the accession. The ICRISAT Genebank at Patancheru, India conserves 849 accessions of proso millet, 749 accessions of barnyard millet, 665 accessions of kodo millet, and 473 accessions of little millet under medium- (4°C and 30% relative humidity) and long-term (−20°C in vacuum-packed standard aluminum foil pouches) storage conditions.

Limited works on germplasm evaluation for various agronomic and nutritional traits, biotic and abiotic stress tolerance have been reported in these crops. Few studies on evaluation and identification of important traits of economic interest are discussed here.

Table 8.1 **Major genebanks conserving germplasms of proso, barnyard, kodo, and little millets worldwide***

Crop/country	Institute	Germplasm accessions		
		Cultivated	Wild	Total
Proso millet				
Australia	Australian Tropical Crops and Forages Genetic Resources Centre (ATCFC)	228		228
Bangladesh	Plant Genetic Resources Centre, BARI (PGRC, BARI)	198		198
Bulgaria	Institute for Plant Genetic Resources "K. Malkov" (IPGR)	489		489
China	Institute of Crop Science, Chinese Academy of Agricultural Sciences (ICS-CAAS)	8451		8451
Czech Republic	Genebank Department, Division of Genetics and Plant Breeding, Research Institute of Crop Production (RICP)	171		171
Germany	Genebank, Leibniz Institute of Plant Genetics and Crop Plant Research (IPK)	165	1	166
Hungary	Institute for Agrobotany (RCA)	243	1	244
India	AICRP on Small Millets (AICRP-Small Millets)	920		920
	International Crop Research Institute for the Semi-Arid Tropics (ICRISAT)	849		849
	National Bureau of Plant Genetic Resources (NBPGR)	994	4	998
Japan	Department of Genetic Resources I, National Institute of Agrobiological Sciences (NIAS)	516		516
Mexico	Estación de Iguala, Instituto Nacional de Investigaciones Agrícolas (INIA-Iguala)	400		400
Poland	Botanical Garden of Plant Breeding and Acclimatization Institute (BYDG)	354		354
	Plant Breeding and Acclimatization Institute (IHAR)	359		359
Russian Federation	N.I. Vavilov All-Russian Scientific Research Institute of Plant Industry (VIR)	8778		8778
Ukraine	Institute of Plant Production n.a. V.Y. Yurjev of UAAS (IR)	1046		1046
	Ustymivka Experimental Station of Plant Production (UDS)	3975	1	3976

Table 8.1 **Major genebanks conserving germplasms of proso, barnyard, kodo, and little millets worldwide*** *(cont.)*

Crop/country	Institute	Germplasm accessions		
		Cultivated	Wild	Total
USA	North Central Regional Plant Introduction Station, USDA-ARS, NCRPIS (NC7)	717	4	721
Barnyard millet				
Japan	Department of Genetic Resources I, National Institute of Agrobiological Sciences	3603	68	3671
India	National Bureau of Plant Genetic Resources	1668	9	1677
	AICRP on Small Millets	868		868
	International Crop Research Institute for the Semi-Arid Tropics	749		749
	The Ramiah Gene Bank, Tamil Nadu Agricultural University, India	232		232
China	Institute of Crop Science, Chinese Academy of Agricultural Sciences	717		717
Kenya	National Genebank of Kenya, Crop Plant Genetic Resources Centre - Muguga	192	16	208
Kodo millet				
Argentina	Banco Activo de Germoplasma de Papa, Forrajeras y Girasol Silvestre		127	127
	Instituto de Botánica del Nordeste, Universidad Nacional de Nordeste, Consejo Nacional de Investigaciones Científicas y Técnicas	390		390
Australia	Australian Tropical Crops and Forages Genetic Resources Centre	54	159	213
Brazil	Embrapa Pecuária Sudeste (CPPSE)	327		327
Colombia	Centro Internacional de Agricultura Tropical (CIAT)		155	155
Ethiopia	International Livestock Research Institute (ILRI)	3	205	208
India	AICRP on Small Millets, Bangalore	1111		1111
	International Crop Research Institute for the Semi-Arid Tropics (ICRISAT)	665		665
	National Bureau of Plant Genetic Resources (NBPGR), New Delhi	2170	10	2180
Japan	Department of Genetic Resources I, National Institute of Agrobiological Sciences (NIAS)	158		158

(Continued)

Table 8.1 **Major genebanks conserving germplasms of proso, barnyard, kodo, and little millets worldwide*** ***(cont.)***

Crop/country	Institute	Germplasm accessions		
		Cultivated	**Wild**	**Total**
Kenya	National Genebank of Kenya, Crop Plant Genetic Resources Centre - Muguga	130		130
New Zealand	Margot Forde Forage Germplasm Centre, Agriculture Research Institute Ltd	281		281
Nigeria	National Centre for Genetic Resources and Biotechnology (NACGRAB)	294		294
USA	Plant Genetic Resources Conservation Unit, Southern Regional Plant Introduction Station, University of Georgia, USDA-ARS	1074	249	1323
Uruguay	Facultad de Agronomía	106	446	552
Little millet				
India	AICRP on Small Millets (AICRP-Small Millets)	928		928
	International Crop Research Institute for the Semi-Arid Tropics (ICRISAT)	473		473
	National Bureau of Plant Genetic Resources (NBPGR)	1253		1253
	Regional Station Akola, NBPGR (NBPGR)	165		165
	The Ramiah Gene Bank, Tamil Nadu Agricultural University, India	108		108
USA	North Central Regional Plant Introduction Station, USDA-ARS, NCRPIS (NC7)	226		226

* Institutes/genebanks with >100 accessions are enlisted.
Source: http://www.fao.org/wiews-archive/germplasm_query.htm.

8.5.1 Germplasm evaluation

8.5.1.1 Agronomic traits

Proso, little, barnyard, and kodo millets germplasm accessions conserved at the ICRISAT Genebank show substantial variation for important agronomic traits (Table 8.2). In proso millet, days to 50% flowering ranges from 26 days to 50 days, plant height from 20 cm to 133 cm, basal tiller number from 1 to 32, and inflorescence length from 22 mm to 400 mm (Upadhyaya et al., 2011). The characterization of proso millet germplasm conserved at the ICRISAT collection revealed that most of the early flowering accessions are from Syria and late flowering accessions are from India; dwarf plant height accessions are from Mexico and tall plant height accessions are

Table 8.2 Diversity in entire and core collections of barnyard, kodo, little, and proso millets conserved at ICRISAT, Patancheru, India

Trait	Mean		Range	
	Entire	Core	Entire	Core
Proso millet				
Days to 50% flowering	34.5	34.9	26–50	28–50
Plant height (cm)	59.4	61.4	20–133	25–133
Basal tillers number	4.0	4.1	1–32	1–32
Flag leaf blade length (mm)	222.7	219.3	80–380	85–380
Flag leaf blade width (mm)	19.5	18.8	6–30	8–30
Flag leaf sheath length (mm)	82.1	80.6	30–170	30–170
Peduncle length (mm)	181.3	179.7	15–400	15–400
Panicle exsertion (mm)	100.1	102.8	0–320	0–320
Inflorescence length (mm)	193.1	193.8	22–400	22–400
No. of nodes	11.4	11.0	2–90	2–90
Inflorescence primary branches number	16.1	15.8	5–29	5–29
Barnyard millet				
Days to 50% flowering	48.8	49.6	30.9–77.2	33.2–73.2
Plant height (cm)	93.2	95.8	44.5–196.5	57.4–196.5
Basal tillers number	7.1	7.0	3.9–20.1	4.2–10.9
Culm thickness (mm)	5.5	5.5	4.7–7.2	4.9–7.2
No. of leaves	6.0	6.1	5.4–7.2	5.4–6.9
Flag leaf blade length (mm)	205.6	207.6	102.8–311.3	127.8–287.9
Flag leaf blade width (mm)	19.6	19.9	7.4–32.0	11.3–32.0
Flag leaf sheath length (mm)	88.1	88.9	59.4–156.5	66.9–156.5
Peduncle length (mm)	144.6	144.4	69.2–277.4	75.4–277.4
Panicle exsertion (mm)	56.5	55.9	29.8–80.6	33.9–77.7
Inflorescence length (mm)	155.5	159.2	81.0–257.8	102.7–240.8
No. of racemes per inflorescence	26.4	26.7	21.9–30.4	22.5–29.8
No. of nodes on primary axis of inflorescence	10.1	10.2	8.7–12.0	9.4–11.3
Length of lowest raceme (mm)	30.0	30.1	25.6–38.5	25.6–38.5
Little millet				
Days to 50% flowering	65.0	67.3	30.9–139.1	35.0–139.1
Plant height (cm)	112.7	115.3	58.3–201.7	60.6–201.7
Basal tillers number	13.0	13.1	10.4–16.9	11.5–16.2
Culm thickness (mm)	6.0	6.0	5.1–7.1	5.3–6.9
Flag leaf blade length (mm)	241.0	245.5	175.2–322.9	191.7–322.9
Flag leaf blade width (mm)	32.1	32.3	22.6–41.3	23.9–41.3
Flag leaf sheath length (mm)	98.5	99.1	81.5–121.3	88.1–114.4
Peduncle length (mm)	159.8	159.8	153.3–166.9	154.6–166.7
Panicle exsertion (mm)	21.2	21.0	6.9–48.2	9.9–48.2
Inflorescence length (mm)	273.4	275.1	198.5–347.6	218.2–330.7

(Continued)

Table 8.2 Diversity in entire and core collections of barnyard, kodo, little, and proso millets conserved at ICRISAT, Patancheru, India *(cont.)*

Trait	Mean		Range	
	Entire	Core	Entire	Core
No. of nodes on primary axis of inflorescence	12.7	12.7	11.7–15.0	11.7–14.0
No. of secondary inflorescence branches	21.9	21.9	17.7–30.0	18.6–30.0
Seed length (mm)	2.3	2.3	2.1–2.4	2.1–2.4
Seed width (mm)	1.6	1.6	1.5–1.8	1.5–1.8
Kodo millet				
Days to 50% flowering	78.7	77.7	56.2–117.4	60.0–110.2
Plant height (cm)	54.6	54.5	44.1–69.3	44.3–63.3
Basal tillers number	15.3	15.2	6.5–30.4	9.2–29.9
No. of leaves	5.7	5.7	5.4–6.7	5.5–6.7
Flag leaf blade length (mm)	191.7	191.9	156.1–226.5	156.1–226.5
Flag leaf blade width (mm)	7.2	7.2	5.9–8.4	5.9–8.4
Flag leaf sheath length (mm)	144.8	144.5	137.7–151.6	137.7–149.7
Inflorescence length (mm)	64.1	64.0	55.4–75.7	57.3–75.7
Sterile primary axis length (mm)	108.2	108.0	96.0–123.2	96.0–122.0
No. of racemes above thumb	3.0	3.0	2.7–3.9	2.7–3.9
Thumb length (mm)	56.5	56.5	50.3–66.0	52.4–63.7
Longest raceme length (mm)	28.6	28.6	25.7–32.4	27.3–32.3

Sources: Upadhyaya et al. (2011, 2014).

from Sri Lanka. Accessions with good exsertion are mostly from Australia and China, and shorter panicle exsertion accessions are from the former USSR, while the longest panicle types are from Nepal (Reddy et al., 2007). Similarly, larger variation of germplasm conserved at ICRISAT for various agronomic traits in barnyard millet (days to flowering from 30.9 days to 77.2 days, plant height from 44.5 cm to 196.5 cm, basal tillers number from 3.9 to 20.1, inflorescence length from 81 mm to 257.8 mm, etc.), little millet (days to flowering from 30.9 days to 139.1 days, plant height from 58.3 cm to 201.7 cm, basal tillers number from 10.4 to 16.9, inflorescence length from 198.5 mm to 347.6 mm, seed length from 2.1 mm to 2.4 mm, etc.), and kodo millet (days to flowering range from 56.2 days to 117.4 days, plant height from 44.1 cm to 69.3 cm, inflorescence length from 55.4 mm to 75.7 mm, etc.) were found (Table 8.2).

Proso millet germplasm accessions conserved at the National Centre for Crop Germplasm Conservation, Beijing, China were evaluated for their agronomic potential, disease resistance, and nutritional content, and elite accessions for specific or multiple traits were identified (Wang et al., 2007). Joshi et al. (2014) reported large variability in the kodo millet landraces collected from seven kodo millet growing districts of Madhya Pradesh, India for agronomic traits and grouped the landraces on the basis of plant height (dwarf, semidwarf, and tall), days to flowering and maturity (early, medium, and late), basal tillers number (low, medium, and high), degree of culm branching, inflorescence length, number of racemes above thumb, length of longest raceme, grain yield per plant, and 1000 grain weight. They reported promising kodo millet genotypes, dwarf types (RPS# 521, 529, 541, 683, 733, 801, and 926), for extra early maturity (RPS# 540, 541, 546, 632, 681, 687, 696, and 700), higher grain yield (RPS# 503, 556, 639, 649, 710, 712, 769, 775, 780, 798, 859, 910, 967, and 977), and higher 1000 grain weight (RPS# 507, 540, 556, 612, 614, 620, 638, 639, 642, 648, 650, 700, 705, 708, 709, 910, 912). Choi et al. (1991) evaluated barnyard millet at Suwon, South Korea, from 1985 to 1990, revealing that barnyard millet was found to be superior to that of other species as a fodder crop (on par with maize) and identified lines IEc 514 and IEc 515 from ICRISAT for high grain and green fodder yields. Gupta et al. (2009a) collected barnyard millet germplasm throughout the Himalayan region mainly from the hill state of Uttarakhand, India and promising donors for plant height (<120 and >200 cm), productive tillers (>4), inflorescence length (>28 cm), raceme number (>50) and raceme length (>3.1 cm), and grain yield (>16 g) were identified.

8.5.1.2 Nutritional traits

In general, small millet grains are the storehouses of many nutrients, phytochemicals, and nonnutritive plant protective functional constituents (Rao et al., 2011; Saleh et al., 2013). Proso, barnyard, kodo, and little millets have the higher amount of protein, crude fiber, minerals, and vitamins as compared to other cereals like rice and wheat. Particularly, proso millet is rich in protein content (12.5%), while barnyard millet is rich in protein (11%), crude fiber (13.6%), and Fe (18.6 mg per 100 g edible portion) (Saleh et al., 2013). It clearly signifies the importance of these crops in terms of nutritional and health perspective. Incorporation of these crops in the daily routine food habits may help in diversifying the food security basket. However, except for proso millet, there are very few or no studies on assessing the extent of genetic variability for grain nutritional traits involving a large number of germplasm in barnyard, kodo, and little millets. In the case of proso millet, Wang et al. (2007) evaluated 6515 germplasm from 14 provinces of China for grain protein and fat content and reported germplasm with high protein (>15%) and fat (>4%).

8.5.1.3 Biotic stress

Proso, barnyard, little, and kodo millets are said to be less affected by pests and diseases; however, there are a few pests and diseases that cause substantial reduction in grain yield of these crops. Limited number of resistant sources for major diseases and to some extent for pests in proso, barnyard, little, and kodo millets have been reported however, and large-scale exploitation of germplasm resources has not been done in these crops.

In proso millet, very few diseases have been reported and the major diseases are head smut, sheath blight, bacterial spot, and so on. Screening 18 proso millet genotypes for sheath blight under artificial inoculation conditions revealed that none of the genotypes were free from sheath blight; however, resistant (<20% disease severity) and moderately resistant (20–30% disease severity) genotypes were reported (Jain and Tiwari, 2013). Breeding of proso millet for resistance to head smut and melanosis (blackening of the grain under the husk, caused mainly by *Pseudomonas syringae* and *Xanthomonas vasicola* pv. *holcicola*) have been reported (Konstantinov and Grigorashchenko, 1986, 1987; Maslenkova and Resh, 1990; Konstantinov et al., 1989). Economically useful mutants with high yield, large grain, good grain quality, and resistance to smut and melanosis were isolated using chemical mutagens such as *N*-methyl-*N*-nitrosourea, dimethyl sulfate and *N*-ethyl-*N*-nitrosourea at the Ukrainian Institute of Plant Production, Breeding and Genetics (Konstantinov et al., 1989). The smut-resistant mutants, like Mutant 5, Mutant 6, 83-10170, and 83-10146, were used in hybridization and the mutant variety Khar'kovskoe 57 with high yield and good quality was released in Ukraine and Dagestan (Konstantinov et al., 1989). Soldatov and Agafonov (1980) tested 300 varieties of proso millet for resistance to melanosis at the Ural'sk Agricultural Experiment Station, Kazakh Soviet Socialist Republic (KSSR), of which 12 varieties were found fairly resistant. Very limited studies on insect, pest screening have been done in proso millet. Shailaja et al. (2009) screened different prerelease and released varieties against infestation of rice moth (*Corcyra cephalonica*), and identified TNAU 151 as comparatively resistant. Promising proso millet germplasm accessions and varieties relatively resistant to shoot fly have been reported in India (Murthi and Harinarayana, 1986) (Table 8.3).

Barnyard millet is mostly affected by smut (*Ustilago panici-frumentacei*) and leaf spot (*Colletotrichum graminicola)*. Grain smut can cause 6.5–60.8% yield loss. Gupta et al. (2009b) screened 257 accessions of barnyard millet, which includes advanced breeding lines for grain smut tolerance and grouped accessions based on reaction against smut infection and identified highly resistance accessions. Screening of barnyard millet genotypes against resistance to diseases led to the identification of resistant/moderately resistant sources for grain smut, head smut, *Helminthosporium* leaf spot or blight, and banded leaf and sheath blight (BLSB) (Table 8.3). Promising germplasm accessions and varieties of barnyard millet relatively resistant to shoot fly have been reported in India (Murthi and Harinarayana, 1986) (Table 8.3).

Head smut and blight are the major diseases in kodo millet. Many researchers have evaluated germplasm/cultivars and identified resistant sources for important diseases like smut and sheath blight (Table 8.3). Shoot fly is the major pest that causes considerable yield loss of up to 40% (Patel and Rawat, 1982). Screening kodo millet genotypes for shoot fly resistance has led to the identification of highly resistant landraces (Table 8.3). Joshi et al. (2014) reported multiple resistant accessions, namely, RPS# 575, 583, 590, 830, 886, 898, and 910 for head smut, sheath blight, and shoot fly.

Little millet is mainly affected by grain smut and sheath blight, and donors for resistance sources have been identified (Table 8.3). Little millet production is quite often affected by pests like shoot fly resulting in heavy loss to the crop. Morphologic characters are found to be associated with resistance to shoot fly in little millet. Tolerant genotypes showed higher trichome length and density, which offer mechanical

Table 8.3 Germplasm/cultivars identified as resistance/tolerance sources for various biotic stresses in proso, barnyard, kodo, and little millets

Crop/biotic stress	Resistant/tolerant sources	References
Proso millet		
Smut	K8763, Saratovskoye 2, Saratovskoye 3, Saratovskoye 6, Veselepodolyankoye 632, Barnaulskoye 80, Gorlinka, "Il'Inovskoe' Kh86, MS1316, Orenburgskoe 9, Khar'kovskoe 86	Ilyin et al. (1993); Konstantinov et al. (1986); Krasavin and Usmanova (1988); Konstantinov et al. (1991); Sharma et al. (1993); Zolotukhin et al. (1998)
Banded leaf and sheath blight	TNAU 137, GPUP 22 and RAUM 8	Jain and Tiwari (2013)
Melanosis	K8789, K8773, K8790, K8740 and K7606, UNIIZ670, Solnechnoe and Krasnoe Toidenskoe 215 (KT215), Orenburgskoe 9, Khar'kovskoe 57	Krasavin and Usmanova (1988); Konstantinov et al. (1989); Konstantinov et al. (1991)
Shoot fly	GPMS # 101, 102, 105, 108, 112, 114, 115, 117, 122, 123, 124, 125, 126, 135, 136, 138, 148, 152, 153, 155, 156, 157, 159, 164, RAUm# 1, 2, 3, MS # 1307, 1316, 1437, 1595, 4872, PM 29-1, BR 6, Co 1	Murthi and Harinarayana (1986)
Rice moth	TNAU 151	Shailaja et al. (2009)
Fall army warm	PI 176653	Wilson and Courteau (1984)
Barnyard millet		
Grain smut	PRB 402, S 841, TNAU# 92, 141, 155, VL# 216, 219, PRB# 901, 903, Co 1	Kumar and Kumar (2009); Kumar (2013); Muthusamy (1981);
Head smut	ABM 4-1, K 1, RAU 8, RBM 7-1, TNAU# 82, 86, 92, 96, 99, 101, 116, 128, 130, PRB# 401, 402, VL# 29, 172, 202, 205, 207, 208, 215, 216, 219, 220, 221, 222	Kumar and Kumar (2009); Kumar (2012)
Leaf spot or blight	TNAU # 116, 130, VL# 221, 222, 172, 29	Kumar (2012)
Brown spot	ABM 4-1, K 1, PRB# 401, 402, S 841, TNAU 82, 86, 92, 96, 99, 101, 116, 128, VL# 29, 172, 202, 205, 207, 208, 209, 216, 215, 220, RAU# 8, 12, RBM# 7, 7-1, VMBC 248	Kumar and Kumar (2009)

(Continued)

Table 8.3 Germplasm/cultivars identified as resistance/tolerance sources for various biotic stresses in proso, barnyard, kodo, and little millets *(cont.)*

Crop/biotic stress	Resistant/tolerant sources	References
Banded leaf and sheath blight (BLSB)	TNAU# 128, 130, VL# 29, 220, RBM 12	Jain and Gupta (2010)
Shoot fly	GECH# 102, 106, 108, 111, 120, 123, 127, 142, 149, 151, 157, 180, 205, 210, 218, 224, 226, 227, 230, 235, 240, 241, 246, 247, 248, 250, 260, 276, 288, VL# 8, 13, 21, 24, 30, 31, 32, ECC # 19, 18, 20, 21, RAU 7, KE 16, K1, PUNE 2386, Bhageshwar Local 2	Murthi and Harinarayana (1986)
Kodo millet		
Head smut	RPS# 539, 575, 581, 583, 590, 804, 818, 820, 830, 859, 886, 898, 910, 977, JK 13, GPLM# 78, 96, 176, 322, 364, 621, 641, 679 720, Acc. no 64, 348, 424	Jain (2005), Joshi et al. (2014); Jain et al. (2013)
Sheath blight	RPS# 502, 503, 508, 510, 516, 529, 531, 535, 543, 548, 550, 556, 566, 575, 577, 579, 585, 593, 607,609, 621, 629, 634, 646, 649, 661, 662, 689, 691, 694, 695, 708, 739, 753, 755, 787, 789, 814, 830, 867, 881, 883, 918, 919, 923, 929, 956, 961	Joshi et al. (2014)
Shoot fly	RPS# 515, 583, 612,628,642, 685, 763, 806, 810, 811, 822, 823, 834, 842, 846, 871, 872, 901, 902, 904, 905, 909, 910, 914, 915, 917, 918, 921, 925, 927, 929, 930, 933, 934, 938, 939, 941, 943, 944, 945, 946, 948, 951, 953, 967, 968, 970, 974, GPUK 3, JK13, GPLM # 6, 11, 20, 21, 29, 32, 39, 42, 45, 50, 60, 106, 110, 113, 117, 119, 120, 121, 131, 142, 155, 158, 160, 170, 172, 173, 178, 180, 185, RPS# 40-1, 40-2, 62-3, 61-1, 69-2, 72-2, 75-1, 102-2, 107-1, 114-1, 120-1, IQS 147-1, Co 2, Keharpur	Joshi et al. (2014); Murthi and Harinarayana (1986)

Table 8.3 **Germplasm/cultivars identified as resistance/tolerance sources for various biotic stresses in proso, barnyard, kodo, and little millets** ***(cont.)***

Crop/biotic stress	Resistant/tolerant sources	References
Little millet		
Grain smut	IPmr 841, 1061	http://www.dhan.org/smallmillets/docs/report/1_Advances_in_Crop_Improvement_of_Small_Millets.pdf
Head smut	GPMR# 65, 82, 67, 105, 70, 73, 80, 83, 92; OLM 36, 40, 203, TNAU# 89, 98, RLM# 13, 14, VMLC# 281, 296, Varisukdhara	Jain (2003); Jain and Tripathi (2007)
Shoot fly	GPMR# 164, 274, 236, 243, 110, 213, 584, 66, 683, 569, 189, 241, 98, 163, 324, 670, 598, 192, 96, 583, 161, 596, 95, 190, GPMR # 7, 17, 18, 20, 22, 26, 46, 53, 78, 84, 92, 98, 101, 104, 106, 107, 112, 114, 115, 116, 117, 124, 132, 134, 136, 141, 148, 149, 163, 169, 170, 171, 172, 175; PRC # 2, 3, 7, 8, 9, 10, 11, 12, RPM# 1-1, 8-1, 12-1, 41-1, RAU# 1, 2, K1, Co 2, Dindori 2-1	Gowda et al. (1996a); Murthi and Harinarayana (1986)

obstruction to young larvae in reaching their feeding sites (Gowda et al., 1996b). Field evaluation of little millet germplasm accessions for shoot fly resistance led to identification of highly resistant accessions (Table 8.3).

8.5.1.4 *Abiotic stress*

Proso, little, barnyard, and kodo millets are also affected by abiotic stresses, though they are generally considered well-adapted to abiotic stresses as compared to most other cereals (Dwivedi et al., 2012). Barnyard millet is reported to be tolerant to drought and waterlogging (Zegada-Lizarazu and Iijima, 2005), while proso millet is susceptible to drought (Seghatoleslami et al., 2008). Lodging is a constraint in many crops, including proso, little, barnyard, and kodo millets, causing substantial losses in grain yield and quality. Use of lodging-resistant cultivars along with good crop husbandry is the most effective way to minimize losses due to lodging. Proso millet lines developed in the United States have had strong selection for lodging resistance (Baltensperger et al., 1995a, 1995b, 2004). Sources for salinity tolerance have been reported (Acc. No. 008211, 008214, and 008226) (Sabir et al., 2011) in proso millet. Heavy metal tolerance (copper and zinc) at seedling stage was found highest in kodo millet followed by proso millet (Arora and Katewa, 1999).

8.6 Use of germplasm in crop improvement

The large size of germplasm particularly in the case of low research priority crops like proso, barnyard, little, and kodo millets reduce use of germplasm in breeding programs due to extremely low funding for research and development as compared to other crops. Developing representative subset of the entire collection of the species is a more economical and efficient way of utilizing germplasm to screen and identify the potential genetic resources for various economically important traits. At ICRISAT, Upadhyaya et al. (2011, 2014) developed core collections in proso, barnyard, little, and kodo millets, which captured genetic variation of the entire collections. These core collections could be effectively evaluated for agronomic and grain quality traits, and for biotic and abiotic stress tolerances to enhance utilization of germplasm in these crops. ICRISAT has distributed a total of >15,000 germplasm accessions of proso (6,047), barnyard (3,932), kodo (2,582), and little millet (2,449) to 25–37 countries (Table 8.4). This includes two and six sets of core collections of proso and barnyard millets, respectively. The most efficient use of germplasm conserved is using them directly as varieties. In barnyard millet, PRJ 1 was released in India during 2003, which is a selection from ICRISAT germplasm accession IEc 542 that originated in Japan.

8.6.1 *Development of core collection*

A core collection consists of a limited set of accessions (about 10%) derived from an existing germplasm collection, chosen to represent the genetic spectrum in the whole collection. Core collection helps to capture the entire diversity to utilize in breeding programs. Core collections have been formed in proso (Upadhyaya et al., 2011), barnyard (Gowda et al., 2009; Upadhyaya et al., 2014), little (Gowda et al., 2008; Upadhyaya et al., 2014), and kodo millets (Upadhyaya et al., 2014) (Table 8.5). At ICRISAT, the entire germplasm collection of proso millets (833 accessions) was stratified into five groups based on races and data on 20 morphoagronomic traits were used for clustering following Ward's method. About 10% (or at least one accession)

Table 8.4 Proso, barnyard, little, and kodo millet germplasm accessions distributed from ICRISAT Genebank, Patancheru, India (updated on Dec. 2014)

Crop	Total number of accessions (countries)	Germplasm samples distributed to			
		ICRISAT	India	Other countries	Total
Proso millet	849 (30)*	216	3,421	2,410 (37)	6,047
Barnyard millet	749 (9)	568	2,483	881 (28)	3,932
Kodo millet	665 (2)	382	1,317	883 (25)	2,582
Little millet	473 (5)	184	1,877	388 (27)	2,449
Total	2,736	1,350	9,098	4,562	15,010

* Number of countries.

Table 8.5 **Core collections in proso, barnyard, kodo, and little millets**

Crop	Accessions used	Traits assessed	Accessions in core collection	References
Proso millet	833	20	106	Upadhyaya et al. (2011)
Barnyard millet	729	24	50	Gowda et al. (2009)
	736	21	89	Upadhyaya et al. (2014)
Little millet	895	21	55	Gowda et al. (2008)
	460	20	56	Upadhyaya et al. (2014)
Kodo millet	656	20	75	Upadhyaya et al. (2014)

was randomly selected from each of 101 clusters to constitute a core collection of 106 accessions (Upadhyaya et al., 2011). Similarly, Upadhyaya et al. (2014) formed core collections in barnyard (89 accessions), little (56 accessions), and kodo millets (75 accessions), representing 11–12% of the entire collection of these crops conserved at ICRISAT Genebank (Table 8.5). These core collections are thus ideal genetic resources for identifying new sources of variation for use in crop improvement and for genomic studies.

8.7 Limitations in germplasm use

A significant number of germplasm accessions have been conserved in proso, barnyard, little, and kodo millets (Table 8.1); however, precise characterization and evaluation of these conserved genetic resources for traits of economic importance is very limited. Large holdings in genebanks and nonavailability of precise evaluation data on traits of economic importance limit the use of germplasm in these crops as well like in major crops (Upadhyaya et al., 2009). Forming subsets like core and mini core collections are the best entry point to search a genetic variability of agronomic and nutritional traits, biotic and abiotic stresses tolerance, and their use in breeding programs.

8.8 Germplasm enhancement through wide crosses

Floral morphology and anthesis behavior of proso, barnyard, little, and kodo millets make them very difficult crops for hybridization. However, emasculation and crossing techniques have been suggested in proso millet (Nelson, 1984) and other small grass florets (Richardson, 1958). To-date there is little or no effort made on interspecific hybridization in proso, barnyard, kodo, and little millets. In little millet, Hiremath et al. (1990) made interspecific hybridization between *P. sumatrense* (cultivated) with

Panicum psilopodium (Wild) and crossability of 23–25% was reported. Hybrids resembled female parent *P. sumatrense* with regard to nonshattering spikelets, intermediate between both the parents with respect to several quantitative characters like height, thickness of stem, leaf width, and spikelet number. Hybrids were highly fertile with 84% seed set and regular bivalent formation in the hybrids strongly suggests that the genomes of *P. sumatrense* and *P. psilopodium* are basically similar and are fully homologous (Hiremath et al., 1990). In the case of barnyard millet, Sood et al. (2014) reported interspecific hybridization between the two cultivated species of barnyard millet, PRJ 1 (*Echinocloa esculenta*) and ER 72 (*Echinocloa frumentacea*). The hybrid of the cross was vigorous with more tillers, high culm branching, and was free from grain smut disease but failed to set seed due to sterility. These studies open up vast avenues for introgression of desirable traits and exploitation of genetic variability for broadening the genetic base of the cultivars.

8.9 Integration of genomic and genetic resources in crop improvement

The foremost challenge for the molecular characterization of proso, barnyard, little, and kodo millets is the availability of very limited genomic resources like DNA markers, lack of genetic/linkage maps, and genome sequences. However, genomic resources of closely related species like foxtail millet where two reference genome sequences are available (Bennetzen et al., 2012; Zhang et al., 2012) can be utilized toward enriching genomic resources in these crops. DNA markers, such as simple sequence repeat (SSR), expressed sequence tag-simple sequence repeat (EST-SSR), ILP (intron length polymorphic), and microRNA-based molecular markers developed using foxtail millet genome sequence information showed >85% of cross-genera transferability among millets including proso, barnyard, little, and kodo millets, as well as nonmillet species (Kumari et al., 2013; Muthamilarasan et al., 2014; Pandey et al., 2013; Yadav et al., 2014). Using the genomic data of switchgrass, Rajput et al. (2014) developed SSR markers for proso millet, which showed that 62% of the switchgrass SSR markers were transferable to proso millet.

Developments in sequencing technologies have made it possible to analyze large amounts of germplasm against low production cost. It enables to screen genebank collections more efficiently for DNA sequence variation, which will be useful for mining sequence variation associated with economically important traits through genome-wide association studies (GWAS). Most recently, Wallace et al. (2015) genotyped the barnyard millet core collection (Upadhyaya et al., 2014) using GBS approach (Elshire et al. (2011) and identified several thousand single-nucleotide polymorphisms (SNPs), and studied the patterns of population structure and phylogenetic relationships among the accessions. The procedure used to identify SNPs following GBS approach in barnyard millet can also be applied easily and rapidly to characterize germplasm collections of other crops as well (Wallace et al., 2015). The GBS approach can play a major role in the crop species like proso, barnyard, little, and kodo millets for which genome sequences are not available.

8.10 Conclusions

Proso, barnyard, little, and kodo millets are nutritious, grown under marginal lands of arid and semiarid regions. Globally, significant numbers of germplasm accessions of these crops are being conserved in genebanks, and reported to have substantial variation for economically important traits. Very limited reports on trait donors for various agronomic and nutritional traits, and biotic and abiotic stress tolerance traits have been reported, and germplasm subsets like core collections in these crops are available for exploitation in crop improvement. Limited availability of genomic resources in proso, barnyard, little, and kodo millets is the major challenge in these crops; however, this could be overcome through use of genomic resources available in taxonomically closest species and high-throughput genotyping technologies. However, these crops continue to be of low priority with limited funding for research and development. Assessing genetic variability of germplasm, use of genetic and genomic resources for breeding high-yielding cultivars, developing crop production and processing technologies, value addition for improving consumption, public–private partnerships, and policy recommendations are needed to upscale these crops as remunerative to farmers.

References

Arora, A., Katewa, S.S., 1999. Germination as a screening index of heavy metal tolerance in three ethno food grasses. J. Environ. Biol. 20, 7–14.

Baltensperger, D.D., Nelson, L.A., Frickel, G.E., 1995a. Registration of 'Earlybird' proso millet. Crop Sci. 35, 1204–1205.

Baltensperger, D.D., Nelson, L.A., Frickel, G.E., Anderson, R.L., 1995b. Registration of 'Huntsman' proso millet. Crop Sci. 35, 941.

Baltensperger, D.D., Nelson, L.A., Frickel, G.E., Heyduck, R.F., Yu, T.T., 2004. Registration of NE-1 proso millet germplasm. Crop Sci. 44, 1493–1494.

Bennetzen, J.L., Schmutz, J., Wang, H., Percifield, R., Hawkins, J., Pontaroli, A.C., et al., 2012. Reference genome sequence of the model plant *Setaria*. Nat. Biotechnol. 30, 555–561.

Bonjean, A.P.A., 2010. Origins and historical diffusion in China of major native and alien cereals. In: He, Z., Bonjean, A.P.A. (Eds.), Cereals in China. CIMMYT, Mexico, DF, pp. 1–14.

Brand-Miller, J., McMillan-Price, J., Steinbeck, K., Caterson, I., 2009. Dietary glycemic index: health implications. J. Am. Coll. Nutr. 28, 446S–449S.

Chandel, G., Meena, R.K., Dubey, M., Kumar, M., 2014. Nutritional properties of minor millets: neglected cereals with potentials to combat malnutrition. Curr. Sci. 107 (7), 1109–1111.

Chase, A., 1929. The North American Species of *Paspalum* (Contributions From the United States National Herbarium)vol. 28U.S. Govt. Print. Off., Washington, DC, 1-310.

Choi, B.H., Park, K.Y., Park, R.K., 1991. Evaluation of genetic resources of foxtail, proso and barnyard millets. Res. Rep. Rural Dev. Admin. Upland Ind. Crops 33 (2), 78–83.

de Wet, J.M.J., 1986. Origin, evolution and systematics of minor cereals. In: Seetharam, A., Riley, K.W., Harinarayana, G. (Eds.), Small Millets in Global Agriculture. Oxford & IBH Publishing Co. Pvt. Ltd., New Delhi, India, pp. 19–30.

de Wet, J.M.J., Prasada Rao, K.E., Brink, D.E., 1983a. Systematics and domestication of *Panicum sumatrense* (Gramineae). J. d'agriculture traditionnelle et de botanique appliquée 30, 159–168.

de Wet, J.M.J., Prasada Rao, K.E., Mengesha, M.H., Brink, D.E., 1983b. Domestication of Sawa millet (*Echinochloa colons*). Econ. Bot. 37, 283.

de Wet, J.M.J., Prasada Rao, K.E., Mengesha, M.H., Brink, D.E., 1983c. Diversity in kodo millet (*Paspalum scrobiculatum*). Econ. Bot. 37, 159–163.

Dwivedi, S.L., Upadhyaya, H.D., Senthilvel, S., Hash, C.T., Fukunaga, K., Diao, X., et al., 2012. Millets: genetic and genomic resources. In: Janick, J. (Ed.), Plant Breeding Reviews, vol. 35, John Wiley & Sons, USA, pp. 247–375.

Elshire, R.J., Glaubitz, J.C., Sun, Q., Poland, J.A., Kawamoto, K., Buckler, E.S., et al., 2011. A robust, simple genotyping-by-sequencing (GBS) approach for high diversity species. PloS ONE 6 (5), e19379.

Gowda, K., Jagadish, P.S., Ramesh, S., Gowda, K.N.M., 1996b. Morphological characters associated with resistance to shootfly in little millet. Karnataka J. Agric. Sci. 9 (1), 63–66.

Gowda, K., Gowda, K.N.M., Ramesh, S., Jagadish, P.S., Gowda, K., 1996a. Field evaluation of little millet germplasm for resistance to the shoot fly, *Atherigona pulla* Wiede. (Diptera:Anthomyiidae). Karnataka J. Agric. Sci. 9 (1), 155–158.

Gowda, J., Rekha, D., Somu, G., Bharathi, S., Krishnappa, M., 2008. Development of core set in little millet (*Panicum sumatrense* Roth ex Roemer and Schuttes) germplasm using data on twenty one morpho-agronomic traits. Environ. Ecol. 26, 1055–1060.

Gowda, J., Bharathi, S., Somu, G., Krishnappa, M., Rekha, D., 2009. Formation of core set in barnyard millet [*Echinochloa frumentacea* (Roxb.) Link] germplasm using data on twenty four morpho-agronomic traits. Int. J. Plant Sci. 4 (1), 1–5.

Gupta, A., Joshi, D., Mahajan, V., Gupta, H.S., 2009b. Screening barnyard millet germplasm against grain smut (*Ustilago panici-frumentacei* Brefeld). Plant Genet. Resour. 8 (01), 52–54.

Gupta, A., Mahajan, V., Kumar, M., Gupta, H.S., 2009a. Biodiversity in the barnyard millet (*Echinochloa frumentacea* Link, Poaceae) germplasm in India. Genet. Resour. Crop Evol. 56 (6), 883–889.

Harlan, J.R., 1975. Crops and Man. American Society of Agronomy and Crop Science Society of America, Madison, WI.

Hiremath, S.C., Dandin, S.B., 1975. Cytology of *Paspalum scrobiculatum* Linn. Curr. Sci. 44, 20–21.

Hiremath, S.C., Patel, G.N.V., Salimath, S.S., 1990. Genome homology and origin of *Panicum sumatrense* (Gramineae). Cytologia 55, 315–319.

Hunt, H.V., Vander Linden, M., Liu, X., Motuzaite-Matuzevicuite, G., Colledge, S., Jones, M.K., 2008. Millets across Eurasia: chronology and context of early records of the genera *Panicum* and *Setaria* from archaeological sites in the Old World. Veget. Hist. Archaeobot. 17, S5–S18.

Hunt, H.V., Campana, M.G., Lawes, M.C., Park, Y.J., Bower, M.A., Howe, C.J., et al., 2011. Genetic diversity and phylogeography of broomcorn millet (*Panicum miliaceum* L.) across Eurasia. Mol. Ecol. 20, 4756–4771.

Hunt, H.V., Badakshi, F., Romanova, O., Howe, C.J., Jones, M.K., Heslop-Harrison, J.S.P., 2014. Reticulate evolution in *Panicum* (Poaceae): the origin of tetraploid broomcorn millet, *P. miliaceum*. J. Exp. Bot. 65, 3165–3175.

Ilyin, V.A., Zolotukhin, E.N., Ungenfukht, I.P., Tikhonov, N.P., Markin, B.K., 1993. Importance, cultivation and breeding of proso millet in Povolzye province in Russia. In: Riley,

K.W., Gupta, S.C., Seetharam, A., Mushonga, J.N. (Eds.), Advances in Small Millets. Oxford and IBM Publishers, New Delhi, India, pp. 109–116.

Jain, A.K., 2003. Occurrence of grain smut of little millet caused by *Macalpinomyces sharmae* in Madhya Pradesh. Plant Protect. Bull. 55, 30–32.

Jain, A.K., 2005. Stable sources of resistance for head smut in kodo millet. Indian Phytopathol. 58, 117.

Jain, A.K., Gupta, A.K., 2010. Occurrence of banded leaf and sheath blight on foxtail and barnyard millets in Madhya Pradesh. Ann. Plant Protect. Sci. 18, 268–270.

Jain, A.K., Tiwari, A., 2013. Evaluation of promising genotypes of proso millet against banded leaf and sheath blight disease caused by *Rhizoctonia solani* Kuhn. Mysore J. Agric. Sci. 47 (3), 648–650.

Jain, A.K., Tripathi, S.K., 2007. Management of grain smut (*Macalpinomyces sharmae*) in little millet. Indian Phytopathol. 60, 467–471.

Jain, A.K., Tiwari, A., Dhingra, M.R., 2013. Resistance against head smut in kodo millet (*Paspalum scrobiculatum* L.). Mysore J. Agric. Sci. 47 (3), 645–647.

Jones, M.K., 2004. Between fertile crescents: minor grain crops and agricultural origins. In: Jones, M.K. (Ed.), Traces of Ancestry: Studies in Honour of Colin Renfrew. Oxbow Books, Cambridge, pp. 127–135.

Joshi, R.P., Jain, A.K., Chauhan, S.S., 2014. Collection and evaluation of kodo millet land races for agro-morphological traits and biotic stresses in Madhya Pradesh. JNKVV Res. J. 48 (2), 162–169.

Konstantinov, S.I., Grigorashchenko, L.V., 1986. Productivity of proso millet forms of different ecological and geographical groups and their susceptibility to melanosis infection. Selektsiya i Semenovodstvo (Kiev) 61, 40–44.

Konstantinov, S.I., Grigorashchenko, L.V., 1987. Inheritance of resistance to melanosis in proso millet hybrids of the first generation. Tsitologiya i Genetika 21 (5), 335–338.

Konstantinov, S.I., Linnik, V.M., Ya, S.L., 1986. Head smut resistant proso variety Khar'kovskoe 86. Selektsiya i Semenovodstvo (USSR) 5, 35–36.

Konstantinov, S.I., Linnik, V.M., Ya, S.L., 1989. Use of smut-resistant induced mutants in breeding proso millet. Selektsiya i Semenovodstvo (Kiev) 66, 25–28.

Konstantinov, S.I., Linnik, V.M., Ya, S.L., Grigorashchenko, L.V., 1991. Breeding proso millet for resistance to diseases. In: Litun, P.P. (Ed), Urozhai i adaptivnyi potentsial ekologicheskoi sistemy polya. Ukrainian Academy of Sciences, Kiev, Ukraine, pp. 112–117.

Krasavin, V.D., Usmanova, R.I., 1988. Proso millet variety Orenburgskoe 9 resistant to head smut. Selektsiya i Semenovodstvo (Moscow) 1, 54–55.

Kumar, B., 2012. Management of important diseases of barnyard millet (*Echinochloa frumentacea*) in mid-western Himalayas. Indian Phytopathol. 65, 300–302.

Kumar, B., 2013. Management of grain smut disease of barnyard millet (*Echinochloa frumentacea*). Indian Phytopathol. 66 (4), 403–405.

Kumar, B., Kumar, J., 2009. Evaluation of small millet genotypes against endemic diseases in mid-western Himalayas. Indian Phytopathol. 62, 518–521.

Kumari, K., Muthamilarasan, M., Misra, G., Gupta, S., Subramanian, A., Parida, S.K., et al., 2013. Development of eSSR-markers in *Setaria italica* and their applicability in studying genetic diversity, cross-transferability and comparative mapping in millet and non-millet species. PLoS ONE 8, 1–15.

Lu, H., Zhang, J., Liu, K., Wu, N., Li, Y., Zhou, K., et al., 2009. Earliest domestication of common millet (*Panicum miliaceum*) in East Asia extended to 10,000 years ago. Proc. Natl Acad. Sci. USA 106 (18), 7367–7372.

Maikhuri, R.K., Rao, K.S., Semwal, R.L., 2001. Changing scenario of Himalayan agro-ecosystem: loss of agro-biodiversity an indicator of environment change in central Himalaya, India. Environmentalist 21, 23–29.

Maslenkova, L.I., Resh, L.P., 1990. Sources of resistance to head smut in proso millet. Nauchno-Tekhnicheskii Byulleten', VASKhNIL, Sibirskoe Otdelenie, Sibirskii Nauchno-Issledovatel'skii Institut Sel'skogo Khozyaistva, vol. 6, pp. 28–33.

Murthi, T., Harinarayana, G., 1986. Insect pests of small millets and their management in India. In: Seetharam, A., Riley, K.W., Harinarayana, G. (Eds.), Small Millets in Global Agriculture. Oxford and IBH publishing, New Delhi, India, pp. 255–270.

Muthamilarasan, M.E., Suresh, B.V., Pandey, G.A., Kumari, K.A., Parida, S.W.K.U., Prasad, M.A., 2014. Development of 5123 intron-length polymorphic markers for large-scale genotyping applications in foxtail millet. DNA Res. 21, 41–52.

Muthusamy, M., 1981. Screening of barnyard millet *Echinochloa frumentacea L.* to smut disease. Proceedings of the National Seminar on Disease Resistance in Crop Plants, December 22 and 23, 1980, pp. 35–36.

NAAS, 2013. Role of Millets in Nutritional Security of India. Policy paper No. 66. National Academy of Agricultural Science, New Delhi, p. 16.

Nelson, L.A., 1984. Technique for crossing proso millet. Crop Sci. 21, 71.

Pandey, G., Misra, G., Kumari, K., Gupta, S., Parida, S.K., Chattopadhyay, D., et al., 2013. Genome-wide development and use of microsatellite markers for large-scale genotyping applications in foxtail millet [*Setaria italica* (L.)]. DNA Res. 20, 197–207.

Patel, R.K., Rawat, R.R., 1982. Note on the estimation of loss in yield of kodo-millet caused by the attack of shootfly. Indian J. Agric. Sci. 52, 880.

Rajput, S.G., Plyler-harveson, T., Santra, D.K., 2014. Development and characterization of SSR markers in proso millet based on switchgrass genomics. Am. J. Plant Sci. 5, 175–186.

Rao, B.R., Nagasampige, M.H., Ravikiran, M., 2011. Evaluation of nutraceutical properties of selected small millets. J. Pharm. Bioallied Sci. 3 (2), 277–279.

Reddy, V.G., Upadhyaya, H.D., Gowda, C.L.L., 2007. Morphological characterization of world's proso millet germplasm. SAT eJ. 3 (1), 1–4.

Richardson, W.L., 1958. A technique of emasculating small grass florets. J. Genet. 18, 69–73.

Sabir, P., Ashraf, M., Akram, N.A., 2011. Accession variation for salt tolerance in proso millet (*Panicum miliaceum* L.) using leaf proline content and activities of some key antioxidant enzymes. J. Agron. Crop Sci. 197, 340–347.

Saleh, A.S.M., Zhang, Q., Chen, J., Shen, Q., 2013. Millet grains: nutritional quality, processing, and potential health benefits. Compr. Rev. Food Sci. Food Saf. 12 (3), 281–295.

Seghatoleslami, M.J., Kafi, M., Majidi, E., 2008. Effect of deficit irrigation on yield, WUE and some morphological and phenological traits of three millet species. Pak. J. Bot. 40 (4), 1555–1560.

Shailaja, S., Jagadish, P.S., Kumar, C.T.A., Nangia, N., Gowda, J., Nagaraja, A., 2009. Evaluation of pre-release and released varieties of proso millet (*Panicum miliaceum* L.) to *Corcyra cephalonica* Stainton (Lepidoptera: Pyralidae) infestation. Environ. Ecol. 27, 445–447.

Sharma, P.N., Sugha, S.K., Panwar, K.S., Sagwal, J.C., 1993. Performance of proso millet (*Panicum miliaceum*) germplasm in relation to smut (*Sphacelotheca destruens*). Indian J. Agric. Sci. 63, 675.

Soldatov, A.F., Agafonov, N.P., 1980. Susceptibility of *Panicum* millet to melanosis in western Kazakhstan. Trudy po Prikladnoi Botanike Genetike i Selektsii 69, 64–69.

Sood, S., Khulbe, R.K., Saini, N., Gupta, A., Agrawal, P.K., 2014. Interspecific hybrid between *Echinochloa esculenta* (Japanese barnyard millet) and *E. frumentacea* (Indian barnyard millet) – a new avenue for genetic enhancement of barnyard millet. Electron. J. Plant Breed. 5 (2), 248–253.

Upadhyaya, H.D., Pundir, R.P.S., Dwivedi, S.L., Gowda, C.L.L., Reddy, V.G., Singh, S., 2009. Developing a mini core collection of sorghum for diversified utilization of germplasm. Crop Sci. 49, 1769–1780.

Upadhyaya, H.D., Sharma, S., Gowda, C.L.L., Reddy, V.G., Singh, S., 2011. Developing proso millet (*Panicum miliaceum* L.) core collection using geographic and morpho-agronomic data. Crop Pasture Sci. 62, 383–389.

Upadhyaya, H.D., Dwivedi, S.L., Singh, S.K., Singh, S., Vetriventhan, M., Sharma, S., 2014. Forming core collections in barnyard, kodo, and little millets using morphoagronomic descriptors. Crop Sci. 54, 1–10.

Vavilov, N.I., 1926. Studies on the origin of cultivated plants. Bull. Appl. Bot. Plant Breed. 16, 1–248.

Wallace, J.G., Upadhyaya, H.D., Vetriventhan, M., Buckler, E.S., Hash, C.T., Ramu, P., 2015. The genetic makeup of global barnyard millet germplasm collection. Plant Genome 8, (1), 1–7.

Wang, L., Wang, X.Y., Wen, Q.F., Cao, L.P., 2007. Identification of protein and fat content for Chinese proso millet germplasm. J. Plant Genet. Res. 8, 165–169.

Wilson, R.L., Courteau, J.B., 1984. Search of plant introduction proso millet for fall armyworm resistance. J. Econ. Entomol. 77, 171–173.

Yadav, C.B., Muthamilarasan, M., Pandey, G., Khan, Y., Prasad, M., 2014. Development of novel microRNA-based genetic markers in foxtail millet for genotyping applications in related grass species. Mol. Breed. 34 (4), 2219–2224.

Zegada-Lizarazu, W., Iijima, M., 2005. Deep root water uptake ability and water use efficiency of pearl millet in comparison to other millet species. Plant Prod. Sci. 8 (4), 454–460.

Zhang, Y., Wang, J., Wang, W., Zeng, P., Han, C., Quan, Z., et al., 2012. Genome sequence of foxtail millet (*Setaria italica*) provides insights into grass evolution and biofuel potential. Nat. Biotechnol. 30, 549–554.

Zohary, D., Hopf, M., 2000. Domestication of Plants in the Old World, third ed. Oxford University Press, Oxford.

Zolotukhin, E.N., Tikhonov, N.P., Lizneva, L.N., 1998. New cultivar of *Panicum miliaceum*, 'Il'Inovskoe. Int. Sorghum Millet Newsl. 39, 133–134.

Subject Index

C

D

G

J

K

O

P

Q

T

U

V

W

Y

Printed in the United States
By Bookmasters